Historiographies of Science

The goal of this series is to provide definitive assessments of the historiography and the future of major fields and approaches within the history of science. Each volume will address the major trends in historical thought within a particular field, the major debates among historians of that field, and promising new directions that may shape future scholarship. Each volume is framed in terms of what a scholar should know about the history of work in that area, if they wanted to make a meaningful and original contribution to that field. Each volume will be written by experts in that field for graduate students and other scholars new to the history of that field. While other areas of history have extensive historiographic literatures, history of science has fewer resources from which to draw. The paucity of historiographical reflections by leading scholars in the history of science makes it more difficult for new scholars to join the field, as they try to assess the traditions of research on their own. These volumes will offer an informed introduction to major issues that will foster new, original research in the history of science. Editors will be asked to select topic areas/ fields that they think have had a substantial and diverse body of scholarship. Each volume will be informed by different methods, theories, and perspectives that can be compared and contrasted in each volume.

Elena Aronova • David Sepkoski •
Marco Tamborini
Editors

Handbook of the Historiography of the Earth and Environmental Sciences

With 20 Figures and 1 Table

Springer

Editors
Elena Aronova
History Department
University of California
Santa Barbara, CA, USA

David Sepkoski
Department of History
University of Illinois
Urbana-Champaign, IL, USA

Marco Tamborini
Institut für Philosophie
Technical University of Darmstadt
Darmstadt, Hessen, Germany

ISSN 2523-7748 ISSN 2523-7756 (electronic)
Historiographies of Science
ISBN 978-3-031-40798-7 ISBN 978-3-031-40799-4 (eBook)
https://doi.org/10.1007/978-3-031-40799-4

University of Illinois at Urbana-Champaign

This Springer imprint is published by the registered company Springer Nature Switzerland AG.
The registered company address is: Gewerbestrasse 11, 6330 Cham, Switzerland

Series Preface

While some areas of history have extensive historiographic literatures, the history of science has fewer resources from which to draw than most. This scarcity of historiographical reflections by leading scholars makes it more challenging for newcomers who must try to assess traditions of historical research as they frame their own contribution to the history of science. As informed introductions to major themes in the writing of the history of science, we hope that this series will both help foster original research in the history of science and further discussion regarding historiographic trends.

The goal of this series is to provide an assessment of the historiography and future of major approaches within the history of science. Each volume addresses the major trends in historical thought within a particular field, the major debates among historians of that field, and promising new directions that may shape future scholarship. Written for graduate students or scholars new to the history of science, each volume is framed in terms of what a scholar should know about the history of work in that area, if they wanted to make a meaningful and original contribution to that field.

The volumes in the historiography of science series are not intended to provide comprehensive reviews of every topic discussed in the history of science. Editors of individual volumes select topic areas and fields that they think have had a substantial and diverse body of scholarship that have been informed by different methods, theories, and perspectives. Because we would like to foster more conversation about historiography, we see the idiosyncrasies of individual chapters not as flawed and partial perspectives but as opportunities to articulate diverse perspectives through an ongoing conversation. These volumes are open for revision through Springer's Meteor publishing platform. Please engage with the authors and editors and help push this historiographic dialogue further.

Michael R. Dietrich

Contents

About the Editors

Elena Aronova is Associate Professor in the Department of History at the University of California, Santa Barbara. She has published on the history of environmental data collection, the history of the International Geophysical Year, the history of seismology, and the historiography of science. She is the author of *Scientific History: Experiments in History and Politics from the Bolshevik Revolution to the End of the Cold War* (University of Chicago Press, 2021), which documents the history of continuous efforts to integrate scientific knowledge and new technologies—from plant genetics to computers—into historical research. She has co-edited two collections of essays: *Science Studies During the Cold War and Beyond: Paradigms Defected* (2016) and *Data Histories* (2017).

David Sepkoski is the Thomas M. Siebel Chair in History of Science at the University of Illinois, Urbana-Champaign. He is the author of many books and articles in the history of the earth and environmental sciences, including *The Paleobiological Revolution: Essays on the Growth of Modern Paleontology* (with Michael Ruse, University of Chicago Press, 2009), *Rereading the Fossil Record: The Rise of Paleobiology as an Evolutionary Discipline* (University of Chicago Press, 2012), and *Catastrophic Thinking: Extinction and the Value of Diversity from Darwin to the Anthropocene* (University of Chicago Press, 2020). Among other awards, he is most recently the recipient of the John Simon Guggenheim Fellowship for 2019–2020.

Marco Tamborini teaches philosophy and history of science at the Technical University of Darmstadt. His research explores the dynamic intersections of the life sciences, engineering, and ethics, with a particular focus on bioinspired and emerging technoscientific fields. His recent book publications include *The Architecture of Evolution: The Science of Form in Twentieth-Century Evolutionary Biology* (University of Pittsburgh Press, 2022), *Entgrenzung: Die Biologisierung der Technik und die Technisierung der Biologie* (Meiner, 2022), and *Biorobotik zur Einführung* (Junius Verlag, 2024).

Contributors

Antony Adler Carleton College, Northfield, MN, USA

Etienne Benson Max Planck Institute for the History of Science, Berlin, Germany

Stephen Bocking Trent University, Peterborough, ON, Canada

Adelene Buckland King's College London, London, UK

Lino Camprubí Universidad de Sevilla/DEEPMED, Seville, Spain

Alberto Celis Universidad de Sevilla/DEEPMED, Seville, Spain

Pey-Yi Chu Pomona College, Claremont, CA, USA

Gillen D'Arcy Wood University of Illinois, Urbana-Champaign, Champaign, IL, USA

Ronald E. Doel Florida State University, Tallahassee, FL, USA

Sebastian Felten Department of History, University of Vienna, Wien, Austria

Silvia Fernanda de Mendonça Figueirôa Depto. de Ensino e Práticas Culturais/DEPRAC, Faculdade de Educação-FE/UNICAMP, Campinas, SP, Brazil

Anna Graber University of Minnesota, Minneapolis, MN, USA

Jarrod Hore UNSW, Sydney, NSW, Australia

Marianne Klemun University of Vienna, Vienna, Austria

Maria Margaret Lopes Programa Interunidades em Museologia-PPGmus, Museu de Arqueologia e Etnologia-MAE/USP, São Paulo, Brazil

Chris Manias King's College London, London, UK

Daniella McCahey Texas Tech University, Lubbock, TX, USA

Irina Podgorny Archivo Histórico de la Facultad de Ciencias Naturales y Museo, Universidad Nacional de La Plata/CONICET, La Plata, Argentina

Lukas Rieppel Brown University, Providence, RI, USA

Martin Rudwick Department of History and Philosophy of Science, University of Cambridge, Cambridge, UK

University of California San Diego, San Diego, CA, USA

Trinity College Cambridge, Cambridge, UK

Matthew Shindell Smithsonian National Air and Space Museum, Washington, DC, USA

Marco Tamborini Institut für Philosophie, Technical University of Darmstadt, Darmstadt, Hessen, Germany

Derek D. Turner Connecticut College, New London, CT, USA

Caitlin D. Wylie Program in Science, Technology, and Society, University of Virginia, Charlottesville, VA, USA

Introduction: The Historiography of the Earth and Environmental Sciences

When we were first approached to edit this volume, the title offered to us was "Handbook of the Historiography of Geology." For a variety of reasons discussed below, we decided on the more capacious title *Handbook of the Historiography of the Earth and Environmental Sciences*, which we felt better captured the state of both current historical and scientific perspectives on the earth sciences. But the deeper motivation for our title is the awareness that human industries, agriculture, transportation, and population expansion have produced a profound change in the earth's climate, environment, and ecology. This transformation—which many consider the onset of a new geological era known as the "Anthropocene"—has inspired historians to explore linkages between the histories of industrial capitalism and imperialism, postcolonial and subaltern studies, Indigenous histories, and environmental history with more traditional histories of geology, paleontology, and climate science.

No matter where one locates the beginning of anthropomorphic climate and environmental change, there is now a broad recognition by historians and practitioners of Science and Technology Studies (STS) that it is impossible to separate the intellectual and technical development of mining, geology and paleontology, oceanography, and other fields in the earth and environmental sciences from the political, social, and economic consequences of European imperialism and extractive capitalism. These connections have not only broadened the historiographic and geographic boundaries of the field, but have also highlighted the need for expanding the community of scholars and readers in the history of the earth and environmental sciences to be more inclusive of perspectives from the Global South, Indigenous communities, and marginalized peoples who have been historically affected by Western scientific and industrial practices. The contributions to this volume are intended to reflect and encourage this need for broader perspective and participation.

The history of the earth sciences is a field that was institutionalized relatively recently: the History of Earth Sciences Society (HESS), along with its associated journal *Earth Sciences History*, was founded in the early 1980s, aimed at promoting and publishing historical work on various areas of the earth sciences, such as geology, geography, geophysics, oceanography, paleontology, meteorology, and climatology (https://pubs.geoscienceworld.org/hess). Beyond the umbrella organizations and journals, a number of subfields have their own professional organizations and communities. Thus, the International Commission of the History of

Oceanography (ICHO), born during the 1968 International Congress on the History of Science in Paris, continues to serve as a global body devoted to linking scholars, writers, and teachers interested in the history of the marine sciences, broadly defined, and internationalizes the conversations in the field. A generation of historians of oceanography was initiated to the field through the ICHO conferences and summer schools, which showcased the new approaches and perspectives in the history of science.

Like other fields in the history of science, the early histories of the earth sciences have been focused on the histories of ideas, their origin and the controversies around them, and their subsequent reception. Martin Rudwick's *The Great Devonian Controversy* (1985) is a history of ideas at its best, a monumental achievement that has become a classic in the history and philosophy of science. Since then, and keeping up with the trends in the history of science, the historians of the earth sciences have been moving beyond the history of scientific theories and knowledge, turning their focus to examining the actual activities and practices of scientists. The "practice turn," which gained prominence in the 1980s and 1990s, emphasizes the social, material, and constructivist aspects of scientific work and developments within the earth sciences. The result of this turn to practice has been a new appreciation of materiality, instrumentation, situatedness, and the labor of creating knowledge about the earth and the environment. The growth and diversification of the history of the earth sciences is a sign of its vitality as a field. At the same time, the expansiveness of literature representing the history of the earth sciences is often daunting for new scholars entering the field. The *Handbook of the Historiography of the Earth and Environmental Sciences* is intended to provide an assessment of the historiography and the future of major subfields and approaches within the history of the earth sciences.

While some areas of history have extensive historiographic literatures (such as the Blackwell's Companion series that has almost 60 volumes dedicated to American history alone), the history of science, and especially the history of the earth and environmental sciences, has very few resources from which to draw. The paucity of historiographical reflections by leading scholars in the history of the earth sciences makes it more difficult for new scholars to join the field, as they try to assess the traditions of research on their own. *Handbook of the Historiography of the Earth and Environmental Sciences* offers an informed introduction to major issues that is intended to foster new, original research in the history of the earth sciences.

The volume is intended for a broad range of readers, including graduate students, other scholars, and scientists, both familiar with and new to the history of the earth and environmental sciences. It is intended to provide a critical assessment of the historiography and the future of major themes and approaches within the history of the earth sciences, understood broadly. Each chapter addresses major trends in historical thought within a particular aspect of the earth sciences (e.g., geology, paleontology, climate, etc.), the major debates among historians of that subfield, and promising new directions or ways of conceiving of the field that may shape future scholarship. Since scholarship in the history of science is increasingly becoming entangled with environmental, economic, bureaucratic, political, gender, and other historical approaches, this collection showcases the breadth and diversity of scholarship on the earth and environmental sciences.

The *Handbook of the Historiography of the Earth and Environmental Sciences* is not intended to give comprehensive coverage of every topic in the historiography of the earth and environmental sciences, nor to give exhaustive "literature reviews." Because these chapters are not literature reviews, but historiographic essays, we selected fields informed by different methods, problems, and perspectives that can be compared and contrasted in each chapter. In developing our list of contributors, we matched important topic areas to scholars with established and emerging expertise in that area and asked for thoughtful essays that convey a clear authorial perspective on past and current scholarship and scholarly debates. Each author in this collection offers their perspective on the historiography of their topic area, framing those topics in terms they found most significant.

Thanks to generous funding provided by the Thomas M. Siebel endowment at the University of Illinois, Urbana-Champaign, the entire collection is published open access. We saw it imperative to secure an open access publication in order for the volume to be accessible to readers without institutional access to Springer collections. Our intention for this volume was to support a conversation among established and emerging historians of the earth sciences regarding their work, its history, and its future.

Themes and Contributions

The volume opens with an essay by Martin Rudwick, whose work has long been considered foundational to the discipline. A former geologist, Rudwick's career tracks the growing professionalization of the history of earth science scholarship, from amateur contributions by retired scientists to a thriving community of professional historians. Rudwick made that transition himself, and his reflections on his long career are especially instructive for younger scholars who were trained after this shift took place. Rudwick provides an overview of the "big themes" in the historiography of the earth sciences by looking back at the development of his career that helped to establish geohistory as a field within the history of science and by sharing his personal reflections on the field's past, present, and future (see Rudwick, "Thinking-and-Doing History of Geology: Reflections on Working in Two Cultures over Eight Decades"). Rudwick's essay serves as a comprehensive introduction to the collection. The rest of the volume is organized in four broad thematic parts: (1) Formations, (2) Institutions and Practices, (3) Perspectives, and (4) Geographies (see Fig. 1).

In geology, the term "formations" designates the basic rock units forming the layers (or strata) of the Earth, a basic element in the spatiotemporal analysis of the Earth's crust on the geological timescale. The title of the opening part of the collection uses the term "formations" in a more capacious sense, as a reference to both the deep time of the geological processes that shaped the oceans (Adler, "Changing Narratives in the History Oceanography"), the planets (Shindell,

BIG THEMES IN HISTORIOGRAPHY OF THE EARTH SCIENCES

Thinking-and-Doing History of Geology: Reflections on Working in Two Cultures over Eight Decades | **Martin Rudwick**

FORMATIONS

Changing Narratives in the History of Oceanography | **Antony Adler**

Climate and the Little Ice Age | **Gillen D'Arcy Wood**

Planets: A History of Observing Worlds and Changing Worldviews | **Matthew Shindell**

Rivers in History: Systems, Agents, and Places | **Etienne Benson**

Histories of Frozen Earth | **Pei-Yi Chu**

INSTITUTIONS AND PRACTICES

Collections and Museums in the Historiography of the Earth Sciences | **Irina Podgorny**

Expeditions and Fieldwork in Earth Sciences | **Marianne Klemun**

Earth Science and Extractive Capitalism in the Age of Empire | **Lukas Rieppel**

Mining and the Formation of Modern Geology | **Sebastian Felten**

Data, Aesthetics, and Visualizations of Deep Time | **Marco Tamborini**

PERSPECTIVES

Indigenous Knowledge and Perspectives | **Stephan Bocking**

Earth Sciences from the Perspectives of Science and Technology Studies | **Caitlin Wylie**

Cold War and Earth Sciences | **Ron Doel**

Fig. 1 The structure of the volume (see discussion in the text)

Historiography of the International Geophysical Year | **Daniella McCahey**

Philosophy of the Earth Sciences | **Derek Turner**

Women Geologists 1780–1840: Re-reading Charlotte Murchison | **Adelene Buckland**

GEOGRAPHIES

History of the Earth Sciences from the South | **Jarott Hore**

Land, Mines, and Bones: African History and Deep Knowledges of the Earth | **Chris Manias**

Historiography of Earth Sciences in Latin America | **Silvia Figueirôa, Margaret Lopes**

Regional Geopolitics, French Geography, and Climatology in "Dry Spain," Eighteenth to Twenty-First Centuries: Birth of the "Mediterranean Climate" | **Alberto Celis, Lino Camprubi**

Mining and Metallurgy in Early Imperial Russia | **Anna Graber**

Fig. 1 (continued)

"Planets: A History of Observing Worlds and Changing Worldviews"), rivers (Benson, "Rivers in History: Systems, Agents, and Places"), the permafrost (Chu, "Histories of Frozen Earth"), and the climate (Wood, "Climate and the Little Ice Age"), as well as the temporal frames of the historiographies of associated disciplines, subdisciplines, and knowledge communities. The five essays in this part decenter traditional spatiotemporal frames for history and reflect on increasingly transnational and interdisciplinary perspectives on the histories of our physical environments.

The history of the earth and environmental sciences is also a history of practices and institutions in specific historical and regional contexts. The chapters in the next part, "Institutions and Practices," track the historiography of these sciences in distinct physical and epistemic locales. These include familiar spaces like the field and the museum (Podgorny, "Collections and Museums in the Historiography of the Earth Sciences"; Klemun, "Expeditions and Fieldwork in Earth Sciences"), and are associated with distinct practices (surveying and collecting, classifying and displaying), as well as specific industrial and economic contexts such as petroleum prospecting and mining for coal and precious metals (Rieppel, "Earth Science and Extractive Capitalism in the Age of Empire"; Felten, "Mining and the Formation of Modern Geology"). These different locales and practices are associated with different epistemological goals, and these chapters are attentive to knowledge production as the exercise of power: over nature, over bodies, over economic systems. Ultimately, as these five chapters demonstrate, the ability to survey, collect, order, display, and visualize natural knowledge about the earth (Tamborini, "Data, Aesthetics, and Visualizations of Deep Time") has been a crucial tool in establishing systems of domination, extraction, and control in imperialistic, colonial, and post-colonial contexts.

The way we study the history of the earth and environmental sciences also reflects different valuations of knowledge and epistemic goals by scholars in the history of science and Science and Technology Studies, as showcased in the next part, "Perspectives." Recent attention to Indigenous knowledge of the earth and its environment (Bocking, "Indigenous Knowledge and Perspectives") decenters narratives of Western imperialism as the exclusive context for knowledge production. Perspectives in STS such as social constructivism and ethnography explore the practices and contexts of knowledge production about the earth and environment as sites of social performance and embodied values (Wylie, "Earth Sciences from the Perspectives of Science and Technology Studies"). Philosophical approaches (Turner, "Philosophy of the Earth Sciences") reveal epistemic and aesthetic assumptions about visual presentations of knowledge, as well as important methodological features of historical science. And studies of earth science in the context of Cold War politics demonstrate the important role that geophysical sciences played in geopolitics (Doel, "Cold War and Earth Sciences") and international scientific cooperation (McCahey, "Historiography of the International Geophysical Year (1957–1958)") in the postwar era.

The contributions in the last part of the volume—"Geographies"—reconsider the narrative of the history of the earth sciences that has been traditionally centered on the study of the accumulation of knowledge within the European and North Atlantic "center" and its spread to the non-European "periphery." The part opens with an essay (Hore, "History of the Earth Sciences from the South") that resituates that narrative by focusing on histories of earth science from and about non-European geographies, especially those in the Southern Hemisphere, and proposes avenues for future research about modern thinking about the planet "from the south" that extends beyond the European revelations of deep time. Other essays in this part illustrate the potential of the place-based approach to the history of geosciences by foregrounding these histories in Africa (Manias, "Land, Mines, and Bones: African History and Deep Knowledges of the Earth") and Latin America (Figueirôa and Lopes, "Historiography of Earth Sciences in Latin America"), by historicizing the extension of the concept of the "Mediterranean climate" to regions as far from the Mediterranean Sea as California and Australia (Celis and Camprubí, "Regional Geopolitics, French Geography, and Climatology in 'Dry Spain,' Eighteenth to Twenty-First Centuries: Birth of the 'Mediterranean Climate'"), and by bringing the historiography of imperial Russian mining into conversation with the historiography of the earth sciences (Graber, "Mining and Metallurgy in Early Imperial Russia").

Taken together, the contributions in this volume suggest that it is time to think a bit differently about the historiography of the earth sciences. Breaking with a traditional disciplinary or chronological organization, this volume showcases innovative ways of framing historiographic perspectives on the earth and environmental sciences broadly conceived, and points to new challenges and contexts for future work by historians and scholars of STS.

Elena Aronova
David Sepkoski

Part I

Big Themes in Historiography of Earth Sciences

Thinking-and-Doing History of Geology: Reflections on Working in Two Cultures over Eight Decades

Martin Rudwick

Contents

Abstract

Editor's note: This chapter was commissioned in recognition of Martin Rudwick's transformative contributions to the history of geology, deep time sciences, and earth history. Unlike other contributions to this volume, it is autobiographical and personal. Rudwick's career is, in itself, a microcosm of modern historiography of earth and environmental sciences, and we are honored to have his personal reflections in this collection. We hope that readers unfamiliar with Rudwick's career and accomplishments in this field will benefit from getting to know him in this way, and that those readers familiar with his scholarship will enjoy this insight into his life's work as a historian.

M. Rudwick (✉)
Department of History and Philosophy of Science, University of Cambridge, Cambridge, UK

University of California San Diego, San Diego, CA, USA

Trinity College Cambridge, Cambridge, UK

E. Aronova et al. (eds.), *Handbook of the Historiography of the Earth and Environmental Sciences*, Historiographies of Science,
https://doi.org/10.1007/978-3-031-40799-4_2

Keywords

History of geology · Martin Rudwick · Historiography of science · Biography

This is not a fully documented research article of the kind that will make the rest of this volume a valuable resource for future work on the historiography of the Earth sciences. David Sepkoski and his fellow-editors have asked me instead to reflect on the intellectual trajectory that my own research has taken, and on some of the incompletely solved problems and inadequately examined issues that in my opinion would be worthwhile for other historians to explore in years to come. I have chosen to do this autobiographically, but I shall mention the details of my own career only in so far as I think they were relevant to what I have been able to do, or at least have tried to do, in my research and publications.

1 I

I start far away from the historiography of the Earth sciences, with two anecdotes from my childhood and adolescence, because I think I can see in retrospect how deeply they influenced my later thinking-and-doing in what became my scholarly field. My upbringing was unremarkable for a child of a middle-class English family during and immediately after the Second World War. My father Joseph Rudwick was the senior science teacher in a leading English public school (in the British, not American, sense), where he taught physics and is said to have been outstandingly good at it (he trained at least one subsequent Nobel prizewinner). To his great disappointment I showed no aptitude for physics and little interest in the phenomena that he found so fascinating. Instead, I was intrigued by all kinds of natural history. For example, my closest boyhood friend (the child of refugees from Nazi Germany) was highly knowledgeable about birds, though I found my ability to emulate him in watching and identifying them was limited by my inherited color-blindness, which had been discovered rather traumatically in an art class at my primary school. But the huge disparity between physics and ornithology alerted me at this early age to the negative effects of the modern anglophone concept of a singular "Science," which, as I soon realized, smothered all the fascinating diversity of these and many other sciences.

I can pinpoint my discovery of the specific science that became the focus of all my later research: It was the moment when I found in the family garden a strange object that I soon learnt was a fossil sea-urchin (it was an *Echinocorys* preserved in flint, derived from the local Chalk). This became the core of my own fossil collection, which grew rapidly as I searched for further specimens in disused Chalk quarries on the nearby Chiltern hills and, on visits to my grandparents, in those in Jurassic rocks on the Cotswold hills. It was further enriched when my godfather Clive Rouse gave me a small collection of superbly preserved fossils that I think he had inherited. He was important to me in childhood as a highly supportive nonparental adult, but also in retrospect as a scholarly historian. He was at the time the leading British expert on the conservation and iconography of wall-paintings in mediaeval churches and secular buildings. Watching him painstakingly uncovering these wonderful images of mediaeval life, and of the mediaeval realities of heaven and hell, saints and sinners, and prelates and peasants, gave me a vivid sense of the sheer strangeness of past periods of history; his evident sympathy for those who long ago had commissioned and created these paintings made a lifelong impression on me. Perhaps without knowing it, he taught me never to look down condescendingly on any of the people of the past, and never to ridicule their ideas. This is an attitude that should surely be a minimal requirement for being a historian of any kind, but it often seems lacking in those who tackle the history of the sciences, including the sciences of the Earth.

I had a largely conventional education. The teaching I was given at secondary level was mostly competent rather than inspiring. The great exception was my biology teacher Douglas Reid, who also actively encouraged my rapidly growing interest in geology. Under his guidance, I taught myself geology largely from books and from studying intensively the fossil collection in the school's museum (under early postwar conditions, the major museums in London were still closed). Beyond the classroom, Reid was a world authority on the biology and systematics of amphipod crustaceans. After I won a scholarship to Cambridge but had to serve further time at a boarding school that I experienced more as a prison, he kept my frustration at bay by giving me a small but serious research task. Using a binocular microscope, it involved dissecting the tiny amphipods that had lived among the barnacles scraped from the hull of a ship docked at Aden (then still a British possession), and using published monographs to identify them and then plot their biogeography. This resulted in my first published scientific paper. I never again attempted this kind of systematic zoology, on either living or fossil animals, but its rigorous discipline was an invaluable training in that it demanded very close attention to morphological detail.

2 II

At Cambridge, I benefitted from the university's distinctive system of undergraduate teaching, in that I could choose to combine geology with mineralogy-and-petrology (which for local historical reasons were taught in a separate department), and with zoology, before specializing in geology and paleontology in my final year. So I was given basic training in a wide range of what would now be called Earth and Life sciences, with the opportunity to concentrate later on paleontology. The advanced teaching of that science was organized along biological lines rather than those determined by stratigraphy, and by the time I graduated the geology department was unusual in being headed by a paleontologist. In fact Oliver Bulman, one of the world's leading experts on graptolites, had formally defined his research field not as paleontology but as "palaeozoology"—an unusual term in the anglophone world at that time—thereby implicitly emphasizing his affinities with the biological sciences rather than the inorganic sciences of the Earth. This highlighted the continuing tension within paleontology between treating fossils primarily as evidence for the long-term history of life, and using them to serve the needs of stratigraphy and as a key to unraveling the complex structures of the Earth's crust: two goals that were and are not always easily compatible.

I remained at Cambridge after graduation and began research toward a PhD, with an academic career in mind. I tried to tackle a problem in the then rather novel field of paleo-ecology. I took an already well-worked piece of stratigraphy and planned to reinterpret it in detail as a historical sequence of changing local environments and their local communities of organisms. I decided to start on the Jurassic rocks in the beautiful countryside of the Cotswolds, and Joscelyn Arkell, at this time the world's leading authority on ammonites and Jurassic stratigraphy (and a Fellow of my College), agreed to be my advisor. But a summer of fieldwork showed me that, in the decades since the area was first studied by geologists, many of the disused quarries had become too overgrown, and the stratigraphy therefore too obscure, to make the project a promising one. However, the fossils I was able to collect, vastly amplified by the collections I went on to explore in the great Sedgwick Museum back in Cambridge, then focused my attention instead on the morphology of the fossils themselves. With this major change of tack, Bulman became my new advisor (he told me that he himself and some other leading paleontologists were also color-blind; this convinced me that my chosen science was one of the few for which this genetic defect was perhaps even an advantage, since it focused one's attention on structures rather than being distracted by colors, which in fossils are usually irrelevant).

I found myself most entranced by the often abundant fossil shells of brachiopods, whose lovely shapes reminded me of sculptures such as those by Henry Moore and Barbara Hepworth, which I had first encountered at an open-air exhibition at the great postwar Festival of Britain. The impact of such esthetic responses on the course of scientific research has not, I think, been adequately studied by historians; certainly it was important in my case. I began to puzzle why brachiopod shells had such diverse and distinctive shapes: not with the aim of reconstructing their evolutionary relationships, but rather of working out how those shapes and structures might have enabled the animals to operate and make a living in the vanished worlds of the deep past. I was greatly influenced by two biological classics to which Reid had first introduced me: D'Arcy Thompson's *Growth and Form* (1917) and E.S. Russell's *Form and Function* (1916). They in turn introduced me to the science of morphology, which had flourished in the nineteenth century but seemed to have been eclipsed by the rise of neo-Darwinism in the early twentieth.

More broadly, and with profound implications for my later intellectual trajectory, these books were my first encounter with the historiography of any of the natural sciences. In particular I was fascinated by Russell's account of Georges Cuvier's early nineteenth-century work on living and fossil mammals and their *functional morphology*—a term rarely used in my own time, at least by invertebrate paleontologists—which I soon adopted as the theme for my own attempts to reconstruct the modes of life of fossil brachiopods. Cuvier became my scientific hero. Yet on reading more widely about the history of the sciences, I found he was generally scorned and dismissed as having been a reactionary opponent of evolutionary ideas. This first made me suspicious of all such attempts to divide the scientists of the past into goodies and baddies, or to judge them by the perceived modernity or otherwise of their ideas. This "presentism" remains, in my opinion, the most pernicious of all the prejudices that still often hamper the research of historians of the sciences.

In the following years, my research on "the functional morphology of fossil brachiopods" (the title of my PhD dissertation) led successively to teaching positions in my department and a research fellowship at my College; a major research grant to study the whole of brachiopod history—from the Cambrian period to the present day—in terms of adaptive evolution and extinction; and my first book *Living and Fossil Brachiopods* (1970), which eventually synthesized all this work. A decisive phase in this research was a year spent flying round the world—with all due modesty I thought of it in retrospect as my personal *Beagle* voyage—during which I studied tropical marine life of all kinds on the Great Barrier Reef; *living* brachiopods, and their then poorly understood suspension-feeding mechanisms, in New Zealand; and some of the world's finest collections of fossil brachiopods of all ages, in Washington DC. I tried to develop a methodological principle that would allow likely functions to be reliably inferred from observable structures, in extinct fossil organisms that cannot be studied alive. This method involved comparing those real structures with the forms which, in any given class of organisms, would have been as effective as physically possible for any candidate function. I called any such optimal structure the "*paradigm*" for that function; I began using that then rather unusual word in my published papers just a year before Tom Kuhn's *Structure of Scientific Revolutions* (1962) made it much more widely familiar with a quite different meaning.

The most striking, and most controversial, example of my use of this paradigm method was my functional analysis of some of the most bizarre or aberrant brachiopods that ever existed, not long (geologically!) before what looked suspiciously like a mass extinction at the end of the Permian period and Palaeozoic era (any such "catastrophic" event was itself a highly controversial idea at this time). Some paleontologists adopted my paradigm method as a useful tool in the reconstruction of life's history; others—mistakenly—took it to depend on assuming an implausibly and universally perfect degree of adaptation or even suspected that I wanted to rehabilitate Paley's natural-theological "argument from design." In any case, this research brought my fossil brachiopods back to life in my mind's eye more vividly than ever; it also provoked much wider thought about the well-known geologists' mantra that "the present is the key to the past."

Hearing C. P. Snow's subsequently famous (or notorious) lecture at Cambridge on *The Two Cultures* (1959) had alerted me to the wider implications of the contrasting cultures of "scientists" and "non-scientists" in the academic world (at this time, the word "culture" was beginning to be widely deployed in contexts far beyond social anthropology). But I could not recognize either myself or my academic friends in Snow's rather crude stereotypes, which were narrowly based on theoretical physicists and literary scholars, respectively, and not, for example, on geologists and historians. However, setting aside the often acrimonious "two cultures" debate, I found myself thinking about both the similarities and the contrasts between what I was doing in paleontology and what I experienced while dipping my toes into that science's history.

3 III

Above all, a decade or more of immersion and involvement in paleontology gave me firsthand experience of the irreducibly *social* dimension of scientific research. I was deeply impressed by Michael Polanyi's *Personal Knowledge* (1958), which articulated just such a conception of the practical and craft-like character of scientific work, and of its necessarily communal structure, based ultimately on relationships of interpersonal trust. So when later I met Kuhn and then read his newly published *Scientific Revolutions*, many of his ideas already seemed to me to be quite familiar and convincing. More specifically, I realized that the Earth sciences and their practitioners had a distinctive culture, although, like M. Jourdain speaking prose, I had been quite unaware of it. To mention just one of its characteristic features, geologists relied on *visual* communication far more than many other scientists: all the way from needing to explain our tentative ideas by sketching diagrams on the back of a proverbial envelope, to needing to design and produce lots of high-quality drawings, maps, and sections, and to use lots of carefully designed photographic images, to present our evidence and conclusions in publishable and persuasive form. Conversely, in contrast to many other kinds of scientist we were often much less dependent on mathematical, quantitative, and statistical techniques, which in my case was just as well since I knew I had no talent in those directions.

Around the same time, I began to experience a contrasting scholarly culture, when I mixed with the small Cambridge group of historians and philosophers of science (HPS). Its intellectual leader, the Kant scholar Gerd Buchdahl, who was strongly encouraging my growing interest in this field, asked if the geology department could offer some lectures on that science's history, and Bulman, knowing my interest in that topic, assigned the task to me. It gave me an incentive to read much more widely in primary sources—I soon learnt that I could not rely much on the few secondary sources then available—and in due course I gave a series of lectures on the history of paleontology. Eventually I developed them into *The Meaning of Fossils* (1972), my first book written as a historian. The title signaled my intention to focus on how these puzzling objects had been interpreted in the course of Western history, and on what they had been taken to *mean*: ranging from the early modern period when there was intensive debate about their organic or inorganic origin (or, as it turned out, both), to the mid-nineteenth century when they were viewed mainly in the context of evolutionary theories (practical limitations prevented me from taking the history any further toward my own time). But I argued that this had been no smooth ascent from error to truth, and that many of the uncertainties and confusions had resulted from the complexities of fossilization and the nature of the fossil record (which were more apparent to a paleontologist than to most historians).

By the time this book was published, I had made the huge decision to try to turn myself into a full-time historian, and I had moved from the geology department to what later became Cambridge's HPS department. Early on, my induction into the culture of historical research was greatly deepened by taking part in a special seminar that brought together some youngish intellectual historians such as Quentin Skinner and John Dunn and some youngish historians of the sciences such as Piyo Rattansi and Bob Young (with myself as a newish recruit). Several outstanding senior scholars such as Joseph Needham, Ernst Gombrich, and Frances Yates were invited to join us in turn to discuss their own research. They dramatically enlarged my perspective by introducing me to the idea that some non-Western cultures had had their own histories of sophisticated scientific activity, and that in the early modern centuries even European cultures had had lively traditions of investigating the natural world—such as astrology, alchemy, and natural magic—that did not deserve to be dismissed as outdated and worthless but investigated as potentially valuable for understanding the past history of natural knowledge.

Thinking a lot about the methodology of my own research gave me further incentives to explore in more detail the history of my own kind of science. I was particularly intrigued by Charles Lyell, whom geologists usually treated as the heroic "uniformitarian" goodie who had vanquished "catastrophist" baddies such as Cuvier. I was struck instead by their contrasting but equally fruitful attempts to understand the history of the Earth and its life. When my fossil work took me to Sicily to hunt for Permian brachiopods, I took advantage of being there to climb almost to the summit of Etna (dormant at the time) to follow Lyell's early fieldwork on the spot, trying to *see* the huge volcano and the rest of Sicily, as he had, as evidence for the vast timescale of the Earth's history and also its sheer eventful complexity. This was the first of my many attempts to "re-tread" the fieldwork of early geologists in detail, as a method potentially as valuable as the "re-staging" of historically decisive experiments in other natural sciences. It later provided me with my contribution to Cecil Schneer's celebrated conference in New Hampshire; this event, and *Toward a History of Geology* (1969), the volume he subsequent edited, marked in effect the birth of this as a coherent field of historical study with its own research community. (Around the same time, I was invited to Yerevan in Armenia—but failed to find funding to get there—for the inaugural meeting of INHIGEO, which aspired to promote the historiography of geology worldwide, but which has been dominated ever since by geologists rather than historians and, during the Cold War years, by those from the Soviet bloc.)

Retreading Lyell's fieldwork on Etna first gave me a taste for using primary sources not intended for publication—in this case, his field notebooks, and letters to his family and colleagues—in contrast to the polished prose of successive editions of his *Principles of Geology*. I began delving into those of his unpublished papers that were accessible (most of those then still at his family home in Scotland were not). In Philadelphia, for example, I found a long letter from William Conybeare, who was usually mentioned in histories of geology, if at all, as a benighted catastrophist. In contrast, I thought his letter was one of the most shrewd of contemporary critiques of Lyell's work, and I transcribed and published it in full. It gave me heightened respect for both sides of their debate, and again a sense that it had been a fruitful interaction within a coherent social group rather than a battle in which one party had vanquished the other.

When in 1975 the centenary of Lyell's death was marked by a conference in London, I chose to give a lecture not specifically on the great man's work but instead on the way that he and other geologists of his era had developed what I termed a "visual language" for their science, combining maps and sections with other visual images of many kinds: These had been derived from diverse and in some cases apparently unrelated precedents such as records of mine workings, picturesque "topographical" landscapes, and offshore coastal views intended for use in navigation. I was much impressed by William Ivins' *Prints and Visual Communication* (1953), and more generally by works on art history, from which I was led, for example, into the history of cartography, and the practical details of prephotographic techniques such as engraving and lithography. My long paper on the early history of geology's "visual language" was published in *History of Science* (1976): Its illustrations became that journal's first-ever publication of so many visual images. But in the short term, my paper had little impact on historians of the sciences, such was their strongly nonvisual culture, still dominated by literary and textual traditions; it had a much greater impact on sociologists, thanks mainly to Bruno Latour's interest in my work. Only many years later, to my bemusement, did historians start to refer retrospectively to my paper as a "pioneer" work on visual representation in the sciences.

By the time this work was published, I had made another major decision, namely, to leave Cambridge. I was appointed the academic successor of the Dutch historian R. Hooykaas, whom I already knew both personally and as the author of important work on the fundamental "principle of uniformity" in sciences such as geology. With his encouragement, I hoped not only to continue and extend his kind of research but also to replicate in Amsterdam the model of interdisciplinary "science studies" that was being developed with much success by the small Unit of that name at Edinburgh, which aimed at combining historical, philosophical, and sociological approaches to the study of the natural sciences (as represented there by Steve Shapin, David Bloor, and Barry Barnes, respectively). My attempt to develop something along these lines in the Netherlands eventually unraveled, mainly due to the fallout from events in the wider world of Dutch politics (in which I was not personally involved). But living on the continent of Europe for several years, learning a new language from scratch and being immersed in an unfamiliar culture, had a profound impact on my intellectual outlook. It made me aware of the linguistic and cultural limitations of the Anglo-American academic world that until then I had taken for granted. More specifically, it made me realize that, in exploring the history of any important scientific issue, it was essential to take into account whatever work had been done by the relevant scientists throughout the scientific world of the time, whatever the country in which they had been living and working, and whatever the language in which they had published their results. I first tried—as far as my limited linguistic abilities allowed—to follow this "transnational" and multilingual policy in my research during the years when I was living and working in Amsterdam, though it did not become explicit in my work until later years.

4 IV

Several years earlier, serendipitous events had given me unanticipated access to a rich hoard of unpublished geological manuscripts, which reinforced my inclination to focus my historical research on the decades in which Cuvier, Lyell, and many other distinguished geologists had done their best work. An indirect descendant of George Greenough, the first president of the Geological Society in London (the first body of its kind in the world), asked me to sort out those of his papers that she had inherited. I found they included letters from almost everyone who was anyone in the geological world of the early nineteenth century, certainly those in Britain but some from much further afield. I used them to publish an account of the foundation and very early history of the Society that Greenough had regarded, with justified pride, as his baby, and also an account of some of his early fieldwork, which impinged on another debate (Plutonists vs. Neptunists) generally depicted, like the Lyellian one, in Manichaean terms of goodies and baddies. Again, it seemed to me to have been instead a fruitful debate within a community of individuals with diverse viewpoints and experience, interacting sometimes with acrimony but more often in a civil manner, and in which neither side, seen in retrospect, had been in any simple sense correct or incorrect.

The value of this viewpoint was reinforced decisively after I found among the Greenough manuscripts a bundle of letters labeled by him as "Great Devonian Controversy." They were from many of the leading geologists of the 1830s, Lyell among them, and they convinced me that I had stumbled on the tip of an outstandingly important historical iceberg. The matter in contention was initially a question of stratigraphy and structural geology of relatively local significance, centered on the rocks of Devonshire in the west of England, but it had implications that soon opened out into the global understanding of the Earth's history, and it also had major practical implications in relation to prospecting for coal. In subsequent years, I transcribed these and many other relevant letters conserved in archives around the world, reuniting both sides of many exchanges, and of course I also studied lots of relevant scientific articles and books published at the time. I used them all to try to reconstruct the course of what turned out to have been complex arguments among a group of geologists—initially in Britain, but soon drawing in others in continental Europe and eventually some as far afield as Russia and America—which included the most distinguished of their time, but also many relatively obscure people who had provided decisive local information. Based on this almost uniquely rich documentation, I tried to construct a detailed narrative of the complex course of the argument, which could be traced month by month and sometimes even day by day (thanks in part to the efficient postal system of the time), over the course of the 1830s and 1840s. This lengthy narrative, and my interpretation of its significance for understanding the dynamics of scientific argument more generally, was eventually published as *The Great Devonian Controversy* (1985). This was, in my opinion, my most important contribution to the historiography of the Earth sciences; certainly it attracted wider critical attention, both positive and negative, than any of my later historical publications.

One feature of my Devonian narrative provoked criticism that to me was unexpected (for example, from Steve Gould, usually a supporter of my work). This was my insistence on describing each successive phase of the complex argument with absolutely no forward reference to the eventual resolution of the controversy or the modern interpretation of the empirical evidence under scrutiny. But I regarded this literary strategy as essential for enabling the readers of my narrative to experience the same confusions and uncertainties as the original actors, the same gradual clarifications and shifting plausibilities of alternative interpretations, and the same eventual resolution of the controversy. I treated the obscurity of the controversy in the modern scientific world as a historiographical advantage: While the narrative was tracing how it had developed over time in unanticipated ways, readers who were well-informed geologists would be just as much in the dark as those who had never previously heard of the idea of "Devonian." But in fact I complemented this rigorously nonretrospective account with a final and substantial part of the book, in which I tried to analyze the controversy "out of earshot of the natives" (a metaphor borrowed from Cliff Geertz's anthropology), with as much retrospective insight as might be needed.

Two further concepts or methods came to govern my interpretation, which I first tried out on a small scale in a study of Darwin's early work (much of it geological), utilizing the superb detailed research by the scholarly "Darwin industry." In place of the usual dichotomy of primary sources—as being either unpublished and private (notebooks, correspondence, etc.) or published and public (articles, books, etc.)—I suggested a *gradient* or "continuum of relative privacy" reflecting the full range of relevant activities of the historical actors: from strictly private notes, through records of conversations, correspondence, reports of discussions at scientific meetings, draft papers, and fully published articles and books. And in place of the deceptive simplicity of distinguishing sharply between "professionals" and "amateurs," I suggested a "gradient of attributed competence" reflecting the varied kinds and degrees of reliance that the most prominent individuals had placed on reports or materials contributed by those they considered less experienced than themselves, irrespective of how any of them had been earning their living (if indeed they had needed to). These two complementary gradients, although obviously dependent on the fallible judgment of the historian, did at least make full use of all classes of available evidence and also recognized the diverse roles of many individuals and groups of widely varied social position and intellectual caliber: ranging from geologists with international reputations to amateur fossil collectors; university professors to country parsons, lawyers, and physicians; landed gentry to lower-class "fossilists" and quarrymen; and a few women of equally diverse social status.

In combination, these gradients revealed the centrality of a "*core-set*" of less than a dozen prominent individuals, whose eventual agreement that the puzzles posed by the controversy had been satisfactorily solved was *in practice* taken to mark the achievement of the consensus that brought the controversy to an end. This concept of a core-set, first proposed by Harry Collins for arguments within the radically different "Big Science" world of modern fundamental physics, was, I claimed, equally valid for the relatively "little" world of early nineteenth-century geology. This implied that the dynamics of scientific research had not differed across those two worlds as much as a superficial comparison might have suggested. And I proposed that the social category of a core-set had as its cognitive equivalent what I termed a "*focal problem*," of which the Devonian puzzle had been a characteristic example. I argued that the cognitive topography, as it were, of any science at any specific time might be that of a generally "quiet background" of steady uncontroversial cumulative progress, while several focal problems, not necessarily related to each other, might at the same time have disturbed that equable situation, and might have led in the long run to major changes in the science. For example, among focal problems agitating the geological world over much the same years as the Devonian argument, but with no close connection to it, had been the reality and causes of recent or current movements of the Earth's crust, whether violent (as earthquakes) or hardly perceptible on a human timescale; the physical causation of the spectacular large-scale folding and faulting of solid rocks, as seen in the Alps and elsewhere; and the origins of widespread "superficial" features such as boulder clay (till), erratic blocks, and scratched bedrock. (This cognitive landscape of quiet background and focal problems did *not* map readily on to Kuhn's distinction between "normal" and "revolutionary" science.)

The consensual outcome of the Devonian controversy was the novel and unanticipated concept of a "Devonian" system and period, with a unique place in the stratigraphical pile and in the history of the Earth and its life. The analysis of the debate in the text of my book was complemented not only by many images reproduced from the primary sources, illustrating the arguments of the original authors, but also by many diagrams of my own design, intended to explicate my historical interpretation in visual form. Among the varied critical responses to the book, these diagrams showed revealingly the *cultural* divide between natural scientists such as geologists and biologists, and many scholars in the humanities and social sciences. Most of the former found my diagrams helpful for following my argument, whereas some of the latter found them bewildering and even unintelligible (one distinguished critic derided the most complex of these diagrams as looking like the well-known and highly diagrammatic map of the London metro or Underground). What I had wanted all my diagrams to illustrate and clarify was my basic claim that the detailed course of the controversy showed not the eventual triumph of a correct interpretation of the evidence over an incorrect one, but rather a dynamic interaction in which each alternative had been modified again and again by empirical and rhetorical pressure from those advocating another viewpoint, and, above all, that the finally satisfactory and consensual outcome was a novel interpretation that *had not been foreseen by anyone* at the outset, and which was therefore a genuinely new piece of knowledge. I suggested that this pattern of interaction might be found to be quite general in the history of the sciences, if other controversies were analyzed in detail in the same way. This was a conclusion that suggested in turn that neither a naïve realism nor a naïve social contructivism was adequate to describe and understand the process of gaining reliable knowledge of the natural world.

5 V

In contrast to the intensely detailed work involved in reconstructing the Devonian controversy, the demands of teaching soon afterward encouraged me to think on a much broader scale about the history of the Earth sciences. I designed a lecture course that would span the centuries that I had covered (at least for paleontology) in my first historical book—starting in the early modern period—but that would also extend forward through the later nineteenth century and the whole of the twentieth up to my own time. Even if I could not hope to contribute substantially to primary historical research on that later period, I wanted to try to understand the often heated debates in the modern Earth sciences over such issues as the reality or otherwise of "continental drift," and of allegedly "catastrophic" episodes in the deep past and their possible causes (my own earlier paleontological research had involved me, if only briefly and marginally, with problems such as the alleged end-Permian mass extinction, the "Cambrian explosion" of metazoan life, and Precambrian "Snowball Earth" episodes). My survey of the *longue-durée* history of the Earth sciences, although necessarily rather superficial, reinforced my sense that the late eighteenth and early nineteenth centuries—the period on which I had already focused most of my historical research—were somehow of pivotal significance in the wider story. I felt reassured to know that I was far from being alone in this opinion. For example, Karl von Zittel, a product of the outstanding German academic world of the late nineteenth century, had famously defined the period around 1800 as "das heroische Zeitalter der Geologie." This was a historical judgment that I found increasingly convincing, the more I immersed myself in the major works of that period and of a few decades on either side of it: spanning the time from, say, Buffon and James Hutton, through that of Cuvier and William Buckland, to that of Lyell and Léonce Élie de Beaumont (I deliberately pick figures from the francophone world to match the anglophones). And I noticed that the very word *géologie* was a neologism proposed by Jean-André Deluc (a francophone Genevan, but resident in England) at just this time, though with a meaning significantly different from what it soon became, when, for example, it was adopted by those who founded the Geological Society and then passed into general use.

Taking the existing historiography as my starting point, I assumed at first that the key concept at issue in these debates must have been that of the magnitude of the timescale of the Earth and its life. It seemed to be almost taken for granted by modern commentators that up until the late eighteenth century the general belief—even among the educated classes in Western countries—had been that the world had existed for only a few millennia (taking Ussher's notorious 4004 BC, or something like it, as the date of cosmic Creation), but that by the mid-nineteenth century Lyell's vision had become widely accepted, extending the Earth's recorded history back across at least tens of millions of years if not more. If this were true, Toulmin and Goodfield's striking title *The Discovery of Time* (1965) would indeed be an apt summary of a major revolution in human thought about the natural world.

I came gradually to the conclusion that this focus on the Earth's timescale, while not wholly mistaken, was deeply misleading. Early on, I found a small but highly significant hint of this in a then unpublished letter (already mentioned): Conybeare told Lyell quite casually that to account for the Earth's past he (Lyell) was welcome to invoke as many "*quadrillions of years*" as he wished, provided he could show that the evidence required that or any similar order of magnitude (that is, not just up to, but far beyond any modern radiometric estimate). Conybeare pinpointed his fundamental disagreement with Lyell not at all in the timescale, but instead in the overall character of the Earth's *history* that had unfolded in the course of unimaginably lengthy time. Had it been a repetitive series of essentially similar events, the products of unchanging causal processes, with "no vestige of a beginning, no prospect of an end" (in Hutton's famous phrasing)? Or had it been a unique and unrepeated sequence of distinctive *Époques de la Nature* (Buffon's equally famous title)? It soon became clear to me that much of the adulation bestowed on Hutton by anglophone geologists was misconceived (and often fueled by British, or even more narrowly Scottish, nationalistic sentiments), and that the primary purpose of his *Theory of the Earth* was not to argue for a vast timescale, but to prove that the Earth is a physical system designed to subsist in what cosmologists called a *steady state*, from and to eternity: In short, it was an *eternalistic* cosmic theory. Buffon's model, which I called "*directional*," was quite different (and much more similar to modern ideas). It plotted—however conjecturally—a series of "milestones" marking an unrepeated linear sequence of major changes that could have brought the Earth from an unfamiliar distant past to the familiar present, and would take it on into an equally unfamiliar distant future, and all this on a suggested timescale that was vast enough to dwarf any human history, even if inadequate by the standards of modern geology. (Buffon's work was well known at this time to francophone historians of science and others able to read French, and well recognized for its outstanding importance, but—significantly—it was not fully translated into English, either during geology's "heroic age" or indeed until the present century.)

On this issue, I found Lyell a figure of unquestionable importance, but a problematic one. My lengthy involvement with his work, using his early field notebooks, letters, lectures, articles, and other publications as primary sources, culminated in Chicago's facsimile reprint (1991) of the first edition of the *Principles*, for which I wrote a long introduction that summarized my interpretation of his overall strategy (on which, more below). I also put much effort into identifying all the sources cited (often in frustratingly cryptic form) in his footnotes, which greatly enlarged my sense of the multinational and multilingual dimensions of the geological world of the early nineteenth century. Lyell himself had an enviable facility with several languages, above all French (which at this time was as much taken for granted, as the primary scientific language, as American English is today). My own attempt to emulate his coverage of the relevant body of geological knowledge led me to realize the relative marginality of what was being published in English at this time, compared with the mass of publications in French; I noticed how often German and Italian authors, for example, had made sure that their works were also published in French translation, in order to make them fully accessible across the scientific world (again, with an obvious parallel to the role of English today).

Alongside the issue of the Earth's timescale, another problem from Lyell and Conybeare's time seemed to me to be similarly misconstrued by many modern commentators, although likewise not wholly mistaken. This was that the geology of their time had been dominated by arguments about the science's relation to the Creation and Flood narratives in the book of Genesis. Gillispie's readable *Genesis and Geology* (1960), while in fact dealing almost exclusively with early nineteenth-century Britain, had popularized the view that the principal culprits in opposing any Lyellian (or modern) magnitude for the Earth's timescale had been the proponents of a supposedly literal interpretation of the opening texts of the Bible. But this was another point at which I felt it was crucially important to recognize the social dimension of the early nineteenth-century debates. There had indeed been many religious people at that time, especially in Britain, who argued for—or just took for granted—a literalistic interpretation of these and other biblical texts: in some cases from plain ignorance of any alternative; in others, in knowing rejection of biblical research that they regarded as both religiously and socially subversive. And these so-called "*scriptural geologists*," while far more prominent in Britain than elsewhere, had clearly been the forerunners of the "creationist" fundamentalists who are flourishing today, most of all in the United States.

However, the common modern assumption of a perennial "conflict" between supposedly monolithic entities labeled "Science" and "Religion" was only credible if one ignored the many religious voices which—even in nineteenth-century England—had rejected biblical literalism in favor of hermeneutics that acknowledged the historical and cultural contrasts between the world of ancient Judaism and their own modern world. So, for example, Conybeare was a highly respected geologist and the principal author of the standard summary of British stratigraphy in the 1820s, but also a prominent intellectual in the Church of England and one who deplored his fellow-clerics' ignorance of scholarly German biblical research. His one-time colleague Buckland, in contrast, began his geological career by advocating a rather literalistic interpretation, attributing many "Superficial" deposits to the biblical Flood, which went down well at Oxford. But over the years, he and other geologists shifted gradually to arguing for a "geological Deluge" that might in reality have been something like a megatsunami around the dawn of human history, possibly near-global in extent, of which the biblical Flood story might just be a faint and even garbled memory. (Modern inferences about the likely global effects of a major asteroid impact in the much more distant past should remind us that there was nothing intrinsically implausible, let alone nonnatural, about ideas of this kind.) Other geologists, in Britain and beyond, agreed with this "*diluvial*" interpretation of the Superficial deposits and other features but reckoned that the field evidence showed that the putative megaevent must have dated from long before any recorded human history, thereby detaching it altogether from the biblical story.

The more closely I looked at these "diluvial" debates in early nineteenth-century geology, the more it seemed to me that their historiography was deeply influenced by the much more recent cultural history of the United States. It seemed no accident that the basic metaphor of "conflict" and the hypostatizing of the unhistorical categories of "Science" and "Religion" (usually with their initial capitals!) were derived mainly from late nineteenth-century American authors. And they seemed to have been reinforced in the United States, then and ever since, by the striking polarization between secularist and religious societal forces—the latter being dominantly fundamentalist in character—which often appear, at least to an outsider, equally dogmatic and mutually intolerant. In contrast, I felt that as a European I had benefitted from working in intellectual environments, both in Britain and the Netherlands, where scientists—even the atheists among them—generally respected scholarly theology as much as any other -ology, because they could hardly avoid recognizing that among their colleagues at least some theologians were at least their intellectual equals, and could not just be dismissed as ignorant, stupid, or deluded. So I felt that as a historian I should be free to explore the course of earlier debates about the relation between the natural sciences and religious beliefs and practices, and to try to evaluate them evenhandedly, without being suspected of having unacknowledged apologetic motives. My own mainstream (i.e., nonfundamentalist) religious commitments only impinged on my historical research in that I did not approach my sources with the indiscriminate hostility to religious ideas that I noticed in some of my American colleagues.

6 VI

A sense of the possibly positive historical role of ideas with roots in biblical scholarship first emerged unexpectedly from my attempts to identify just what it was that had made my favorite historical period so "heroic" for the Earth sciences, if in fact this could not be attributed simply to an enlargement of the Earth's timescale. As a research project that I intended to be preparatory to a more comprehensive review of the Earth sciences in that period—or perhaps unconsciously as a way of postponing that formidably daunting task—I began to assemble examples of early attempts to imagine *visually* what the Earth and its life might have looked like before there were any human beings to observe and record it. The history of such images, which eventually became the basis for my album of *Scenes from Deep Time* (1992), seemed at first a straightforward story in which Cuvier's authoritative reconstructions of the skeletons of extinct mammals from their fossil bones had been extended by him and other naturalists into depictions of those and other animals, bodily restored and living in their likely habitats. At first, these reconstructions were not formally published, and were disseminated only in marginal and even jocular forms, as if to indicate that they were not to be taken too seriously, but before long they were being put forward more confidently as legitimate inferences from the fossil remains of the organisms and from preserved evidence of their environments.

By the middle of the nineteenth century, this new genre of imagined "scenes from deep time" was well established. Magnificent sequences reliably based on fossil evidence, such as those by the botanist Franz Unger (or rather, by his highly accomplished artist, though under his direction), depicted the long-term *history* of life as geologists' research had made it known; some such reconstructions extended even to life-sized three-dimensional models, such as the famous Crystal Palace monsters. For what the newly named "palaeontologists" had established beyond reasonable doubt, in collaboration with the newly named "stratigraphers," was that the Earth had indeed had a true *history*, in the Buffonian sense of a linear sequence of unrepeated phases or periods, each with its own distinct forms of life. Its constituent events and processes might have been identical in kind and more or less equable in intensity throughout, as Lyell the "*uniformitarian*" asserted, or they might have been occasionally more sudden and violent than had ever been experienced in the relatively brief span of recorded human history, as his "*catastrophist*" critics such as Élie de Beaumont claimed. The philosopher William Whewell, proposing that pair of labels, recognized that Lyell's sect was very exclusive, since he himself was its only prominent advocate (though joined later, to some extent, by the young Darwin). But even Lyell, like his catastrophist critics, acknowledged and indeed emphasized that the now flourishing science of geology was centrally concerned with reconstructing the long-term *history* of the Earth.

While piecing together the origins of this novel genre of "scenes from deep time," I realized how closely their pictorial conventions were derived from earlier artistic traditions of depicting *human* history. Delving, however superficially, into art history had taught me how scenes based on the Bible had long been depicted in just the same style as those from Classical history: Both sources were of course central to the Western cultural sense of the distant human past. But the well-established genre of biblical illustrations, as displayed spectacularly, for example, in Johann Scheuchzer's huge *Physica Sacra*, had also clearly provided inspiration and precedents for the first attempts to depict even earlier phases of the Earth's history, recruiting the newly appreciated evidence of fossils. No iconographical change of gear, as it were, was required in order to move from depicting the successive "days" of Creation in conventional fashion—culminating typically in the Garden of Eden, stocked with modern animals as well as Adam—to showing wholly extinct species convincingly "resurrected" (Cuvier's metaphor) from their fossil remains. In the early nineteenth century, this tradition of employing the same pictorial conventions for human history both biblical and secular, and of using both as a joint template for scenes from the deepest prehuman past, was flourishing more than ever. For example, John Martin's large paintings of the biblical Flood and of the destruction of Roman Pompeii, both depicted in the same melodramatic *martinien* style, were both profitable (once they were converted into inexpensive prints for the expanding middle-class market). And the same genre was being deployed equally effectively to depict the newly discovered prehuman history based on fossil evidence. Henry De la Beche famously depicted *Duria antiquior* ("an earlier Dorset") swarming with ichthyosaurs and plesiosaurs (for the financial benefit of the "fossilist" Mary Anning, who had found some of the best specimens), and Gideon Mantell commissioned Martin to depict his newly discovered iguanodons, resurrected in the long-vanished landscape of an earlier Sussex. These were just some of the first examples of a pictorial genre that has of course continued ever since—now enhanced by digital technology—to be highly influential in conveying to the general public a sense of the long-term history of the Earth and its life.

That "scenes from deep time" could also be effective for expressing more fundamental arguments in geology became apparent in the case of De la Beche's widely circulated print entitled "Awful Changes." This showed—in the style of a contemporary political caricature—a "Professor Ichthyosaurus" explaining a *fossil* human skull to a lively audience of ichthyosaur students and other reptiles all known in reality only as (Jurassic) fossils. De la Beche's sequence of draft sketches for this caricature fortunately survived at the back of one of his field notebooks. They confirmed what should have been apparent from the design and wording of the circulated final image: This was no light-hearted parody of a novel pictorial genre, but a deeply serious criticism of the theoretical goals of Lyell's newly published *Principles*. It was an imagined scene from the distant posthuman *future*, when—as Lyell himself had suggested unambiguously in print—organisms more or less similar to those being found as fossils in ancient strata might quite possibly *return*, as the Earth continued to turn on an unimaginably lengthy cycle of geographical and environmental change. As one of De la Beche's sketched alternative designs hinted more clearly, Lyell was reviving Hutton's cyclic theory, but reconciling it with the newer evidence of an apparently linear or "directional" fossil record, by shrinking the entirety of the Earth's known history into just one part of one far lengthier cycle. Another alternative design would have made the same point with different imagery: It showed Father Time holding a chemical balance with Europe and Africa in one pan and America in the other, and with the pans (or crustal plates) oscillating endlessly like a child's see-saw on a timescale shown on a clock calibrated in "millions of centuries." (The final caricature, reproduced to illustrate my article analyzing the whole sequence of sketches, prompted *Isis* to decorate its front cover—for the very first time!—with a pictorial image.)

This interpretation of Lyell's *Principles*, as having at its heart a controversially steady-state and cyclic theory of the Earth's history, might seem to have been incompatible with the more familiar reading of his massive work, as being designed primarily to advocate a thoroughly naturalistic view of the world; and specifically to refute the belief, still widespread among the reading public, that the Earth's history was reliably summarized in the Creation stories in Genesis. Here again I felt that if the social dimension were taken into account, the historiographical either/or would become a both/and. Lyell himself made it clear that he was aiming his work primarily at his scientific equals at the Geological Society in London and at "geologists" everywhere else in the world, but at the same time he also hoped it would influence and help to persuade intelligent and educated "general readers" to adopt a more scientific and secular view of the world. His *Principles* was one of several major works published in Britain around this time, which explicitly shared—or at least aspired to—this kind of dual readership or, more precisely, a continuum of readers from the most to the least well-informed about the science.

My interpretation of Lyell's model for the Earth's history was confirmed over the years, at least in my opinion, by more detailed study of the *Principles*. I took seriously Lyell's own claim that his third and final volume—often neglected or even ignored by historians who focused too exclusively on Lyell's impact on the young Darwin—was designed as the culmination of the whole massive work. It used the exhaustive inventory of currently active processes ("present causes"), as set out in the two earlier volumes, as having just provided the basic "alphabet and grammar" needed to decipher the otherwise unknown natural "language" in which the Earth's long-term history had been recorded. Lyell reconstructed the more recent part of that history in *retrospective* sequence, backward from the present into increasingly strange "Tertiary" (later, Cenozoic) former worlds. He even quantified the lapse of time (though not in years) by the rate of change in their molluscan faunas, adopting Whewell's suggested labels of Pliocene, Miocene, and Eocene; and he hoped that in future this kind of chronology would be extended back into the still more remote worlds recorded in "Secondary" and "Transition" (later, Mesozoic and Palaeozoic) rocks and fossils. Far from abandoning or negating the *historical* goals of the whole work, this retrospective strategy simply followed Cuvier's earlier prescription that geologists should adopt a methodological policy of penetrating from the known present back into the unknown past, in order to then construct a forward narrative of the Earth's true history. However, Lyell redefined and radically reinterpreted the apparently still older rocks that other geologists called "Primary"—with no fossils at all—as being either "*metamorphic*" or "*plutonic*" in origin. This left the Earth with no record whatever of its earliest history, and it therefore left Lyell's controversial cyclic theory as impossible to prove or disprove.

7 VII

My survey of geological publications in the decades around 1800, however superficial, had already made me recognize the centrality, as perceived at the time, of another individual whose achievements had certainly matched Lyell's. This was Cuvier, whose prominence in the early nineteenth-century scientific world has been obscured by his later—and unjustified—double dismissal, both as an opponent of the idea of organic evolution and as a catastrophist in geology. I was first alerted to this historiographical distortion by clues in unpublished sources. For example, among Cuvier's voluminous correspondence conserved in the Paris archives I found a letter in which Lyell had warmly urged his French hero—as Cuvier evidently was—to publish an English edition of his famous *Discours Préliminaire*, in order to correct the misleading impression of his ideas conveyed by Robert Jameson's successive editions in English. And in Lyell's manuscript notes for his early public lectures, I noticed an appreciative and evidently sincere eulogy of Cuvier, news of whose unexpectedly early death had then just reached London.

Clues such as these just extended what E.S. Russell's classic work had taught me decades earlier: Cuvier had been a figure of towering importance not only in the history of biology but also in the history of geology. The more I immersed myself in Cuvier's published work—and later, in his unpublished papers in the Paris archives—the more I realized how disastrously Jameson had distorted Cuvier's ideas for his English-speaking readers. Not only did Jameson claim, mistakenly, that Cuvier had intended to promote the kind of "scriptural" geology popular in Britain at that time, with all its social and political implications. No less seriously, he had misrepresented Cuvier's attitude to geology by giving the English edition the title "Theory of the Earth." In fact, as his papers showed, Cuvier had publicly poured withering scorn on what that phrase denoted, namely, a distinctive scientific genre.

Any work entitled "Theory of the Earth"—of which there were many, Hutton's now being just the best known in English—was a would-be comprehensive explanation of *all* the Earth's major features and processes: comparable in its ambitions to, say, Newton's theory of gravitation in cosmology (I have suggested the neologism "*geotheory*" for this kind of work, to distinguish it from ordinary geological theories of more limited scope and more modest ambitions). It was no accident that Cuvier did *not* give his own major geological essay the title of "Theory of the Earth"; instead he called it a *discours préliminaire*, which implied that its affinities, and its author's sympathies, lay with the celebrated introduction to the great *Encyclopédie* of the Enlightenment.

I tried to complement the existing fine analyses of Cuvier's biology (most notably Bill Coleman's) with a study of his geological ideas. The link was of course Cuvier's research on fossil mammals (and, later, other fossil tetrapods). Using his flair for comparative anatomy, his work had begun with his sensational identification of the South American *Megatherium* as a gigantic edentate, and with his confirmation that both the Siberian mammoth and the North American mastodon were distinct from either of the living species of elephant. This brought the controversial issue of *extinction* to the center of debate among naturalists. As Cuvier claimed, these fossil remains of large terrestrial mammals, otherwise unknown, offered the strongest evidence in favor of the reality of extinction: They counted against the most plausible alternative, namely, the "living fossil" argument that any fossil species unknown alive might still be flourishing in some unexplored continental interior or in the ocean depths (as exemplified by the modern discoveries of *Metasequoia* and the coelacanth). Once extinction was accepted as a real feature, and one not always attributable to human activities (as the fate of the dodo on Mauritius was), it also became easier to envisage the natural world as having had a true past *history*, by showing that (to adopt a famous later saying) it was a foreign country where they did things differently. Cuvier's analysis of form and function in living animals, making adaptation an even more central concept than before, then became his key to the reconstruction of the skeletons of putatively extinct mammals from their scattered fossil bones, and hence to their spectacular "resurrection" as animals as lively in the mind's eye as the living mammals in the menagerie next to his house at the natural history museum in Paris.

Cuvier's fruitful collaboration with the director of Sèvres, the mineralogist Alexandre Brongniart, in their joint survey of the "Tertiary" stratigraphy of the Paris Basin—it was Cuvier's only significant geological fieldwork—led them, on the evidence of distinct assemblages of fossil mollusks, to interpret the sequence of formations as the record of an alternation of marine and freshwater conditions. The details of the sequence suggested that these changes in environment had been sudden, and Cuvier took them to be examples of the major physical changes that most geologists believed had occasionally punctuated the Earth's immensely long history. They saw evidence for such "*révolutions*" in, for example, the huge scale of folding in solid rocks and the occurrence of fossils of obviously marine origin, high on mountain ranges such as the Alps and Pyrenees. The inference that such major events must have been sudden was rooted in the highly respected Newtonian maxim that imputed causes should always be proportionate to their observed effects, though it must have gained in plausibility from its metaphorical analogy with the repeated *coups d'état* and other dramatic political events that Cuvier and his contemporaries had witnessed from 1789 onward. Anyway, it was taken for granted that these drastic *révolutions* in the natural world were all due to *natural* causes of some kind.

As already mentioned, Cuvier recognized that the most effective method for the new science of "geology" was to work backward from the observable present world into the unfamiliar world of the evidently remote prehuman past. The crucial boundary appeared to be marked by traces of the most recent major *révolution*, such as the deposits later called "Superficial," many of which seemed to be unrelated to observable causal processes. Cuvier, like many of his contemporaries, thought this major event might have been none other than the Flood recorded in Genesis. This "diluvial" inference initially reassured the religious conservatives in French society and caused consternation among their secularist critics: Both factions wanted to recruit this outstanding naturalist to their own cause. But in fact Cuvier, who had earlier adopted the role of a Classical scholar in order to demonstrate that all the animals described in antiquity were of species still known alive, also used that kind of literary scholarship to review the evidence for a possibly worldwide "Deluge," even earlier and at the dawn of human history. This was, for Cuvier, a typical cross-cultural Enlightenment exercise. He evaluated the Flood story in Genesis in just the same way as all the similar stories—which he reviewed at much greater length—from the Classical world and all other known ancient civilizations, as far afield as China. He took all of them to be faint and perhaps distorted memories of a real and drastic natural event in the earliest history of humankind: not a unique event, but the most recent in a lengthy series of *révolutions* that had punctuated the immeasurably long prehuman history of the Earth.

My attempt to understand and do justice to Cuvier's geological research, and most importantly his outstanding work with fossil vertebrates, took me increasingly often to Paris—always a personal pleasure—and to the Cuvier archives, where I compared his draft texts and lectures with the published versions, and found many revealing revisions. I also read as widely as I could in his extant correspondence. As Dorinda Outram had demonstrated, this reflected his spectacular rise to scientific prominence throughout Europe, even during the long years of war, partly as a consequence of his powerful institutional positions at what was then the very center of the scientific world. Eventually I decided to make new translations into modern English, not only of Cuvier's immensely influential *Discours*—in the hope of replacing, for anglophone readers, Jameson's misleading nineteenth-century editions—but also of the more important parts of his earlier published papers on fossil mammals, which he had reprinted in the great four-volume *Ossemens Fossiles* to which his *Discours* was *préliminaire*. My *Georges Cuvier, Fossil Bones and Geological Catastrophes* (1997) was thus an anthology rather than a biography, but each text was introduced with a brief review of its historical significance. I hoped the book would help to revive interest in and appreciation of Cuvier's research among historians of the Earth sciences, and also be of some use to historians of biology.

8 VIII

The major effort required to prepare the translations for my Cuvier anthology caused further postponement of my long-term plan to write a comprehensive account of the "heroic age of geology" in its wider historical context. In retrospect the long delay was worthwhile, because it encouraged me to read more widely in the geological books published in the main scientific languages of that time, and at least to skim through the equally multilingual scientific periodicals—which in these decades were rapidly expanding in range—to assess their coverage of my theme and to read more closely the articles most relevant to it. Duly daunted by the magnitude of what I reckoned was needed, I tried to decide on parameters that would make it manageable.

I wanted to trace the emergence of the new science of "geology," primarily within what was recognized (if not by this name) as the international "scientific community" of the time. This would mean focusing on those who were often referred to, and not just in French, as "*savants*," a term that I adopted in order to avoid the anachronistic "professional" overtones of the modern word "scientist." Conversely I would have to give less attention to the impact of their work on wider social circles, and vice versa; I had to accept that in the intellectual and political climate of the early twenty-first century my account, when published, would probably be dismissed disparagingly as both internalist in scope and elitist in sympathies. My review of materials from the later eighteenth century seemed to me to confirm Zittel's judgment that there had been major shifts of thinking about the Earth at just that time, but I wanted to establish a baseline for my account that would avoid an indefinite regress into the evidence of earlier decades and centuries. I decided on a literary device based by analogy on modern stratigraphical practice, namely, to choose fairly arbitrarily a date that could act as a "golden spike" driven into the seamless flow of history, and to try to survey what the world of the Earth sciences had looked like, at that specific point in time, to any well-informed participant. I chose the year 1787, in which the self-styled "physical geographer" Saussure became the first savant, and one of the first human beings, to stand on the summit of Mont Blanc, the highest spot in Europe; it was also almost the last year of the *ancien régime* and almost on the eve of political revolutions that soon convulsed the entire Western world. Although my literary device was criticized by some later commentators (including one eminent geologist who ought to have understood my metaphorical "golden spike"), I think it worked quite effectively as a prelude to my survey of the varied sciences of the Earth as they were being practiced at just that time.

Next, then, I had to sketch the implicit map of natural knowledge, as it was understood around 1787, to see where any sciences dealing with the Earth had been located. This involved cutting through a lot of presentist historiographical thinking about the relations between the sciences, which tended to treat terms such as "physics," "biology"—and notably "geology"—as suprahistorical and timelessly permanent in meaning. In contrast, a primary distinction was made at the time between "*natural history*" and "*natural philosophy*," between sciences dealing with the *description* and *classification* of natural entities of all kinds, and those that focused on the *causes* and *explanation* of events and processes in nature. (The history of the classification of the sciences, or of the changing map of knowledge, is in my opinion a seriously underexplored research topic, and I recognized my own efforts as provisional and perhaps even amateurish.)

In the late eighteenth century, "natural history" retained its traditional breadth of meaning, covering the description and classification not just of animals and plants but also minerals and fossils, and all other natural entities, even nebulae: Most importantly, the word "history" itself still simply meant description and had no necessary connotations of any dimension of time. At this period, there were three distinct natural-history sciences that dealt with features of the Earth. "*Mineralogy*" was a science of specimens, collected in the field but studied indoors and conserved in museums or private "cabinets"; it described and classified not only minerals in the modern sense but also rocks and fossils. "*Physical geography*" was a spatial science, studied primarily in the field and then published as textual descriptions, landscape views, and maps; it described and classified topographical features such as mountains and coastlines, rivers, and volcanoes. "*Geognosy*" was a structural science, studied wherever rocks could be seen in three dimensions, in coastal cliffs, inland quarries, and underground in mines; it described and classified rock structures such as folds and faults, and piles of rock "formations"; the latter were usually described from the top downward, with no reference to temporal sequence. ("Geognosy" was *not* used to denote a particular theory about the causal origin of the rocks, contrary to what was often asserted in older secondary sources; it was not a "theory of the earth" opposed to Hutton's).

Like the word "history" in natural history, natural philosophy retained the traditional breadth of meaning of "philosophy" (significantly, the world's oldest scientific periodical was called "philosophical"); it included sciences such as "*physics*," which dealt with the causal analysis of natural phenomena of all kinds, not just those covered by the modern science. Specifically, "*physique de la terre*" (for which there was at the time no common term in English, and which was not synonymous with modern "geophysics") dealt with phenomena such as earthquakes and with all the causal processes involved in the formation of mountains and valleys, volcanoes, and other features described and classified by any of the natural-history sciences. All such causal analyses (like analyses of chemical reactions, for example) necessarily involved a temporal dimension of "before and after," but this did not carry any indication of their place in *the history of the Earth itself*. For this, I adopted the modern scientific term *geohistory*, in order to avoid having to distinguish it repeatedly from "history" in the everyday sense of *human* history. (Regrettably and confusingly, "geohistory" has recently begun to be used in some circles with the totally different meaning of the *historiography* of the Earth sciences: In the correct sense, I was doing geohistory in my first career, when I plotted the place of brachiopods in marine environments from the Cambrian period onward, but *not* in my second career, as a historian of the Earth sciences.)

My review of the practice of these four sciences of the Earth in the late eighteenth century—for which I relied on outstanding research by historians such as François

Ellenberger, Rhoda Rappaport, and Ken Taylor—highlighted the rarity and marginality of truly *geohistorical* interpretations of what eighteenth-century savants reported, in contrast to the ubiquity of their natural-historical descriptions of what they observed, and/or their natural-philosophical explanations of the likely causal origins of the same features. In contrast, Desmarest's reconstruction of three distinctive *époques* in the geohistory of Auvergne, as inferred from its ancient volcanoes and their lava flows, and Giraud-Soulavie's rather similar analysis of those in Vivarais, stuck out as conspicuous exceptions, unusual for their detailed reconstructions of successive phases in the vastly extended prehuman *history* of central France. What I found difficult to express adequately, because it is so alien to the thinking of modern Earth scientists, was the marginal role, even late in the eighteenth century, of such genuinely *historical* thinking about the Earth. There were plenty of ideas about the time dimension, most notably in the distinctive genre of "Theory of the Earth," but what they lacked was a clear sense of the *unpredictable contingency* of historical events as experienced in the human sphere. So it should not have been surprising to find that truly *geohistorical* thinking first emerged in the Earth sciences in the form of deliberate and explicit analogies with *human* history.

This became apparent in the metaphors that savants such as Soulavie chose to use to explain what they were doing. They drew on analogies with several kinds of (human-) historical scholarship that were flourishing in their day, adding the word "*nature's*" to its key terms. From *chronology*, or the compilation of accurate dates for events in human history, they derived the idea of "nature's annals," which could be divided by the decisive events that were perceived as "nature's epochs" (using a word that then, as in Buffon's work already mentioned, denoted events rather than periods of history). From *chorography*, or the compilation of comprehensive accounts of the present character and past history of specific countries or localities, they derived the concept of "nature's documents" conserved in "nature's archives." From *epigraphy* and *numismatics*, they recognized by analogy the necessity of learning "nature's languages" and deciphering "nature's inscriptions," not least those found on "nature's coins" such as fossils. And from *antiquarian* studies (including, in modern terms, archaeology), and particularly the then novel techniques of careful excavation, they learnt how to interpret "nature's monuments" as evidence for reconstructing the vanished worlds of the deep past. All these fruitful analogies depended of course on what was being done by scholars in the humanities, who at this time enjoyed even greater prestige and public approval than the naturalists; most striking of all—and therefore a favorite source of analogies—were the sensational excavations at the sites of Herculaneum and Pompeii, first discovered in the mid-eighteenth century and throwing dramatic new light on the vanished everyday world of the ancient Romans.

Traced forward through the turbulent Revolutionary and Napoleonic years—which disrupted the sciences throughout the Western world, even bringing them, in France, briefly to a standstill—these analogies with human historiography gradually became so familiar that in the end they seemed almost commonplace. Cuvier famously proclaimed himself a "a new kind of antiquarian," who was using fossil bones to "resurrect" the animals of a vanished prehuman world. A generation later,

Lyell, while retreading some of Desmarest's fieldwork in Auvergne, wrote in his notebook a striking essay on the "Analogy of geology & history"; later, in his *Principles*, he concluded his lengthy review of the previous history of geology by comparing the science's future prospects with what the great German scholar Barthold Niebuhr had recently achieved with Roman history. By Lyell's time, however, the analogy with human historiography was being so much taken for granted by geologists that it was less commonly explicit in their scientific work, though it remained useful and prominent in what they wrote with a wider public in mind.

9 IX

I was not of course the first historian to notice, and indeed emphasize, the centrality of this transposition of human-historical ways of thinking into the Earth sciences, but I came to regard it as a major theme, if not indeed *the* major theme, in my interpretation of the history of these sciences during their most "heroic" age. A pivotal point, in my opinion, was Buckland's celebrated forensic analysis of the fossil bones in Kirkdale Cave in northern England, which earned him the Royal Society's highest award and praise from naturalists everywhere. Although much less spectacular than many well-known bone caves in continental Europe, Kirkdale was unprecedented for Buckland's detailed reconstruction of a whole ecosystem of extinct animals in a vanished world. And the caricature attached to Conybeare's jokey poem about it—portraying Buckland himself crawling into the cave, to the surprise of the resident hyenas—highlighted his achievement of effective science-based time-travel: The geologist himself was crawling into a scene from deep time, able to describe this vanished world as if he were an eyewitness. Buckland saw good reason to believe that the event that had annihilated the hyenas, mammoths, and other megafauna had been recent enough to be equated with the biblical Flood, but his Cuvierian "resurrection" of a whole ecosystem remained no less valid in the light of the alternative inference that the geological "Deluge"—construed as some kind of megatsunami—had long preceded any human records.

As I surveyed the international world of geology in the heyday of Buckland, Lyell, and many other heavyweights, I came to believe that historiographically it had been seriously distorted by the modern sidelining of this "diluvial" theory, on no better grounds than the historical association of the idea of a geological Deluge with literalist readings of the Flood story in Genesis. When INHIGEO met at Neuchâtel to review the historical origins of the glacial theory (and also, quite separately, the origins of modern ideas of Alpine tectonics), we were taken to see, and were duly impressed by, huge cottage-sized erratics such as the famous *Pierre à Bot*. Yet there was little discussion of the earlier and fruitful interpretation of erratics, by von Buch and other prominent geologists, in terms of some kind of diluvial event. This explanation was of course eventually superseded by the glacial theory—more lastingly by Charpentier's detailed evidence for glacial extension, than by Agassiz's bizarre "Snowball-Earth" idea—but not before it had served to highlight the sheer scale of a major disruption *of some kind* in the geohistorically recent past. To point out the value of the Deluge idea in the early nineteenth century, as a fairly good—but not perfect—explanation of erratic blocks, scratched bedrock, and suchlike features widely scattered around Europe and North America, is not a surreptitious way of supporting those whose modern representatives attribute the erosion of the Grand Canyon to Noah's Flood.

That the Earth's climate might in the recent past have been far colder than at present was utterly unexpected, because it ran counter to the abundant evidence of formerly hotter conditions, the most obvious being the widespread Coal formation and its tropical-looking fossil flora. The glacial theory undermined *both* of the most long-lasting concepts of geohistory, namely, that the Earth had either been slowly cooling in Buffonian manner from some kind of incandescent origin, or that it had maintained some kind of Huttonian or Lyellian steady state from as far back as any evidence survived. Either way, the glacial theory reinforced the view that, on the contrary, the planet had truly had a *history* as unpredictable—even in retrospect—as human history; from this, it followed that geohistory would always have to be reconstructed, as it were "bottom up," from surviving piecemeal evidence of the deep past, rather than being deduced "top down" from the supposedly unchanging laws of physics.

Having brought my account of "the reconstruction of geohistory"—which I used as my subtitle—all the way from Saussure's "golden spike" ascent of Mont Blanc to Agassiz's sensational "Snowball Earth" conjecture about the geohistorically recent past, and having been obliged by the unanticipated total length of the narrative to split it in two, I published it at long last as *Bursting the Limits of Time* (2005) and its sequel *Worlds before Adam* (2008); the titles alluded, respectively, to Cuvier's programmatic purple prose and to the marginality of the whole of human history in relation to the vast complexity of prehuman geohistory. The narrative was massive, densely illustrated, and heavily footnoted, because I hoped it would alert other historians to a profusion of underexplored sources, many of which I felt I had merely skimmed, and which I hoped would provide starting points for new lines of research, which I could not possibly anticipate.

After a period of authorial exhaustion, I was persuaded by nonacademic friends to write one last narrative, setting my detailed account of geology's most "heroic" half-century in its context, by condensing it to form just the central part of an expanded history, extending back to the seventeenth century and forward to the present day. I used my earlier experience of teaching some very smart undergraduates a course with precisely this scope, adapting it to suit general readers of equal intelligence. This was *haute vulgarization*—as the French admirably define it—intended to be accessible to the same "general public" that Cuvier had hoped would read his *Discours* and Lyell his *Principles*, and which I had not previously attempted to reach. *Earth's Deep History* (2014) won the relevant prizes (for a book written for general readers) from both American and British history-of-science societies, so I like to think I got the intellectual level about right. Reviewing the earlier period, I was most impressed by historical work (especially Tony Grafton's) on the science of *chronology*, the then flourishing scholarly project within which Ussher's dating of Creation at 4004 BC, now widely ridiculed, made good sense, and which deserved to be recognized as conceptually continuous with what twentieth-century scientists—blissfully unaware of the historical roots of the word—have named *geochronology*. For that later period, extending to the present day, I was equally impressed by some of the fine published historical research on major twentieth-century issues such as "continental drift" and "neo-catastrophism," particularly when, as, for example, in Naomi Oreskes' fine work on the widespread *rejection* of what later became plate tectonics, they were not driven by the triumphalism that so often mars historical accounts of recent scientific theorizing. And finally, after my last book was published, I was persuaded to write a couple of articles putting my own early paleontological research, and that of my then colleagues, students, and critics, in its proper context as part of the twentieth-century movement of "palaeobiology." This was my first and only attempt to explore from primary sources (including my own recollections) the terrain of fairly recent science, while wearing my historian's hat.

To conclude this overlong account of my work on the historiography of the Earth sciences: I feel privileged to have benefitted from a first career in which I was deeply immersed in the culture of one of those sciences, which I think gave me an invaluable "feel" for their empirical materials and problems; and then to have tackled the mental and emotional adjustment needed to shift gear into investigating the roots of such problems from the perspective of the quite different culture of intellectual history.

The historiography of the Earth sciences has hitherto been a Cinderella on the wider stage of the historical understanding of natural knowledge of all kinds. In my opinion, it will only gain its proper prominence when the Earth sciences themselves are fully recognized as crucial for understanding our human impact on the future of our planet, and as having had—not least through their gradual disclosure of the Earth's immensely lengthy and eventful past *history*—a profound impact on our evaluation of what in less politically correct times was called "Man's place in Nature."

Competing Interest Declaration The author(s) has no competing interests to declare that are relevant to the content of this manuscript.

Martin J.S. Rudwick: List of Publications

[History of Earth Sciences only]
[Single-author book titles in ***bold italics***]

Hutton and Werner compared: George Greenough's geological tour of Scotland in 1805. *British Journal of the History of Science* 1: 117–35, 1 fig. 1962

The foundation of the Geological Society of London: its scheme for cooperative research and its struggle for independence. *British Journal of the History of Science* 1: 325–55, 1 fig. 1963

A critique of uniformitarian geology: a letter from W.D. Conybeare to Charles Lyell, 1841. *Proceedings of the American Philosophical Society* 111: 272–87. 1967

Lyell on Etna, and the antiquity of the earth. *In* Cecil J. Schneer (ed.), *Toward a History of Geology*. M.I.T. Press. 288–304, 6 figs. 1969. [repr. Rudwick, *Lyell and Darwin*, 2005, art. IV]

Uniformity and progression: reflections on the structure of geological theory in the age of Lyell. *In* D.H.D. Roller (ed.), *Perspectives in the History of Science and Technology*. University of Oklahoma Press, 209–27. 1971. [repr. Rudwick, *Lyell and Darwin*, 2005, art. I]

The Meaning of Fossils: Episodes in the History of Palaeontology. 3 + 287 pp., 52 figs. Macdonalds/American Elsevier. 1972. [2nd ed., Science History, 1976; new impression, University of Chicago Press, 1985; Spanish ed., Blume (Madrid), 1987]

Darwin and Glen Roy: a "great failure" in scientific method? *Studies in the History and Philosophy of Science* 5: 97–185, 8 figs. 1974. [repr. Rudwick, *Lyell and Darwin*, 2005, art. X]

Poulett Scrope on the volcanos of Auvergne: Lyellian time and political economy. *British Journal of the History of Science* 7: 205–42, 4 figs. 1974. [repr. Rudwick, *Lyell and Darwin*, 2005, art. III]

Charles Lyell F.R.S. (1797–1875) and his London lectures on geology, 1832–33. *Notes and Records of the Royal Society of London* 29: 231–63. 1974. [repr. Rudwick, *Lyell and Darwin*, 2005, art. VIII]

Caricature as a source for the history of science: De la Beche's anti-Lyellian sketches of 1831. *Isis* 66: 534–60, 12 figs. 1975. [repr. Rudwick, *Lyell and Darwin*, 2005, art. VII]

The emergence of a visual language for geological science, 1760–1840. *History of Science* 14: 149–95, 25 figs. 1976. [repr. Rudwick, *New Science of Geology*, 2004, art. V]

Historical analogies in the early geological work of Charles Lyell. *Janus* 64: 89–107. 1977 [repr. Rudwick, *Lyell and Darwin*, 2005, art. V]

Charles Lyell's dream of a statistical palaeontology. *Palaeontology* 21: 225–44. 1978. [repr. Rudwick, *Lyell and Darwin*, 2005, art. VII]

Transposed concepts from the human sciences in the early work of Charles Lyell. *In* L.J. Jordanova and R.S. Porter (eds.), *Images of the Earth: Essays in the History of the Environmental Sciences*. British Society for the History of Science, Monograph 1: 67–83. 1979. [2nd ed., 1995, pp. 77–91.]

Senses of the natural world and senses of God: another look at the historical relation of science and religion. *In* A.R. Peacocke (ed.), *The Sciences and Theology in the Twentieth Century*. Routledge & Kegan Paul. 241–61. 1981.

Charles Darwin in London: the integration of public and private science. *Isis* 63: 186–206, 2 figs. 1982. [repr. Rudwick, *Lyell and Darwin*, 2005, art. IX]

The Great Devonian Controversy: The Shaping of Scientific Knowledge among Gentlemanly Specialists. xxxiii + 494 pp., 74 figs. University of Chicago Press. 1985.

The shape and meaning of earth-history. *In* David C. Lindberg and Ronald L. Numbers (eds.), *God and Nature: Historical Essays on the Encounter between Christianity and Science*. University of California Press, 296–321, 1 fig. 1986. [repr. Rudwick, *New Science of Geology*, 2004, art. II]

A year in the life of Adam Sedgwick and company, geologists. *Archives of Natural History* 15: 243–68, 7 figs. 1988. [repr. Rudwick, *New Science of Geology*, 2004, art. XI]

Introduction. *In* [facsimile reprint of] Charles Lyell, *Principles of Geology*. University of Chicago Press. 1: i–lviii. 1990.

Bibliography of Lyell's sources. *In* [facsimile reprint of] Charles Lyell, *Principles of Geology*. University of Chicago Press. 3: Appendices 113–60. 1991.

Scenes from Deep Time: Early Pictorial Representations of the Prehistoric World. xiii + 280 pp., 106 figs. University of Chicago Press. 1992. [Japanese ed., Shinhyoron, 2009]

Minerals, strata and fossils. *In* Nicholas Jardine, James Secord and Emma Spary (eds.), *Cultures of Natural History*, Cambridge University Press. 266–286, 10 figs. 1996. [repr. Rudwick, *New Science of Geology*, 2004, art. III]

Cuvier and Brongniart, William Smith, and the reconstruction of geohistory. *Earth Sciences History* 15: 25–36, 4 figs. 1996. [repr. Rudwick, *New Science of Geology*, 2004, art. VII]

Georges Cuvier, Fossil Bones, and Geological Catastrophes: New Translations and Interpretations of the Primary Texts. xvi + 301 pp., 25 figs. University of Chicago Press. 1997.

Recherches sur les ossements fossiles: Georges Cuvier et la collecte des alliés internationaux. *In* Claude Blanckaert *et al.* (eds.), *Le Muséum au Premier Siècle de son Histoire* [bicentennial volume]. Muséum National d'Histoire Naturelle (Paris) [Archives]. 591–606, 4 figs. 1997. [trans. Rudwick, *New Science of Geology*, 2004, art. VIII]

Lyell and the *Principles of Geology*. *In* D.J. Blundell and A.C. Scott (eds.) *Lyell: The Past is the Key to the Present*. Geological Society, London [Special Publication 143 (Lyell centennial vol.)]. 3–15, 3 figs. 1998. [repr. Rudwick, *Lyell and Darwin*, 2005, art. II]

Georges Cuvier's paper museum of fossil bones. *Archives of Natural History* 27: 51–68, 8 figs. 2000. [repr. Rudwick, *New Science of Geology*, 2004, art. IX]

Jean-André de Luc and nature's chronology. *In* C.L.E. Lewis and S.J. Knell (eds.), *The Age of the Earth*. Geological Society [Special Publication 190]: 51–60, 1 fig. 2001. [repr. Rudwick, *New Science of Geology*, 2004, art. VI]

The New Science of Geology: Studies in the Earth Sciences in the Age of Revolution. 336pp., 66 figs. Ashgate. 2004. [repr. 13 articles, 1976–2001, with new Introduction]

Travel, travel, travel: geological fieldwork in the 1830s. *In* Rudwick, *New Science of Geology*, 2004, art. XII, 10 pp. 2004. [not previously published]

Bursting the Limits of Time: The Reconstruction of Geohistory in the Age of Revolution. xxiv + 708 pp., 179 figs. Chicago: University of Chicago Press. 2005.

Picturing nature in the age of Enlightenment. *Proceedings of the American Philosophical Society* 149: 279–303, 20 figs. 2005.

Lyell and Darwin, Geologists: Studies in the Earth Sciences in the Age of Reform. 337pp, 37 figs. Ashgate. 2005. [repr. 10 articles, 1969–98, with new Introduction]

Author's response [to] Gabriel Gohau, Simon Schaffer and James Secord: Review symposium, The Geohistorical Revolution [on Rudwick, *Bursting the Limits of Time*, 2005]. *Metascience*, 16: 359–95. 2007.

[with Hugh Torrens] Walk with the Founding Fathers: a field trip on the Isle of Wight [A Field Guide]. 34pp. History of Geology Group, Geological Society, London. 2007.

Worlds Before Adam: The Reconstruction of Geohistory in the Age of Reform. xxiv + 614 pp., 165 figs. University of Chicago Press. 2008.

Biblical Flood and geological deluge: the amicable dissociation of geology and Genesis. *In* M. Kölbl-Ebert (ed.), *Geology and Religion: A History of Harmony and Hostility*, London, The Geological Society [Spec. Pub. 310], 103–10, 3 figs. 2009.

Author's response [to Celâl Şengör: A Rankean View of Historical Geology and its Development [essay review of Rudwick, *Bursting the Limits of Time*, 2005]. *Earth Sciences History*, 28: 108–42. 2009.

In the footsteps of Charles Darwin: the Parallel Roads of Glen Roy: Field Guide. 36 pp. History of Geology Group, Geological Society, London. 2009.

Geohistory and the historicity of Genesis [in the work of Deluc]. *In* J. L. Heilbron and René Sigrist (eds.), *Jean-André Deluc: Historian of Earth and Man*. Slatkine Érudition, 242–60. 2011.

The Silurian of 'Siluria' and the idea of a Palaeozoic era [A Field Guide]. 20 pp., 12 figs. History of Geology Group, Geological Society, London. 2013. [International Congress of the History of Science, Technology and Medicine, Manchester, 2013]

Earth's Deep History: How it was Discovered and Why it Matters. ix + 360 pp., 94 figs. University of Chicago Press. 2014. ['Simplified Chinese' ed., SDX Joint Publishing Company, 2020]

Insiders and outsiders: INHIGEO seen from the sidelines. *In* W. Mayer, R.M. Clary, L.F. Azuela, T.S. Mota, and S. Wołkowicz. (eds.), *History of geoscience: celebrating 50 years of INHIGEO. Geological Society of London, Special Paper* 442: [8pp, 2 figs.] 2016.

Two careers across two cultures. *In* Frank James (ed.), Some significances of the Two Cultures debate. *Interdisciplinary Science Reviews* 41: 268–277. 2016.

Functional morphology in paleobiology: origins of the method of 'paradigms'. *Journal of the History of Biology* 51 (1): 135–178. 2017.

Functional morphology in paleobiology: the fate of the method of 'paradigms'. *Journal of the History of Biology* 51 (3): 479–533. 2017.

Part II

Formations

Changing Narratives in the History of Oceanography

Antony Adler

Contents

Abstract

The formal scientific study of the oceans only gained prominence in the late nineteenth century. The earliest accounts of the development of oceanography focused on notable scientists whose contributions laid the groundwork for later oceanographic exploration. In the second half of the twentieth century, ocean sciences expanded considerably as increased government funding and technological advancements enabled more extensive research and exploration. Subsequently, scholarship on the history of oceanography grew as well, exploring themes that included the overlap of ocean science and empire, the transformative impact of World War II and the Cold War on oceanographic funding and research, and the rise of international collaboration. As the historiography of the history of oceanography evolves, scholars have increasingly adopted interdisciplinary research methods to develop more comprehensive understanding of humanity's evolving relationship with the ocean.

A. Adler (✉)
Carleton College, Northfield, MN, USA
e-mail: aadler@carleton.edu

E. Aronova et al. (eds.), *Handbook of the Historiography of the Earth and Environmental Sciences*, Historiographies of Science,
https://doi.org/10.1007/978-3-031-40799-4_8

Keywords

Oceanography · 19th century · Cold war · Marine stations · Technology · Exploration · Ocean history

1 Introduction

Any attempt to write the history of ocean science requires an initial decision: which areas of study fall under the contemporary umbrella term "oceanography?" Oceanography, after all, is a science dedicated to studying three-fourths of our planet. Oceanographic sciences are now often divided into different subdisciplines: physical oceanography, chemical oceanography, biological oceanography, geological oceanography, geophysics, atmospheric science, as well as fisheries science. Like the branching of an evolutionary tree, these now distinct subdisciplines share common points of origin. In the nineteenth century they all might have fallen under the now defunct term of thalassography. Similarly, the term "hydrography" was commonly employed in the mid nineteenth century to denote the practice of measuring some of the physical features of the marine environment that later interested oceanographers, such as ocean depths and currents; however, the primary aim was to facilitate navigation rather than develop theories to explain the natural forces observed. The commonality linking practitioners of these subfields is that their subjects of study are all situated in the marine environment. They therefore tend to use the same instruments and platforms – most importantly, research vessels – though some branches of marine science are more dependent on the infrastructure of land-based laboratories near the coast.

In writing a historiography of the history of oceanography, I follow the advice of ocean sciences historian Eric Mills:

> As historians of oceanography we can either confine ourselves to the past few decades, when a definable field of oceanography existed, or make the more daring and significant choice of looking back to earlier, more puzzling times. Definitions turn out to be cramping, inhibitory, and worst of all, an unadventurous aid to our scholarship. History of oceanography is what historians of oceanography write about. (Mills 1990, p. 2)

This broad definition has been embraced by those who now identify themselves as historians of oceanography. As the editors of the proceedings of the Fifth International Congress on the History of Oceanography wrote in 2002: "Oceanography is a hybrid, a mixed science – probably more so than any other branch of the natural sciences. [...] [I]t is an aggregation of approaches, involving the full spectrum of the physical, biological and earth sciences, in their formal, empirical and applied manifestations, intended to yield an understanding of a portion of our particular planet" (Benson and Rehbock 2002, p. ix). The leadership of the International Commission for the History of Oceanography (ICHO) agreed in 2020 that a mission statement should define the organization as "a global body devoted to linking scholars, writers, and teachers interested in the history of the marine sciences, broadly defined."[1] ICHO's recent project to compile a shared online bibliography has thus far resulted in a catalogue of over nearly 1300 entries. An all-encompassing historical treatment of everything historians of oceanography consider to fall within the scope of the "history of oceanography" would be so broad as to resist coherence. However, historians' embrace of such a broad field of scholarship is itself a recent historical development.

At the outset, I should acknowledge the necessary limitations of the present work. While there is now an established body of scholarship comparing the history of oceanography in different national contexts, much remains to be done. The majority of existing studies focus on developments in the anglophone world or on Western Europe. Very little has been published (in English) on the rich history of ocean sciences in Russia (Mikhailov et al. 2002). This absence is doubly notable given Russian scientists' early entry into these fields. For example, Russia commissioned one of the earliest custom-built vessels for marine exploration, the *Andrei Pervozvanny*, launched in the 1899 (Lajus 2021). Even a transnational history of American and Soviet ocean sciences during the Cold War remains to be written. Similarly, with the notable exception of a few case studies which have explored the history of marine science in Japan (Ericson 2020) and China (Neushul and Wang 2000; Wang 2007; Luk 2020), the existent scholarship remains largely focused on developments in Europe and North America. Finally, I have chosen not to include as yet unpublished dissertations in this discussion.[2]

In this short chapter, some histories of the ocean sciences must inevitably be touched upon more briefly than others. The history of fisheries science, for instance, deserves its own treatment (for useful histories of fisheries science see Smith 1994; Rozwadowski 2002; Hubbard 2006; Finley 2011; Bolster 2014, and most recently, Banoub 2021).[3] There have been a few other historiographies of oceanography to which the reader's attention is called. The major foundational historiographical essay on this subject still remains Eric Mills' "The Historian of Science and Oceanography After Twenty Years," published in the journal *Earth Science History* (Mills 1993). Margaret Deacon provided a bibliographical introduction to the subject in the second edition of her book, *Scientists and the Sea 1650–1900*, published in 1997. Readers are also advised to consult Samantha Muka's recent comprehensive historiography of the history of marine biology (Muka 2021). Some of the works discussed by Muka coincide with my own attempt in the present essay. Muka provides greater

[1] The International Commission for the History of Oceanography was created in 1968 as a sub-group within the Division of History of Science of the International Union of History and Philosophy of Science (IUHPS/DHS). The first president of the commission was André Gougenheim, former director of the French Hydrographic Service.

[2] An extensive bibliography of scholarship on the history of the marine sciences – including doctoral theses – is compiled by ICHO and accessible online: https://www.zotero.org/groups/2526220/history_of_marine_science_icho_library

[3] For an overview of the adjacent subdiscipline of marine environmental history readers are advised to consult Schwerdtner-Máñez and Poulson (2016).

in-depth attention, however, to the history of marine stations and specifically to biological research. Painting with a broad brush, I trace what I see as the major currents that have shaped the development of the history of oceanography as currently practiced by historians of science.

2 First Systematic Frameworks

The first histories of oceanography were written by ocean scientists in the late nineteenth and early twentieth centuries. These appeared as introductory chapters of scientific expedition narratives. Intended to contextualize the work of the expeditions, these historical overviews aimed at securing a place for their authors within an emerging discipline of oceanography by framing their reports as the leading edge of an accumulated body of knowledge. Thus, Scottish naturalist Wyville Thomson, who would go on to be chief scientist on the *Challenger* expedition, provided a summary of developments in physical and marine zoological science as prologue to his report on the cruises of the *Lightening* and *Porcupine* (Thomson et al. 1873). He explained the beginnings of scientific interest in the depths as resulting from the requirements of submarine telegraphy, and paid homage to the naturalist Edward Forbes, proponent of the methodology of sampling using a dredge: "every year adds enormously to our stock of data, and every new fact indicates more clearly the brilliant results which are to be obtained by following his methods" (Thomson et al. 1873, p. 6). Similarly, John Murray, who took over responsibility for the publication of the *Challenger* expedition reports after Thomson's death, provided a comparable historical summary in the first volume, published in 1880. Murray subsequently repeated this account in the introduction to his co-authored report on the voyage of the *Michael Sars* (Murray et al. 1912).

The primacy of the *Challenger* expedition in many histories of oceanography stems in part from the intentional historical aggrandizement of the project by some of the scientists involved. No other oceanographic expedition has been the subject of so many histories. (To name but a few: Linklater 1972; Corfield 2003; Zuroski 2017; Lopes 2018; Macdougall 2019; Jones 2023). John Young Buchanan, the expedition chemist, for example, boldly claimed in 1919 that "the science of oceanography was born at sea in lat. 25° 45' N., long. 20° 14' W., on 15 February 1873," when *Challenger* scientists conducted their first deep-sea dredging. Such claims did not go unchallenged, however, as even some contemporaries noted that scientific interest in the oceans had begun much earlier.[4]

The first English-language monograph dedicated specifically to the history of oceanography was published in 1923, but again it was written by a marine scientist. Writing in the preface to his book, *Founders of Oceanography and Their Work: An Introduction to the Science of the Sea*, William Herdman informed his readers:

[4]See the review of Buchanan's book in *The Geographical Journal* 56, 70–71 (1920). https://doi.org/10.2307/1781303

"I have myself lived through the period that has seen the development of the Natural History of the Sea into the Science of Oceanography, and have known intimately most of the men who did the pioneer work" (Herdman 1923, p. v). The basis for this work was a series of twenty public lectures. Herdman could boast of a direct personal connection to the most famous of nineteenth-century oceanographic voyages. As a student he had worked as an assistant to Charles Wyville Thomson – lead scientist of the *Challenger* expedition. Shortly after graduating from the University of Edinburgh in 1879, Herdman joined the '*Challenger* Office' under the direction of Sir John Murray and was given the task of organizing and studying the Tunicata collection returned by the *Challenger* expedition (his reports were then published between 1882 and 1888). As he went on to explain, the first half of his book provided biographies of "some of the leading men who have made our science," with whom he claimed to have firsthand acquaintance. The second half discussed subjects in oceanography about which he was himself "most interested" (Herdman 1923, p. vi).

Despite Herdman's autobiographical focus, Mills credits him for developing, what Mills described as "one of the first attempts to provide a systemic framework for the history of oceanography" (Mills 1983, p. 1). According to Herdman, the history of oceanography in the nineteenth century could be divided into three periods: "the period of Edward Forbes, the great Manx naturalist; the period of Wyville Thomson, ending with its climax, the 'Challenger' expedition; and the post-'Challenger' period of Sir John Murray and modern oceanography" (Herdman 1923, p. 8). Thus Herdman's biographical chapters primarily discussed the works of Forbes, Thomson, Murray, but also described the contributions of the Swiss-born American naturalists Louis Agassiz and Alexander Agassiz, of Prince Albert 1st of Monaco, and finally, of the Prussian founder of the Naples Zoological station, Anton Dohrn. These scientists continue to figure prominently in subsequent overviews of the history of the marine sciences, with recent accounts contextualizing their work within the social and political environment of their times (Adler 2017).

Almost a half century elapsed before another attempt was made to devise a systematic framework for the history of oceanography. Public interest in ocean exploration surged during the 1960s. The development of technologies like saturation diving equipment and submersibles capable of exploring extreme depths seemed to presage humanity's colonization of the oceans. This was also a period when oceanographers hoped to forge large-scale international cooperative programs for ocean exploration. For example, UNESCO organized an international oceanography conference in Copenhagen in July 1960; this led to the creation of the Intergovernmental Oceanographic Commission (IOC). It was during this period, when ocean research was enjoying international government support and a newly vaunted science of oceanography was garnering widespread public interest, that marine scientists became more interested in documenting the history of what was increasingly presented as a single research field.

The next iteration of what Mills dubbed a "systemic framework" for understanding the history of ocean sciences was developed by German oceanographer Georg Wüst in 1964. Wüst foregrounded the importance of fieldwork as a central method in addressing research questions and developing theories in a burgeoning new science.

"Progress," in his view, had been the result of "great oceanic expeditions," improvements in instrumentation and methodology, and the development of theory (Wüst 1964, p. 3). Wüst grouped "the most outstanding" deep-sea expeditions as falling within four eras: the era of exploration (1873–1914), the era of national systematic and dynamic ocean surveys (1925–1940), the period of new marine geological, geophysical biological and physical methods (1947–1956), and finally, the era of international research cooperation after 1957.

Wüst's periodization, while useful for understanding broad developments in oceanography, is in some ways limited. His analysis focused on the history of the study of the open ocean and prioritized physical oceanography while ignoring other fields of marine science, despite crossovers in research interests, methods, and shared scientific instrumentation. Like many other early histories of oceanography, Wüst's periodization took as its starting point the voyage of *HMS Challenger*, thus excluding earlier scientific work at sea. The end point of 1964 – when Wüst's article was published – also proves confining from a retrospective twenty-first-century perspective. Government funding for oceanography, particularly in the United States, reached its apogee during the Cold War before a subsequent decline. From a more long-term perspective, the period Wüst regarded as the moment of maturation of the discipline, is better described as a temporary and historically contingent period. Yet, Wüst's approach leaves little room for the possibility of historical contingency. The development of oceanography in different regions is presumed to have to follow a single necessary path. He accordingly argues that oceanography "in many parts of the world ocean, particularly the Indian Ocean and the South Pacific [...], [had] not advanced beyond the second and third stages of research." He predicted that the Indian Ocean would become "the next field of systematic international cooperation on a large scale" (Wüst 1964, p. 20). At the time of his writing, the International Indian Ocean Expedition (IIOE) was taking place and he was involved in advocating for this project. Thus his predictions for the development of oceanography illustrate the manner in which a scientist's own scientific work may shape their historical outlook (Wüst 1960). Both Wüst and Herdman, taking a long view of the history of oceanography, find periods of constancy and points of change, but their accounts remain, on the whole, deterministic, assuming an unvarying path of scientific progress.

Only a few years after Wüst published his article, the first meeting of the international congress for the history of oceanography was organized in Monaco in 1966. Though the majority of attendees were oceanographers, the meeting marked the first organized attempt to bring together scholars interested in the *history* of ocean sciences. It was also notable for the inclusion of a paper by Harold Burstyn, "The Historian of Science and Oceanography" (Burstyn 1968). For his part, Burstyn does not recall the meeting favorably, remembering it as "dominated by superannuated oceanographers."[5] He found them so hostile to his perspective that he limited his presentation to reading the abstract of his paper. Nevertheless, in his essay –

[5]Personal communication, Burstyn to Adler, 14 December 2021.

published in 1968 in the congress proceedings – he argued that the history of oceanography was uniquely well suited for understanding the interplay of science and society.[6] As he wrote: "no other field [...] has been so dependent for its development on the growth of the modern state, no other field has either been 'big science' or no science at all" (Burstyn 1968, p. 670). Accordingly, he argued, two tasks fall to the historian of science: tracing the history of theoretical and methodological developments in oceanography and tracing the impact of social, economic, and political forces upon those developments.

In his own work, Burstyn did just this by describing both the scientific and political influences which led to the creation of the *Challenger* expedition, interpreting it as an early form of "big science" characterized by "big money, large multi-disciplinary teams, a division of labour, team commitments, the building of coalitions, strong leadership, hierarchical organization, and the making of budgetary compromises with an eye on public opinion" (Burstyn 2001, p. 54). If the organization of early oceanographic expeditions was a precursor model for the organization of other forms of science in the mid-twentieth century, he urged, this was all the more reason for historians of science to give the history of oceanography greater attention. His 1966 congress paper – and subsequent writing – was a call to action for historians of science. Histories of oceanography might have been compiled by scientists, but the subject was ready for the attention of historians as a means to learn more about broader developments in the history of science. Thus, the history of oceanography, like other branches of the history of science, was to be transformed by the social constructivist turn of the late 1960s and early 1970s and the institutionalization of the history of science as an academic field.

3 Defining Subject and Scope

An initial problem for the history of oceanography, as an emerging subfield in the history of science, was to identify the scope of its subject matter. In aid of this, four important monographs, cataloguing research vessels and oceanographic instrumentation, were published in the 1980s and following decade. Stewart B. Nelson's *Oceanographic Ships: Fore and Aft* (1982) covers American research vessels up to 1971, while Tony Rice's *British Oceanographic Vessels 1880–1950* does the same for the British fleet up to the period shortly after World War II. It is noteworthy that Rice explicitly states in his foreword that the chronology covered follows Wüst's periodization (Rice 1986, p. 6). Two catalogues of oceanographic instrumentation were also published: Anita McConnell's *No Sea Too Deep: The history of Oceanographic Instruments* (1982) and Christian Carpine's *La pratique de l'océanographie au temps du Prince Albert 1er* (2002), a compilation of catalogues

[6] Eric Mills later described this paper as a "bellwether" due to its significance in helping to establish a more methodological approach for the study of the history of oceanography. See Mills (2008).

that first appeared in the *Bulletin de l'Institut Océanographique de Monaco* between 1987 and 1999.[7] Both McConnell's and Carpine's catalogues document the process by which sampling instrumentation gradually became more standardized and efficient. Scientists and professional instrument makers refined sampling instrumentation by adopting newly invented construction materials like brass, rubber, and celluloid. McConnell's book, though primarily a chronological summary of developments in instrumentation, also offers an argument about the factors responsible for technological progress in oceanography. As she explained: "It is contended that significant improvements in underwater apparatus came at those times when: (a) a navy considered that its surface operations would be thereby advanced; or (b) commercial interests demanded it. These improvements depended for their execution on: (c) the availability of suitable materials, including ships and technological processes, and (d) adequate manpower to work intensively in the chosen region. Whereas (a) and (b) are alternatives, (c) and (a) are essential factors" (McConnell 1982, p. 148).

Historian Margaret Deacon (daughter of the celebrated British oceanographer George Deacon) deserves credit for greatly expanding the chronological scope of the history of the marine sciences.[8] In a 1965 paper she demonstrated that members of the Royal Society took up questions related to oceans as early as the late seventeenth century (Deacon 1965). She subsequently expanded this study in *Scientists and the Sea 1650–1900* (Deacon 1997). Not only did Deacon reject the mythology of a single nineteenth-century point of origin for oceanography, she also provided a broader geographical scope by including French, German, and Scandinavian scientists in her account.[9] Most importantly, Deacon's book offers a history that traces multiple and diverse motives driving ocean research over time. To see the *Challenger* expedition as the origin point for marine science, she argues, is "to misjudge its significance," to neglect "the work of earlier scientists," and to underestimate "the strength of the parallel movement in other countries" (Deacon 1971, p. 369). Deacon later even asserted that the international success of the *Challenger* expedition was in some ways counterproductive for marine biology in Britain: "[*Challenger*'s] very success made life more difficult for would-be oceanographers in Britain, by giving the impression that everything necessary had already been done" (Deacon 1993 p. 24). She also showed that while the British government was generous in its funding of the *Challenger* expedition, it was considerably less generous in funding deep-water research in the latter half of the nineteenth century. Though her focus

[7]Nelson was Senior Advisor to the Oceanographer of the US Navy; Rice worked as a biologist at the British Natural Environmental Research Council Institute of Oceanographic Sciences; McConnell was Curator of Oceanography and Geophysics at the London Science Museum; Carpine was Curator at the Musée Océanographique in Monaco.

[8]Deacon worked in the Science Studies Unit at the University of Edinburgh before becoming Research Assistant at the British National Maritime Museum.

[9]In her study Deacon delineated the end of British oceanographic primacy, claiming Scandinavia had become "the new headquarters of marine science" by the end of the nineteenth century (Deacon 1971, p. 390).

remains on Europe, Deacon's work provides important historical contextualization, highlighting the connection of tidal science with navigation, the crossovers between terrestrial and marine biological research questions, and the early internationalization of ocean science.

One of the first comprehensive histories of oceanography can be found in *The Edge of an Unfamiliar World: A History of Oceanography* (Schlee 1973), by historian Susan Schlee. Schlee calls attention to the parallel development of large-scale government-sponsored research programs in Britain and the United States. Curiously, there are no references to Deacon's work in Schlee's book, though in many ways the works complement one another. Perhaps the omission was due to the limited circulation of the first edition of Deacon's book. While Deacon examined the causes that led to increased scientific interest in oceans in the seventeenth century, Schlee's narrative begins in the late 1800s and explores the development of the American Coast and Geodetic Survey and the Hydrographic Office of the Navy. Deacon and Schlee's works together provide a helpful foundation upon which subsequent scholarship continues to build.

Some of the topics covered by Deacon and Schlee are revisited in Helen Rozwadowski's (2005) *Fathoming the Ocean: The Discovery and Exploration of the Deep Sea*. Reflecting the methodological and focal shifts in the history of science since the publication of these earlier works, Rozwadowski's account of the history of British and American oceanography does far more to embed the history of scientific discoveries in social and cultural history. Rozwadowski highlights multifaceted aspects of Victorians' discovery of the sea and re-examines the oceanographic research vessel as a negotiated social space. Seagoing scientists of the nineteenth century were not only devising a new science, she argues, they were also "forging a functional workplace" (Rozwadowski 2005, p. 195).

Historian Michael Reidy too expanded earlier accounts of the history of ocean sciences while focusing on the British context; like Rozwadowski, he shows that a multiplicity of actors were involved in the diffusion of scientific knowledge about the seas. Observing that oceanography and empire grew in tandem, he argues for greater attention to the intersection of science and statecraft. As Reidy explains, "the British Admiralty, maritime community, and scientific elite collaborated to bring order to the world's seas, estuaries, and rivers. These experts transformed the vast emptiness of the oceans into an ordered bounded grid, inscribed with isolines of all kinds – tidal, magnetic, thermal and barometric – in areas uncharted and on coasts unseen" (Reidy 2008, p. 6).

Transnational comparison has been a feature of histories of oceanography since the early 1980s. The oceanic turn in various fields of scholarship over the past several decades has encouraged historians to focus on transnational connections across oceanic spaces. Stimulated by Fernand Braudel's multi-volume history of the Mediterranean world (1949), maritime historians increasingly turned their attention to the movement of peoples, commodities, and ideas across ocean spaces. This interpretive framework, first applied to the Mediterranean and Atlantic Worlds, has since been applied to the Indian Ocean and to the Pacific Ocean (Pearson 2007; Matsuda 2012). Following this trend, and recognizing the transnational character of much oceanographic science, historians have brought an increasingly transnational perspective to their work.

In his foundational work on the history of biological and physical oceanography, Eric Mills compares developments in England, Germany, France, and the United States. This approach allows him to highlight distinctive national characteristics in the development of the marine sciences. His history of biological oceanography largely focused on the development of quantitative plankton studies in Germany, Scandinavia, the United Kingdom, and the United States (Mills 2012). His history of physical oceanography, *The Fluid Envelope of Our planet. How the Study of Ocean Currents Became a Science*, points to what Mills terms the "paradox of French marine science," the apparent absence of scientific interest in physical oceanography in France in the late nineteenth century at a time when dynamic oceanography was taking off in Scandinavia, Germany, and the United States (Mills 2011, p. 163). Mills shows that physical oceanography research was prompted first by concerns about the decline of commercial fisheries and then, in the second half of the twentieth century, by the requirements of submarine warfare. While Mills reveals different national styles of oceanography, he also highlights the emergence of a transnational scientific community interested in the same research problems and exchanging methods and theories.

Attention to the shaping of oceanography by different national contexts remains an important feature of much scholarship to this day. While most of the English-language corpus has focused on the history of oceanography in the anglophone world, there are exceptions. To give but a few examples: Bo Poulson wrote about Danish oceanographer Johannes Schmidt and his work with the Carlsberg Foundation (Poulsen 2016); Vera Schwach described the development of marine science in Norway (Schwach 2013); Franziska Torma examined popular interest in marine biology in Germany in the early twentieth century (Torma 2012); and Colin Summerhayes and Cornelia Lüdecke detailed the history of German undersea mapping in the South Atlantic during the 1930s (Summerhayes and Lüdecke 2012).[10]

4 Technology, Place, and Space

The International Congress for the History of Oceanography held several international conferences after the initial meeting in Monaco in 1966. Subsequent meetings took place in Edinburgh, Scotland (1972); Woods Hole, Massachusetts (1980); Hamburg, Germany (1990); La Jolla, California (1993); Quindao, China (1999); Kaliningrad, Russia (2003); and Naples, Italy (2008).[11] Over time these congress meetings became

---

[10] See the chapter ► "Historiography of the International Geophysical Year (1957–1958)" by Daniella McCahey in this volume on developments in the Soviet Union.

[11] The first three meetings marked commemorative dates in the history of oceanography. The Monaco meeting corresponded with the centennial of the start of Prince Albert Ist's career at sea, the Edinburgh meeting marked the centennial of the voyage of the *Challenger* expedition, and the Woods Hole meeting marked the fiftieth anniversary of the Woods Hole Oceanographic Institution.

more international and aimed to encourage the application of historical methods to the study of oceanography. A scan of the countries represented in the early history of oceanography congress meetings is informative. Some regions were better represented at the first three congress meetings than others (initially France, the U.K., and the United States). This weighting can be partially explained by the meeting locations. To be noted: attendees at these first meetings were almost exclusively affiliated with science departments or oceanographic institutions. Attendees did span the Cold War divide. There were eight attendees from the United States at the 1966 congress and five from the U.S.S.R. Four contributors to the third congress proceedings were from the U.S.S.R.

A meeting held in 1993 at the Scripps Institution of Oceanography in La Jolla marked a turning point, being organized by two historians of science: Keith Benson and Philip Rehbock. "It was our specific goal to facilitate this transition by encouraging the participation of greater numbers of non-oceanographers, without reducing in the slightest the opportunities for oceanographers to participate as fully, and as valuably, as they always have. Thus, without diminishing the legitimacy and importance of the motives of the oceanographer-historian – reminiscence, chronology, precedence, acknowledgement, public recognition – we wish also to stimulate an analysis of the history of oceanography that gives full credit to its scientific complexity and social relationships" (Benson and Rehbock 2002, p. x). Nevertheless, despite this declaration of purpose, the papers assembled in the ICHO congress proceedings are so wide-ranging that it is not possible to discern a specific historiographical thread running through them. In his forward to the proceedings of an eighth and final congress held in 2008, Keith Benson, then-president of ICHO, lamented that attendance had greatly diminished from previous meetings. He attributed this to an international decline in funding for oceanography. In tandem with the decline in funding for marine science, he argued, "support for ancillary social science research, such as the history of oceanography" had also declined (Benson 2013, p. 9).

More united in approach than the ICHO congress papers were some of the proceedings produced in conjunction with the Maury Conferences for the History of Oceanography – a series of meetings sponsored by the Office of Naval Research. These included at least one book which provides thematic framing for the collected essays: *The Machine in Neptune's Garden: Historical Perspectives on Technology and the Marine Environment.* Borrowing its title from a classic work of environmental history by Leo Marx,[12] the collection highlights the various ways in which technological developments have been integral to the process of gaining knowledge about the oceans. The editors also situate the papers in historiographical context, placing the history of oceanography at the intersection of environmental history and the history of science and technology. As the editors explain in their introduction: "The history of science and technology attends to the intellectual, political, and

[12] Marx's *The Machine in the Garden* (1964) explored the tension between nature and technology in American culture, focusing on the nineteenth and early twentieth centuries. Marx demonstrated how the enduring conflict between the idealized pastoral landscape, symbolizing a simpler, harmonious life, and rapid industrialization driven by machines shaped American identity, and influenced perceptions of progress, the environment, and our relationship with technology.

social contexts in which instruments were developed and deployed, so telling the history of the ocean floor, or any other part of the sea, must involve sophisticated consideration of science and technology to understand how knowledge of the marine environment was created" (Benson et al. 2004, p. xiv). It is critical, they argue, that historians incorporate attention to the history of technology in their histories of oceanography because "oceanographers' reliance upon the machine to define and construct knowledge of the marine world has created oceanography's interdisciplinary nature" (Benson et al. 2004, p. xxii). This approach focused historians' attention on the ways technology mediates scientists' data gathering in the ocean environment; "instruments" may take the form of sampling devices, but can also include everything from radioactive isotope tracers to mathematical models.

Tony Rice has shown that technological limitations and innovations are key to understanding the revolution in oceanography that occurred in the late nineteenth century. Whereas vessels under sail had difficulty hauling a dredge or remaining in position for an instrument deployment, steam engines provided greater maneuverability. This was imperative for the deployment of instruments to greater depths. Rice explains that in the case of *HMS Challenger*, "without the donkey engine mounted on the main deck it would have been quite impossible to achieve the plethora of net hauls, soundings and thermometer and water bottle deployments that were made at the 362 official stations worked during the voyage" (Rice 2009, p. 30). Attention to the history of technology and instrumentation also reveals the transfer of techniques, methodologies, and research questions from one scientific discipline to another. The history of marine microbiology, for instance, reveals that sterilization techniques developed in land-based laboratories and applied to the study of terrestrial microbiology were deployed on oceanographic expeditions at the end of the nineteenth century to address questions about the prevalence of microbes in the deep sea (Adler and Dücker 2018; Dolan 2020).

However, in addition to the vessel itself, understood as an instrument, different technologies require platforms from which they can be deployed. Science always takes place somewhere specific (Adler 2014). Beginning in the late 1980s, historians and sociologists of science became increasingly interested in understanding the physical spaces in which scientific work is conducted – most importantly the laboratory. Attention to laboratory settings thus came to shape the work of scholars of the marine sciences. Work on the history of marine biological stations is expansive (and still growing), having developed in tandem with shifting attention by historians of science to both the role of the laboratory in generating scientific knowledge and to the unique characteristics of science conducted in the field (see, for example, Pauly 1988; Benson 1988; de Bont 2015; Adler 2016; Matlin et al. 2020). Some of these studies have looked primarily at the development of biology, or marine ecology (Egerton 2014), but others have explored the histories of these spaces for what they reveal about the social and gendered dynamics of scientific knowledge-making (Muka 2014; Steiner 2018; Kohlstedt 2022).

As marine stations rose to prominence, work in these locations came to be seen as a rite of passage for aspiring marine scientists. This was certainly the case for the Naples Marine Station in Italy and the Marine Biological Laboratory in Woods Hole,

Massachusetts, USA. The first histories of these institutions were produced by people connected with them either professionally or through family connections. Susan Schlee, for instance, who chronicled the history of the Woods Hole Institute's first research vessel *Atlantis* (1978), was married to a marine geologist living in Woods Hole. As Helen Raitt and Beatrice Moulton, co-authors of the first history of Scripps Institution of Oceanography, write in their introduction:

> [I]t is the rare scientist who will take time to write more than journal articles or research papers. The lure of the laboratory and the undiscovered is always stronger than the lure of past history and dusty archives, so that it falls to wives and observers like us to dust off the scrapbooks and attempt to write some of it down. (Raitt and Moulton 1967)[13]

In the twentieth century, as ocean sciences fissured into more narrowly defined subdisciplines, the institutions that supported oceanographic work also became more specialized. Scientists increasingly regarded marine biology as a separate branch of study from physical oceanography or marine geology. In the United States, the Woods Hole Oceanographic Institute on the East Coast and the Scripps Institution of Oceanography on the West Coast became the two most important institutional centers for oceanographic research. The ascendency of these two institutions is closely tied to the types of marine research they hosted – physical oceanography, acoustics, and marine geology. The importance of these forms of oceanographic data to the U.S. Navy was made starkly evident by their utility for battle strategy during the Second World War and the Cold War. As a result, the United States Navy became the primary source of funding for oceanographic research in the postwar era. Funding was dispersed through the Office of Naval Research. This partnership lasted throughout the Cold War, but with the relaxing of tensions after the fall of the Soviet Union, funding declined.

Declines in government funding starting in the 1970s led both oceanographers and historians to look back upon the preceding period as a "golden age of global oceanography" (Shor 1978, p. 35). For historians interested in the development of oceanography, the Cold War became a topic of particular fascination. Examination of the relationship between scientists and the navy gave rise to many questions of interest to historians of science: what is the relationship between science and the state? How have military concerns shaped developments in science? What is the relationship between science and diplomacy? When and how do institutional

[13]Helen Raitt chronicled a Scripps voyage on which she joined her husband, marine geophysicist Russell Raitt. She was the first woman to go on such an extended oceanographic voyage (Raitt 1956). Similarly, a later history of Scripps was produced by Elizabeth Noble Shor, wife of marine geophysicist George G. Shor (1978). Muriel L. Guberlet, who wrote a popular history of oceanography, was then-widow of marine biologist John Earl Guberlet (1964). While the foremost authority on the life and work of the pioneer oceanographer Prince Albert 1st of Monaco was Jacqueline Carpine-Lancre, wife of Christian Carpine, curator at the oceanographic museum in Monaco. This trend of a direct familial connection to subject of study has continued. However, the gendered restrictions on who can undertake oceanographic fieldwork has changed; I am too married to an oceanographer.

bureaucracies and scientific ideals of objectivity collide? Answers to these questions have varied, as we will see in the next section.

5 The Cold War: Scientists and the State

The partnership between government agencies, military, and scientific communities in the second half of the twentieth century has been the subject of much investigation by historians science. The relationship between civilian scientific institutions and state agencies was not easily established; historians have long debated whether scientific research was distorted or aided by this influence.[14] In the immediate postwar period, scientists and university administrators were wary of political censorship and restrictions on freedom of inquiry (Sapolsky 2014, p. 40). Cold War-era government funding had the greatest influence on those branches of science that had been most closely enrolled in the war effort during World War II – physics, engineering, and the earth sciences. Among these, ocean sciences – particularly in the United States – were the most indelibly impacted. After the war, the Office of Naval Research (ONR) with its responsibility for marine affairs was the first federal agency to support academic science until supplemented by the National Science Foundation in 1950 (Knauss 2000). Historians of oceanography, like historians of developments in other scientific disciplines during the Cold War have grappled with the question of how geopolitics and military priorities shaped, or were shaped by, the marine sciences.[15] The following section outlines some of the conclusions that scholars have taken in their analysis of this period in the development of the oceanography.

Gary Weir, in *An Ocean in Common: American Naval Officers, Scientists, and the Ocean Environment*, traces the relationship between civilian oceanographers and military officers largely from the perspective of the Navy. He shows that Navy officers attempted to partner with oceanographers as early as the 1920s. However, it was World War II that gave civilian scientists the opportunity to establish themselves as cross-cultural brokers between Naval officers and scientific institutions. As Weir explains: "After four years of war and frequent interaction with the navy, many pivotal wartime translators and their apostles easily crossed the cultural divide into senior civil service jobs as research scientists or administrators. [...] [M]any of oceanography's translators became permanent insiders, living simultaneously and comfortably in two cultures" (Weir 2001, pp. 271–272). Weir shows how this relationship was sustained, first through the Naval Research Laboratory, then the United States Navy Hydrographic Office, and finally through the Office of Naval Research. He shows how this new partnership led to

[14] The origins of this debate can be traced to back to claims made by historian of physics Paul Forman in the late 1980s. Daniel Kevles was critical of Forman's (1985) claim that Cold War science policy had distorted the trajectory of American research, arguing instead that physics was simply "what physicists do – or have done" (Kevles 1990, p. 263).

[15] See Ronald E. Doel's chapter in this volume on the ► "Cold War and Earth Sciences."

the development of instrumentation of import to both naval strategists and ocean scientists – the most famous being the bathythermograph, an instrument for measuring water temperature variation at depth. Such instrumentation was not only important for ocean circulation modeling, it was also vital to the Navy for sonar deployment and submarine detection (Weir 2001, pp. 126–144).

While Weir examined the Cold War history of oceanography in the United States by looking at how this civilian-military partnership was established, Jacob Hamblin observed how this partnership overlapped with efforts in international diplomacy. A seeming paradox of Cold War oceanography is that while the Western bloc competed with the Soviet Union for global military supremacy – particularly in the realm of undersea warfare – scientists frequently sought international collaboration in support of their efforts to explore the oceans. These efforts led to the First International Congress of Oceanography held in New York in 1959. In *Oceanographers and the Cold War: Disciples of Marine Science*, Hamblin traces the increasing bureaucratization of oceanography in the second half of the twentieth century. While the stated benefit of cooperation in marine science was the "easing of tensions," international collaboration also allowed scientists to "maintain their international reputations" while continuing to carry out classified research for the military. Hamblin argues that scientists' efforts of international cooperation posed no impediment to their military work. Both served the interests of oceanographers (Hamblin 2005, p. 261).

In a follow-up book, Hamblin examined the history of debates surrounding the disposal of radioactive waste in the oceans during the same period (Hamblin 2008). Many ocean scientists had concerns over the safety of ocean disposal that conflicted with the more sanguine views of other government advisors. Efforts to win the scientists' approval eventually led to the creation of new patronage opportunities with research funding offered through the Atomic Energy Commission. However, the issue of ocean dumping did not recede, becoming a matter of international diplomacy when the Soviets, while denying their own radioactive waste disposal, accused Western powers (particularly Britain and the United States) of poisoning the oceans.

Like Hamblin, Samuel Robinson urged that developments in international diplomacy were crucial to understanding the history of twentieth-century oceanography. An important complement to U.S.-centric accounts of this period, Robinson's book traces the growth of oceanographic science in Britain, focusing on the expansion of the National Institute of Oceanography under the direction of physical oceanographer George Deacon (Robinson 2018). While in many ways the story of scientific and military collaboration in Britain parallels developments in the Unites States, Robinson also points to important differences. He shows that the history of British twentieth-century oceanography, entwined with the history of military surveillance and intelligence gathering, must be understood in connection to Cold War strategic collaboration between Britain and the United States. However, whereas U.-S. surveillance planners prioritized global-scale reconnaissance, British military planners were preoccupied with the threat of conventional attacks against home bases. Tensions arose between British physical oceanographers and British fisheries scientists. Fisheries scientists had worked in collaboration with Soviet scientists under the auspices of ICES (the International Council for the Exploration of the Seas), while oceanographic surveys of the North Sea operated by NATO gathered data of strategic military value that were not to be shared. Yet, by the 1960s, British oceanographers were beginning to shift from military to civilian sources of patronage. The rise of the environmental movement defined new problems and offered new opportunities, with the National Institute of Oceanography eventually reframing its research mission as aiding restoration ecology.

Naomi Oreskes has an extensive publication record spanning a wide range of topics. She has written about the history of the theory of plate tectonics, climate change denial, and the nature of scientific consensus. Additionally, Oreskes has dedicated considerable attention to describing how military funding influenced the trajectory of oceanographic research in the twentieth century. With Ronald Rainger she co-authored an essay which described how individual animosities and intellectual disagreements led Harald Sverdrup, then-director of the Scripps Institution of Oceanography, being denied security clearance at the outbreak of World War II. This study demonstrated the emerging impact of military funding on American science in the twentieth century (Oreskes and Rainger 2000). She also published a study examining the division of scientific labor in oceanography during World War II. The invention of the bathythermograph launched a major data collection project. While oceanographers identified their work as an inherently masculine endeavor, analyzing bathythermograph records became women's work. Oreskes argued that this research focus can be attributed to the practical value of ocean temperature data for anti-submarine warfare and oceanographers' desire to secure patronage from the navy. She explains that while "measuring the temperature of ocean water was considered fundamental science, at the end of World War II it had greater value for maintaining ties with the U.S. Navy than for producing scientific novelties" (Oreskes 2000, p. 375). Fields like marine acoustics and ocean circulation modeling flourished from the late 1940s onward because they promised to have a direct application to submarine warfare and undersea communication. While Navy funding sometimes shaped the questions oceanographers sought to explore, it was nevertheless regarded by the scientists as a good patron, Oreskes argues, as it made no attempt to determine the outcome of investigations it sponsored (Oreskes 2021).

In sum, oceanography played a crucial role in Cold War military strategy for the United States, other NATO member states (Turchetti 2012), and the Soviet Union. Extensive research on underwater acoustics, ocean currents, and other ocean phenomena was carried out to enhance submarine warfare capabilities, sonar systems, and anti-submarine warfare techniques. Concurrently, significant technological advancements emerged, including deep-sea submersibles, bathymetric mapping systems, and underwater communication technologies. Historians of Cold War oceanography have studied the interplay between military funding, scientific research, and the development of these technologies (Doel et al. 2006). These studies examined the role of military patronage in accelerating scientific discovery and its influence in shaping the direction of research. Historians have highlighted the importance of studying the histories of marine institutions and the careers of individual scientists in the context of the Cold War, exploring institutional frameworks and career trajectories of specific researchers to gain a more comprehensive understanding of the dynamics, motivations, and impact of scientific endeavors during this period.

6 Where Do we Go Next?

What are some possible future directions for the history of oceanography? This account suggests the field is now quite expansive. While opportunities for future work still include the telling of unrecorded events in the development of oceanographic science, there is also a need to reexamine some of the established narratives of our field. As with other topics in the history of science, many early histories of oceanography were written by individuals who began their careers as scientists. These works are sometimes especially valuable because they inform those not trained in marine science of the discoveries and the scientists deemed most important by their contemporary colleagues. Yet, these histories often foregrounded developments in theory and focused on giving credit for first discovery. The quintessential example is the essay by Wüst (1964) previously discussed.[16] Historians can still do more to reexamine established narratives of the history of oceanography by bringing to bear perspectives informed by the social sciences and the humanities.

A paper by historians Penelope Hardy and Helen Rozwadowski highlights the continued relevance of the past to the present. In their 2020 article, prompted by debates over the removal of Confederate statues in the United States and by ongoing efforts of scholars to grapple with the legacies of racism in American history, Hardy and Rozwadowski call for a critical reexamination of Matthew Fontaine Maury. Sometimes labeled the "father" of oceanography in early hagiographic biographies, Maury fought for the Confederacy during the Civil War and praise of his legacy can be read as aiding the rehabilitation of the Confederacy: "Celebrating [Maury's] scientific contributions supported Lost Cause ideology by eliding his role in the Confederacy and support for the institution of slavery" (Hardy and Rozwadowski 2020, p. 11). This example illustrates how historians of oceanography can break away from uncritically deployed tropes or constrained accounts of developments in theory or discovery. Hardy's and Rozwadowski's study reminds us that the history of science – oceanography included – is demonstrably entwined with power and politics, with implications for social relationships in our present society.

More recent interdisciplinary works in social sciences and humanities have looked at the ways in which engagement with the ocean as a space set apart from terrestrial environments has encouraged new forms of human imagination. Some of these works have taken an ethnographic approach to explore how contemporary oceanographers conceptualize their subjects of interest and co-produce representations of nature through the mediation of scientific instruments (Helmreich 2009;

[16]For more recent example of studies that examine the development of specific oceanographic instruments see Wernand (2010) on the secchi disc or Polmar and Mathers (2021) on the Bathyscaph.

Lehman 2018). Others have focused on the history of human perception into the depths, while outlining the history of technologies that made watery spaces visible via diving equipment, photography, or aquaria (Cohen 2022; Elias 2019; Muka 2023). I situate my own contribution to the history of oceanography here as well. Adopting the Science and Technology Studies concept of "sociotechnical imaginaries," my book *Neptune's Laboratory: Fantasy, Fear, and Science at Sea* (2019) details the international development of oceanography, with attention to ways that the ocean sciences shaped – and were shaped by – anticipated futures for humanity. From the nineteenth century to the present, hopes about resources the ocean might offer to sustain humanity, and fears about civilizational collapse via conflict or environmental disaster, have played an important role in determining the research topics that gain government support and the research questions that become matters of public interest (Adler 2019).

The most promising future approach may come from embracing greater interdisciplinarity. Some contemporary historians of oceanography, Rozwadowski in particular, have called for an all-encompassing attention to "ocean history," approaching the ocean as a geographically varied, three-dimensional, biologically diverse, and culturally conceived environment (Rozwadowski 2013). Many have sought to blend analytical approaches from the history of science and environmental history. While historians of oceanography have treated the history of ocean science as shaped by requirements of "the field," they have been slower to acknowledge the heterogeneous character of the marine environment. In advocating for a more inclusive approach, Rozwadowski writes: "ocean history does seem to be rather naturally oriented toward consideration of categories of space. While history traditionally addresses a particular geographic location, a specific time period, and finite groups of historical actors, environments such as oceans seem to require a different approach. Perhaps history of gyres, or trade-wind belts, or seamounts, or the deep sea, or tides might produce insights that histories of specific places such as Long Island Sound or the Grand Banks cannot" (Rozwadowski 2013, p. 138).

A more encompassing view of the spaces in which marine scientists have operated encourages historians of oceanography to borrow approaches from geography – such as attention to verticality (Adler 2020, 2022). As historian Christopher Pastore has recently demonstrated, using geography as a framework for analysis can offer a corrective to long-established tropes, helping to avoid oversights brought about by simply following the scientists. Pastore shows that the great voyages of exploration of the eighteenth and nineteenth centuries – held up as points of origin for the history of marine science – "overshadow the less heroic ways that scientific knowledge about the sea was produced closer to shore, informally, and in many cases centuries earlier." Instead of looking so exclusively at expedition vessels for the origins of marine science, we should investigate the "tidal pools, shallow bays, and coral reefs" that garnered the interest of seventeenth-century practitioners of natural history (Pastore 2021, p. 123). The "science of shallow waters" thus pushes back the established chronology of marine science and reveals a more diverse group of actors as central to the production of early marine scientific knowledge. Efforts to broaden our scope might also include identifying historical agents in the non-human world. The potential of this approach is demonstrated in a recent essay that examines the manner in which marine creatures shaped the economic and cultural worlds of the pearl fishery in the Gulf of Mannar (Fernando 2022).

Writing in 2008, Eric Mills addressed the then-current state of the field in an editorial published in the newsletter of the International Commission for the History of Oceanography. According to Mills, historians of the marine sciences still found their work frequently overlooked by historians of science and the discipline of history continued to be dominated by political history. At the same time, Mills expressed optimism that the history of oceanography would continue to develop and that younger historians would study the "interfaces between the marine sciences and public policy, military establishments, exploration and the environment" (Mills 2008, p. 3). Surveying the state of the field in 2022, I think Mills' optimism remains warranted. The era of large international history of oceanography congresses dominated by practicing ocean scientists is over, but in the early 2020s historians of ocean sciences are well represented (often in ICHO-sponsored panels) at history of science, history of technology, and environmental history conferences alike.

As it matures as a subdiscipline of the history of science, the history of oceanography is benefitting from the willingness of scholars to embrace interdisciplinarity. Historians of oceanography are not as concerned with the divisions in ocean science important to biological oceanographers, physical oceanographers, or chemical oceanographers. In fact, informed by the social sciences and humanities, they are increasingly interested in more encompassing approaches to knowledge-making, including formal, informal, and tacit knowledge. They recognize that not all scientific knowledge about the sea has been produced by scientists. And that historical research need not be focused on a geographically confined marine ecosystem. We can examine the ocean sciences as practiced across a variety of environments. Histories of twentieth-century oceanography would now be regarded as incomplete if they failed to give attention to the role played by the marine sciences in international diplomacy or the efforts of private industry in underwater exploration (Martínez-Rius 2020).

Increasingly, historians of the marine sciences conceptualize their work as part of a broader project to historicize oceans – spaces long imagined to be timeless and unchanging. Collapsing fisheries, plastics pollution, and the global effects of climate change have eroded misconceptions of the ocean as immune to human activity (Rozwadowski 2019). Developing a richer understanding of the history of oceanography becomes ever more imperative as an aid to an ocean literacy now essential in shaping the human and planetary future.

References

Adler, Antony. 2014. The ship as laboratory: Making space for field science at sea. *Journal of the History of Biology* 47: 333–362.

———. 2016. The hybrid shore: The Marine Station movement and scientific uses of the Littoral, 1843–1910. In *Soundings and crossings: Doing science at sea, 1800–1970*, ed. Katharine Anderson and Helen M. Rozwadowski, 145–178. Sagamore Beach: Science History.

———. 2017. Legitimizing marine field science: Albert Ist of Monaco. In *Understanding field science institutions*, ed. Patience A. Schell, Christer Nordlund, Karl Grandin, and Helena

Ekerholm, 157–191. Sagamore Beach: Science History Publications/Watson Publishing International.
———. 2019. *Neptune's laboratory: Fantasy, fear, and science at sea*. Cambridge, MA: Harvard University Press.
———. 2020. Deep horizons: Canada's underwater habitat program and vertical dimensions of marine sovereignty. *Centaurus* 62: 763–782. https://doi.org/10.1111/1600-0498.12287.
———. 2022. Science: Histories, imaginations, spaces. In *The Routledge handbook of ocean space*, ed. Kimberly Peters, Jon Anderson, Andrew Davies, and Philip Steinberg, 34–45. New York: Routledge.
Adler, Antony, and Erik Dücker. 2018. When Pasteurian science went to sea: The birth of marine microbiology. *Journal of the History of Biology* 51: 107–133.
Banoub, Daniel. 2021. *Fishing measures: A critique of desk-bound reason*. St. John's, Nfld: Memorial University Press.
Benson, Keith R. 1988. Why American Marine Stations?: The teaching argument. *American Zoologist* 28: 7–14.
———. 2013. Forward. In *Places, people, tools: Oceanography in the Mediterranean and beyond. Proceedings of the eighth international congress for the history of oceanography*, ed. Christiane Groeben, 9–10. Naples: Giannini Editore.
Benson, Keith R., and Philip F. Rehbock, eds. 2002. *Oceanographic history: The Pacific and beyond*. University of Washington Press.
Benson, Keith R., Helen M. Rozwadowski, and David K. van Keuren. 2004. Introduction. In *The machine in Neptune's garden: Historical perspectives on technology and the marine environment*, xiii–xxviii. Sagamore Beach: Science History Publications/Watson Publishing International.
Bolster, W. Jeffrey. 2014. *The Mortal Sea: Fishing the Atlantic in the age of sail*. Cambridge, MA/London: Belknap Press of Harvard University Press.
Braudel, Fernand. 1949. *La Mediterranee et le monde mediterraneen a l'époque de Philippe II*. Paris: Colin.
Burstyn, Harold L. 1968. The historian of science and oceanography. In *Premier Congrès International d'Histoire de l'Océanographie, Monaco – 1966*, ed. Jacqueline Carpine-Lancre and John Leighly, vol. 2, 665–676. Monte Carlo: Bulletin de l'Institut Océanographique 2. Imprimierie Nationale de Monaco.
———. 2001. 'Big science' in Victorian Britain: The challenger expedition (1872–6) and its report (1881–95). In *Understanding the oceans: A century of ocean exploration*, ed. Margaret Deacon, Tony Rice, and Colin Summerhayes, 27–48. London/New York: University College London Press.
Cohen, Margaret. 2022. *The underwater eye: How the movie camera opened the depths and unleashed new realms of fantasy*. Princeton: Princeton University Press.
Corfield, Richard. 2003. *The silent landscape: The scientific voyage of HMS challenger*. Washington, DC: Joseph Henry Press. https://doi.org/10.17226/10725.
de Bont, Raf. 2015. *Stations in the field: A history of place-based animal research, 1870–1930*. Chicago: University of Chicago Press.
Deacon, Margaret. 1965. Founders of marine science in Britain: The work of the early fellows of the royal society. *Notes and Records of the Royal Society of London* 20 (1): 28–50.
Deacon, Margaret B. 1971. *Scientists and the sea, 1650–1900: A study of marine science* (1st ed). London: Academic Press.
Deacon, Margaret B. 1993. Crisis and compromise: The Foundation of Marine Stations in Britain during the late 19th century. *Earth Sciences History* 12: 19–47.
———. 1997. *Scientists and the sea, 1650–1900: A study of marine science*. 2nd ed. Brookfield: Ashgate.
Doel, Ronald E., Tanya J. Levin, and Mason K. Marker. 2006. Extending modern cartography to the ocean depths: Military patronage, cold war priorities, and the Heezen-Tharp mapping project, 1952–1959 [Included in virtual special edition celebrating the journal of historical geography at 40]. *Journal of Historical Geography* 32 (2006): 605–626.

Dolan, John. 2020. The origins of oceanography in France. *Oceanography* 33: 126–133. https://www.jstor.org/stable/26937752.

Egerton, Frank N. 2014. History of ecological sciences, part 51: Formalizing marine ecology, 1870s to 1920s. *The Bulletin of the Ecological Society of America* 95: 347–430. https://doi.org/10.1890/0012-9623-95.4.347.

Elias, Ann. 2019. *Coral empire: Underwater oceans, colonial tropics, visual modernity*. Durham: Duke University Press.

Ericson, Kjell David. 2020. The Misaki Marine Biological Station's dual roles for zoology and fisheries, 1880s–1930s. In *Why study biology by the sea?* ed. Karl S. Matlin, Jane Maienschein, and Rachel A. Ankeny, 87–115. Chicago: University of Chicago Press.

Fernando, Tamara. 2022. Seeing like the sea: A multispecies history of the Ceylon pearl fishery 1800–1925. *Past & Present* 254: 127–160. https://doi.org/10.1093/pastj/gtab002.

Finley, Carmel. 2011. *All the fish in the sea: Maximum sustainable yield and the failure of fisheries management*. Chicago: University of Chicago Press.

Forman, Paul. 1985. Behind quantum electronics: National security as basis for physical research in the United States, 1940–1960. *Historical Studies in the Physical and Biological Sciences*. 18: 149–229.

Guberlet, Muriel L. 1964. *Explorers of the sea: Famous oceanographic expeditions*. New York: The Ronald Company.

Hamblin, Jacob Darwin. 2005. *Oceanographers and the cold war: Disciples of marine science*. 1st ed. Seattle: University of Washington Press.

———. 2008. *Poison in the well: Radioactive waste in the oceans at the Dawn of the nuclear age*. Rutgers: Rutgers University Press.

Hardy, Penelope K, and Helen M Rozwadowski. 2020. Maury for modern times: Navigating a racist legacy in ocean science. *Oceanography* 33 (3): 10–15. https://doi.org/10.5670/oceanog.2020.302.

Helmreich, Stefan. 2009. *Alien Ocean: Anthropological voyages on microbial seas*. Berkeley: University of California Press.

Herdman, William A. 1923. *Founders of oceanography and their work*. London: Edward Arnold.

Hubbard, Jennifer Mary. 2006. *A science on the scales: The rise of Canadian Atlantic fisheries biology, 1898–1939*. Toronto: University of Toronto Press.

Jones, Erika. 2023. *The challenger expedition: Exploring the ocean's depths*. London: Royal Museums Greenwich.

Kevles, Daniel. 1990. Cold war and hot physics: Science, security, and the American state, 1945–56. *Historical Studies in the Physical and Biological Sciences* 20: 239–264.

Knauss, John A. 2000. The emergence of the National Science Foundation as a supporter of ocean sciences in the United States. In *50 years of ocean discovery: National Science Foundation 1950–2000*, 3–8. Washington, DC: National Academy Press.

Kohlstedt, Sally Gregory. 2022. Collaboration, gender, and leadership at the Minnesota Seaside Station, 1901–1907. *Journal of the History of Biology*. https://doi.org/10.1007/s10739-022-09679-4.

Lajus, Julia. 2021. Materiality of marine sciences in late Imperial Russia and early Soviet Union: Research vessels, instruments, laboratory practices. *Art* 14: 245–265. https://doi.org/10.4000/artefact.10129.

Lehman, Jessica. 2018. From ships to robots: The social relations of sensing the World Ocean. *Social Studies of Science* 48: 57–79. https://doi.org/10.1177/0306312717743579.

Linklater, Eric. 1972. *The voyage of the challenger*. Garden City, New York: Doubleday & Company Inc.

Lopes, Maria Margaret. 2018. The challenger Deep-Sea expedition (1872–1876) in Brazil: The circulation of news and knowledge. In *Viaggiatori: The scientific dialogue linking America, Asia and Europe between the 12th and the 20th century. Theories and techniques traveling in space and time*, ed. Fabio D'Angelo, 118–132. Naples: Associazione culturale Viaggiatori.

Luk, Christine Yi Lai. 2020. The first Marine Biological Station in modern China: Amoy University and amphioxus. In *Why study biology by the sea?* ed. Karl S. Matlin, Jane Maienschein, and Rachel A. Ankeny, 68–86. Chicago: University of Chicago Press.

Macdougall, Doug. 2019. *Endless novelties of extraordinary interest: The voyage of H.M.S. challenger and the birth of modern oceanography*. Illustrated edition. New Haven: Yale University Press.

Martínez-Rius, Beatriz. 2020. For the benefit of all men: Oceanography and Franco-American scientific diplomacy in the cold war, 1958–1970. *Berichte zur Wissenschaftsgeschichte* 43: 581–605. https://doi.org/10.1002/bewi.202000015.

Matlin, Karl S., Jane Maienschein, and Rachel A. Ankeny, eds. 2020. *Why study biology by the sea?* Chicago: University of Chicago Press.

Matsuda, Matt K. 2012. *Pacific worlds: A history of seas, peoples, and cultures*. Cambridge: Cambridge University Press.

McConnell, Anita. 1982. *No sea too deep: The history of oceanographic instruments*. Bristol: Hilger.

Mikhailov, N.N., R. Tatusko, and Sydney Levitus. 2002. *Russian marine expeditionary investigations of the World Ocean*. Silver Spring: U.S. Dept. of Commerce, National Oceanic and Atmospheric Administration, National Environmental Satellite, Data, and Information Service. ftp://ftp.nodc.noaa.gov/pub/data.nodc/woa/PUBLICATIONS/atlas56.pdf.

Mills, Eric L. 1983. Problems of deep-Sea biology: An historical perspective. In *The sea*, ed. Gilbert T. Rowe, vol. 8, 1–79. Wiley.

———. 1990. What is history of oceanography. *History of Oceanography Newsletter* 2: 2–3.

———. 1993. The historian of science and oceanography after twenty years. *Earth Sciences history* 12: 5–18. https://www.jstor.org/stable/24137418.

———. 2008. Editorial: The history of oceanography after forty years. *History of Oceanography Newsletter* 20: 3–4.

———. 2011. *The fluid envelope of our planet: How the study of ocean currents became a science*. Toronto: University of Toronto Press.

———. 2012. *Biological oceanography: An early history, 1870–1960*. University of Toronto Press.

Muka, Samantha K. 2014. Portrait of an outsider: Class, gender, and the scientific career of Ida M. Mellen. *Journal of the History of Biology* 47: 29–61. https://doi.org/10.1007/s10739-013-9354-z.

Muka, Samantha. 2021. Historiography of marine biology. In *Handbook of the historiography of biology. Historiographies of science*, ed. Michael R. Dietrich, Mark E. Borrello, and Oren Harman, 435–459. Cham: Springer International Publishing. https://doi.org/10.1007/978-3-319-90119-0_24.

———. 2023. *Oceans under glass: Tank craft & the sciences of the sea*. Chicago: Chicago University Press.

Murray, Sir John, Johan Hjort, Haakon Hasberg Gran, and Bjørn Helland-Hansen. 1912. *The depths of the ocean: A general account of the modern science of oceanography based largely on the scientific researches of the Norwegian steamer Michael Sars in the North Atlantic*. London: Macmillan.

Nelson, Stewart B. 1982. *Oceanographic ships, fore and Aft*. Washington, DC: U.S. Gov. Printing Office.

Neushul, Peter, and Zuoyue Wang. 2000. Between the devil and the Deep Sea: C. K. Tseng, Mariculture, and the politics of science in modern China. *Isis* 91: 59–88. https://doi.org/10.1086/384626.

Oreskes, Naomi. 2000. 'Laissez-Tomber': Military patronage and women's work in mid-20th-century oceanography. *Historical Studies in the Physical and Biological Sciences* 30 (2): 373–392. https://doi.org/10.2307/27757836.

Oreskes, Naomi. 2021. *Science on a Mission: How military funding shaped what we do and don't know about the ocean*. Chicago: University of Chicago Press.

Oreskes, Naomi, and Ronald Rainger. 2000. Science and security before the atomic bomb: The loyalty case of Harald U. Sverdrup. *Studies in History and Philosophy of Science Part B: Studies in History and Philosophy of Modern Physics* 31 (3): 309–369. https://doi.org/10.1016/S1355-2198(00)00019-8.

Pastore, Christopher L. 2021. The science of shallow waters: Connecting and classifying the early modern Atlantic. *Isis* 112 (1): 122–129. https://doi.org/10.1086/713566.

Pauly, Philip J. 1988. Summer resort and scientific discipline: Woods hole and the structure of American biology, 1882–1925. In *The American development of biology*, ed. Ronald Rainger, Keith R. Benson, and Jane Maienschein, 121–150. University of Pennsylvania Press. https://www.jstor.org/stable/j.ctv5138tr.9.

Pearson, Michael. 2007. *The Indian Ocean*. London: Routledge.

Polmar, Norman, and Lee J. Mathers. 2021. *Opening the great depths: The Bathyscaph Trieste and pioneers of undersea exploration*. Annapolis: Naval Institute Press.

Poulsen, Bo. 2016. *Global marine science and Carlsberg – The Golden connections of Johannes Schmidt (1877–1933)*. Leiden: Brill.

Raitt, Helen. 1956. *Exploring the deep pacific*. New York (State): Norton.

Raitt, Helen, and Beatrice Moulton. 1967. *Scripps institution of oceanography, first fifty years*. Los Angeles: Ward Ritchie Press. http://scilib.ucsd.edu/sio/hist/gc29r2.pdf.

Reidy, Michael S. 2008. *Tides of history: Ocean science and Her Majesty's Navy*. Chicago: University of Chicago Press.

Rice, Anthony L. 1986. *British oceanographic vessels 1800–1950*. London: Ray Society.

———. 2009. Marine science in the age of sail. *Zoologica Scripta* 38: 25–31. https://doi.org/10.1111/j.1463-6409.2007.00305.x.

Robinson, Samuel A. 2018. Ocean science and the British cold war state. In *Palgrave studies in the history of science and technology*. Cham: Springer International Publishing. https://doi.org/10.1007/978-3-319-73096-7_1.

Rozwadowski, Helen M. 2002. *The sea knows no boundaries: A century of marine science under ICES*. Copenhagen: The International Council for the Exploration of the Sea.

———. 2005. *Fathoming the ocean: The discovery and exploration of the Deep Sea*. Cambridge, MA: Belknap Press of Harvard University Press.

———. 2013. The promise of ocean history for environmental history. *The Journal of American History* 100: 136–139. https://doi.org/10.1093/jahist/jat069.

———. 2019. *Vast expanses: A history of the oceans*. London: Reaktion Books.

Sapolsky, Harvey M. 2014. *Science and the navy: The history of the office of naval research*. Princeton University Press.

Schlee, Susan. 1973. *The edge of an unfamiliar world; A history of oceanography*. New York: Dutton.

———. 1978. *On almost any wind: The Saga of the oceanographic research vessel Atlantis*. Ithaca: Cornell University Press.

Schwach, Vera. 2013. The sea around Norway: Science, resource management, and environmental concerns, 1860–1970. *Environmental History* 18: 101–110. https://doi.org/10.1093/envhis/ems107.

Schwerdtner-Máñez, Kathleen, and Bo Poulson, eds. 2016. *Perspectives on oceans past: A handbook of marine environmental history*. Dordrecht: Springer Nature.

Shor, Elizabeth Noble. 1978. *Scripps institution of oceanography, 1903–1978*. San Diego: Tofua Press.

Smith, Tim D. 1994. *Scaling fisheries: The science of measuring the effects of fishing, 1855–1955*. Cambridge: Cambridge University Press.

Steiner, Katharina. 2018. Copepods and Fisherboys: Advanced marine biological research and street poverty in Naples c. 1890. In *Urban histories of science: Making knowledge in the City, 1820–1940*, ed. Oliver Hochadel and Agustí Nieto-Galan, 80–101. New York: Routledge.

Summerhayes, Colin, and Cornelia Lüdecke. 2012. A German contribution to South Atlantic seabed studies, 1938–39. *Polarforschung* 82: 93–101.

Thomson, Charles Wyville, William Benjamin Carpenter, John Gwyn Jeffreys, and Charles Fitz-Gerald Thomson. 1873. *The depths of the sea. An account of the general results of the dredging cruises of H.M. SS. "Porcupine" and "lightning" during the summers of 1868, 1869, and 1870, under the scientific direction of Dr. Carpenter ... J.G. Jeffreys ... and Dr. Wyville Thomson ... with ... Illustrations and maps*. London: Macmillan.

Torma, Franziska. 2012. Snakey waters, or: How marine biology structures global environmental sciences. *Rachel Carson Center Perspectives* 2: 13–21.
Turchetti, Simone. 2012. Sword, Shiel and Buoys: A history of the NATA sub-committee on oceanographic research, 1959–1973. *Centaurus* 54: 205–231. https://doi.org/10.1111/j.1600-0498.2012.00258.x.
Wang, Zuoyue. 2007. China goes to the poles: Nationalism and internationalism in Chinese polar oceanography. In *Extremes: Oceanography's adventures at the poles*, ed. Keith R. Benson and Helen M. Rozwadowski, 269–302. Sagamore Beach: Science History Publications.
Weir, Gary E. 2001. *An ocean in common: American naval officers, scientists and the ocean environment*. College Station: Texas A&M University Press.
Wernand, Marcel R. 2010. On the history of the secchi disc. *Journal of the European Optical Society – Rapid Publications* 5 (0). https://doi.org/10.2971/jeos.2010.10013s.
Wüst, Georg. 1960. Proposed international Indian Ocean oceanographic expedition, 1962–1963. *Deep Sea Research* 6: 245–249.
———. 1964. The major Deep-Sea expeditions and research vessels 1873–1960: A contribution to the history of oceanography. *Progress in Oceanography* 2: 3–52.
Zuroski, Emma. 2017. Situating the local in a global expedition: HMS challenger expedition in New Zealand, 1874. *Journal of the Royal Society of New Zealand* 47: 107–111.

Climate and the Little Ice Age

Gillen D'Arcy Wood

Contents

Abstract

An ever-increasing "natural archive" of historical climate proxy data has allowed for the recent emergence of a new historiography of climate change. This chapter reviews the significance of recent titles in climate history focused specifically on the Little Ice Age – the period of deteriorated global weather conditions from the fourteenth to the nineteenth centuries that witnessed frequent, destabilizing episodes of drought, cold, famine, and disease across both eastern and western hemispheres. Prior to the broad availability of historical climate data this century, academic historians were able to ignore issues of climate and environment within an anthropocentrist framework for historical analysis. The new climate history of the Little Ice Age, by contrast, advocates an improved understanding of climate instability as a key driver of social change at both local and hemispheric scales, in ways that radically revise canonical narratives of the Late Medieval and Early Modern periods. The chapter concludes with consideration of two other new

Parts of this chapter have been reproduced with permission of the Journal of Early Modern Cultural Studies 8.2 (2008): 33–55.

G. D'Arcy Wood (✉)
University of Illinois, Urbana-Champaign, Champaign, IL, USA
e-mail: gdwood@illinois.edu

E. Aronova et al. (eds.), *Handbook of the Historiography of the Earth and Environmental Sciences*, Historiographies of Science,
https://doi.org/10.1007/978-3-031-40799-4_9

climate change paradigms relevant to current historiography: deep time and the Anthropocene.

Keywords

Little ice age · Climate change · Historiography · Early modern period · Modern period · Deep time · Anthropocene

1 Introduction: The New Climate *Zeitgeist*

For most of the twentieth century, climate represented the untouchable third rail of professional historiography. Enlightenment thinkers of the colonial period, following Montesquieu, linked climatic zones to racialist tropes, helping to legitimize the subjection of African, Asian, and Pacific peoples. Climatic determinism, with its roots in Hippocratic theory, has thus long been stigmatized as an historical prop to European imperialism and an intellectual embryo of biological race theory. Until recently, to propose climate conditions as an influence on human history was to raise the unwelcome specter of a disgraced ideology.

Climate's stocks as a historiographic paradigm have risen in the twenty-first century, however. In an era of increasing extreme weather, rising seas, and global food system stress, the necessity of integrating climate and environmental change into the work of history has become increasingly urgent. Crude climatic determinism remains anathema, but historical materialism without environmental awareness now appears vulnerable to a dangerous determinism of another kind: a grand narrative of class, government, empire, and trade whose anthropocentric terms are no less a product of Enlightenment liberal ideology than the current ecological crisis itself.

Across a variety of academic fields, climate and human history are no longer estranged. In the sciences, the once marginal field of historical climatology, pioneered in the 1960s and 1970s by Hubert Lamb, now spawns institutes. Given that standardized data for temperature and rainfall are available in most regions only from the mid-nineteenth century on, historical climatologists have mined a multitude of proxy sources – combining a "natural" archive of ice cores, tree rings, and fossilized pollens with a more conventional human archive of agricultural records, weather diaries, etc. – to fill out the climate record across a broad spatio-temporal range: from a single famine year in, say, seventeenth-century Europe, to the beginnings of sedentary human culture 10,000 years ago, and back through the dim, paleoclimatological eras of human evolution. In each case, the propensity of the world climate system to dramatic short and long-term fluctuations – with corresponding ripple effects across human populations – has been a major and consistent finding.

In traditional fields within the humanities, too, a historicized ecological consciousness has become the new norm. Since the 1970s, environmental history has evolved from a parochial American focus to a key interdisciplinary paradigm in both regional studies and world history. The now-thriving field boasts major journals, a professional society, and – the ultimate test of viability – faculty lines. In an adjacent discipline, a new generation of anthropologists have recognized the untenability of the traditional intradisciplinary divide between physical anthropology – which takes seriously the climatic adaptations of early humans – and cultural anthropologists who are, in Carole Crumley's words, "suspicious ... of the determinisms: racial, environmental, social." Anthropologists, in tune with the overwhelming majority of professional historians, have long aggressively prosecuted their commitment, on principle, to "denying environment a meaningful role in human history" (*Historical Ecology* 2–3).

A founding presumption of the new climate historicism is that anthropogenic climate change did not begin sometime in the 1950s or 1980s, but rather with the fossil fuel technologies, colonial resource extraction, and global economic growth associated with the early modern period. The task of climate history for this period (which coincides with the so-called Little Ice Age) is to illuminate how specific instances of human ecological agency – land use, agricultural technology, and food production, the opening of trade routes, urban growth, as well as armed conflict, subsistence, and public health crises – were acculturated and rationalized. To fully comprehend the colonial enterprise post-1500, for example, requires we trace the mutual influences between European states and their annexed global environments to show how the export of North Atlantic biota and agricultural, mining, and manufacturing infrastructures within the intercontinental trading network facilitated western contact with (and imprint upon) "exotic" climates and ecosystems, not just peoples.

The new climate history is thus, by its nature, interdisciplinary, requiring a crossing (or leap) into the biophysical sciences and a technical literacy in one or more of the disciplines of climatology, geology, geography, and the biological and environmental sciences. As will be seen by the cluster of groundbreaking recent titles in climate history discussed here, the historiographic potentials are impressive. Humanists by training and sensibility, climate historians are able to explain, in ways physical scientists are not, what the hard data of historical climatology *meant* in the minds and lived experiences of the people who endured or benefited from a specific meteorological regime, and how societies have adapted to and shaped environmental change. Such climate-aware historicism involves the study of extreme weather and environmental change as objects of knowledge and desire analyzed through "thick" description of specific episodes of socio-ecological stress, such as the droughts, floods, famines, and plagues of the Little Ice Age. The goal of this new eco-critical approach is to reconstruct the rationalizations of historical actors in their relation to specific ecosystems and climate regimes, while tracing the ways various providential, instrumentalist, and proto-ecological views of the world and its natural resources first gained currency and found expression. Vitally important – and this is the true revolution behind the new climate paradigm – are critical histories of socio-cultural responses to "exogenous" natural phenomena beyond human control: for instance, the extreme weather events, natural disasters, and flow-on social convulsions of the middle centuries of the last millennium. Despite major research advances, however, the new climate history paradigm remains highly contested. As we shall see, the marriage of climate science and academic historiography never did run smoothly.

2 Anthropocentrism: The Last Ideology

In 2014, the editors of the *Journal of Interdisciplinary History* (*JIA*), a prominent scholarly forum for academic historians, devoted its winter issue to the Little Ice Age (LIA), the period roughly from 1300 to 1850, during which much of the globe suffered regular spasms of extreme weather characterized by well-below-average temperatures and erratic rainfall. The establishment of the LIA over the last 50 years as an informal but consequential period in late Holocene climate has been made possible through the cross-feeding of relevant historical sources with proxy temperature and rainfall data – from tree rings, sedimented pollens, speleothems, etc. – accumulated by dozens of researchers across multiple fields. These data collectively attest to the widespread impacts of extended cold, drought, and flooding on human communities across Europe, Asia, and the Americas in the centuries after 1300, with links both direct and indirect to devastating disease outbreaks and civil unrest. As such, the LIA represents an exciting frontier of interdisciplinary, and potentially collaborative, research among historians and Earth scientists.

Given the established consensus surrounding LIA data, and the rich research opportunities on offer, it comes as a surprise to recall that the lead article in the 2014 *JIA* special issue contested the very existence of the LIA. "We find no statistical evidence of any major breaks, trends, or cycles in European weather of the sort that one could associate with an LIA," averred its authors, Morgan Kelly and Cormac Ó Gráda. The claim was startling, cavalier, and readily falsifiable. To make their "case," the authors, both well-known economic historians, deployed forms of statistical obfuscation, straw man argument, and data cherry-picking readily recognizable from far less reputable forums than the *JIH* – specifically the relentless stygian content stream of climate disinformation sponsored by industry in the decades since global warming first emerged as a political threat to fossil fuel capital investment in the late 1980s. In the measured words of climate historian Sam White, who was offered the opportunity to rebut the Kelly/Ó Gráda article in the same issue, the authors' argument for a phantom LIA had "serious flaws," not least their presumption, as non-specialists, that putative flaws in the interpretation of weather data had "somehow evaded hundreds of climate scientists" working for decades in the field (327, 334). The article, White concluded (with a detectable note of exasperation), "is a peculiar – and entirely unfounded – criticism of robust and widely accepted climate science" (351).

A chief peculiarity surrounding the 2014 publication of the Kelly/Ó Gráda article is that the debate it pretended to join – and to which the *JIA* editors lent their assistance in staging – did not exist. The fact of the LIA was and is uncontested by the overwhelming majority of climate scientists and historians. A further historical irony is that the contrarian article's appearance was swamped in the early 2010s by a near-tsunami of new historical scholarship on the LIA that has definitively established the 1300–1850 period of climate deterioration, and climate historiography itself, in the mainstream of professional historical practice (though it continues to receive a "cold welcome" from significant quarters of the academic establishment, for reasons adduced below).

My selective bibliography of recent LIA scholarship for discussion here comprises six widely reviewed books published in the years immediately before and after the controversial 2014 issue of the *JIA*: Sam White's *The Climate of Rebellion in the Early Modern Ottoman Empire* (2011) and *A Cold Welcome: The Little Ice Age and Europe's Encounter with North America* (2017); Geoffrey Parker's monumental *Global Crisis: War, Climate Change and Catastrophe in the Seventeenth Century* (2013); Gillen D'Arcy Wood's *Tambora: The Eruption that Changed the World* (2014); Bruce Campbell's *The Great Transition: Climate, Disease and Society in the Late-Medieval World* (2016); as well as significant chapters on the LIA in John L. Brooke's ground-breaking "big history," *Climate Change and the Course of Global History* (2014).

Before I engage with these books' collective description of the LIA and their consequences for late medieval/early modern historiography, it will be worthwhile to assess more largely what the contrived LIA "debate" in the *JIH* reveals about the longstanding distrust of academic historians toward climate science and the new climate-historical paradigm. My purpose is not to relitigate the case so easily debunked by White in 2014, but to draw lessons regarding an intellectual culture that recently offered a prestigious platform to discredited tropes and methods of climate denial.

In *The Great Transition*, Bruce Campbell identifies anthropocentrism as the critical impediment to a fully descriptive historiography of social change: "the long dominance of anthropocentric analyses grounded in either Marxist theory or neoclassical economics means that scant attention has as yet been paid to the historical roles of environmental processes and forces" (27). The complex, LIA-driven socio-ecological transformations of the fourteenth and fifteenth centuries, he argues by way of example, "stubbornly refuse to fit comfortably within explanations that give causal primacy to mostly territorially specific human processes and relationships" (395). To properly account for the sweeping social crises of the LIA in Europe requires instead a "holistic approach" that "breaches conventional historical boundaries of time and place" and, most importantly, contests the anthropocentric privileging of human agency to the exclusion of all environmental drivers (396).

According to Campbell, the persistent error in traditional historiography has been "to create a binary divide between human and environmental agencies," where the natural world functions as a mere backdrop to the ever-fascinating drama of human self-determination. Traditional historiographic practice, both liberal and Marxist, "label[s] human actions as 'endogenous', and therefore to be understood and explained, and environmental variables as 'exogenous', and consequently less worthy of examination" (396). Under the long-prevailing anthropocentrist regime, the understanding of natural systems and their embedded relation to human affairs has not been a goal, or even a working concept in mainstream historiography. Even the post-1970s emergence of environmental history did not produce a revolution in methodology or disciplinary realignment commensurate with our twenty-first-century cluster of mounting ecological crises or the biophysical research data tools that have emerged to track them.

Deconstruction of this suffocating binary logic is thus first-order work for the new climate historian, but represents an enormous challenge. A key problem lies in re-orienting research in the fields of modern history conspicuously endowed with rich archives and socio-economic data, where the opportunities for human-centered analysis are more or less inexhaustible and continue to be professionally rewarded. "The trouble has arisen," as White expresses it, "from integrating the role of climate into the current historiography wherever an abundance of evidence has already provided adequate social, economic, and political interpretations of events." Call it the climate/history paradox: how it is that decades of brilliant and highly satisfying scholarship in modern social and political history have nevertheless "left us without clear and well-documented examples of how and why climate events could shift the course of human history" (*Climate of Rebellion* 13–14).

White readily acknowledges, moreover, that the climate history paradigm has not been well served by its popular, journalistic iterations, whose research "has often been uneven and the authors … far too ready to reach hasty and dramatic conclusions."[1] Climate history, in the popular marketplace, has too often overcorrected for the anthropocentric bias of professional historiography by "fail[ing] to consider social or political context or make adequate room for human agency." For this and other reasons, the climate paradigm has met with entrenched "skepticism from academic historians" with the result that, at his time of writing in 2011, White might reasonably complain that "climate has yet to play a major role in the mainstream historiography of most regions" (13). Even as biophysical and demographic data pointing to the severity of the LIA accumulated through the 1970s and 1980s, the information remained largely inert – ignored by academic historians who resisted integrating climatic contexts into their working paradigms of historical change. It's some measure of the impact of White and his fellow new climate historians that his 2011 statement is somewhat less demonstrably true now. The new climate history of the past decade has earned newfound respectability for the field "by bringing climate into human history in a more nuanced way, neither simplifying its role nor neglecting its impact" (14).

The 2010s cohort of new climate historians, sensitive to the age-old charge of environmental determinism, have consciously adopted a vocabulary indebted to ecology – with its focus on complexity, contingency, and chaotic interaction – in place of simplistic frameworks of cause and effect. Bruce Campbell has eloquently framed the guiding principle of the new LIA scholarship:

> To do justice to a complex past and the dynamism of the natural world in which people lived, worked and reproduced it is necessary to understand how climate and society, ecology and biology, microbes and humans, acting separately and in combination with each other, shaped the course of history … As [the Great Transition] unfolded, interactions and feedbacks, both environmental and human, occurred at a range of nested temporal and spatial scales – from the short-term to the long-term and the micro to the macro – as change cascaded through the prevailing socio-ecological system … (396)

[1] White cites the following popular climate histories as examples: Fagan (1999) and his sequel book on the Little Ice Age (2000), Linden (2006), Diamond (2005), and Keys (2000).

A key paradigm shift in this eco-historical approach is elasticity of scale, in both time and place. To comprehend the evolution and impact of the Black Death – to take Campbell's central example – requires the subtle imbrication of microbial, human, trans-national, and atmospheric scales, of local and global impacts, and a working conception of trans-human, biological time and space. The sudden penetration of bubonic *Y. pestis*, via multiple vectors, into human communities in Asia in the mid-fourteenth century exemplifies the historical unfolding of the Great Transition at large, characterized by "a cascading process of contingent chaotic development … [where] each stage of each component in the transition was unique, with outcomes that were rarely predetermined and always prone to unpredictability" (396). A feature of this process is its nonlinear evolution. An inconspicuous bacterium makes the leap to a rodent host. Soon millions are dead and human societies are transformed. The lesson of the Black Death is that historical agents not only reside at different spatio-temporal scales, but transgress and transcend them, with the potential for metastasizing impacts at each phase.

The development of new environmental histories that transcend anthropocentrist traditions requires that scholars abandon a reflexive indifference toward the biophysical sciences, allowing for perception of the full currents of social change in periods of climate and environmental stress. For the climate historian, the challenge is to deftly manage the necessary analytic vocabularies for each spatio-temporal level, as well as the "cascading" interactions between them. The fact that acquisition of these technical vocabularies requires immersion in multiple scientific fields sets a high threshold for effective climate history, and helps explain why adoption of the paradigm has been slow.

But other disincentives are at work, too. The example of Kelly/Ó Gráda, who simply ignored climate science and the paleoclimate literature in their critique of the LIA, suggests that the age-old "two cultures" problem, putting the sciences and social sciences and humanities at odds, dies hard, even under the existential pressure of a global climate change emergency. The Earth system sciences, dedicated to descriptions of the biophysical world, have traditionally been irrelevant to anthropocentrist historiography, inflected by neoclassical economics, Marxist theory, or other human-centered approaches. The fact that so-called science studies – dedicated to the critique of scientific methods and institutional histories – itself became a popular field of humanistic inquiry in the late twentieth century points to an ingrained suspicion of the biophysical sciences, of which Kelly/Ó Gráda is a piquant latter-day example.

The methodological burden for the new climate history is thus a heavy one. Its successful practice requires not only the historian's self-education about the natural world and its processes, but also a critical awareness of the anthropocentric biases that have inhibited integration of that knowledge in the past and that continue to govern much research in the discipline. Scholars have been habituated to articulating human history and cultural evolution as a legible story of progress or decline governed by human-built institutions and enacted by moral agents, with the natural world relegated to scenic backdrop. Social history in the new climate paradigm, by contrast, is ecological in character. It operates on the assumption that human affairs exhibit the same patterns of systems connectivity, complexity, and nonlinear transformation that we observe in the organic world, from the genetic profiles of bacterial disease to continental weather systems.

To be clear, the research sciences have not always offered a progressive, importable paradigm for the humanities. For most of the twentieth century, across many fields, scientific practice was reductionist in approach, with the goal of isolating natural organisms and their observed behaviors within ever smaller ambits. But the advent at the turn of the twenty-first century of large-scale digital technologies and data-merging capability – from genome sequencing, to climate modeling, to satellite imaging of land and sea – has enabled scientists to emerge from discrete disciplinary problem-solving to address the larger "Earth system" picture, including the tightly coupled, vital relationship between humans and their environments.

Nowhere in the new climate historiography of the 2010s is the implication that modern historical change is innocent of social power inequities and the institutionalized racism that underwrote wholesale exploitation of subaltern populations in the colonial period, only that these crucial human factors are embedded within a larger, and heretofore largely ignored, socio-environmental nexus. The goal of the new climate history, rather, is to demonstrate, over and over, the historical and material interdependence of human and natural systems – something the academic humanities until now have largely failed to do. "The new climate science," as John L. Brooke argues, "has destroyed the refuge of skeptical historians, who traditionally discount the impact of natural forces because they presumably operated as an unknowable constant. These natural forces are now 'knowable,' and they were not 'constant'" (391).

According to the new scientific paradigm, *biocomplexity* – the chaotically variable interaction between organic elements on multiple scales – is the defining characteristic of all ecosystems, including human societies. The new climate history – inspired by the science of biocomplexity – tests narrative frameworks for describing the nonlinear interaction of historical actors and events across multiple scales of time and space, from the molecular to the intercontinental, from a single growing season in the mid-1600s to millennia to deep time. Such an approach was inconceivable before about 1990. For historians of the LIA, their entire research agenda has been enabled by the productivity of new methods and tools that revolutionized paleoclimate research over the last half century, most notably the development of physical proxy records for past climates.

3 History by Proxy: The Little Ice Age

In his Prologue to *Global Crisis*, entitled "Did Someone Say 'Climate Change'?," Geoffrey Parker calls the professional history community to account for "denying a link between climate and catastrophe" (xviii). Climate skepticism might have been justified into the 1980s, when radiocarbon dating was in its infancy, the first carbon dioxide readings of ice cores were only just being recorded, general circulation models of the atmosphere were crude, and paleoclimate models rare in the scientific literature. Even pioneering climate historian Emmanuel Le Roy Ladurie, writing in the 1960s, could argue only for a "slight, perhaps negligible" influence of climate on history (*Times of Feast* 119). But with the emergence since the 1990s of a readily available and increasingly comprehensive historical climate archive, the pressure to incorporate this vast wealth of new environmental information into the historical narrative has grown overwhelming (albeit not, as we have seen, irresistible). In the decade-and-a-half prior to the publication of *Global Crisis*, some 3500 papers with the keywords "ice cores" alone were published, more than the entire paleoclimate bibliography in 1980.

In addition to ice cores drawn from polar ice sheets and alpine glaciers, which record changing levels of air temperature, rainfall, and volcanic anomalies at an annual resolution, Parker nominates three other significant proxies for past climates whose robustness and reach have developed rapidly in the past 30–40 years. *Palynology*, the study of pollens in lake sediments, reconstructs the state of vegetation at the time the pollens' deposit, from which temperature, rainfall, and seasonality can be deduced. While palynology is a recent science, the potential of tree-ring growth as a proxy climate record has been appreciated for a century and more, and now boasts a wide-ranging global database. The science of *dendrochronology* is, of course, limited to the lifetime of a given tree, which cannot extend its proxy record, as ice cores may do, to timescales of hundreds of thousands of years. But trees, terrestrial and ubiquitous, are especially informative in reconstructing past spring and summer temperatures, and their period of annual growth. Finally, the study of stalagmites and other cave deposits – collectively called *speleothems* – provides a further key to past climate conditions. Oxygen isotope variations in the calcite sheath of a speleothem record changes in temperature and rainwater composition over geological timescales. The recent explosion in the speleothem record includes extensive new data on the LIA in Europe, as well as detailed reconstruction of fluctuations in the Asian monsoon over the last half-a-million years.

Collectively, this "natural archive" offers thousands of data points of potential relevance to the historian, which, when combined with the traditional human archive of official chronicles, personal letters and diaries, agricultural records, and instrumental data (inter alia) represents a new historiographic paradigm of formidable explanatory power. The plenitude of combined natural and human sources now available, argues Parker, in reference to the seventeenth-century world crisis in particular, "allow[s] historians to integrate climate change with political, economic and social change with unprecedented precision. Accounts of climatic conditions in Africa, Asia, Europe, and the Americas in the mid-seventeenth century abound, while millions of measurements of tree-rings, ice-cores, pollen deposits and stalactite formations are available" (xxii). Unsurprisingly, given the novelty of their data and methods, the new climate historians, Parker conspicuous among them, have produced radically original accounts of their historical periods, which together form a comprehensive account of global strife in the LIA and the relentlessly adverse climate conditions that helped precipitate it.

The engagement of these historians with the frontiers of Earth and environmental science extends further to consideration of the physical causes – astral and geological – of the LIA. Debate on this question in the relevant scientific fields continues to be lively, but a working consensus has converged around an upsurge in low-latitude volcanic activity after 1250 combined with natural orbital cycles and erratic fluctuations in solar irradiance evidenced by sunspots (which were themselves the object of fascination among astronomers in the period). As Campbell observes, "sunspot activity had effectively ceased" by the last decade of the thirteenth century, ushering in the so-called Wolf Solar Minimum (198). This combined with the cumulative thermal deficit of a suite of tropical volcanic eruptions – in 1258, 1269, 1276, and 1286 – which spread a reflective aerosol veil around the planet, ensured a sharp downward trend in northern hemisphere and global temperatures beginning in the 1290s, as well as flow-on disruptions to weather patterns and rainfall. A global climate deterioration post-1300 is reflected clearly in the narrowed growth rings of trees in both the Old and New Worlds.

In *The Great Transition,* Campbell points to the consecutive harvest failures of 1315–1316 and then 1329–1330 in Europe as the first of the many social crises engineered by the climate reversals of the LIA. "Incessant heavy rain" across the continent caused crop failure and resulting famine, which "brought the weaknesses and vulnerabilities of the struggling societies to the fore" (258). But this was a mere prelude to the global catastrophe of the mid-fourteenth century when climatic stressors initiated release of bubonic plague into the human community. After a period of average temperatures recorded in the global tree record, 1342–1343 witnessed an "exceptionally strong negative environmental forcing on a global scale [that] coincides almost exactly with the point when in western Asia *Y. pestis* finally spilled over from sylvatic to commensal rodent populations and then began to spread far and wide with devastating consequences for human populations" (285). Campbell's emphasis on ecological conditions surrounding the emergence of the plague runs counter to traditional historiographic focus on demographic pressures, with Malthusian arguments linking crop failure, civic instability, and population decline. The evidence from climate proxies insists, by contrast, that "the demographic catastrophe" of the Great Death – which killed up to a third of Europe's population – "owed more to ecological than to economic circumstances" (319). A further episodic decline in solar irradiance in the mid-fifteenth century – the so-called Spörer Solar Minimum – reinforced the negative socio-ecological spiral triggered by the Black Death "as populations and economic output became locked onto a downward path" (394).

Sam White, in *The Climate of Rebellion in the Early Modern Ottoman Empire*, takes up the LIA narrative from the late sixteenth century when rebel armies "defied the imperial government for more than a decade, laying waste to wide stretches of the [Ottoman] empire" (185). Early western historiography of the Ottomans, orientalizing in tendency, focused on the decadence and corruption of sultanate governance to explain the Celali Rebellion of the 1590s, while more recent revisionist scholarship has downplayed altogether the importance of social disorder, preferring to emphasize "the resilience of the empire" (188). Missing from these accounts is the glaring fact that the 1590s witnessed the longest drought in the Eastern Mediterranean in 600 years. The value of the new climate history is nowhere more evident than in such a case, where existing historical paradigms, set alongside newly available climate data, are revealed to be conspicuously inadequate. The core of White's book is to extend our understanding of the "general crisis" of the mid-seventeenth century eastward, where "the troubles of Ottoman lands prove[d] so extreme and so prolonged, even relative to other victims" (188). Extreme cold froze the Bosphorus in 1621, and the first in a series of ill-fated sultans was deposed the following year. The Ottoman Empire's dependence on a single annual harvest of wheat and barley – itself dependent on mild temperatures and generous spring rainfall – proved a tragic Achilles' heel once the LIA tightened its grip.

In his follow-up LIA monograph on the seventeenth century, *A Cold Welcome,* White shifts focus to the Americas, for which "oxygen isotopes in Greenland ice cores indicate that temperatures throughout the early 1600s remained at their lowest point of the Little Ice Age" (141). The first decade of the seventeenth century coincided with the tentative installment of English settlements at Jamestown on the Virginia coast, which barely survived the extreme weather, and only then at the cost of the majority of settlers' lives and a brutal struggle between the Europeans and indigenous communities (who suffered continent-wide from the same crop-killing cold and drought). But as White's broad and original survey of early European forays into the Americas shows, the well-known Jamestown disaster was far from the only instance of a halting colonial enterprise shaped by sharply adverse Little Ice Age conditions:

> The first Spanish expeditions into the Southeast, one after another, fell victim to storms, hunger, disease, and freezing winters. The French Huguenot attempts to settle Florida suffered famine and ended with a hurricane. England's first colony at Roanoke disappeared after drought, hunger, and storms. Spain's outpost in Florida fared so poorly in its environment that Philip III debated whether it was even worth saving. The Popham colony ended after a single winter of extreme cold in Maine. The Spanish Empire's expeditions to New Mexico faced icy winters, drought, or both. Several of the first French colonies in Canada and New England were decimated by scurvy during long winters. (290)

In *A Cold Welcome,* White showcases the capacity of the new climate history to identify patterns that are invisible to traditional historiography with its artificially limited focus on individual nation-state or regional narratives. The beauty of climate, as it were, is that it has no respect for geopolitical or other man-made boundaries, and compels the historian to think instead across latitudes and along isothermic lines, in this case from Canada, to Florida, to New Mexico, to Cahokia, Illinois. What results is a deconstructed form of history, with humanity displaced from its complacent center – the "endogenous" to which all else must be "exogenous."

Like White's two volumes, which extend the hemispheric boundaries of LIA studies east and west, Geoffrey Parker's epic *Global Crisis* combines a revisionist focus on the "general crisis" in Europe in the mid-seventeenth century with comparative analyses, particularly from East and South Asia. A cluster of a dozen volcanic eruptions along the Pacific Rim between 1638 and 1644 triggered the fitful collapse of the East Asian monsoon and a spike in droughts and floods associated with El Niño through the 1640s and 1650s, "extremes of weather seldom witnessed before and never (so far) since" (112). This trans-hemispheric climate deterioration, a protracted episode of global cooling without parallel in the entire Holocene record (i.e., the last 12,000 years), coincided with "more rebellions and revolutions than any comparable period in world history," from Catalonia and Naples to England, to the Mughal Empire, to dynastic overthrow in China.[2] Parker documents, moreover, how climatic and political instability reflected for the ordinary people of these regions in "a sharp deterioration in the overall quality of life," measured in bread prices, disease outbreaks, the traumas of war, and even height averages (668). Other impacts were long-term, even epochal. The catastrophic outbreak of plague in Italy and Spain after 1649, for example, "set the seal on the decline of the Mediterranean as the heart of the European economy forever" (670). Those who best survived the climate catastrophe – the Dutch and, to a mixed degree, England – assumed economic and political leadership of a reorganized, barely recovering continent.

The "long tail" of climate change impacts on human history is likewise the focus of my 2014 book *Tambora: The Eruption that Changed the World*, which extends the new LIA studies into the nineteenth century. The eruption of the Tambora volcano in 1815 in the then Dutch East Indies was, by multiple measures, the most climatically significant of the Holocene, depressing temperatures and altering storm tracks worldwide for 3 years. From this single, disruptive geo-climatological event, I trace a series of "teleconnected" influences on nineteenth-century history, from the disastrous evolution of epidemic cholera in Bengal, to the institution of narcotic crop production in Yunnan (later the "Golden Triangle"), to the United States' first economic depression, to the first emergence of the modern welfare state in Western Europe. This study, which builds on pioneering research into the post-Tambora period by John Post in the 1970s, draws equally fully from the human archive of poems, paintings, and diaries as the "natural archive" of ice cores, tree rings, and other climate proxies. What emerges from this case-study approach to the Tambora event – whose extreme climatic impacts lasted only 3 years – is powerful evidence for the long-lasting influence on human affairs of even a limited episode of global climate change.

Published the same year as *Tambora*, John L. Brooke's *Climate Change and the Course of Global* takes the opposite approach on the same assumptions. Instead of building outward from a single climate disaster, he surveys the *longue durée* of human development through the lens of climate change from our evolutionary origins to the present era of global warming sometimes labeled the Anthropocene. In his chapters on the LIA, Brooke echoes the emergent consensus surrounding a new climate-historical narrative for the late Middle Ages, namely, that "the fundamental forces at work in the fourteenth century crisis were exogenous to the socioeconomic system … the result of severe natural disasters, not the dynamics of the demographic and economic systems." Taken together, these "exogenous shocks must be seen as foundational to the modern world" (380). One of these foundations was, of course, the transatlantic slave trade, which only "grew to monstrous proportions … when the full drought impacts of the Little Ice Age struck

[2] For a book-length study of China's political instability in the Little Ice Age, with direct reference to climate deterioration, see Brook's (2010).

the Sahel" in the mid-seventeenth century (443). In refashioning the heretofore eurocentric paradigm of the "general crisis" via the lens of global climate and the LIA, Brooke, like his fellow new climate historians, expands the map of modern history to a scale commensurate with the twenty-first-century paradigm of the Earth System sciences. Famine in Ireland and the slave trade in West Africa, for example, may now be naturally comprehended with a single frame, no longer segregated by artificial spheres of historiographic interest. Working from this paradigm, Brooke is able to trace teleconnections between events both small and large, local and hemispheric – connections invisible to conventional historiography and its regionalist, anthropocentric biases.

4 Coda: Climate, Deep Time, and the Anthropocene

As might be expected given the all-consuming focus on climate change this century – both in academic research and the popular media – the LIA is not the only new historical paradigm to emerge that relies on climate science methods and metrics for its period framing and narrative priorities. Two others are the "deep time" or "big history" approach exemplified by Brooke's 2014 book, and the hot debate surrounding the "Anthropocene," a historical framework defined by scaled human interventions in natural systems, including the carbon and nitrogen cycles, agriculturation of land, and the wholesale engineering of freshwater resources. Exactly when these human interventions reached a tipping point at which a "Human Age" on Planet Earth might be said to have commenced – as recent as the industrial period and Nuclear Age, or as ancient as neolithic deep time? – is not a resolvable question, but the emphasis of the Anthropocene paradigm on the interface between human and natural systems automatically reframes the epistemic principles of historiographic inquiry for our age in ways analogous to recent revisionist scholarship on the Little Ice Age.

Stephen Jay Gould has called the popularization of Deep Time in the aftermath of Charles Lyell's *Principles of Geology* (1830–1833) the most significant cultural revolution since Galileo. Darwin's *On the Origin of Species* (1859), inspired by Lyell, helped imprint a new, abyssal figure of time on the popular imagination. History was the first emergent discipline to feel the shuddering impact of the new Deep Time paradigm, and professional historians of the late-nineteenth century responded with Darwinian ingenuity. The biblical timescale of 6000 years would be preserved by substituting the creation of the Earth with a refurbished chronology invoking "The Birth of Civilization," with the Mediterranean as its cradle.

Human history thus retained its essential biblical geography and chronology, in terms of scale, but with the tropes of *Genesis* replaced by mathematics, writing, cities, and the rule of law as the new narrative of origins (this civilization-based narrative formed the basis of my high school history curriculum as late as the 1980s). Via this secularized origin story, the professional history establishment of the twentieth century combined ingenuity with simple denial to keep at bay the Victorian deep time revolution that continues to define and drive the modern bio-physical sciences. The humanities' hygienic separation from the natural and physical sciences, of which deep time is the founding, enabling principle, has been rigorously policed across generations.

Of course, no respectable scholar today would champion a triumphalist "History of Western Civilization" model beginning in the Mediterranean 5000 years ago. But with the rise of competing liberal, Marxist, and postcolonial narratives of modernity in the second half of the twentieth century, history's sacralized timeline has been further radically foreshortened to the period of European global expansion and fossil fuel industrialization since 1500. Where once the period before Sumer and the invention of writing was cast into the oblivion of "pre-history," now a post-1500 modernity narrative reigns supreme. Twenty-first-century Anglo-American history departments are accordingly populated by specialists and sub-specialists of "modern" European imperialism, globalization, and its aftermath of subaltern rising. The work in this field is often brilliant and revelatory but it belongs, nevertheless, to a larger, suffocating hegemony of ideas – what deep time historian Daniel Smail has called "the sterile presentism that grips the historical community" (*On Deep History* 9). In short, the intellectually restless Victorians excavated deep time, but academic historiography has managed, in the 150 years since, to reinter it. Modernity – glittering, savage, and sublime – is the privileged subject; what predates it is forever, irredeemably, pre-modern.

Presentism is likewise an objection leveled against the idea of an Anthropocene, the most narrow definition of which focuses on the exponentional growth of human impacts on natural cycles and ecosystems since the 1950s, a period definition popularized by John McNeil as "The Great Acceleration." But advocates of the Anthropocene have also met with resistance from a more familiar quarter, namely, social historians uncomfortable with the importation of scientific models and methods into the practice of narrative history. I conclude this chapter with a description of a refashioned "two cultures" debate from the early 2010s that coincides with the rash of book-length publications on the LIA featured here, a debate that suggests the LIA might already be in the process of being subsumed within a larger conceptual framework for the understanding of climate and human history.

In late October, 2012, Hurricane Sandy formed in the warm waters of the Carribean, before wheeling north along the Atlantic seaboard as a massive, highly unusual "superstorm" with wind gusts of 90 mph. Climate change set the table for Sandy. Historically high sea surface temperatures intensified the winds and increased rainfall by 35%, while 6 in. of sea level rise over the last century meant Sandy's destructive surge rode in at greater loft, enough to flood the subway stations on Manhattan's low-lying southern tip. By the time Sandy's fury was spent, an average 40 ft of beachfront had been swallowed up along the entire 150-mile Jersey Shore. 350,000 homes were damaged or destroyed, while losses stood at $75b in the United States alone (Trenberth et al. 2015).

Social historians Andreas Malm and Alf Hornborg cite the example of Sandy – and its disproportionate impacts on Haiti and the New York/New Jersey area – as an example of the "differentiated vulnerability" of nations in the age of climate change (Malm and Hornborg 2014). Climate change – expressed in weather catastrophes –

is a hemispheric political actor with no respect for nation-state boundaries. But Hurricane Sandy highlighted the close correlation between climate change and global inequity. In the aftermath of Sandy, New Jersey called in the National Guard, and pumped billions of dollars into restoring beaches, while Haitians at a more exposed latitude suffered hundreds of casualties, lost entire built communities, and were dependent on foreign aid for survival. The lesson of Sandy: given the same extreme conditions, the misery of global warming will be overwhelmingly visited upon poor, under-resourced populations in the global south.

Malm and Hornborg's Hurricane Sandy example is part of their larger argument against the Anthropocene as a historical paradigm (published in the inaugural issue of *The Anthropocene Review* in 2014). For Malm and Hornborg, Anthropocene discourse represents a further unwelcome intrusion of the physical sciences into the critical domain of social science and the humanities. Theorists of the Anthropocene, they argue, favor coarse-grained, deep-time narratives, emphasizing human species evolution from hunter-gatherer communities to sedentary agriculture through to recent industrialization, with an exponential increase in environmental impacts at each developmental stage. This species-level narrative homogenizes humanity, with no recognition that industrialization, most particularly, was shaped and monetized by a core of North Atlantic nations at the expense of the global southern periphery – exploited for its material resources, cheap labor, and unprotected markets. Through the multi-millennial lens of the Anthropocene, these subaltern populations disappear from view.

Malm and Hornborg's dispute with Anthropocene Studies belongs to a familiar post-colonial critique of first-world environmentalism. What we witnessed with Hurricane Sandy, they argue, was not a new historical phenomenon – the global threat of climate change – but rather the latest episode in a centuries-old tale of European globalization. With a sleight-of-hand characteristic of bourgeois grand narratives, the Anthropocene flattens the nuanced history of culture and power, of global inequity, of differentiated responsibility and suffering, into a planetary drama that quietly forgives all carbon debts.

Significantly, however, support for the Anthropocene paradigm has come from post-colonial writers themselves. In his famous "Climate of History" article of 2009, Dipesh Chakrabarty argued that the geological scale of climate change alters the baseline terms for critique. Theorizing climate change demands a species-level category of human agency and a "deep time" elucidation of world history. Chakrabarty insists, moving forward, on both fine-grained historical differentiation *and* a planetary, deep-time consciousness – but his critics, including Malm and Hornborg, have caricatured his argument as binary, involving a zero-sum choice between critique and science.

Chakrabarty's call for a new climate change paradigm has been echoed by, among others, the Indian novelist Amitav Ghosh, in a series of lectures published under the title *The Great Derangement: Climate Change and the Unthinkable* (2016). In Ghosh's revisionist history of modernity, China enjoyed the benefits of coal for hundreds of years before James Watt's invention of the steam engine; nineteenth-century Burma was a Petro-State; while industrialization was never a Grand Plan enacted upon an inert global south. Steam-powered mechanization and shipping were, rather, opportune technologies to further Britain's inchoate economic agenda of capital accumulation, which enabled only the temporary, brutal supremacy of the West over major-power rivals India and China, who have, since the late twentieth century, caught up rapidly. The argument for "differentiated responsibility" still holds, Ghosh argues, but "the complexity of the carbon economy's genealogy holds a lesson also for those in the global south who would draw a wide and clear line between 'us' and 'them' in relation to global warming" (114). For Ghosh, Malm and Hornborg's simple dichotomy of industrialized North and impoverished Global South risks reiterating a colonialist narrative in which white Europeans, for good or ill, must always be the active agents, the central players.

Ghosh makes a related point regarding the slipperiness of agency in the age of climate change: global-scale environmental degradation, including climate change, has been a catastrophic outcome of industrialization, but was never its purpose. It is "the unintended consequence," Ghosh writes, "of the very existence of human beings as a species." Responsibility is differentiated, yes, but is also a shifting, complex product of time, opportunity, infrastructure, and resource endowment. Historical time, through this lens, refracts and expands telescopically. Liberal historiography emphasizes the technological turning points of modernity – be it the printing press, Watts' steam engine, or the nuclear reactor. But in the revisionist terms of a long Anthropocene historiography, for the industrial revolution to have occurred at all required cultural antecedents – in technology and social organization – stretching back thousands of years into deep time. "Every human being who has ever lived," says Ghosh, "has played a part in making us the dominant species on this planet … The events of today's changing climate … represent the totality of human actions over time" (115).

Much of current writing on the "long" Anthropocene is occupied with detailing this contingent, even chaotic deep time evolution. For Chakrabarty and Ghosh, an Anthropocene historiography presumes the co-dependence of human and natural systems, and a critical realignment commensurate with the existential, species-level threat posed by environmental systems collapse. Through the funhouse mirror of the Anthropocene, the human agent is distorted, elongated, and even at times unrecognizable. By contrast, the Malm and Hornborg critique of the Anthropocene seems itself anthropocentric. Indiscriminate geophysical forces – and the chaos they bring – do not belong within an analytic framework where human agency and government are the first principles and sole unquestioned truths.

Amitav Ghosh calls climate change our "Great Derangement" – an apt catastrophist epithet for our time. In the nineteenth-century natural sciences, French catastrophism, embodied in the writings of Baron Cuvier, stood opposed to the British gradualist school of Hutton and Lyell, which insisted the key to the past lay in the present – that the materials and processes of the Earth, if not its superficial appearances, were constant over geological time. But Cuvierian catastrophism is once again ascendant under the Anthropocene paradigm. In our accelerated climate change regime, the future will *not* look like the past. In fact, our present is already a "no-analogue state," as Earth scientists term it (Steffen et al. 2004). The mind-bending challenge for climate scientists and climate historians alike is to describe the ever-shifting conditions of this no-analogue world we inhabit.

To engage the discourse of the Anthropocene, historians cannot avoid the challenge of Deep Time. A century and more since the secularization of Biblical chronology, consciousness of alternative bio- and geological timeframes has belatedly begun to make inroads in the academic humanities, for example, through the sheer persuasive weight of the burgeoning paleoclimate archive. Scientific literacy is indispensable to the scholarly engagement with Deep Time, as the 2010s proliferation of original research in the Little Ice Age period described in this chapter attests. The generation of historians following in its wake enjoys a unique license to embrace deep geological timescales, and the conceptual modalities of the Anthropocene, while remaining fully disciplined period scholars of modernity. With cutting-edge tools to hand, the climate-conscious twenty-first-century historian will integrate a traditional archival, text-based approach with the material scope, data-mining powers, and interpretive reach of multiple scientific disciplines – the better to fashion original narratives of the global past tailored to a messy planetary future.

References

Brook, Timothy. 2010. *The troubled empire: China in the Yuan and Ming dynasties*. Cambridge, MA: Harvard University Press.

Brooke, John L. 2014. *Climate change and the course of global history: A rough journey*. Cambridge: Cambridge University Press.

Campbell, Bruce M.S. 2016. *The great transition: Climate, disease and society in the late-medieval world*. Cambridge: Cambridge University Press.

Chakrabarty, Dipesh. 2009. The climate of history: Four theses. *Critical Inquiry* 35: 197–222.

Crumley, Carole. 1995. *Historical ecology: Cultural knowledge and changing landscapes*. Santa Fe: School of American Research.

Diamond, Jared. 2005. *Collapse*. New York: Norton.

Fagan, Brian. 1999. *Floods, famines and emperors*. New York: Basic Books.

Ghosh, Amitav. 2016. *The great derangement: Climate change and the unthinkable*. Chicago: University of Chicago Press.

Kelly, Morgan, and Cormac Ó Gráda. 2014. The waning of the little ice age: Climate change in early modern Europe. *Journal of Interdisciplinary History* 44 (3): 301–325.

Keys, David. 2000. *Catastrophe*. New York: Ballantyne.

Le Roy Ladurie, Emmanuel. 1971. *Times of feast, times of famine: A history of climate since the Year 1000*. Trans. Barbara Bray. New York: Doubleday & Company.

Linden, Eugene. 2006. *Winds of change*. New York: Simon & Schuster.

Malm, Andreas, and Alf Hornborg. 2014. The geology of mankind: A critique of the Anthropocene narrative. *The Anthropocene Review* 1: 62–69.

Parker, Geoffrey. 2013. *Global crisis: War, climate change and catastrophe in the seventeenth century*. New Haven: Yale University Press.

Smail, Daniel Lord. 2008. *On deep history and the brain*. Berkeley: University of California Press.

Steffen, W. et al. 2004. *Global Change and the Earth System: A Planet Under Pressure*. IGBP Publications.

Trenberth, Kevin E., John T. Fasullo, and Theodore G. Shepherd. 2015. Attribution of climate extreme events. *Nature: Climate Change* 5: 725–730.

White, Sam. 2011. *The climate of rebellion in the early ottoman empire*. Cambridge: Cambridge University Press.

———. 2014. The real little ice age. *Journal of Interdisciplinary History* 44 (3): 327–352.
———. 2017. *A cold welcome: The little ice age and Europe's encounter with North America*. Cambridge, MA: Harvard University Press.
Wood, Gillen D'Arcy. 2014. *Tambora: The eruption that changed the world*. Princeton University Press: Princeton.

Planets: A History of Observing Worlds and Changing Worldviews

Matthew Shindell

Contents

Abstract

As cultural and scientific objects, the planets have been understood differently by people across cultures and time periods. The physical understanding of the planets, their origins, natures, movements, and relationship to Earth has changed considerably over time. An historical examination of these changes, beginning with a few examples from the ancient world and then focusing primarily on the European and American contexts from the medieval world to the present, illustrates that the planets are windows into the worldviews and preoccupations of the culture in which planetary ideas and understandings originated. There have also been changes to how historians understand the practice and purpose of planetary

M. Shindell (✉)
Smithsonian National Air and Space Museum, Washington, DC, USA
e-mail: shindellm@si.edu

E. Aronova et al. (eds.), *Handbook of the Historiography of the Earth and Environmental Sciences*, Historiographies of Science,
https://doi.org/10.1007/978-3-031-40799-4_10

observations and theories during these periods, and these changes are addressed in each section.

Keywords

Planets · Exoplanets · Planetary science · Earth science · Astronomy · Astrology · Archaeoastronomy · Ancient science · Medieval science · Modern science · Spaceflight · Robotic exploration

1 Introduction

Humans have lived with objects of the natural world for hundreds of thousands of years, and their understandings of these objects have changed as new worldviews and knowledge systems have become dominant. Few objects have changed as dramatically as the planets, which not only transformed from wandering lights in the night sky into other worlds, but also found themselves enlisted in arguments about the material substance of the universe, the history and evolution of the Earth and solar system, the origin and extent of life, and the possible futures of humankind. The history of human understanding of the planets, though discontinuous across time and geography, offers a window into changes and variations in how people in different cultures and time periods have viewed themselves and their relationship to their own world. This chapter offers a brief history of human ideas about the planets. The author considers archeological speculations about the long prehistory of planetary observation, presents some examples that illustrate the prevalence in the ancient world of astronomical-astrological links between the planets and governance, appreciates the long medieval Islamic and Christian intellectual preoccupation with the Aristotelian-Ptolemaic world system, follows the slow transformation of the Earth into a planet and the planets into worlds in the seventeenth and eighteenth centuries, examines the natural histories and cosmic epics that gave the planets a common origin and evolutionary path, and concludes with our present era of robotic exploration of our own solar system and the perhaps countless planetary systems beyond it. In each section, the author also discusses relevant changes in historical approaches to planetary observation.

2 Prehistoric Humans and the Planets

We know nothing at all about the first human observations of the planets, let alone how our prehistoric ancestors understood their motions. If the planets can be said to have been "discovered," then their moment of discovery is buried irrecoverably deep in the past. Archaeology tells us that Homo sapiens have existed for at least 300,000 years, and we don't know when, where, or at how many times and in however many places they may have noticed the planets moving in the night sky first over the African continent, and later around the world. For that matter, we don't know if the various other human species who coexisted with Homo sapiens, or who preceded them for two million years, noticed the planets moving – the present wisdom is that archaic humans such as Neanderthals were intelligent and lived socially complex lives from very early on (Harari 2018; Graeber and Wengrow 2021). For all we know, Homo sapiens were recipients as well as originators of prehistoric planetary knowledge. Since we only have written records representing the past 5500 years, this leaves a vast amount of time during which humans not so different from the ones who have written this book might have been observing and developing ideas about the planets and passing these ideas on via oral tradition and even artifacts that mostly no longer exist. It therefore makes little sense to talk about the "discovery" of the planets (at least those that can be seen with the unaided eye), but instead to draw attention to their ubiquity and endurance in human cultures around the world.

We can't say much about the practices or purposes of early humans regarding planetary observation. We can say that their reasons for observing the planets and the questions they brought to these observations were almost certainly very different from ours. Early human communities may have observed the motions of the planets, in addition to the Sun and Moon, as ways of tracking seasonal changes, predicting animal migration patterns, and perhaps to determine the times of year when multiple groups would gather together for communal events. Archaeologists have found calendars of lunar cycles carved into bone that date back at least 34,000 years (Aveni 1995, 23). Archaeoastronomers believe that the architects of Stonehenge in England (~3000 BCE) and Nabta Playa in Southern Egypt (5000–4000 BCE) designed these megalithic structures to unite the social lives of community members with celestial events. Earlier literature attempted erroneously to find elements of modern astronomical knowledge or practice in these structures, describing them as "ancient observatories" used to predict eclipses or the movements of the planets. More recent work insists that prehistoric peoples gathered, socialized, and perhaps worshipped in these spaces designed to celebrate celestial motions and alignments. Anthony Aveni suggests that these were "sacred" observatories: "consecrated space[s] for watching the sky" in which "cosmic encounters were celebrated because they served to call people together to conduct rites to their gods" (Aveni 1997, 91). If these were spaces of religious, social, and perhaps even political significance, facilitating the maintenance of communal ties and collective identity, their removal from the world of objects and relationships in which they once participated makes them now seem mysterious wonders of a bygone world.

For these reasons, we don't find much history of science literature treating prehistoric astronomy. We do, however, find a wealth of literature from archaeoastronomers who have attempted to reconstruct a history from what scant archaeological materials do exist, combined with computer technologies that allow them to replay cosmic occurrences in different locations at different times. One area that is ripe for historical analysis is the history of these contemporary practices and their knowledge claims about the past. In many instances, the answers we find in the deep past reflect more the questions and methods researchers bring to them than the prehistoric peoples we want to interrogate (Villalobos and Barnhart 2014).

3 The Planets in Ancient Cultures

By the time the first literate cultures emerged, knowledge about the night sky, its changes, and the connection of celestial events to natural and social occurrences was probably well established. We find this knowledge in creation stories as well as the astronomy-astrology practiced by cultures around the world. For example, the *Popol Vuh* of the ancient Maya not only recounts the creation of the "sky-earth" (universe) and the exploits of the hero twins Hunahpu and Xbalanque, but also tells how these and other gods and demigods came to reside in the night sky and how the motions of the Sun, Moon, and planets reenact their exploits (Tedlock 1996; Bazzett 2018). Monuments throughout the ancient Mayan world record the dates on which communities reenacted portions of the *Popol Vuh* in annual rituals that legitimated the divine rule of Mayan kings (Freidel et al. 2001; Aldana 2010), and archaeoastronomers believe that these reenactments were timed and city architecture planned so that planetary motions could participate in the renewal of the city's religious charter. It is primarily for this reason, argues historian of science Gerardo Aldana, that royal families patronized the production of astronomical knowledge (Aldana 2010). The few Mayan codices that survive (of the thousands that existed precontact) speak to a well-developed practice of calendar-making that was useful not only in timing ritual events, but also as an agricultural almanac (Schele and Miller 1986; Aveni et al. 2003). Aldana reminds us to consider the context in which the Maya scribes worked rather than impose our own understandings and assumptions about the value of astronomical knowledge to agrarian societies in any attempt to reconstruct their intellectual project (Aldana 2022).

Royal patronage of Mayan astronomy created an expert class of astronomer-astrologer scribes. The knowledge that these scribes produced was meant to be understood within the context of an animate universe maintained by the actions of humans and gods. It was not esoteric knowledge for understanding the world, but for participating in it. These qualities of ancient astronomy-astrology are shared by many cultures around the world, even if the specific beliefs about the universe and the place of humans within it differ. In many ancient cultures, planetary motions and other celestial events were understood to yield information vital for good governance. In ancient China, for example, a creation story explained how the heavens and Earth became separate from each other (Major 1993). Only a heavenly monarch (thearch) attuned to the natural order of the universe could maintain harmony between these now separate parts of the world (Tseng 2011). A legitimate emperor acted as mediator between human society and the universe (Xiaochun 2000). An elite class of scribes practicing *tian wen* (heavenly writing) provided astrological guidance to the emperor that allowed him to rule in accordance with the universe (Sivin 2009; Cullen 2017). As with the Maya, the necessity of astrological information in the service of the state led to the patronage of astronomy-astrology.

The ancient culture that has long fascinated historians of science interested in the origins of astronomy is that of the Babylonian empire. Here again we see a culture who believed they lived in a cosmic state, in which their social world was connected to the celestial world, and in which knowledge of the heavens was necessary for good governance. The Babylonians believed that the appearance and motions of the planets delivered messages from their gods. Again, an elite class of astronomer-astrologer priests supported by royal patronage observed the heavens in search of good and ill omens of events to come. This information was necessary for state planning, and also for determining what ritual practices were necessary to communicate back to the gods (Rochberg 2004). Unlike later understandings of astrology, in which planetary influence causes earthly occurrences, Babylonian astrology understood planetary omens as warnings of what would happen if appropriate actions were not taken (Rochberg 2010).

Historians of science such as Noel Swerdlow have drawn attention to the mathematical instruments the Babylonian priests developed for predicting planetary motion and in them seen how "in the belief in omens lies the birth of science" (Swerdlow 1998, 5). The historian of astrology Nicolas Campion likewise insists that what happened in ancient Mesopotamia was "a small scientific revolution" (Campion 2008, 75). Historians and anthropologists have deciphered the roughly 7000 omens found in the 70 tablets of the *Enuma Anu Enlil*, as well as the reports the astrologer scribes sent to their royal patrons to piece together the complex system the scribes used to divine and predict. While the mathematics they employed does resemble science in many ways, it was employed for purposes that most today would consider incompatible with science. However, both the argument that this was science and the modern prejudice against it are indicative of the problem of demarcating science. On the one hand, searching for evidence of science in earlier eras rather than understanding what practitioners and their patrons understood themselves to be doing yields a skewed representation of ancient knowledge production. However, it is equally limiting to attempt to uphold a modern definition of science and to dismiss or diminish anything from the past (or present) that does not resemble it (Pigliucci and Boudry 2013). As one of the recognized founders of the history of science, George Sarton wrote in his own treatise on ancient science, "It is better to leave out for the nonce the consideration of science as science, and to consider only definite problems and their solutions" (Sarton 2011). Following the planets through time and geography does not help us to determine at what point science began, but it does help us to see how and in what contexts many of the practices we now associate with different forms of science emerged.

4 Greek Philosophers Shape the Cosmos

Practices of observing, recording, and attempting to predict the motions of the planets seem to have emerged in the context of cosmic states. So far, at least in the examples given above, this did not equate to any investigation of the structure or nature of the planets or the universe in which they moved. This was not a failure of these ancient projects, it simply wasn't the goal. One of the earliest places where we see a culture engaged with the structure and materiality of the universe is ancient Greece. In the emerging tradition of Greek natural philosophy, the planets were understood not as signs or omens, but as physical objects that affected specific events and both collective and individual fates. This idea emerged in the writings of Greek philosophers still well known today (partly because of their influence over the following centuries, and partly because of the enduring interest of modern historians). Plato's *Timaeus* imagined a new creation story based on the observed nature of the universe. He posited a living universe animated by a soul, constructed by a divine craftsman. Its movements are perfect and orderly. Humans, made from the imperfect residue of the universe, can observe the perfect heavens and aspire to restore themselves more nearly to perfection. The works of Plato's student Aristotle established the four corruptible sublunar elements – Earth, water, air, and fire – and proposed a fifth element with a separate physics, the aether, that comprised the heavens. Plato and Aristotle introduced a new geometrical structure to the universe. The planets and the fixed stars all moved around the Earth in perfect circular motion (Lindberg 2008).

Plato and Aristotle provided the philosophical underpinnings of an understanding of the planets that would endure for centuries, but it was the Roman astronomer, Claudius Ptolemy, in the work that became known as the *Almagest*, who gathered together the astronomical and astrological tools that would be employed by those who followed (Gingerich 1993; Ptolemy 1998; Neugebauer 1975). Ptolemy's work was no mere synthesis of earlier Greek ideas and practices; he benefited from increased contact with other cultures thanks to Alexander the Great's imperial ambitions. Ptolemy wrote his works from the Roman Egyptian city of Alexandria, where he enjoyed the city's museum and libraries and the Greek, Egyptian, and Babylonian knowledge they contained. Ptolemy's system, while not perfect, was mathematically more sophisticated than previous approaches and did succeed in making planetary astronomy predictive while preserving the quality of uniform circular motion. Thus, his mathematical model could be justified through existing arguments from Greek natural philosophy (Evans 1998). Ptolemy's other great contribution to the history of planetary astronomy was a four-book compendium on astrology, traditionally called the *Tetrabiblos*. These texts not only presented those aspects today associated with astrology – the influence of planets and zodiac signs on individuals (i.e., horoscopes) – but also addressed how the planets affected all aspects of the world, including the people, animals, and things that would be found in different regions of the world, only a small portion of which was believed to be inhabited. The *Almagest* and the *Tetrabiblos* had profound effects on understandings of the planets for centuries.

5 Planets and the Medieval Monotheist World

To the medieval natural philosopher, the planets – their positions, qualities, and movements through the zodiac – were of paramount importance. The planets participated in an intricate "machina mundi" (world machine) that structured the medieval European worldview. The Muslim, Jewish, or Christian God was the ultimate cause of everything within this world machine, and God's will was enacted through the movements of the heavens. All of creation was connected. To some extent, these connections can be understood mechanically – in a world built from nested spheres, the movement of one sphere affected that of the others. But in other respects, the connections through which one thing affected another were occult; as much as the universe was a machine, it was also a tapestry of interwoven influences – relationships that were based on immaterial connections, attractions, and similarities. Natural philosophers adopted an Aristotelian-Ptolemaic worldview, based on the physics and elements of Aristotle and the methods, observations, and astrology of Ptolemy. They understood all change in the sublunar world to be due to the effects of the movements of the stars and planets. As the fourteenth-century Parisian scholar Themon Judaeus wrote, "every natural power of this inferior sensible world is governed by the heavens" (Grant 1997, 162). Left to their own devices, without the planets to induce birth, growth, maturation, degeneration, death, and decay, the elements would separate and all complex forms would cease to be. As the Islamic natural philosopher Ibn Rushd (Averroes) observed, "if celestial motion were destroyed, the motion of all inferior beings would be destroyed and so also would the world" (Grant 1994, 588).

Each planet had different effects on the world, and these effects varied depending on in which sign it was observed and what other planets were nearby. Why were the planets all different? Certainly, they each behaved differently. Mercury and Venus seemed tied to the Sun in their motions, with Venus shining bright sometimes in the early morning and other times in the evening. Mars could change in both brightness and color, from the color of wheat to that of blood, and these changes often occurred right before the planet seemed briefly to reverse direction in its path among the fixed stars. Jupiter and Saturn moved more slowly and steadily in the sky, but even they occasionally seemed to change direction. These visible differences no doubt informed speculation about the differences of the planets, although how they inspired early thinkers to associate different qualities with them is lost to time. But medieval natural philosophers also followed Ptolemy in inferring different effects based on what they knew of the Sun and Moon, both of which were also considered to be planets. The Sun was the example par excellence of planetary influence. While the Earth was the center of the universe, the Sun was the center of the heavens, and the most noble planet. It supplied heat and light. Its movement through the ecliptic brought the seasons. The Moon not only appeared different to the Sun, but it also caused different changes – its movements, phases, and appearance brought the tides, marked the times for planting, harvest, and religious ceremonies. It followed that the other five planets would also have differing qualities and influences (Grant 1994).

With regard to the history of the earth sciences, planetary knowledge was essential to understanding the characteristics of peoples around the world, as well as the nature of the lands they inhabited. These influences were understood to reach deep into the Earth's interior, producing different metals and precious stones, or bringing forth vapors that could corrupt the air (in this way they could be blamed for bringing forth diseases and plague) (Wey Gómez 2008; Horrox 1994). The planets were also understood to affect the airs and waters directly, producing different weather events and defining regional climates. A tradition of astrological weather forecasting developed during this time period and persisted, in altered form, in modern almanacs (Fleming 2005; Lawrence-Mathers 2019).

In addition to affecting change in the sublunar world, the planets were also understood to affect the human body. Planetary knowledge belonged as much to the physicians as it did to the astronomers. Just as the Moon influenced the tides, it was also believed to influence the fluid humors of the body, as defined by the ancient Greek physician Hippocrates and more recent Islamic physicians such as Ibn Sina (Avicenna). The four humors corresponded to the four sublunar elements, and the body was conceptualized as a microcosm of the universe. Each planet had its own associations with the different senses and body parts. Planetary positions could be used to diagnose illness, and to determine when and what remedy to apply. They could also enhance or diminish the efficacy of various cures, as well as the medical competence of the physician (Isserles 2017; Geller 2014).

6 Finding Progress in Medieval Planetary Astronomy

Historians of modern science once mistakenly accused medieval natural philosophers of blindly following ancient authorities rather than embarking on their own investigations of nature. The Middle Ages were presented as a time of intellectual stagnation, in which nothing new was achieved. The period was presented as a foil to the "Scientific Revolution" that swept away dogmatic commitment to the Aristotelian-Ptolemaic worldview (Kuhn 1957, 1996; Lindberg and Westman 1990). More recent scholarship, however, has considered medieval scholarship on its own terms and has found it to be contiguous with the rise of modern science (Grant 1996; Falk 2020). The ancient texts coveted by scholars in the Islamic and Christian worlds during the medieval period were the subjects of intense scrutiny, criticism, and improvement, and this work lies at the center of at least two great renaissances of interest in the structure and purpose of the heavens. First in the Islamic world, the Abbasid dynasty promoted the cultivation of different types of learning, including astronomy (Kraemer 1986). A. I. Sabra has described this period as a "cultural explosion" featuring the translation, assimilation, and naturalization of ancient science and philosophy (Sabra 1987, 1996). This included the construction of libraries and centers of learning, and an ambitious project that sought to acquire manuscripts from around the known world, and employed Muslim, Persian, Jewish, and Christian scholars to translate and make sense of the "sciences of the ancients" (Goldstein 1986, 2001).

This work also involved empirical observation and the development of mathematical tools and instruments, such as refining the African astrolabe and developing the equatorium (Goldstein 1986; Morrison 2013). Historians have only barely addressed the importance of the Maragheh observatory and the associated "Maragheh school" in which the founder Nasir al-Din al-Tusi and other prominent astronomers of medieval Islam worked to improve Ptolemaic planetary theory through the compilation of more extensive astronomical tables (zij) and new mathematical models of planetary motion. They made important new theoretical insights in planetary theory in their studies of the discrepancies found in Ptolemy's *Almagest* and his *Planetary Hypotheses*. The contributions of these scholars, as well as earlier work by al-Battani and Ibn Sina would continue to influence astronomy, astrology, and medicine through the seventeenth century. Interest in the Maragheh astronomers has primarily been for their place in the history of mathematical astronomy and its influence on the development of the Copernican system (Neugebauer 1975; Swerdlow and Neugebauer 1984; Morrison 2014). Islamic natural philosophers have otherwise been relatively neglected compared to their European peers and successors, even as Islamic scholars like Ibn Sina took the first steps toward bringing the geocentric model of the cosmos in line with the new monotheism, considering the planetary influences as a natural means through which God controlled occurrences on Earth (Morrison 2013). New scholarship promises to bring renewed attention to medieval Islamic science and medicine (Küçük 2019; Ragab 2018; al-Musawi 2015). The reception of these Greek and Arabic texts in Christian Europe beginning in Spain at the end of the eleventh century initiated a second great renaissance of learning, at the center of which were planetary astronomy and astrology (Ryan 2011).

7 Decentering the Earth

We are now almost up to the point where most histories of modern planetary physics begin. This period, long known as the Scientific Revolution, is described as the time during which science became more empirical, experimental, mathematical, and untethered to ancient or religious authority. At the center of this revolution is Copernicus's heliocentric universe, set forward in his 1543 *De revolutionibus orbium coelestium* (On the Revolutions of the Heavenly Spheres), which removed the Earth from the center and made it move like a planet in orbit around the Sun (Kuhn 1957). This work did not immediately unite terrestrial and planetary physics; the heavens were still spherical, and planetary motion was still perfectly circular. Nor did it do away with the idea of planetary influence, and scholars have in fact suggested that Copernicus's work should be understood as engaged in the defense of astrology (Westman 2011, 2016). Copernicus's work did speak to a generally shared consensus among astronomers that the Ptolemaic system with all its modifications was becoming increasingly cumbersome. Tycho Brahe celebrated Copernicus's mathematical approach to planetary motion, but did not believe that the Earth could move – it was too heavy, and he had never seen any evidence in the relative positions of the fixed stars that indicated that the Earth was moving, or that the sphere of the fixed stars could be far enough away that no parallax would be observed (Blair 1990). Brahe proposed a compromise that combined elements of the geocentric and heliocentric models and preserved Aristotle's distinction between the heavens and the sublunar world – the planets orbited the Sun in a heaven that was fluid, not crystalline, while the Sun, Moon, and fixed stars orbited the central Earth. What became known as the Tychonic system coexisted with the Ptolemaic and Copernican conceptions of the universe. For many astronomers, such as Christopher Clavius, neither Ptolemy nor his rivals knew the true structure of the universe (Lattis 1994).

Our view of this period as revolutionary is influenced not only by the historical literature's focus on prominent "Copernicans" such as Galileo Galilei and Johannes Kepler, but also on the heroic treatment these figures received. While once depicted almost as modern scientists ahead of their time, more recent treatments of these figures have placed them in their proper contexts (Biagioli 1994; Heilbron 2012; Voelkel 2001). Galileo did introduce the telescope to astronomy, his observations of the Moon and planets did imply that the planets were similar to the Earth, and he did argue in his famous *Dialogue Concerning the Two Chief World Systems* that the planets were "world globes" perhaps subject to the same changes experienced by the Earth. But Galileo struggled to explain how such worlds would orbit a central Sun. His correspondence demonstrates that he did try to come up with an explanation as early as 1615. It was an explanation that drew upon astrological notions of the Sun's occult influence, passages from the Bible, and his observations of sunspots: The movement of the sunspots demonstrated that the Sun rotated; perhaps they were also evidence that the Sun was consuming a power bestowed upon it from the fixed stars; perhaps this power then emanated from the Sun in a "spirited, tenuous, and fast substance" and penetrated everything in the universe, causing the celestial objects to rotate around the center; and perhaps this entire system was what was spoken of in Psalm 19 (Finocchiaro 1989, 65). This was by no means the type of modern explanation we today associate with Galileo and modern science, but one appropriate to an astronomer educated in the scholastic tradition. Only later would Galileo's planetary observations and Kepler's laws of planetary motion be divorced from Aristotelean logic and neo-Platonic geometry and incorporated into a new physical explanation of the planets. The transformation of planets into worlds and the Earth into a planet was a slow one.

8 A Family Resemblance

New philosophies and ultimately new mathematics emerged in the Enlightenment that supported a family relationship between the planets. Rene Descartes put forward one of the earliest fully mechanical descriptions of the universe that described how the planets, including the Earth, might all be formed from the same matter. Descartes's universe was a plenum filled with matter, some so small as to be imperceptible, and held the solar system together in a vortex of matter in motion. God no longer enacted his will through the motion of the planets, but instead created the universe and acted only as the first cause of its perpetual motion (Westfall 1977). Such a model lent itself to imagining other planetary systems orbiting other stars, each a vortex. And it also lent itself to speculation about life on other planets throughout a larger universe. Bernard Le Bovier de Fontenelle in France and Christiaan Huygens in the Netherlands – early champions in the extraterrestrial life debate – both suggested that the Cartesian vortices implied innumerable such worlds, and gave us no reason to doubt their inhabitation (Dick 1996; Crowe 2012).

It was the English physicist, mathematician, and alchemist Isaac Newton who provided the mechanical philosophy that ultimately came to define modern science.

To matter and motion, Newton proposed adding a third category – natural forces. The mathematical description of the universal force of gravity in his 1687 *Philosophiæ Naturalis Principia Mathematica* (Mathematical Principles of Natural Philosophy) was Newton's landmark achievement. Newton's forces acted at a distance, without direct contact, which eliminated the need for a plenum but also roused the skepticism of mechanists who saw this as a return to the influences and occult forces of Renaissance natural philosophy (Henry 2020). In Newton's own view, he had replaced the plenum with God's sensorium, in which His laws could act universally without interruption across time and space (Brooke 1991). God had likewise bestowed the qualities of matter (including gravity) at the time of creation, a belief historians have tied as much to Newton's alchemy as to his Christianity (Westfall 1993; Henry 2020). Brahe had objected to the Earth's motion in part due to its weight. Newton's universal law of gravity allowed the Earth and other planets to move precisely because of their mass and their resulting gravitational relationship to the Sun.

Enlightenment geologists in search of a comprehensive "theory of the Earth" began to take seriously its history as a planet and its relationship to the other planets in the solar system. The Earth was a member of a family of planets, and it was susceptible to universal natural forces. It had originated and developed over time based on these forces. Benoît de Maillet's posthumously published *Telliamed* (1748) was one of the first such theories. Descartes had proposed that the Earth had once been a global ocean whose waters had slowly been stripped by the motion of the vortex. Maillet saw evidence for this in the sedimentary rocks containing fossilized marine shells he had seen on mountains and deserts. This gradual drying of the Earth through natural planetary processes had created a long and ongoing sequence of change in which surface features were first deposited and then uncovered as the sea level lowered (Taylor 1998).

Georges-Louis Leclerc, Comte de Buffon published his own theory of the Earth during this same period. Newton's theory of gravity, unlike Descartes's plenum and vortices, was agnostic about whether planetary systems would be common or rare. Buffon proposed that the Earth and planets might have originated not due to beneficent design, but by cosmic accident. A comet might have grazed the Sun, its collision producing hot globules of solar matter that then congealed into planets. While Buffon did not believe that this was the true history of the solar system, he did believe that his theory demonstrated that the uniformities of the solar system could be accounted for by invoking a single natural cause (Numbers 1977). Buffon's hypothetical story of solar system formation made all the planets the results of the same event, and all subject to the same directional process. Each planet had to cool from its initially molten state. After the crust solidified, great deluges created an ocean so deep that all the Earth's surface was under water, with the global ocean eventually retreating to reveal the continents. As with Maillet, this global ocean explained the presence of sedimentary rocks on mountain tops. Buffon was also interested in fossils, and the Earth's long period of cooling and drying out was his way of explaining why fossils resembling the bones of tropical animals could be found in northern latitudes (Bowler 2009).

9 Nebular Hypotheses and New Planets

Other thinkers elaborated on Buffon's idea of a single original cause for the regular motions of the solar system. Pierre-Simon, marquis de Laplace imagined that the planets all orbited in the same direction and in roughly the same plane because they had all originally been part of a single rotating body. The Sun's atmosphere might once have been hotter, extending as far as the distant orbit of Saturn, and rotating along with the Sun. As the atmosphere cooled, it contracted and began to rotate faster, throwing off rings that then condensed into planets. As in Buffon's theory, the planets began as hot molten spheres before a cooling process shaped and transformed them over time (Brush 1987, 1996b).

The work of the German-born British astronomer William Herschel and his sister Caroline introduced a new idea to solar system formation. The Herschels set out to identify and classify the different types of nebulae they observed to understand how a gravity-driven process condensed clouds of nebular fluid into stars, which then attracted additional nebular material to produce planets, and then ultimately exploded to provide the material for new stars and planetary systems (Schaffer 1980a, b). William Herschel also became an authority on the planets due to his telescopic discovery of the planet Uranus in 1781. Herschel's new planet was a surprise discovery but it conformed to predictions about the distances between planetary orbits – a mathematical expression known as the Titius-Bode law. This seeming confirmation of the law set astronomers looking for a planet between Mars and Jupiter, where a planet was predicted but none observed. In 1801 they discovered Ceres and what became known as the asteroid belt. When Neptune was discovered in the mid-nineteenth century, it did not conform to the law's prediction.

10 Cosmic Epics and Dying Worlds

While laws governing the formation and distance between planets were proving unreliable, nineteenth-century British intellectuals nonetheless celebrated the progress of astronomy, along with all other evidence of progress they could identify in their ascendance to world domination. Historians Bernard Lightman and Nasser Zakariya have pointed out how directional cosmic narratives and "evolutionary epics" that traced the evolution of the physical and biological world did so in service of the Victorian ideal of progress, and remain a staple of popular science into the present (Lightman 2007; Zakariya 2017). One of the most famous of the Victorian epics, and also the best historically studied, is Robert Chambers's 1844 *Vestiges of the Natural History of Creation* (Chambers 1994; Secord 2000). In fact, the book was a natural history of progress that wove together often racist and classist ideas about human intellect, culture, and civilization into a story of the natural unfolding of the universe one could read as a natural argument for the ascendance of the British Empire. The initial chapters of the book cover the formation of the planets, which was based on Laplace's nebular hypothesis. Chambers described the planets as "bound up in one chain – interwoven in one web of mutual relation and harmonious agreement" (Chambers 1994, 12).

All of the planets were understood to have a common beginning, were acted upon by the same set of natural forces, and would all ultimately come to the same end. The American astronomer Percival Lowell, whose outspoken advocacy of canals and intelligent canal builders on Mars has been extensively treated by historians of science, saw on Mars the future of the Earth (Hoyt 1996; Dick 1996; Markley 2005; Lane 2011; Nall 2019). Lowell drew upon his own version of the nebular hypothesis as it applied to the evolution of the planets. Every planet in Lowell's model went through six stages: first was the sun stage, in which the planet was hot enough to radiate light; second was the molten stage, in which the planet was still incredibly hot but beginning to cool; third was the solidifying stage, in which the planet developed a solid surface; fourth was the terraqueous stage, where there was liquid water at the surface; fifth was the terrestrial stage, when the oceans receded; and sixth was the dead stage, with neither water nor atmosphere. Mars, because it was smaller than the Earth, had already reached the terrestrial stage. The Moon, smaller than Mars, was what the red planet would become once its water and atmosphere were fully depleted. Mars – a barely habitable desert world on which survival required global feats of engineering – was thus an example of the next step in Earth's own evolution, as natural forces caused deserts to increase and multiply.

The nineteenth century also stands out as the period in which geographical practices were first systematically applied to the planets. One can hardly imagine the Martian canal controversy without considering the authority of maps and globes. Despite their shortcomings, geographer K. Maria D. Lane argues that topographical maps of Mars became valuable cartographic icons in staking and communicating scientific knowledge claims during this period (Lane 2005, 2011). The process of mapping, and the use of a terrestrial vocabulary of seas, channels, islands, and continents (not to mention canals), reinforced the idea that Mars was Earth-like. While astronomers disagreed publicly in their interpretations of the features they saw on Mars, nearly all took for granted the idea that Mars was physically Earth-like, and most agreed that it was a living world, even if this life might be limited to vegetation. The maps connected planetary astronomy to the imperial practice of exploration, the context in which nineteenth-century geographical practice was rooted.

11 Twentieth Century

The nebular hypothesis was challenged in the twentieth century by ideas reminiscent of Buffon's chance one-off occurrence. Perhaps the solar system had formed when the Sun interacted with another star passing nearby. Tidal forces might have torn material from the Sun and this material might have congealed into the planets. Many of Herschel's nebulae were revealed to be very large galaxies, making his natural history of stars and planetary systems untenable. But spectroscopic evidence of the composition of the stars and planets (accumulating since the mid-1800s) provided ammunition for arguments that planetary formation could still result from a preexisting solar nebula (Brush 1996a, c; DeVorkin 2000). The discovery of radioactivity and the utilization of radioactive decay and isotope ratios in new geochemical and geophysical methods – techniques accelerated by World War II and Cold War projects in atomic and nuclear science – put limits on the age of the Earth and solar system, and yielded insights into the deep history of the Earth and solar system (Brush 1996b; Doel 1996; Shindell 2014, 2019).

In the USA, the earth and planetary sciences (the latter hardly existed before the Cold War) benefited from increased military and national patronage after World War II. This patronage had dramatic effects on the direction of these fields, even as new technologies and methods yielded new evidence about planetary structure, dynamics, and geomagnetism, and new understandings of global planetary systems (Doel 2003; Oreskes 1999, 2021). This included the founding of the National Aeronautics and Space Administration in the context of the Space Race between the USA and the Soviet Union. The rise of the rocket and the prospect of studying the planets from outside of Earth's atmosphere, whether in orbit around Earth or in missions to fly by, orbit, or even land on another world attracted the interest of scientists from multiple disciplines. While within their home disciplines interest in the planets may have in some cases been marginalized, these scientists soon found a home in the burgeoning interdisciplinary field of planetary science (Tatarewicz 1990; Doel 1996).

The direction of US planetary science was shaped by technologies and funding opportunities made available by the state. NASA built a space science community through funding to universities for research and graduate fellowships – "a constituency that would help support the agency and its goals" (Lambright 2014). Scientists who wished to ask questions about the planets had to learn how to ask them in the new technical language of spaceflight. They worked with NASA engineers, mission planners, and the same aerospace contractors engaged in solving problems for the military and CIA to design instruments for planetary exploration. They worked within the constraints and timelines of missions that were often set with little regard for science. David DeVorkin has argued that many of the scientists who participated in the early years of rocket-powered science "placed themselves in the dual role of contributors to science and inventors of a new way to conduct science," and developed "a tool-building technical culture" (DeVorkin 1992). Peter Westwick's work has illuminated a trend of interdependency between civilian and military contracts at the NASA center behind most of the robotic science and exploration, Caltech's Jet Propulsion Laboratory (JPL). Before JPL first began its work to support lunar and planetary exploration in the late 1950s, most of the the lab's work was classified. Civilian contracts during the space race brought that percentage down to almost negligible levels, as government funds poured in for robotic planetary exploration. But even if JPL was mostly invested in developing civilian space technologies during these years, the lab nonetheless existed within a larger aerospace community in and around Los Angeles, California, that military contracts shaped into a "gun belt metropolis" (Westwick 2007).

Historians have approached twentieth-century planetary science and exploration in a variety of ways. Some, such as Schorn, have placed planetary exploration within a longer history of planetary astronomy or, like Sheehan, into an even longer history of planetary observation (Schorn 1998; Sheehan 1992). Histories funded by the NASA History Office have tended to focus on technical and administrative mission histories, including early pre-Apollo robotic missions to the Moon, the Viking program, and the Galileo project (Hall 1977; Ezell and Ezell 1984; Meltzer 2012). The History Office has also produced a very useful reference book on the history of planetary spacecraft (Siddiqi 2018). More recently, NASA published an edited volume commemorating 50 years of robotic exploration (Billings 2021). Other missions, such as the Voyager grand tour of the outer planets, have been chronicled without NASA support (Pyne 2010; Bell 2015). Some historians such as Koppes and Westwick have focused on the Jet Propulsion Laboratory as the institution at the center of robotic planetary exploration, and the interface between the national space agency and the academic community (Koppes 1982; Westwick 2007). Launius and McCurdy have treated a broader history of robots as space explorers (Launius and McCurdy 2008).

Mars has been more thoroughly explored than any other planet in our solar system (aside from Earth, of course). And it has also received the most attention from historians and popular authors. Erik Conway has explored the history of JPL in respect to engineering and operating missions to explore Mars (Conway 2015). Crossley, Markley, and Shindell have worked to place our present fascination with Mars – manifested not only in robotic exploration but also in imagined futures of human expeditions or long-term habitation – in a longer, literary, and intellectual history of imagining Mars and its connections to Earth (Markley 2005; Crossley 2011; Shindell 2023). Meanwhile, sociologists and anthropologists have made Mars exploration a site of ethnographic analysis and participant observation, providing perspectives on both the culture of planetary exploration and the role of culture in planetary practices (Vertesi 2015, 2020; Messeri 2016; Mirmalek 2020).

12 The Future of the Planets

Some of these most recent works speak to early twenty-first-century planetary science and exploration. It is yet unclear how our present moment should be understood, and no great work of the history of twenty-first century planetary physics has yet been written. Perhaps the meaning and significance of post-Cold War exploration will be determined by what comes next in the exploration of the planets. The US, Chinese, and Indian space programs all have plans of returning humans to the Moon in programs that emphasize scientific exploration as well as technology development. In most of these plans, Mars is considered the next logical destination after the Moon for human exploration (as it was for some in the Apollo era) (Logsdon 2015; Lambright 2014). If humans do return to the Moon and set foot on other planets and moons in the decades to come, how will this affect our view of the planets? No doubt our knowledge of them will be enhanced as it was with the return of samples from the Moon during Apollo, as well as the robotic return of samples from comets and asteroids. As for the potential significance of these findings, let alone the cultural changes that might result from becoming a multiplanet species, we can only speculate based on plans and ambitions.

We can already speak to the significance of the narratives produced by scientific and speculative visions of the planets. We live on a planet that seems to be unique in our own solar system in its ability to support life, and more alive with dynamic processes. There is reason to believe that life may have emerged on Mars or elsewhere in the solar system (though we've yet to find it). The notion that another planet could be made habitable likewise seems reasonable (though perhaps overly ambitious, at least with present technologies). The discovery of thousands of planetary systems, and even Earth-sized planets around other stars, has shown us that our situation may not be as unique as once believed, and that the formation of planets around stars is a regular occurrence (Impey 2023). The arrangements of these systems, many with gas giants larger than Jupiter orbiting close to their star, have changed our ideas about the order in which our own planets may have formed and the gravitational relationships that might have reordered them and placed Earth in the so-called "habitable zone." We can find this story, more or less, in the recently released decadal survey for planetary science and astrobiology (National Academies of Sciences 2022). New space-based tools bring these exoplanet worlds into greater resolution, and yet they remain well out of reach.

Our ability to tell versions of this story, to imagine the historical relationship between the parts of our solar system, and to speculate about our future in this solar system and others is all based on the use of powerful technologies that mediate our relationship to the planets. Propulsion technologies lift rockets and spacecraft beyond Earth. Computer technologies guide these spacecraft and control their operations in space. Robotic technologies combined with instruments of seeing allow us to collect data remotely, and even to move from site to site on planetary surfaces. And yet this is still a very human enterprise. For every robot at work in the solar system, there are hundreds if not thousands of humans on Earth who worked to make the mission possible or continue to work to support and operate it, and teams of scientists working to interpret received data and formulate new questions. Bureaucratic funding structures and politics at the local and federal levels still define what questions these experts ultimately get to pursue in space. Publications like the decadal survey allow this expert community to exert their influence upon the priority setting process (Asner and Garber 2019). And, when it comes to imagining the broader significance of planetary science and exploration, we continue as a culture to transpose ideas of who we are, our utopian and dystopian visions of our future, and our understandings (celebratory and critical) of the past that shaped our present.

13 Conclusion

This present era of robotic exploration of our solar system began little more than half a century ago; the blink of an eye in the face of thousands of years of observing the sky. Advances in robotics, computing, and imaging have given us unprecedented access to the planets. Humans and robots have even brought pieces of the Moon, comets, and asteroids home to Earth for analysis. It is tempting to believe that our current understanding of the planets is more permanent – more based on "real" empirical evidence – than those of our historical antecedents. Many have found it tempting to tell the history of planetary exploration as the story of how we came to truly know our solar system. Indeed, it does seem true that the technologies of exploration, including the laboratory methods of earth and planetary science, have constrained theories about the origin and evolution of the solar system. The truth may be, however, that our science has barely scratched the surfaces of the worlds we've explored, and that our present understanding may one day seem rooted in our own cultural biases – defined through the epistemological lenses of capitalism, expansionism, the pursuit of resources, and a technological militarism that has defined twentieth- and twenty-first-century life. It is wise, perhaps, for historians of contemporary science to think of ourselves and our moment as just the latest chapter in a very long and (dis)continuous history of locating ourselves and our significance in our universe.

References

Aldana, Gerardo. 2010. *The apotheosis of Janaab' Pakal: Science, history, and religion at classic Maya Palenque*. Boulder: University Press of Colorado.

———. 2022. *Calculating brilliance: An intellectual history of Mayan astronomy at Chich'en Itza*. Tucson: University of Arizona Press.

al-Musawi, Muhsin J. 2015. *The medieval Islamic republic of letters: Arabic knowledge construction*. Notre Dame: University of Notre Dame Press.

Asner, Glen R., and Stephen J. Garber. 2019. *Origins of 21st-century space travel: A history of NASA's decadal planning team and the vision for space exploration 1999–2004*. Washington, DC: National Aeronautics and Space Administration.

Aveni, Anthony F. 1995. *Ancient astronomers*. Washington, DC: Smithsonian.

———. 1997. *Stairways to the stars: Skywatching in three great ancient cultures*. New York: Wiley.

Aveni, Anthony F., Harvey M. Bricker, and Victoria R. Bricker. 2003. Seeking the sidereal: Observable planetary stations and the ancient Maya record. *Journal for the History of Astronomy* 34: 145–161.

Bazzett, Michael, Trans. 2018. *The Popol Vuh*. Minneapolis: Milkweed Editions.

Bell, Jim. 2015. *The interstellar age: The story of the NASA men and women who flew the forty-year Voyager Mission*. New York: Dutton.

Biagioli, Mario. 1994. *Galileo, courtier: The practice of science in the culture of absolutism*. Chicago: University of Chicago Press.

Billings, Linda, ed. 2021. *50 years of solar system exploration: Historical perspectives*. Washington, DC: National Aeronautics and Space Administration.

Blair, Ann. 1990. Tycho Brahe's critique of Copernicus and the Copernican system. *Journal of the History of Ideas* 51 (3): 355–377.

Bowler, Peter J. 2009. *Evolution: The history of an idea*. Berkeley: University of California Press.

Brooke, John Hedley. 1991. *Science and religion: Some historical perspectives*. New York: Cambridge University Press.

Brush, Stephen G. 1987. The nebular hypothesis and the evolutionary worldview. *History of Science* 25 (3): 245–278.

———. 1996a. *Fruitful encounters: The origin of the solar system and of the Moon from Chamberlin to Apollo*. Vol. 3. A history of modern planetary physics. New York: Cambridge University Press.
———. 1996b. *Transmuted past: The age of the earth and the evolution of the elements from Lyell to Patterson*. Vol. 2. A history of modern planetary physics. New York: Cambridge University Press.
———. 1996c. *Nebulous earth: The origin of the solar system and the core of the earth from Laplace to Jeffreys*. Vol. 1. A history of modern planetary physics. New York: Cambridge University Press.
Campion, Nicholas. 2008. *A history of western astrology*. Vol. 1. The ancient world. New York: Bloomsbury Academic.
Chambers, Robert. 1994. *Vestiges of the natural history of creation and other evolutionary writings*. Edited by James A. Secord. Chicago: University of Chicago Press.
Conway, Erik M. 2015. *Exploration and engineering: The Jet Propulsion Laboratory and the quest for Mars*. Baltimore: Johns Hopkins University Press.
Crossley, Robert. 2011. *Imagining Mars: A literary history*. Middletown: Wesleyan University Press.
Crowe, Michael J. 2012. *The extraterrestrial life debate, 1750–1900*. Mineola: Dover Publications.
Cullen, Christopher. 2017. *Heavenly numbers: astronomy and authority in early imperial China*. New York: Oxford University Press.
DeVorkin, David H. 1992. *Science with a vengeance: How the military created the US space sciences after World War II*. New York: Springer.
———. 2000. *Henry Norris Russell: Dean of American Astronomers*. Princeton: Princeton University Press.
Dick, Steven J. 1996. *The biological universe: The twentieth century extraterrestrial life debate and the limits of science*. New York: Cambridge University Press.
Doel, Ronald E. 1996. *Solar system astronomy in America: Communities, patronage, and interdisciplinary science, 1920–1960*. New York: Cambridge University Press.
———. 2003. Constituting the postwar earth sciences: The military's influence on the environmental sciences in the USA after 1945. *Social Studies of Science* 33 (5): 635–666.
Evans, James. 1998. *The history and practice of ancient astronomy*. New York: Oxford University Press.
Ezell, Edward Clinton, and Linda Neuman Ezell. 1984. *On Mars: Exploration of the Red Planet, 1958–1978*. Washington, DC: National Aeronautics and Space Administration.
Falk, Seb. 2020. *The light ages: The surprising story of medieval science*. New York: W. W. Norton.
Finocchiaro, Maurice A., ed. 1989. *The Galileo affair: A documentary history*. Berkeley: University of California Press.
Fleming, James Rodger. 2005. *Historical perspectives on climate change*. New York: Oxford University Press.
Freidel, David, Linda Schele, and Joy Parker. 2001. *Maya cosmos: Three thousand years on the Shaman's path*. New York: HarperCollins Publishers.
Geller, Markham Judah. 2014. *Melothesia in Babylonia: Medicine, magic, and astrology in the ancient near east*. Boston: Walter de Gruyter GmbH & Co KG.
Gingerich, Owen. 1993. *The eye of heaven: Ptolemy, Copernicus, Kepler*. New York: American Institute of Physics.
Goldstein, Bernard R. 1986. The making of astronomy in early Islam. *Nuncius* 1 (2): 79–92.
———. 2001. Astronomy and the Jewish community in early Islam. *Aleph* 1: 17–57.
Graeber, David, and David Wengrow. 2021. *The Dawn of everything: A new history of humanity*. New York: Farrar, Straus and Giroux.
Grant, Edward. 1994. *Planets, stars, and orbs: The medieval cosmos, 1200–1687*. New York: Cambridge University Press.
———. 1996. *The foundations of modern science in the middle ages: Their religious, institutional and intellectual contexts*. New York: Cambridge University Press.

———. 1997. The medieval cosmos: Its structure and operation. *Journal for the History of Astronomy* 28 (2): 147–167.

Hall, R. Cargill. 1977. *Lunar impact: A history of project ranger*, NASA history series 4210. Washington, DC: National Aeronautics and Space Administration.

Harari, Yuval. 2018. *Sapiens: A brief history of humankind*. New York: Harper Perennial.

Heilbron, John L. 2012. *Galileo*. Oxford: Oxford University Press.

Henry, John. 2020. Primary and secondary causation in Samuel Clarke's and Isaac Newton's theories of gravity. *Isis* 111 (3): 542–561.

Horrox, Rosemary. 1994. *The black death*. Manchester: Manchester University Press.

Hoyt, William Graves. 1996. *Lowell and Mars*. Tucson: University of Arizona Press.

Impey, Chris. 2023. *Worlds without end: Exoplanets, habitability, and the future of humanity*. Cambridge: MIT Press.

Isserles, Justine. 2017. Bloodletting and medical astrology in Hebrew manuscripts from Medieval Western Europe. *Sudhoffs Archiv* 101 (1): 2–41.

Koppes, Clayton R. 1982. *JPL and the American space program: A history of the Jet Propulsion Laboratory*. New Haven: Yale University Press.

Kraemer, Joel L. 1986. *Humanism in the renaissance of Islam: The cultural revival during the Buyid Age*. Leiden: E. J. Brill.

Küçük, Harun. 2019. *Science without Leisure: Practical naturalism in Istanbul, 1660–1732*. Pittsburgh: University of Pittsburgh Press.

Kuhn, Thomas S. 1957. *The Copernican revolution: Planetary astronomy in the development of western thought*. Cambridge: Harvard University Press.

———. 1996. *The structure of scientific revolutions*. Chicago: University of Chicago Press.

Lambright, W. Henry. 2014. *Why Mars: NASA and the politics of space exploration*. Baltimore: Johns Hopkins University Press.

Lane, K. Maria D. 2005. Geographers of Mars: Cartographic inscription and exploration narrative in late Victorian representations of the red planet. *Isis* 96 (4): 477–506.

———. 2011. *Geographies of Mars: Seeing and knowing the red planet*. Chicago: University of Chicago Press.

Lattis, James M. 1994. *Between Copernicus and Galileo: Christoph Clavius and the collapse of Ptolemaic cosmology*. Chicago: University of Chicago Press.

Launius, Roger D., and Howard E. McCurdy. 2008. *Robots in space: Technology, evolution, and interplanetary travel*. Baltimore: Johns Hopkins University Press.

Lawrence-Mathers, Anne. 2019. *Medieval meteorology: Forecasting the weather from Aristotle to the Almanac*. Cambridge: Cambridge University Press.

Lightman, Bernard. 2007. *Victorian popularizers of science: Designing nature for new audiences*. Chicago: University of Chicago Press.

Lindberg, David C. 2008. *The beginnings of western science: The European scientific tradition in philosophical, religious, and institutional context, prehistory to A.D. 1450*. Chicago: University of Chicago Press.

Lindberg, David C., and Robert S. Westman. 1990. *Reappraisals of the scientific revolution*. Cambridge: Cambridge University Press.

Logsdon, John M. 2015. *After Apollo?: Richard Nixon and the American space program*. New York: Palgrave Macmillan.

Major, John S. 1993. *Heaven and earth in early Han thought: Chapters three, four, and five of the Huainanzi*. Albany: State University of New York Press.

Markley, Robert. 2005. *Dying planet: Mars in science and the imagination*. Durham: Duke University Press.

Meltzer, Michael. 2012. *Mission to Jupiter: A history of the Galileo project*. Washington, DC: National Aeronautics and Space Administration.

Messeri, Lisa. 2016. *Placing outer space: An earthly ethnography of other worlds*. Durham: Duke University Press.

Mirmalek, Zara. 2020. *Making time on Mars*. Cambridge: MIT Press.

Morrison, Robert. 2013. Islamic astronomy. In *The Cambridge history of science, volume 2: Medieval science*, ed. David C. Lindberg and Michael H. Shank, 109–138. Cambridge: Cambridge University Press.
———. 2014. A scholarly intermediary between the Ottoman Empire and Renaissance Europe. *Isis* 105 (1): 32–57.
Nall, Joshua. 2019. *News from Mars: Mass media and the forging of a new astronomy, 1860–1910*. Pittsburgh: University of Pittsburgh Press.
National Academies of Sciences, Engineering. 2022. *Origins, worlds, and life: A decadal strategy for planetary science and astrobiology 2023–2032*. National Academies Press. https://doi.org/10.17226/26522.
Neugebauer, Otto. 1975. *A history of ancient mathematical astronomy*. Berlin: Springer.
Numbers, Ronald L. 1977. *Creation by natural law: Laplace's nebular hypothesis in American thought*. Seattle: University of Washington Press.
Oreskes, Naomi. 1999. *The rejection of continental drift: Theory and method in American earth science*. New York: Oxford University Press.
———. 2021. *Science on a mission: How military funding shaped what we do and don't know about the ocean*. Chicago: University of Chicago Press.
Pigliucci, Massimo, and Maarten Boudry, eds. 2013. *Philosophy of pseudoscience: Reconsidering the demarcation problem*. Chicago: The University of Chicago Press.
Ptolemy. 1998. *Ptolemy's Almagest*. Trans. G. J. Toomer. Princeton: Princeton University Press.
Pyne, Stephen J. 2010. *Voyager: Seeking newer worlds in the third great age of discovery*. New York: Viking.
Ragab, Ahmed. 2018. *The medieval Islamic hospital: Medicine, religion, and charity*. Cambridge: Cambridge University Press.
Rochberg, Francesca. 2004. *The heavenly writing: Divination, horoscopy, and astronomy in Mesopotamian culture*. New York: Cambridge University Press.
———. 2010. *In the path of the Moon: Babylonian celestial divination and its legacy*. Boston: Brill.
Ryan, Michael A. 2011. *A kingdom of stargazers: Astrology and authority in the late medieval crown of Aragon*. Ithaca: Cornell University Press.
Sabra, A.I. 1987. The appropriation and subsequent naturalization of Greek science in medieval Islam: A preliminary statement. *History of Science* 25 (3): 223–243.
———. 1996. Situating Arabic science: Locality versus essence. *Isis* 87 (4): 654–670.
Sarton, George. 2011. *Ancient science through the Golden age of Greece*. New York: Dover Publications.
Schaffer, Simon. 1980a. Herschel in Bedlam: Natural history and stellar astronomy. *The British Journal for the History of Science* 13 (3): 211–239.
———. 1980b. 'The great laboratories of the universe': William Herschel on matter theory and planetary life. *Journal for the History of Astronomy* 11: 81–110.
Schele, Linda, and Mary Ellen Miller. 1986. *Blood of kings: Dynasty and ritual in Maya art*. New York: George Braziller.
Schorn, Ronald A. 1998. *Planetary astronomy: From ancient times to the third millennium*. College Station: Texas A&M University Press.
Secord, James A. 2000. *Victorian sensation: The extraordinary publication, reception, and secret authorship of vestiges of the natural history of creation*. Chicago: University of Chicago Press.
Sheehan, William. 1992. *Worlds in the sky: Planetary discovery from the earliest times through Voyager & Magellan*. Tucson: University of Arizona Press.
Shindell, Matthew. 2014. From the end of the world to the age of the earth: The cold war development of isotope geochemistry at the University of Chicago and Caltech. In *Science and technology in the global cold War*, ed. Naomi Oreskes and John Krige. Cambridge: MIT Press.
———. 2019. *The life and science of Harold C. Urey*. Chicago: University of Chicago Press.
———. 2023. *For the love of Mars: A human history of the red planet*. Chicago: University of Chicago Press.

Siddiqi, Asif A. 2018. *Beyond Earth: A chronicle of deep space exploration, 1958–2016*. Washington, DC: National Aeronautics and Space Administration.
Sivin, Nathan. 2009. *Granting the seasons: The Chinese astronomical reform of 1280, with a study of its many dimensions and a translation of its records*. New York: Spring.
Swerdlow, Noel M. 1998. *The Babylonian theory of the planets*. Princeton: Princeton University Press.
Swerdlow, Noel M., and Otto Neugebauer. 1984. *Mathematical astronomy in Copernicus's De Revolutionibus*. Berlin: Springer.
Tatarewicz, Joseph N. 1990. *Space technology & planetary astronomy*. Bloomington: Indiana University Press.
Taylor, Kenneth L. 1998. Earth and heaven, 1750–1800: Enlightenment ideas about the relevance to geology of extraterrestrial operations and events. *Earth Sciences History* 17 (2): 84–91.
Tedlock, Dennis, Trans. 1996. *Popol Vuh: The definitive edition of the Mayan book of the dawn of life and the glories of gods and kings*. Subsequent edition. New York: Touchstone.
Tseng, Lillian Lan-ying. 2011. *Picturing heaven in early China*. Cambridge: Harvard University Press.
Vertesi, Janet. 2015. *Seeing like a rover: How robots, teams, and images craft knowledge of Mars*. Chicago: University of Chicago Press.
———. 2020. *Shaping science: Organizations, decisions, and culture on NASA's teams*. Chicago: University of Chicago Press.
Villalobos, Gerardo Aldana y, and Edwin L. Barnhart, eds. 2014. *Archaeoastronomy and the Maya*. Oxford: Oxbow Books.
Voelkel, James R. 2001. *The composition of Kepler's Astronomia Nova*. Princeton: Princeton University Press.
Westfall, Richard S. 1977. *The construction of modern science: Mechanisms and mechanics*. New York: Cambridge University Press.
———. 1993. *The life of Isaac Newton*. New York: Cambridge University Press.
Westman, Robert. 2011. *The Copernican question: Prognostication, skepticism, and celestial order*. Berkeley: University of California Press.
———. 2016. *Copernicus and the astrologers*. Washington, DC: Smithsonian Libraries.
Westwick, Peter J. 2007. *Into the black: JPL and the American space program, 1976–2004*. New Haven: Yale University Press.
Wey Gómez, Nicolás. 2008. *The tropics of empire: Why Columbus sailed south to the indies*. Cambridge: The MIT Press.
Xiaochun, Sun. 2000. Crossing the boundaries between heaven and man: Astronomy in ancient China. In *Astronomy across cultures: The history of non-western astronomy*, ed. Helaine Selin and Sun Xiaochun, 423–454. Boston: Springer.
Zakariya, Nasser. 2017. *A final story: Science, myth, and beginnings*. Chicago: University of Chicago Press.

Rivers in History: Systems, Agents, and Places

Etienne Benson

Contents

Abstract

Several influential river histories published in the past several decades have used systems-theoretic ideas and language to break down boundaries between nature and culture, between environment and technology, and between the humanities and the natural sciences. These works have served as models not only for historical research on rivers but for research on landscapes and environments in general. At the same time, a number of recent river histories have eschewed talk of systems, instead approaching rivers as historical agents and geographical places. Through a selective survey of recent works in the history of science and environmental history, this chapter offers a critical assessment of the use of systems theory and systems language to understand the entangled natural-cultural histories of landscape features and the Earth as a whole.

Keywords

Rivers · History · Hydrology · Geomorphology · Environment · Technology · Systems

---

E. Benson (✉)
Max Planck Institute for the History of Science, Berlin, Germany
e-mail: ebenson@mpiwg-berlin.mpg.de

E. Aronova et al. (eds.), *Handbook of the Historiography of the Earth and Environmental Sciences*, Historiographies of Science,
https://doi.org/10.1007/978-3-031-40799-4_12

1 Introduction

Is a river a system? Is it useful for historians to conceptualize rivers, mountains, coastlines, and other landscape features in systemic terms? What about the Earth as a whole? In recent years, many historians have explicitly or implicitly answered these questions in the affirmative. Systems talk has provided a way of building bridges and blurring boundaries between the natural and the cultural, between the technological and the social, and between the natural sciences and the humanities. At the same time, the translation of contingent and specific historical phenomena into the seemingly universal language of systems has been criticized for underemphasizing difference, decentering power and inequality, and reducing historical agency to a by-product of system dynamics. Nonetheless, there has been little sustained reflection on the implications of adopting systems talk for work in environmental history and the history of the environmental sciences. This chapter uses the historiography of rivers and river science as a starting point for such a reflection.

As dynamic landscape features with complex relations to human economies, societies, political structures, and technological infrastructures, rivers have seemed particularly well suited to analysis in terms of systems. In *Confluence* (2011), for example, Sara Pritchard presents the twentieth-century Rhône River as the epitome of a broader class of "envirotechnical systems," which she defines as "historically and culturally specific configurations of intertwined 'ecological' and 'technological' systems, which may be composed of artifacts, practices, people, institutions, and ecologies" (Pritchard 2011, 19). Importantly, Pritchard argues that the Rhône is not simply influenced or affected by the construction of dams, irrigation canals, power stations, and other sociotechnological systems, rather the river becomes an envirotechnical system in its own right through its relations to these systems (Pritchard 2011, 20). Moreover, the natural and human-made elements of this system influence each other so profoundly that the line between natural and human-made begin to shift and blur. This is not an accident, as Pritchard shows. On the contrary, shifting and blurring the natural/cultural divide to create a particular "envirotechnical regime" (Pritchard 2011, 23) was the explicit project of engineers and politicians seeking to strengthen the French state in the years following World War II.

While its articulation of envirotechnical analysis was novel, Pritchard's *Confluence* was not the first historical work to apply systems-theoretic terms to the study of rivers. Richard White's *The Organic Machine* (1995), in particular, had served as a model for taking a systems approach to the history of rivers and other landscape features since its publication in 1995. In this slim but wide-ranging volume, White seeks to describe historical changes in the human relationship to the Columbia River in a way that does not reinforce the conceptual divide between nature and culture. To do so he develops the concept of the "organic machine," which he defines as "an energy system which, although modified by human intervention, maintains its natural, its 'unmade' qualities" (White 1995, 2). Central to White's natural-cultural vision of the Columbia is a capacious conception of "work" that borrows and expands upon the scientific definition of that term. It is their common capacity to transfer energy from one place to another, White argues, that allows water, fish, steamboats, and humans to be understood as part of a single system—an organic machine—that bridges the nature/culture divide.

Like many other works of environmental history in the closing decades of the twentieth century, *The Organic Machine* was influenced by a school of ecological science that had emerged under the label of "ecosystem ecology" a few decades earlier. Articulated in the 1930s by the British ecologist Arthur Tansley, the concept of the ecosystem came of age in the USA with the help of the Cold War–era Atomic Energy Commission, eventually becoming central to the discourse of the environmental movement and environmental policymaking (Hagen 1992; Anker 2001; Martin 2018). When environmental history emerged in the USA in the 1970s as a new label for scholarship on the relationship between humans and their material surroundings, ecosystem ecology was at the peak of its influence, and its systems-theoretic view of relationships among living things and their inorganic environments shaped environmental historians' descriptions of nature. In his 1980 book *Land Use, Environment, and Social Change*, for example, White anticipated his later use of systems language by describing Island County, Washington, as a "system" constituted of "[m]en and women, the land and its plants and animals, and the social and economic institutions men had created" (White 1980, 75). Similarly, in *Changes in the Land*, one of the founding texts of American environmental history, William Cronon advocated for the "system-oriented perspective which an ecological approach allows" (Cronon 1983, 13).

That is not to say that all environmental historians wholeheartedly embraced systems talk. Donald Worster, for example, expressed skepticism about ecosystem ecology in *Nature's Economy*, a wide-ranging history of "ecological thought" in the West since the eighteenth century (1977, rev. ed. 1994). In Worster's view, ecosystem ecology's focus on the quantification of flows of energy and matter made it the handmaiden of an "imperial" view of nature as an exploitable resource and an object of human domination. In a subsequent work, *Rivers of Empire* (1985), he reserved the term "system" largely for social and technological systems: irrigation systems, economic systems, systems of government, and so forth. Rivers and landscapes could be and had been subjected to such systems, but for Worster they were not properly viewed as systems in their own right. The term "system" was reserved for things that humans had built, usually for the purpose of dominating nature, and therefore a systems perspective could not, by definition, encompass both the natural and the human-made.

Despite such reservations, historians increasingly embraced systems talk across a wide range of subfields from the 1970s onward. Among historians of technology, for example, the work of Thomas P. Hughes on electrical networks and other large technical systems was particularly influential. In *Networks of Power* (1983), Hughes argued that historians had hitherto focused too much attention on individual machines or devices—the light bulb, the automobile, and so forth—and too little on the systems of which those machines or devices were components, and on the environments within which those systems were designed to function. Hughes' systems approach helped set the parameters of the history of technology in the following decades, while also influencing the emerging interdisciplinary field of infrastructure studies (Star 1999). By the 1990s, both environmental historians and historians of technology were therefore prepared to see their subjects of study in terms of systems—the former via the adoption of metaphors from ecosystem ecology, the latter via the study of large technical systems. This embrace of systems talk made it increasingly easy for historians to envision both modern human societies and nonhuman aspects of the natural world as components of a single natural-cultural system.

Facilitated by systems talk, the subfield of envirotechnical history emerged in the late 1990s at the intersection of environmental history and the history of technology. While envirotech scholarship over the past several decades has embraced a variety of approaches and conceptual frameworks (for surveys see Stine and Tarr 1998; Reuss and Cutcliffe 2010; Pritchard 2014; Zeller 2017; Pritchard and Zimring 2020), its leading practitioners have drawn heavily on systems language. Examples include Timothy LeCain's argument that "the subterranean mine can best be understood as a hybrid system that combined natural and technological systems" (LeCain 2009, 53); Finn Arne Jørgensen's study of "historical attempts to develop technological and organizational systems that can keep bottles from becoming waste" (Jørgensen 2011, 6); James Morton Turner's "systems-based approach, focused on the interconnected stocks and flows of materials across the natural and built environment" in the history of lead-acid batteries (Turner 2015, 49); Julie Cohn's description of electric power networks as "complex and dynamic envirotechnical systems" (Cohn 2017, 338); and Jeremy Vetter's study of field science in the US West as a product of "intertwined environmental and technological systems" (Vetter 2016, 5).

As these examples suggest, the language of systems has been central to the way historians have attempted to integrate the study of technology and environment under the rubric of envirotech. Indeed, the very existence of this subfield of historical scholarship arguably rests on the commensurability between nature and technology that has been established by viewing both as systems of a particular sort. The implications of adopting this perspective have, however, largely gone unexamined. Rivers, because they have served as paradigmatic examples, provide an especially suitable subject for exploring these implications. Has viewing rivers as systems placed constraints on the kinds of questions or themes that have historians pursued? Where does the very idea of rivers as systems come from, and what are the alternatives? These questions are especially important to ask at a time when integrated systems models of Earth and human society are being advocated across the natural and human sciences.

2 How Rivers Became Systems

Historians of science have rarely singled out rivers as a subject of study, instead situating the study of rivers in relation to broader themes in the history of the Earth sciences. There is a rich literature, for example, on theories of the Earth proposed by European naturalists in the early modern period, in which the subject of rivers figured prominently. The question for early modern naturalists was, on its face, simple: Where did the water in rivers come from, and where did the water that rivers deposited into the oceans go? As Yi-Fu Tuan (1968) and others have documented, the answers offered were various, but by the end of the seventeenth century a consensus had emerged that water was neither created nor destroyed but rather recirculated in some fashion from the oceans to the sources of rivers and streams. This emerging consensus was linked to natural-theological speculations about the formation and functioning of the Earth by authors such as Thomas Burnet and John Ray (Tuan 1968; Rappaport 1997; Barnett 2019). Two pathways were generally proposed for the recirculation of water: one via evaporation from the oceans and precipitation over land, and the other via subterranean channels that brought water inland in liquid or vapor form. In the eighteenth century, partly as a result of empirical studies showing that precipitation was sufficient to supply rivers—along with a lack of empirical studies demonstrating subterranean circulation—the former became widely accepted. As a number of scholars have argued, this was the period when the concept that would later become known as the "hydrological cycle" was born (Tuan 1968; Biswas 1970; Linton 2010).

Historians of science have also examined research on rivers as part of the history of geology in the eighteenth and nineteenth centuries, when the identification and sequencing of geological strata became central to the emerging discipline (Greene 1982; Rudwick 1985; Secord 1986). Here the focus shifted from flows of water to the sediments that such flows were capable of transporting. Erosion and deposition were among the phenomena at the heart of the catastrophism-uniformitarianism debate, which pitted advocates of sudden, dramatic change in the geological history of the Earth (as might be caused by catastrophic flooding) against advocates of slow, incremental change (as might be caused by the ordinary flow of a river). Catastrophists could build on several centuries of research into the Noachian flood, as described in the Bible, to support their claims that the Earth's surface had been primarily shaped by sudden convulsions. For a mid-nineteenth-century uniformitarian such as Charles Lyell, however—building on the earlier work of James Hutton and John Playfair—stratigraphic evidence suggested that change had been gradual, effected by processes visible to the geologist in the present (Sepkoski 2020). Rivers were among the most important of the geomorphological and stratigraphic agents responsible for such gradual change.

In comparison to the nineteenth century and earlier, rivers play a much smaller role in the historiography of geology in the twentieth century, which has focused on the rise of geophysics and the development of the theory of plate tectonics (e.g., Oreskes 1999)—themes that, even if they do not exclude an interest in rivers, nonetheless direct attention elsewhere. Instead, twentieth-century research on rivers has been mainly the domain of historians of hydrology, who have often—in contrast with historians of the emergence of the hydrological cycle in the early modern period—been more interested in the implications of river science for practical water management than for general theories of the Earth. Giacomo Parinnello's work on hydrology in the Po River basin in the nineteenth century and early twentieth century is a case in point. Parrinello (2017) shows how the creation of a hydrological monitoring network for the river basin both advanced the understanding of river science and served the immediate needs of the state, which sought to ensure adequate water supplies and minimize damage from flooding. Other scholars have explored links between water resource management, engineering, and river science in other national and colonial contexts (e.g., Reuss 1998; D'Souza 2006; Ertsen 2016; Macfarlane 2020).

The story of how scientists came to conceptualize rivers as systems follows directly on this history, but it is a story that to date has received little attention from historians of science. Instead, the history of river science in the twentieth century has largely remained the domain of scientists reflecting on the history of their own fields, often in ways that focus narrowly on theories and techniques. These scientist-historians have situated the history of rivers not mainly in relation to geology or hydrology, per se, but rather in relation to a field which combines elements of both—namely geomorphology, the study of the Earth's surface and the forces that shape it, among which flowing water is one of the most important (Tinkler 1985). For these historians of geomorphology, the study of rivers was a central locus for a dramatic transformation of their discipline in the mid-twentieth century, when qualitative, evolutionary models of landscape development were replaced by quantitative, systems-theoretic models heavily influenced by hydraulic engineering and thermodynamics. It was this shift, concentrated in the decades immediately following World War II, that introduced the idea of the river as a thermodynamic system—that is, a system that can be characterized in terms of transfers of energy.

Initially studied largely by partisans of this development, this history has both been critiqued by scientist-historians who question its assumptions (Sack 1991, 1992) and, more recently, situated in broader social context by historians of science and STS scholars (Lave 2012; Parrinello 2017). Their work has questioned the characterization of this shift as revolutionary, noting its deep roots in the history of the Earth sciences and the management of river systems and its incomplete adoption and multiple revisions since the mid-twentieth century. Nonetheless, there are clear connections between the changes documented in the history of river science and broader shifts in the Earth and environmental sciences during this period, when systems approaches were widely adopted. The history of Cold War–era ecosystem ecology is perhaps the best studied example (Hagen 1992; Bocking 1997; Coleman 2010; Martin 2018); other examples include climate modeling (Edwards 2010) and Earth system science (Steffen et al. 2020). Taken together, these works show that the Earth sciences also had their "age of system" (Heyck 2015), which in many ways continues today. The turn to systems-theoretic approaches to rivers might not have been as revolutionary as its most fervent advocates claimed, but it was certainly a distinctive characteristic of the post-World War II era.

The idea that rivers can be analyzed as engines for performing the work of moving water and sediment—where "work" is understood in the physicists' sense, as the energy transferred from one place to another—emerged from a historical moment when new techniques for the quantitative study of the Earth intersected with a massive expansion of dams and other water engineering works. In the case of rivers, the mid-twentieth turn toward thermodynamics-inspired systems is closely connected to the boom in water engineering works that took place at the same time. In the USA, where systems-theoretic approaches were pursued most enthusiastically, the period from the 1920s to the 1960s was one of extensive dam construction, as environmental historians have amply demonstrated (Worster 1985; Harvey 1994). As in other parts of the world (Evenden 2015; Parrinello 2018), many of these dams were built for power generation as well as water supply, and the engineers who managed them took into account not only fluctuations in consumer demand for electrical power but also natural phenomena such as precipitation, snowmelt, infiltration, and evaporation. They became concerned with what Ashley Carse (2012), writing about the Panama Canal, has described as "nature as infrastructure." Moreover, as hydropower facilities were integrated into larger power networks, water power became fungible with energy generated from other sources, particularly fossil fuels. The growth of irrigation networks, some of them incorporating electric pumping stations, similarly encouraged engineers to approach rivers as components of large technical systems (Fiege 1999).

Published several decades after this scientific turn, White's *The Organic Machine* built on and creatively expanded the systems view of the river, reinterpreting "energy" as something that linked the exertions of laboring human and animal bodies, the flow of water and sediment in the river, and the electricity generated by nuclear power plants, among other things. In a certain sense, this approach was nothing new. As Norton Wise and others have documented (Smith and Norton Wise 1989; Daggett 2019), concepts of energy and work articulated through thermodynamics have provided bridges between the analysis of physical and economic systems since at least the nineteenth century. But the mobilization of these terms for historical scholarship found a new resonance in the 1990s. Perhaps not surprisingly given its origins in the application of an engineering approach to natural phenomena, this way of writing about rivers proved to be a very powerful one for dismantling borders between nature and culture, and for studying the intertwined development of natural and technical systems.

3 Rivers as Agents and Places

This way of studying river history has not, however, been the only one that historians have pursued. In fact, a great deal of historical scholarship on rivers in recent decades has declined to adopt a thermodynamics-inspired view of rivers as systems, even as it has largely accepted White's argument that rivers ought to be approached as hybrids of nature and culture (see, e.g., Mauch and Zeller 2008). As Matthew Evenden (2018) has argued, the rapidly expanding field of river historiography since the 1990s has abandoned or transcended some of the preoccupations of White's *The Organic Machine* (1995), including its concern with blurring the nature/culture distinction, its privileging of embodied labor as a way of knowing nature, and—of particular relevance here—its use of flows of energy as an analytical tool. Instead, according to Evenden, much recent work has focused on situating rivers in urban contexts or has used border-crossing rivers as the locus for comparative or transnational environmental histories. Implicit in this analysis is a further point: that much of this literature approaches rivers not as *systems* but rather as historical *agents* or geographical *places*. Describing some works that take these alternative approaches will help clarify what is distinctive about the systems approach and why some scholars have declined to adopt it.

Much of the recent scholarship on river history eschews systems talk in favor of biographical or life-history approaches that figure rivers as nonhuman agents of historical change. Marc Cioc's *The Rhine: An Eco-Biography* (2002), for example, follows White's example in showing how the Rhine, like the Columbia, had been transformed into an "organic machine" or "techno-river" through a century and a half of human engineering (Cioc 2002, 6, 14). In contrast to White, however, Cioc is not particularly interested in using the language of systems to deconstruct the nature/culture divide. The "biography" of the book's title situates the industrial remaking of the river in the nineteenth and twentieth centuries as an episode in the river's longer life history, whose beginning dates to before the arrival of the first humans in Europe. The Rhine may presently be an organic machine, in other words, but it was not always such, and may yet become something else—and throughout and despite these changes, it will remain the Rhine. This life-history approach contrasts with the systems approach, in which rivers are seen as integral parts of wider socio-technical-environmental systems. Instead, Cioc centers the Rhine as an entity with its own historical identity and agency, and thereby highlights how deeply that identity and agency have been compromised by human intervention. A similar approach can be seen in David Blackbourn's characterization of the development of the Oder River, which is framed as part of a centuries-long (but never quite complete or successful) project of the "conquest of nature" in Germany (Blackbourn 2006).

These histories follow a narrative arc of increasing human domination, but not all histories that emphasize nonhuman agency do so, particularly when they are set in periods other than the modern one. Whereas Cioc's and Blackbourn's stories both end with a human-dominated river that remains under the human yoke even after the folly of efforts at complete domination have been recognized and partly reversed, Ling Zhang's *The River, the Plain, and the State* (2016) positions the Yellow River as an uncontrollable entity with just as much historical agency as society and the state. Focused on the eleventh century, when the Yellow River dramatically shifted its course into the Hebei Plain, Zhang's book describes an agent that (or perhaps *who*) resolutely refused to submit to human domination. Indeed, the Yellow River often seemed to respond to human interventions with indignation and retributive violence. Unabashed about describing the river's actions in anthropomorphic terms, Zhang shows how it resisted and frustrated the imperial aims of the Northern Song Dynasty, pulling it into a "hydraulic mode of consumption" that ultimately undermined the imperial state. Although influenced by American environmental history, Zhang explicitly rejects ecology as a model for historical scholarship because, she argues, "ecological thinking keeps at its ontological core the flow and transformation of some essential matter, such as energy" (Zhang 2016, 15)—a reduction, she argues, that risks misdescribing the emergent, entangled relations among particular historical entities.

As Zhang's study suggests, historians who approach rivers as historical agents have found the relationship between rivers and the state to be of particular interest. In doing so, they sometimes hark back to Karl Wittfogel's work on "oriental despotism," which argued that authoritarian states had been built on a foundation of water control, particularly the organization of large irrigation systems (Wittfogel 1957). Both Zhang and Worster explicitly take up Wittfogel's problematic work and attempt to revise and revitalize it for new purposes. But the theme of rivers and states is a broader one that need not take Wittfogel's environmentally determinist approach either as model or foil. As David Biggs shows in *Quagmire* (2010), the Mekong River resisted a variety of efforts at control from at least the 1860s, when France colonized the river's delta region and began attempting to reengineer it through canals and drainage systems, through the 1970s, when the Viet Minh succeeded in driving out the USA, which by then had its own legacy of failed river engineering projects. The river's agency, in this account, neither builds states nor destroys them; instead, it serves as the occasion for what Biggs, citing one of his historical actors, calls "works of Penelope" (Biggs 2010, 37)—that is, projects with no end.

These and other recent works refuse to conceptualize rivers as systems because doing so, they claim, obscures or misrepresents their historical agency, thereby making it difficult to see both how nonhuman entities have actively shaped history (particularly evident in Zhang's and Biggs' studies) and how human interventions have undermined or compromised their ability to do so in particular historical periods (the focus of Cioc's study, as well as Blackourn's). This reflects an interesting return to environmental history's roots, albeit with a twist. As Paul Sutter has argued, the turn toward hybridity in environmental history marked by the publication of *The Organic Machine* in the mid-1990s was, among other things, an effort to move beyond an unreflective conceptualization of "nature's agency" that had characterized the first wave of American environmental history (Sutter 2013). But for many environmental historians, the replacement of those crude notions of agency with a systems-based approach that threatened to evacuate history of agency entirely was a step too far. Instead of rejecting agency because of its association with a certain kind of liberal subjecthood, they have tried to articulate an alternative notion of "nature's agency" that is relational and emergent (Nash 2005). An example of this move can be found in Roderick Wilson's *Turbulent Streams*, a study of Japan's rivers from the seventeenth century to the early twentieth century that builds explicitly on White's work but also shifts the analytic focus from "systems" to what Wilson describes as "riparian relations" (Wilson 2021, 11).

Another set of river historians has taken a different approach, one that largely sidesteps the question of agency but similarly distances itself from the view of rivers as systems. This approach is fundamentally geographical, and some of it has in fact been written by scholars who identify themselves primarily as historical geographers rather than as environmental historians. This approach considers rivers as places or, sometimes, as connectors or borders between places; it also investigates the history of the imagination of rivers as distinct geographical entities (Cunha 2019). Studies of urban rivers have been particularly prominent in this line of research (see, e.g., Evenden and Castonguay 2012). When the term "river system" is used in this context, which is not uncommon, it is mainly as a way of referring to a network of streams connected within a single watershed, rather than to any kind of systems-theoretic perspective on rivers. In other words, the word "system," when it appears, is meant to indicate only that a particular river is being studied in relation to other streams. This place-based approach to rivers encompasses much of the work that Evenden (2018) covers in his survey of river historiography, including both urban and transnational river histories, and is part of a broader movement to bring the same level of attention to bodies of water—rivers, lakes, oceans, and so forth—that has long been accorded to solid land.

Whereas the systems approach tends to flatten hierarchies—everything is, after all, part of the same system of cycles, exchanges, and feedback loops—historical scholarship that conceives of rivers as places often brings questions of power and inequality to the forefront. This is the case, for example, in Ari Kelman's study of the New Orleans' waterfront, *A River and Its City* (2003), which shows how various residents of the city have struggled for access to and control over nature and public space from the early nineteenth century to the mid-twentieth century. By approaching the Mississippi River as an urban site, Kelman's book highlights the ways that the river's historical changes—perhaps most dramatically, the floods of 1927—intersect with differences in power among various groups and individuals in New Orleans. As *A River and Its City* documents, following flooding along the lower Mississippi in the spring of 1927, New Orleans' commercial elites decided to dynamite the Poydras levee, some 12 miles downstream of the city, to reduce the chance of an unintentional breach of one of the city's own levees. This decision had catastrophic affects for the poor and sparsely populated river parishes of St. Bernard and Plaquemines, whose residents had no say in the decision. In this way, the city's response to the threat of flooding both reflected and exacerbated the deep differences of power in the community.

A nuanced attention to the way social distinctions and power differentials in a city interact with the physical behavior of a river—or in this case, multiple rivers—is also evident in Debjani Bhattacharyya's *Empire and Ecology in the Bengal Delta* (2018). Like Kelman, Bhattacharyya focuses on the intersection of land and water, including the shifting boundary between the two, but she is particularly attentive to how the very distinction itself, between the city and river, is a subject of historical conflict. Moreover, as the title suggests, *Empire and Ecology in the Bengal Delta* is centrally concerned with the imposition of colonial power over a landscape—or, as she puts it, with "the life of the delta at the intersection of a powerful riverine environment and imperial human intentions" (Bhattacharyya 2018, 38). Under the East India Company in the late eighteenth century, a process of imposing European property law onto the fluid landscape of the Bengal delta was initiated; that process generated a legal and economic framework that then progressively remade the material reality of the delta. This process consolidated some forms of power, Bhattacharyya argues, but also reshuffled hierarchies in the city by generating new sources of wealth and power on newly solidified land.

Viewing rivers as places rather than as systems shifts attention from the functional interconnections between humans and nature—something that these books, building on earlier scholarship, take more or less for granted—to the ways that power (and not just energy or work) is distributed over space. In doing so, these works contribute to a broader rethinking of power and politics in environmental history. If the first generation of self-identified environmental historians was concerned mainly with showing how humans exercised power over nature, almost always at nature's expense, and the second generation was intent on showing how nature and culture were entangled or hybrid (Sutter 2013), then these works show how power within human communities is structured at least in part by the properties and actions of nonhuman entities. In other words, it is not just that "the conquest of nature ... will always include the conquest, the domination or exploitation" of some people by others (Williams 1980), or that nature and culture are constitutively entangled. It is also that nonhuman entities and forces—whether the flooding of the Mississippi, the shifting sandbars of the Bengal Delta, or the quagmires of the Mekong—affect which humans have power and how they exercise it. Power and privilege are, in this sense, emplaced in ways that are hard to articulate in a systems approach.

4 Conclusion

Is a river a system? Many scientists and engineers since the mid-twentieth century have thought so. And is it useful for historians to follow those scientists in viewing rivers as systems? That question cannot be answered with a simple yes or no. Some historians have undoubtedly found it productive. The systems perspective on rivers has worked as a solvent to break down barriers between nature and culture, metaphorically expanding on thermodynamic and systems-theoretic themes to show how "natural" and "cultural" domains ultimately trade in the same basic currency. For White, that currency was energy, albeit a more expansive conception of energy than that offered by physics (see also Demuth 2019); for other scholars, including Pritchard, the identity of the currency is less explicit, but the argument for viewing "technology as natural" and "nature as technological" (Pritchard 2011, 21–22) suggests that whatever form it may take, the currency is shared. Since the 1990s, viewing rivers and other landscape features as systems, envirotechnical or otherwise, has been one of the most effective tools available to environmental historians for moving beyond the too-easy distinctions between nature and culture that characterized the first wave of scholarship in the field. In that sense, it has certainly been useful for historians to view rivers as systems.

As the very selective survey of river historiography presented above also suggests, however, that strategy has not been universally embraced by environmental historians. If systems talk has been a solvent for nature and culture as they were commonly defined half a century ago, recent scholarship has often tried to precipitate out distinctive entities from the natural-cultural solution, albeit ones that do not neatly line up on either side of the nature/culture divide as it has been conventionally conceived. The biographical or agential approach, for example, may recognize that a river such as the Rhine became a "techno-river" over the course of the nineteenth and twentieth centuries while emphasizing that this is only one episode in a much longer life history, over the course of which the Rhine has exercised historical agency to various degrees and in various ways. Similarly, the geographical approach may acknowledge the complex web of feedback loops connecting all of the entities in a particular place, but it also tends to emphasize the persistence of certain hierarchies and relations of power, which themselves are constituted by and not merely imposed upon the river. Rivers may be systems, this scholarship suggests, but they are not only systems, and utility of concepts such as organic machine and envirotechnical system may be especially limited for the study of rivers in nonmodern and non-industrialized contexts.

Much of this review has focused on works of environmental history for the simple reasons that self-described historians of science have contributed relatively little to recent river historiography. The scientific disciplines most directly concerned with rivers, including hydrology, geomorphology, and aquatic ecology, have received little attention from modern historians of science, and historians of earlier periods have rarely focused on rivers in their own right. Instead, most recent historiography of the Earth sciences has focused on issues and perspectives that, while they are certainly capable of incorporating rivers, do not center them. One reason is that much of this work has sought to understand how scientists came to construct understandings of the Earth that operate on the largest of temporal and spatial scales—that is, studies of the emergence of planetary and deep-time perspectives (Rudwick 1985; Oreskes 1999; Edwards 2010; Selcer 2018; Barnett 2019). This emphasis is understandable at a moment when humanity's global impacts have become increasing obvious and worrisome. But it has also led historians to neglect smaller scales, as well as the sciences that operate at those scales. Rivers certainly can be and have been studied on the global scale—for instance, in attempts to calculate the total discharge or sediment load of all the world's rivers—but in doing so their specificities as rivers tend to be lost.

River history therefore provides an opportunity for historians of science to approach the history of the Earth sciences anew, at scales and in topologies that are not neatly captured by the global/local dichotomy. In doing so, the systems perspective surely has its place, and historians of science are well positioned to show how that perspective can illuminate the entanglement of environmental and technological systems with systems of knowledge. In fact, some historians of science have begun to call for moving beyond the global/local dichotomy to focus on the "regional" (Vetter 2011) or "meso" (Güttler 2019) scales. Fittingly, the scholars calling for this scalar refocusing have also made envirotechnical analysis central to their work, emphasizing the importance of environmental characteristics of particular regions as well as regional-scale technological infrastructures. At the same time, building on Pritchard's (2011) analysis of envirotechnical systems and regimes, they have expanded the borders of envirotechnical analysis to include the study of what one might call envirotechnical epistemes. Vetter (2016), for instance, has described the particular mode of knowledge production characteristic of the "railroad age" in the US West, while Güttler (2020) has explored the diverse forms of knowledge produced through and about the Frankfurt Airport in the twentieth century. Rivers have not yet been the focus of such work, but—given their scale, their importance, and the extent to which they have been integrated into human infrastructures—they are very well suited to it. There are also promising and as-yet unexplored opportunities here to connect envirotechnical work on rivers-as-systems to work by historians of science on experimental systems (e.g., Rheinberger 1997).

But a systems approach does not exhaust the ways in which historians of science might yet contribute to river historiography. Viewing rivers as historical agents, for example, opens up a distinct set of questions about the history of knowledge about rivers. Unsurprisingly, the history of science as a field continues to privilege the human knower, understood either as an individual or a member of a larger collective. Over time, the variety of knowers admitted into the historiography has grown, as has attention to the material conditions, resources, and tools required for knowledge to be produced and circulated. In that sense, the heroic liberal subject of a certain kind of progressive history of science has gradually been decentered. Still, historians of science, including historians of the Earth sciences, have been slow to engage with the kinds of arguments presented by environmental historians about nonhuman agency. How do changing envirotechnical conditions shape what is known about the Earth? Can events in the life history of a river—from daily fluctuations in flow to a catastrophic thousand-year flood—redirect not only water and sediment but also the course of knowledge? These questions suggest the potential for a non-anthropocentric history of the Earth sciences that has yet to be written.

Approaching rivers as places also raises questions that are hard to articulate when one approaches them as systems, particularly questions about power and difference. Here, too, historians of science can build on the work of environmental historians while expanding the focus to include the production and circulation of knowledge. Rivers and riverfronts are not just sites where competing interests are pursued and where hierarchies are established and sometimes challenged, they are also sites where various actors produce knowledge, often in competing or even incommensurate ways. Work on civic science, popular epidemiology, community-based science, and environmental justice has shown how the concerns and modes of knowledge production of the people who inhabit a place can differ profoundly from those of scientists who visit such a place solely in a professional capacity (Brown 1990, 1992; Di Chiro 1996, 2008; Fortun 2001; Wylie 2018). How do differences in status and power among groups of people inhabiting riverine landscapes affect the generation and communication of knowledge? River history offers a productive site for exploring that question in dialogue with place-based histories of science (e.g., Galison and Thompson 1999; Kohler 2002; Livingstone 2003).

The language of systems—whether used in a strictly systems-theoretic or thermodynamic sense or in a looser and more colloquial sense—is a powerful way of breaking down borders between nature and culture that today seem increasingly inappropriate for understanding the history of the Earth and the Earth sciences. Systems talk is ubiquitous in the contemporary Earth sciences for good reason; it helps scientists working in a variety of disciplines, on disparate phenomena, and across a variety of scales build models together. And historians who have creatively appropriated and extended that language for their own purposes have shown that it can be similarly useful for humanistic understandings of the planet's past. Envirotech scholarship has benefited greatly from the adoption of the vocabulary of systems theory; one might even say that the subfield largely owes its existence to that vocabulary, which opened up new ways of bringing natural and technical systems into relation. But it is important to remember that this is just one of several idioms available, and that, like any idiom, it is better at articulating some things than others. Historians of science have much to gain by studying rivers as envirotechnical systems that are also systems of knowledge, but they should not lose sight of the fact that rivers can also be agents and places, and that studying them as such may bring to the surface histories that would otherwise remain submerged.

References

Anker, Peder. 2001. *Imperial ecology: Environmental order in the British Empire, 1895–1945*. Cambridge, MA: Harvard University Press.

Barnett, Lydia. 2019. *After the flood: Imagining the global environment in early modern Europe*. Baltimore: Johns Hopkins University Press.

Bhattacharyya, Debjani. 2018. *Empire and ecology in the Bengal Delta: The making of Calcutta*. Cambridge: Cambridge University Press.

Biggs, David Andrew. 2010. *Quagmire: Nation-building and nature in the Mekong Delta*. Seattle: University of Washington Press.

Biswas, Asit K. 1970. *History of hydrology*. New York: American Elsevier Publishing Company.

Blackbourn, David. 2006. *The conquest of nature: Water, landscape, and the making of modern Germany*. New York/London: W.W. Norton & Company.

Bocking, Stephen. 1997. *Ecologists and environmental politics: A history of contemporary ecology*. New Haven: Yale University Press.

Brown, Phil. 1990. *No safe place: Toxic waste, leukemia, and community action*. Berkeley: University of California Press.

———. 1992. Popular epidemiology and toxic waste contamination: Lay and professional ways of knowing. *Journal of Health and Social Behavior* 33: 267–281.

Carse, Ashley. 2012. Nature as infrastructure: Making and managing the Panama Canal watershed. *Social Studies of Science* 42: 539–563.

Cioc, Mark. 2002. *The Rhine: An eco-biography, 1815–2000*. Seattle: University of Washington Press.

Cohn, Julie. 2017. Data, power, and conservation: The early turn to information technologies to manage energy resources. *Information & Culture* 52: 334–361.

Coleman, David C. 2010. *Big ecology: The emergence of ecosystem science*. Berkeley: University of California Press.

Cronon, William. 1983. *Changes in the land: Indians, colonists, and the ecology of New England*. New York: Hill and Wang.

D'Souza, Rohan. 2006. *Drowned and dammed: Colonial capitalism, and flood control in eastern India*. New Delhi: Oxford University Press.

da Cunha, Dilip. 2019. *The invention of rivers: Alexander's eye and Ganga's descent*. Philadelphia: University of Pennsylvania Press.

Daggett, Cara New. 2019. *The birth of energy: Fossil fuels, thermodynamics, and the politics of work*. Durham: Duke University Press.

Demuth, Bathsheba. 2019. *Floating coast: An environmental history of the Bering Strait*. New York: W.W. Norton & Company.

Di Chiro, Giovanna. 1996. Nature as community: The convergence of environment and social justice. In *Uncommon ground: Rethinking the human place in nature*, ed. William Cronon, 298–320. New York: Norton.

———. 2008. Living environmentalisms: Coalition politics, social reproduction, and environmental justice. *Environmental Politics* 17: 276–298.

Edwards, Paul N. 2010. *A vast machine: Computer models, climate data, and the politics of global warming*. Cambridge, MA: MIT Press.

Ertsen, Maurits W. 2016. *Improvising planned development on the Gezira Plain, Sudan, 1900–1980*. New York: Palgrave Macmillan.

Evenden, Matthew. 2015. *Allied power: Mobilizing hydro-electricity during Canada's Second World War*. Toronto: University of Toronto Press.

———. 2018. Beyond the organic machine? New approaches in river historiography. *Environmental History* 23: 698–720.

Evenden, Matthew, and Stephane Castonguay, eds. 2012. *Urban rivers*. Pittsburgh: University of Pittsburgh Press.

Fiege, Mark. 1999. *Irrigated Eden: The making of an agricultural landscape in the American West*. Seattle: University of Washington Press.

Fortun, Kim. 2001. *Advocacy after Bhopal: Environmentalism, disaster, new global orders*. Chicago: University of Chicago Press.

Galison, Peter, and Emily Ann Thompson, eds. 1999. *The architecture of science*. Cambridge, MA: MIT Press.

Greene, Mott T. 1982. *Geology in the nineteenth century: Changing views of a changing world*. Ithaca: Cornell University Press.

Güttler, Nils. 2019. "Hungry for knowledge": Towards a meso-history of the environmental sciences. *Berichte zur Wissenschaftsgeschichte* 42: 235–258.

———. 2020. *Alles über das Fliegen: eine politische Wissensgeschichte des Frankfurter Flughafens*. Wien/Berlin: Turia + Kant.

Hagen, Joel. 1992. *An entangled bank: The origins of ecosystem ecology*. New Brunswick: Rutgers University Press.

Harvey, Mark W.T. 1994. *A symbol of wilderness: Echo Park and the American conservation movement*. Albuquerque: University of New Mexico Press.

Heyck, Hunter. 2015. *Age of system: Understanding the development of modern social science*. Baltimore: Johns Hopkins University Press.

Hughes, Thomas Parke. 1983. *Networks of power: Electrification in Western society, 1880–1930*. Baltimore: Johns Hopkins University Press.

Jørgensen, Finn Arne. 2011. *Making a green machine: The infrastructure of beverage container recycling*. New Brunswick: Rutgers University Press.

Kelman, Ari. 2003. *A river and its city: The nature of landscape in New Orleans*. Berkeley: University of California Press.

Kohler, Robert E. 2002. *Landscapes & labscapes: Exploring the lab-field border in biology*. Chicago: University of Chicago Press.

Lave, Rebecca. 2012. *Fields and streams: Stream restoration, neoliberalism, and the future of environmental science*. Athens: University of Georgia Press.

LeCain, Timothy J. 2009. *Mass destruction: The men and giant mines that wired America and scarred the planet*. New Brunswick: Rutgers University Press.

Linton, Jamie. 2010. *What is water? The history of a modern abstraction*. Vancouver: University of British Columbia Press.

Livingstone, David N. 2003. *Putting science in its place: Geographies of scientific knowledge*. Chicago: University of Chicago Press.

Macfarlane, Daniel. 2020. Nature empowered: Hydraulic models and the engineering of Niagara Falls. *Technology and Culture* 69: 109–143.

Martin, Laura J. 2018. Proving grounds: Ecological fieldwork in the Pacific and the materialization of ecosystems. *Environmental History* 23: 567–592.

Mauch, Christof, and Thomas Zeller, eds. 2008. *Rivers in history: Perspectives on waterways in Europe and North America*. Pittsburgh: University of Pittsburgh Press.

Nash, Linda. 2005. The agency of nature or the nature of agency? *Environmental History* 10: 67–69.

Oreskes, Naomi. 1999. *The rejection of continental drift: Theory and method in American earth science*. New York: Oxford University Press.
Parrinello, Giacomo. 2017. Charting the flow: Water science and state hydrography in the Po watershed, 1872–1917. *Environment and History* 23: 65–96.
———. 2018. Systems of power: A spatial envirotechnical approach to water power and industrialization in the Po Valley of Italy, ca. 1880–1970. *Technology and Culture* 59: 652–688.
Pritchard, Sara B. 2011. *Confluence: The nature of technology and the remaking of the Rhône*. Cambridge, MA: Harvard University Press.
———. 2014. Toward an environmental history of technology. In *The Oxford handbook of environmental history*, ed. Andrew C. Isenberg, 227–258. Oxford: Oxford University Press.
Pritchard, Sara B., and Carl A. Zimring. 2020. *Technology and the environment in history*. Baltimore: Johns Hopkins University Press.
Rappaport, Rhoda. 1997. *When geologists were historians, 1665–1750*. Ithaca: Cornell University Press.
Reuss, Martin. 1998. *Designing the bayous: The control of water in the Atchafalaya Basin 1800–1995*. Alexandria: Office of History, U.S. Army Corps of Engineers.
Reuss, Martin, and Stephen H. Cutcliffe, eds. 2010. *The illusory boundary: Environment and technology in history*. Charlottesville: University of Virginia Press.
Rheinberger, Hans-Jörg. 1997. *Toward a history of epistemic things: Synthesizing proteins in the test tube*. Stanford: Stanford University Press.
Rudwick, M.J.S. 1985. *The great Devonian controversy: The shaping of scientific knowledge among gentlemanly specialists*. Chicago: University of Chicago Press.
Sack, Dorothy. 1991. The trouble with antitheses: The case of G.K. Gilbert, geographer and educator. *Professional Geographer* 43: 28–37.
———. 1992. New wine in old bottles: The historiography of a paradigm change. *Geomorphology* 5: 251–263.
Secord, James A. 1986. *Controversy in Victorian geology: The Cambrian-Silurian dispute*. Princeton: Princeton University Press.
Selcer, Perrin. 2018. *The postwar origins of the global environment: How the United Nations built Spaceship Earth*. New York: Columbia University Press.
Sepkoski, David. 2020. *Catastrophic thinking: Extinction and the value of diversity from Darwin to the Anthropocene*. Chicago: University of Chicago Press.
Smith, Crosbie, and M. Norton Wise. 1989. *Energy and empire: A biographical study of Lord Kelvin*. Cambridge: Cambridge University Press.
Star, Susan Leigh. 1999. The ethnography of infrastructure. *American Behavioral Scientist* 43: 377–391.
Steffen, Will, Katherine Richardson, Johan Rockström, Hans Joachim Schellnhuber, Opha Pauline Dube, Sébastien Dutreuil, Timothy Lenton, and Jane Lubchenco. 2020. The emergence and evolution of earth system science. *Nature Reviews Earth & Environment* 1: 54–63.
Stine, Jeffrey K., and Joel A. Tarr. 1998. At the intersection of histories: Technology and the environment. *Technology and Culture* 39: 601–640.
Sutter, Paul. 2013. The world with us: The state of American environmental history. *Journal of American History* 100: 94–119.
Tinkler, Keith J. 1985. *A short history of geomorphology*. London: Croom Helm.
Tuan, Yi-fu. 1968. *The hydrologic cycle and the wisdom of God: A theme in geoteleology*. Toronto: University of Toronto Press.
Turner, James Morton. 2015. Following the Pb: An envirotechnical approach to lead-acid batteries in the United States. *Environmental History* 20: 29–56.
Vetter, Jeremy. 2011. *Knowing global environments: New historical perspectives on the field sciences*. New Brunswick: Rutgers University Press.
———. 2016. *Field life: Science in the American West during the railroad era*. Pittsburgh: University of Pittsburgh Press.

White, Richard. 1980. *Land use, environment, and social change: The shaping of Island County, Washington*. Seattle: University of Washington Press.
———. 1995. *The organic machine*. New York: Hill and Wang.
Williams, Raymond. 1980. *Problems in materialism and culture: Selected essays*. London: Verso.
Wilson, Roderick I. 2021. *Turbulent streams: An environmental history of Japan's rivers, 1600–1930*. Boston: Brill.
Wittfogel, Karl August. 1957. *Oriental despotism: A comparative study of total power*. New Haven: Yale University Press.
Worster, Donald. 1985. *Rivers of empire: Water, aridity, and the growth of the American West*. New York: Pantheon.
Wylie, Sara Ann. 2018. *Fractivism: Corporate bodies and chemical bonds*. Durham: Duke University Press.
Zeller, Thomas. 2017. Aiming for control, haunted by its failure: Towards an envirotechnical understanding of infrastructures. *Global Environment* 10: 202–228.
Zhang, Ling. 2016. *The river, the plain, and the state: An environmental drama in Northern Song China, 1048–1128*. Cambridge: Cambridge University Press.

Histories of Frozen Earth

Pey-Yi Chu

Contents

Abstract

This chapter provides an overview of histories of frozen earth while arguing that scholarship must attend to the historicity of *permafrost* as a scientific idea. It begins by discussing histories written before the establishment of permafrost science as a discipline. In this stage, terms for frozen earth had not yet been standardized, and interest in it was connected to major questions of nineteenth-century geology, especially the theory of ice ages. The chapter then considers histories written during the period of consolidating permafrost science, when state goals of colonization, industrialization, and militarization motivated research into frozen earth. Frozen earth became *permafrost*, and scientists used history to debate the discipline's goals and orientation. Finally, the chapter analyzes histories written amid rising global awareness of anthropogenic climate change and the thawing of frozen earth. If, previously, histories of frozen earth were written primarily by scientists, in this stage scholars in the humanities and social sciences became increasingly engaged as well. They have explored the meanings of frozen earth in a variety of cultural contexts. This chapter argues that human engagements with frozen earth invested the phenomenon with multiple essences and

P.-Y. Chu (✉)
Pomona College, Claremont, CA, USA
e-mail: pey-yi.chu@pomona.edu

E. Aronova et al. (eds.), *Handbook of the Historiography of the Earth and Environmental Sciences*, Historiographies of Science,
https://doi.org/10.1007/978-3-031-40799-4_13

names, of which *permafrost* was but one. While some conceptions of frozen earth recognized its inherent dynamism and heterogeneity, *permafrost*, a legacy of settler colonialism, suggested permanence and stability. Recovering alternative ways of understanding frozen earth can point to ways of living with change rather than maintaining expectations of stillness.

Keywords

Frozen earth · Permafrost · Cryosphere · Cryology · Arctic · Sub-arctic · Siberia

1 Introduction

In 2015, the historian of science Sverker Sörlin posited the emergence of "cryo-history," histories of ice and snow, during an epoch of retreating cold. The shrinking of glaciers and reduction of sea ice that accompanied global warming had drawn worldwide attention to the mutability of landscapes sometimes assumed to be frozen and timeless. Sörlin observed that ice had a history not only in a geological sense but in social and cultural terms as well. He called for cryo-histories – derived from *kryos* or κρυος, Ancient Greek for cold, frost, or ice – that examine human relationships with cold and its manifestations (Sörlin 2015). Sörlin's essay served as much as an indicator as a harbinger. It revealed how, in response to anthropogenic climate change, scholars in the humanities and social sciences had already been turning their attention to the cold (Piper 2009; Carey 2010; McCannon 2012; Martin-Nielsen 2013; Josephson 2014).

The attention encompassed permafrost, the strata of frozen earth beneath the surface of the planet in high-latitude and high-altitude regions. Textual records of perennially frozen earth date to the sixteenth century; systematic investigation into the phenomenon began during the nineteenth century; and the term *permafrost* appeared in the twentieth century as an English loan translation of the Russian expression, *vechnaia merzlota* (вечная мерзлота). The development of knowledge about frozen earth was accompanied by debates about its essence. As scholars within and beyond the natural sciences investigated the phenomenon, they have been drawn variously to its materiality and fluidity. While some have emphasized its behavior and impacts as a fixed object, others have focused on its instability as an idea and as an element of nature. The question of whether and how to engage with the indeterminacy of permafrost therefore runs as an underlying divide in histories of frozen earth.

This divide emerged not simply from the challenges to positivism posed by science studies in the second half of the twentieth century. Whereas positivists held that words, including scientific terms, corresponded to a single observable phenomenon, science studies scholars understood them to contain multiple meanings that depended on one's interpretive framework (Hess 1997). In the case of *permafrost*, however, the term's many referents, even from a positivist perspective, are suggested by its definition: "ground (soil or rock and included ice and organic material) that remains at or below 0 °C for at least two consecutive years" (van Everdingen 1998). *Permafrost* can refer to soil or rock, with or without ice and organic material; it may be frozen in the sense of hardened and held by ice or nonfrozen in the sense of containing liquid water; and it may have recent or prehistoric origins. *Permafrost* as an idea therefore overlaps imperfectly with encounters with frozen earth in nature. Moreover, the term and its definition do not capture ways of recognizing and interacting with frozen earth outside of European scientific tradition. The multidimensional quality of frozen earth arises both from the varied substances that constitute it and the distinct contexts in which knowledge about it was produced.

While providing an overview of histories of frozen earth, this chapter argues that historiography must attend to the historicity of *permafrost* as a scientific idea. This imperative has both practical and philosophical dimensions. Since the term *permafrost* first appeared in 1943, research into the historical record dating before the mid-twentieth century requires knowledge of other linguistic markers of frozen earth (Kabakchi and Butters 1989). Such markers may appear in not only European but also Uralic, Tungusic, Turkic, Inuit, and Yupik languages. They point to different facets of frozen earth, revealing the phenomenon's relationships to science as well as to the everyday life and worldviews of diverse cultures.

Histories of frozen earth must recover, interrogate, and perhaps help renew these relationships as part of resisting what climate scientist Mike Hulme has called climate reductionism. As Hulme argued, climate reductionism operates by isolating and elevating climate as the dominant factor determining human survival and flourishing on the planet. But doing so crowds out the unpredictability, dynamism, and contingencies of human agency (Hulme 2011). Moreover, by suggesting a direct, causal relationship between climate change and negative social effects, climate reductionism risks reinforcing what philosopher Kyle Whyte has called crisis epistemology. Crisis epistemology casts dangers such as climate change as unprecedented and urgent in hopes of spurring purposeful, decisive action. By portraying problems and their solutions as novel and certain, however, crisis epistemology justifies measures that may perpetuate historical harms, especially to Indigenous peoples. In contrast, Whyte promotes an epistemology of coordination based on Indigenous values of kinship and reciprocity that considers past legacies and cultivates care, cooperation, and repair (Whyte 2020). Such work, prosaic rather than heroic, is necessary for what feminist scholar Donna Haraway has called "staying with the trouble," or fostering creative interdependence and survival amid crisis (Haraway 2016). Histories that illuminate the many-sidedness and mutability of frozen earth, including and beyond the twentieth-century idea of *permafrost*, can contribute to this effort.

This chapter examines histories of frozen earth chronologically, using the development, consolidation, and popularization of permafrost science as a guiding organizational principle. It begins by discussing histories written before the establishment of permafrost science as a discipline. Such histories resembled literature reviews in which naturalists and explorers recounted earlier encounters with phenomena related to frozen earth as they advanced fresh ideas and observations of their own. In this stage, terms for frozen earth had not been standardized, and interest in it was connected to major questions of nineteenth-century geology, especially the theory of ice ages. The chapter then considers histories written during the period of consolidating permafrost science. These often aimed to explain the discipline's progression and relationships to other fields and to educate the broader public. In this stage, scientists also used history to bolster or contest the reputations of key individuals and institutions and to debate the discipline's goals and orientation. Finally, the chapter analyzes "cryo-histories" – to recall Sörlin's term – written amid rising global awareness of anthropogenic climate change and the concomitant thawing of frozen earth. If, previously, histories of frozen earth were written primarily by scientists, in this stage scholars in the humanities and social sciences became increasingly engaged as well. They have explored the meanings of frozen earth in a variety of cultural contexts, Siberian and North American, Indigenous and settler. The historiography sets the stage for broader and deeper exploration of how people have imagined and lived with the dynamic phenomenon of frozen earth on a changing planet.

2 Frozen Earth Before Permafrost Science

Frozen earth coalesced as an object of professional scientific interest over the course of the nineteenth century. In prior centuries, scattered mentions of frozen earth had appeared in the writings of explorers, naturalists, and colonial military governors. During the nineteenth century, European men of science sought to compile and synthesize these nuggets of information in monographs and treatises dedicated to frozen earth. On the one hand, the nineteenth-century texts constitute primary sources about frozen earth research. On the other, they represent one kind of historiography of knowledge about frozen earth. Given their position as both primary and secondary sources, they played an important role in defining and negotiating the parameters of frozen earth as a subject of study. For the first historians of frozen earth, writing history constituted part of scientific investigations. Curiosity about frozen earth was strongly connected to debates and discussions about the possibility of previous ice ages. As they gathered and analyzed the record of observations about frozen earth, they wrestled with the phenomenon's essence and definition. Their debates lent a range of meanings to the ontology of frozen earth, investing it with indeterminacy. This section does not provide a comprehensive survey of published sources about frozen earth prior to the establishment of permafrost science; rather, it highlights those that delved into the history of knowledge about the phenomenon.

Baltic German explorer Alexander von Middendorff is often credited as being the first scientist to provide an overview and detailed observations of frozen earth. He did so in a series of reports presenting the results of his expedition to northern and eastern Siberia from 1842–1845 (von Baer et al. 1848; von Middendorff 1867). However, Middendorff's studies were guided and informed by the work of Karl Ernst von Baer. Thus, the historiography of frozen earth arguably begins with Karl

Ernst von Baer's *Materials Toward Knowledge of the Imperishable Soil-Ice of Siberia* (*Materialien zur Kenntniss des unvergänglichen Boden-Eises in Sibirien*).

In the history of science, Baer is best known for definitively establishing the existence of eggs in mammals (Sarton 1931). But Baer had interests beyond embryology, including biogeography, meteorology, and geology (Tammiksaar 2000). In the late 1830s, he set out to collect all information about frozen earth available in European sources. These ranged from reports of English expeditions to Arctic North America in the sixteenth century to descriptions of northeastern Siberia by Imperial Russian explorers in the early nineteenth century. With these materials, Baer composed a treatise in which he recounted the history of observations of frozen earth and proposed an agenda for future investigations. He also provided an overview of the phenomenon as he understood it, including its appearances, conditions of existence, geographical distribution, and effects upon rivers and groundwater. Completed by 1843, the text ran to over 200 typewritten pages and represented the first written systematic study of frozen earth. Its bibliography serves as an invaluable guide to sources about frozen earth up to Baer's time. By historical accident, Baer's *Materials* remained unpublished until 2000 in Russian translation and 2001 in the original German (Ber 2000; von Baer 2001). The Russian version includes an appendix with Russian translations of some of the primarily German-language sources that Baer consulted. The German version, edited by historical geographer Erki Tammiksaar, provides a more comprehensive bibliography, but without the extracts.

Baer's monograph set a precedent for and exerted a hidden influence upon subsequent writings on frozen earth. Firstly, it reveals that connections between science and empire enabled the historiography of frozen earth. As a Baltic German, Baer belonged to the noble class of Estland, Livland, and Courland (present-day Estonia and Latvia), which had been colonized by Germans beginning in the thirteenth century. After the region was annexed by Russia in the eighteenth century, Baltic Germans became part of the Russian Empire's elite. Called "subcontractors of empire" by one historian, they notably occupied positions in state service, the military, and academia (Sunderland 2014). Baer's circle in Saint Petersburg, where he took a position as a member of the Academy of Sciences in 1834, included officers of the Imperial Navy of German origin. Navigators like Friedrich von Lütke and Ferdinand von Wrangell straddled the worlds of science and empire, their voyages advancing Russian Imperial claims while contributing to natural history and cartography (Daum 2019). The phenomenon of perennially frozen earth came to Baer's attention through Wrangell, who encountered it while surveying the northeastern coast of Siberia in the early 1820s. Subsequently, Baer organized an expedition to eastern Siberia – ultimately carried out by another Baltic German, Alexander von Middendorff – to investigate frozen earth in situ. The initiative was approved by the Imperial government due partly to the relevance of frozen earth to mining and agricultural colonization (Sukhova and Tammiksaar 2015). Baer relied on intelligence from previous expeditions to write his monograph, which had the goal of informing and guiding Middendorff's investigations. Activities tied to colonial expansion furnished inspiration and sources for Baer as well as those who built on his work.

Baer's monograph also posed key scientific puzzles related to frozen earth that were subsequently elaborated upon and debated by scholars of both earth and human history. Baltic Germans' transnational connections informed these intellectual concerns. Like others of his estate, Baer studied at German-speaking universities and published and corresponded in German, one of the three main scientific languages of nineteenth-century Europe (Whelan 1999; Gordin 2015). He studied at Würzburg and lived for 17 years in Prussia, where he became acquainted with the likes of Alexander von Humboldt and Louis Agassiz (Tammiksaar 2000). Plugged into an international network of scholars, Baer situated frozen earth in the context of new theories that were transforming understandings of geohistory. These included, foremost, the theory of ice ages advanced by Agassiz, which Baer connected to the enigma of extinct mammoths as well as frozen earth. Since at least the ninth century, mammoth tusks had been found and traded throughout Eurasia because of their ivory, a luxury commodity. By Baer's time, Georges Cuvier had posited that mammoths, parts of whose skeletons and even carcasses could be encountered in Siberia, were "a lost elephant species" – extinct (Cohen 2002; McKay 2017). Central to these debates was a nearly complete mammoth skeleton retrieved from Arctic Siberia by Imperial Russian naturalist Michael Friedrich Adams that had been found with skin and hair (Adams 1808; Woods 2023). What caused the mammoths' extinction, and how did their parts remain so well preserved? Given that elephants were known to inhabit hot climates in Africa and Asia, how did mammoths end up in cold Siberia?

The nature of frozen earth and the process of its formation constituted part of the puzzle of how mammoths were preserved and possibly transported. At issue was the precise situation at the mouth of the Lena River, along the Arctic coast of Siberia, where the mammoth retrieved by Adams had been located. Adams, relaying the testimony of Osip Shumakhov, the Tungus hunter who discovered the mammoth, suggested that the mammoth had been encased in ice (Tilesio 1815). Such a situation supported a prevailing theory, the drift hypothesis, that hefty objects such as boulders could be transported over long distances, carried in ice floes propelled by water (Krüger 2013). Perhaps mammoths had also been transported from warmer climes via such a mechanism. Drift hypothesis derived from diluvialism, a counterpoint to ice age theory that posited that the earth's features had been shaped by powerful floods.

On the other hand, subsequent reports by Arctic explorers suggested that mammoths had perished in place rather than being carried from afar. Members of an Imperial Russian expedition in 1815–1818, led by Baltic German navigator Otto von Kotzebue, found an abundance of mammoth tusks on the coast of Alaska. The tusks appeared to have fallen out of cliffs containing massive deposits of ice beneath a layer of soil. Kotzebue and Adelbert von Chamisso, a naturalist on the expedition, referred to the formations variously as ice-ground (*Eisgrund*), ice-rocks (*Eisfelsen*), and ice-mountains (*Gebirgsart [Eis]*, *Eisbergen*) (von Kotzebue 1821; von Chamisso 1821). To Baer, such masses of ice potentially served as evidence of the extensive glaciation posited by Agassiz's ice age theory. However, later explorers of the area – called Eschscholtz Bay after Johann Friedrich von Eschscholtz, another naturalist accompanying Kotzebue – contested the impression of buried glaciers of pure ice. After visiting Eschscholtz Bay in 1826, British explorer Frederick William Beechey observed cliffs consisting predominantly not of ice but of frozen mud and gravel (Beechey 1831). Eager for more detailed and reliable descriptions, Baer organized expeditions to Siberian sites of mammoth remains by not only Middendorff but also geologist Friedrich Karl Schmidt (von Baer 1866). They attested that the mammoths had been buried in frozen clay and sand rather than encased in pure ice. Their observations supported the view that the creatures once inhabited the far north, their carcasses having been preserved by the frozen earth (Schmidt 1869, 1872). The ice-rich frozen earth observed by the Kotzebue expedition, the mechanism of its formation, and its relationship to mammoths and the ice ages remained a source of scientific debate (Shur et al. 2022). By linking these together, Baer articulated a puzzle that persisted in the history and historiography of science.

Finally, Baer's *Materials* carries historiographical weight because it introduced terminology for frozen earth that was subsequently debated, appropriated, and obfuscated. Baer advanced two terms related to frozen earth: *Boden-Eis* and *Eis-Boden*. *Boden-Eis* referred to deposits of ice in the earth, ranging from soil or rock congealed by ice to the ice cliffs observed by the Kotzebue expedition. By *Eis-Boden*, however, Baer meant that portion of the earth's crust where the balance of heat exchange between air and land sustained the ongoing presence of *Boden-Eis*. Whereas *Boden-Eis* referred to the substance of frozen earth, which required the presence of ice, *Eis-Boden* referred to a space, anticipating the twentieth-century concept of cryosphere discussed below.

Because Baer's *Materials* remained unpublished, his distinction between *Boden-Eis* and *Eis-Boden* as well as his spatial understanding of *Eis-Boden* became forgotten. Middendorff briefly cited and excerpted Baer's *Materials* (von Middendorff 1875), and those excerpts were cited by another researcher of frozen earth, Baltic German explorer Eduard von Toll (von Toll 1895). Both writers used Baer's terms while altering their meanings. Middendorff used *Boden-Eis* to refer only to large, visible deposits of ice in the earth, which Toll proposed to call *Steineis* in lieu of *Boden-Eis*. As for *Eisboden*, Middendorff established its meaning as soil and rock with a negative temperature, even if ice was not present or visible, which Toll and subsequent writers followed. Instead of denoting a space, *Eisboden* became the substance of frozen earth, which subordinated the materiality of ice and the notion of a space of heat exchange. This shift led to a fixed rather than dynamic understanding of the phenomenon that became controversial and consequential in the next phases of the historiography of frozen earth.

3 Disciplinary Histories

While nineteenth-century histories of frozen earth revolved around the puzzle of ice age theory, during the twentieth century, economic development provided the dominant context for writings about frozen earth. As states pursued mineral extraction, agricultural settlement, and industrial construction in the Eurasian and North

American Arctic and sub-Arctic, frozen earth became perceived as an obstacle. The strategic importance of these projects to countries such as the Soviet Union and the USA prompted governments to nurture the institutionalization of a science of frozen earth. Organizations such as the USSR Academy of Sciences and the US Army Corps of Engineers created institutes and laboratories dedicated to the phenomenon (Obruchev 1947; Wright 1986). Simultaneously, scientific terminology for frozen earth became consolidated through publications and national and international conferences. The English term *permafrost* was coined from the Russian *vechnaia merzlota* in 1943, and the first International Conference on Permafrost took place in 1963 (French and Nelson 2008; Dietz 1963). All of these political and intellectual developments shaped scientific understandings of frozen earth into the twenty-first century.

The combination of state interest and the professionalization of science invested histories of frozen earth with political dimensions. These histories were written primarily by practitioners of the new discipline of permafrost science as they strived to define and celebrate their field, establish intellectual genealogies, and sustain funding. Partly what made their narratives political was their explicit or implicit promotion of the strategic economic and military imperatives that motivated state support of permafrost science. Because permafrost science both served and benefited from development projects, scientists and engineers did not interrogate their colonial implications, framing them rather as impetus for scientific progress (Sumgin and Demchinskii 1938; Legget 1954; Brown 1970). Some histories, however, intervened additionally in politics within the discipline of permafrost science itself (Meister and Saltykov 1958). As permafrost science was institutionalized, individuals set forth competing terminologies and frameworks for the discipline while advancing their careers. Their professional, intellectual, and sometimes personal disagreements were exacerbated by different understandings of the essence of frozen earth. Frozen earth's indeterminacy therefore persisted from the nineteenth to the twentieth centuries. Previously, as frozen earth emerged as a scientific object, its indeterminacy was revealed by the multiple names given to the phenomenon. Now, as a field of science was established around frozen earth, its indeterminacy was revealed by its contested definitions.

As the society that pioneered the science of frozen earth, the Soviet Union also witnessed the most politically charged debates about the phenomenon's essence. These debates centered on what to adopt as the most accurate scientific term for and definition of frozen earth, revealing disagreement about its ontology. Disputes about terms and definitions also carried broader implications for the goals and methods of the emerging discipline and its position in relation to other earth sciences. Should frozen earth be understood as a concrete body and aggregate structure whose behavior could be tested, measured, and modified for the purposes of stable engineering (Sumgin 1934)? Or should it be understood as a continually evolving condition or process responsive to changes in its surrounding environment (Tolstikhin 1935)? Furthermore, should frozen earth be defined and classified according to its temperature, material composition, age, or spatial location (Parkhomenko 1938; Sumgin 1938)?

Soviet scientists first debated these questions as they and the country as a whole were subjected to enormous pressure to transform an agricultural society into an industrial power. Industrialization and agricultural collectivization formed part of the program of building socialism under the dictatorship of Josef Stalin during the 1930s. They unfolded amid tensions with fascist regimes, human-made famines, and campaigns of repression in which people accused of betraying the party-state were arrested, sentenced to forced labor, and executed. Such a political environment distorted and raised the stakes of scientific debate. On the one hand, the political culture of the ruling Communist Party encouraged scientists to engage in critique and discussion (Kojevnikov 1998; Pollock 2006). On the other hand, it fostered intolerance of pluralism and eagerness to identify scapegoats for failures. Consequently, scientific discussions about frozen earth became enmeshed in accusations of sabotage and individuals' alleged lack of patriotism (Elenevskii 1937; Obruchev 1940).

The focal point of Soviet scientists' debate about terminology was the concept of *vechnaia merzlota*, the origin of the English term *permafrost*. *Vechnaia merzlota* was promulgated by Mikhail Sumgin, who published an influential monograph in 1927 that contained the expression in its title (Sumgin 1927). Sumgin spearheaded the USSR Academy of Sciences' research program on frozen earth, organizing conferences, editing a journal, and becoming recognized as the founder of the science of frozen earth (Obruchev and Chekotillo 1943). His promotion of *vechnaia merzlota* enabled the term to gain traction in scientific and popular discourse (Sumgin 1935).

But Sumgin did not invent the expression. Rather, *vechnaia merzlota* emerged from a longer history of practical engagement with frozen earth in the Russian Empire. The word *merzlota* first appeared in print in an 1838 issue of *Mining Journal* in connection with strata of frozen earth encountered while searching for gold deposits in Siberia ("Nabliudeniia" 1838). Likewise, *vechnaia merzlota* first appeared in a report published in 1873 by Russian geographer Petr Kropotkin about an expedition to survey a route to Siberian gold mines. While fulfilling an economically significant task, Kropotkin made geological observations of the territory that brought him into contact with frozen earth (Kropotkin and Poliakov 1873). Then, starting in the 1890s, the Russian Imperial government undertook construction of the Trans-Siberian railroad with a broader vision of bringing agricultural development to Siberia (Marks 1991). *Vechnaia merzlota* appeared increasingly in works of engineering (Mushketov 1895; Boev 1897; Matseevich 1907; Passek 1911; Bogdanov 1912; L'vov 1916) and soil science (Polynov 1910; Prasolov 1911; Sukachev 1911) as people sought to understand frozen earth for the sake of economic projects. Therefore, alongside research into ice age theory written primarily in German language, the Russian Empire also produced a literature on frozen earth in Russian language oriented to economic development. *Vechnaia merzlota* as a concept emerged from the latter trend, which persisted as the Russian Empire was replaced by the Soviet Union. By promulgating *vechnaia merzlota*, Sumgin served as a bridge across the revolutionary divide.

The problem lay in the fact that other scientists found both the term *vechnaia merzlota* and Sumgin's definition of it inaccurate and misleading. They argued that the adjective *vechnaia* – "eternal" – did not correspond to the fluidity, transience, and evolution of natural phenomena, especially on geological timescales (Baranov 1933; Tolstikhin 1935). They also objected to the ambiguity of the noun *merzlota*, whose multiple meanings rendered it imprecise. Depending on usage, critics pointed out, *merzlota* could refer to frozen soil, a substance; the condition of frozenness; or a process of freezing that encompassed the crystallization of water (Meister and Shvetsov 1955). Beyond the expression itself, Sumgin provoked controversy with his definition: soil or ground that maintained a negative temperature continuously for at least 2 years and up to tens of thousands of years (Sumgin 1937). Thanks to Sumgin's initiatives to establish a science of frozen earth, his definition was not only adopted in the USSR but also spread to the USA. It still forms the basis for the definition of permafrost today.

Despite its spread and adoption, however, Sumgin's definition was continually contested. From the 1930s through the 1950s, different Soviet scientists, including Ivan Baranov, Nestor Tolstikhin, Sergei Parkhomenko, and Petr Shvetsov, raised the same core objections. Firstly, Sumgin defined the substance of frozen earth as soil (*pochva*) or ground (*grunt*). But his critics argued that soil and ground belonged to different categories of concepts. Soil referred to specific earth materials, whereas ground referred generally to that part of the earth's surface that served as a foundation for construction, independent of its material content. Moreover, soil extended a few meters from the earth's surface, whereas frozen earth was known to extend to depths of over 100 m. Any definition of frozen earth must therefore include rock.

Secondly, Sugmin defined frozen earth according to its negative temperature. But his critics argued that whether or not something was frozen depended not on negative temperature alone. Rather, it depended on the presence of ice, the phase change of water from liquid to solid, which crucially changed the characteristics of the substance in question. In nature, the presence of minerals and impurities in soil and rock meant that water sometimes remained liquid even at some negative temperatures. Thus, even if soil or rock had a negative temperature, it could not be said to be frozen without the presence of ice.

Finally, Sumgin distinguished frozen earth according to the period of its existence, with 2 years as the minimum for characterizing "eternal" frozen earth. But his opponents argued that the length of time for which soil or rock remained frozen depended on other factors. These included its spatial location and material composition as well as its surrounding environment and climate, all of which affected the degree to which it absorbed and radiated heat. Definitions and classifications of frozen earth ought to be based on such primary, independent variables rather than a secondary, dependent variable such as its period of existence. Sumgin's critics also pointed to the incongruity of labeling frozen earth as "eternal" if it had existed for only 2 years (Baranov 1933; Tolstikhin 1935, 1941; Parkhomenko 1938; Shvetsov 1955).

In lieu of the misleading *vechnaia merzlota*, Sumgin's critics developed an alternative view of frozen earth that highlighted its dynamism and connection to the rest of the planet. They took inspiration from the concept of the cryosphere advanced by Polish scientist Antoni Dobrowolski in his 1923 work *The Natural*

History of Ice (Dobrowolski 1923; Zukriegel 1935). Dobrowolski drew attention to ice as an important yet underappreciated substance on the planet, ubiquitous and uniquely influential in both shaping the physical environment and conditioning earthly life. Ice could be found on land, as with glaciers; on water, as with sea ice; and in the air, as in the form of tiny particles. Compared to other solid substances, ice had a low melting point, well within the range of temperatures near the earth's surface. This property made it highly unstable, perpetually melting amid warmer temperatures and growing amid cooler temperatures. The instability of ice in turn affected the surrounding world, sculpting the terrain and creating or blocking passages across space on the macro- and micro-scales. Some changes had a seasonal rhythm while others took place across longer periods, depending on the latitude and altitude of ice. The spatial distribution of ice varied according to its location along the earth's axis: nearer to the poles it was found at and below the surface, and toward the equator it was found at high elevations. This space where ice existed and exerted influence constituted an envelope around the planet called the cryosphere.

According to this framework, frozen earth must be seen as not an isolated phenomenon but part of the cryosphere, the realm of ice. Frozen earth comprised the intersection of the cryosphere and the lithosphere, the earth's crust or outer shell; it might therefore be called the "cryolithozone." Soviet scientists tied the concepts of cryosphere and cryolithozone to a tradition of geographical thought originating in the Russian Empire. They traced the idea of an envelope of the earth characterized by ice and snow from Mikhail Lomonosov in the eighteenth century to Vladimir Vernadskii in the twentieth. Dobrowolski, a Polish scientist born in the Russian Empire, could be seen as belonging to this intellectual genealogy (Shvetsov and Dostovalov 1959).

Advocates of the cryosphere framework distinguished their physical-geographical view of frozen earth from the engineering and soil science views that gave rise to Sumgin's *vechnaia merzlota*. But although its origins differed from the geotechnical view, the cryosphere framework was also shaped by the needs of empire, specifically continental empires. As historian of science Deborah Coen has argued, an intellectual orientation toward thinking across scales emerged as a characteristic of earth science in nineteenth-century Eurasian continental empires (Coen 2018). Such an intellectual orientation was connected to two imperial needs: the need to identify and develop natural resources, and the need to tie diverse peoples and places into one political whole. Both goals required recognizing local diversity – whether in the social realm of language and religion or the physical realm of minerals and climate – within a broader unity. The cryosphere framework, which Soviet scientists traced to patterns of thought in the continental Russian Empire, similarly viewed diverse local manifestations of ice as part of a planetary envelope. Moreover, proponents of the cryosphere framework took interest in the technical challenges of operating on and working with ice and compounds of ice and earth. They also sought the unification and standardization of knowledge in central institutions, whether on the national or international level.

An analogous debate between advocates of *vechnaia merzlota* and advocates of a cryosphere framework played out in North America. Sumgin's concept of *vechnaia merzlota* was promulgated by Siemon Muller, a Russian-American geologist who wrote a foundational English-language monograph on frozen earth based on Russian-language literature. The monograph was commissioned during World War II by the US Army, which encountered frozen earth as an obstacle while building military infrastructure in Alaska and Canada. Muller drew extensively on the work of Sumgin and his colleagues in the USSR Academy of Sciences. He adopted Sumgin's idea of *vechnaia merzlota*, including its definition, while giving the Russian expression an English translation, *permafrost*, shorthand for "permanently frozen ground" (US Geological Survey 1943; Muller 1947).

Although the term was quickly accepted among military engineers, it provoked objections among scientists, notably geologist Kirk Bryan (Barnes 1946; Bryan 1946b). Bryan, like Sumgin's critics, noted the inaccuracy of suggesting permanence as a characteristic of frozen earth. Recognizing the dynamism of frozen earth, he advocated for terminology that more effectively accommodated the processes of its formation and evolution. To this end, frozen earth ought to be called *pergelisol*, from Latin roots *per* ("throughout"), *gelare* ("to freeze"), and *solum* ("soil"). From this name, the term *pergelation*, describing the process of its formation, could be derived. Situating frozen earth within the broader effects of the freezing and thawing of water and ice on the earth's surface, Bryan proposed to call its study "cryopedology" (Bryan 1946a). His view echoed the cryosphere framework and the idea of "geocryology" as the study of frozen earth developed in the USSR by Sumgin's critics (Meister and Saltykov 1958).

Despite objections to both, the terms *vechnaia merzlota* and *permafrost* stuck. Various circumstances prevented a fuller, transnational discussion of alternative scientific names for frozen earth. These included political obstacles to scientific exchange during the early Cold War, language barriers, and the primacy of geotechnical concerns over theoretical questions. Beginning in the 1960s, however, interest in the cryosphere framework was revitalized by growing research into climate science. Glaciologists became increasingly aware of the importance of ice in regulating the earth's climate system. They saw the need to understand its behavior in relation to not only the hydrosphere but also the atmosphere and lithosphere. Appreciating the planetary significance of ice drew attention to value of the cryosphere as an organizing concept (Jones 2008; Barry et al. 2011). The cryosphere framework received validation in 2008, when the International Association of Cryospheric Sciences was established within the International Union of Geodesy and Geophysics. In the twenty-first century, some scientists continued to aspire to integrate frozen earth into the cryosphere framework (Dobiński 2006, 2011).

4 Cryo-Histories

Histories of frozen earth entered a new phase in the twenty-first century. From investigating the ice ages and boosting Arctic industrialization, writings about frozen earth became concerned primarily with understanding anthropogenic climate change. In 2000, struck by humanity's impacts on the planet, scientists Paul Crutzen and Eugene Stoermer posited a distinct geological epoch defined by human activities that they called the Anthropocene. The Anthropocene was characterized by a massive expansion in humans' use of energy through the burning of fossil fuels (Crutzen and Stoermer 2013; Steffen et al. 2007, 2011). By elevating the concentration of greenhouse gases in the atmosphere, humans were causing average temperatures to rise, altering the Earth's climate system. As a result, rising sea levels, retreating glaciers, and reduced sea ice as well as heat waves, drought, and tropical storms were bringing harm to human and beyond-human life (Pörtner et al. 2022). The concept of the Anthropocene and the realities of global warming prompted scholars beyond the sciences to rethink humanity's relationship with the rest of nature (Chakrabarty 2009).

This rethinking included frozen earth, and it drew in humanists and social scientists, not only permafrost scientists themselves. As historians, anthropologists, human geographers, and political scientists investigated frozen earth, they foregrounded its human dimensions. Their engagement with frozen earth contributed to the emerging fields of Arctic social sciences and environmental humanities. On one hand, they explored the intellectual history of frozen earth, the ways that people have known, imagined, and made sense of the phenomenon. On the other hand, they explored cultural practices surrounding frozen earth, the ways that people have interacted and built livelihoods in connection with frozen earth. Rather than take for granted the progress of scientific knowledge or state-sponsored economic development, humanists and social scientists critically interrogated such processes. They were especially attuned to the colonial aspects of the history of science and modernization. When it came to frozen earth itself, they emphasized its pluralism and dynamism. They uncovered multiple essences of frozen earth that stemmed from diverse understandings of the phenomenon. They also emphasized its fluidity rather than its eternity or permanence. Such an emphasis coincided with rising global awareness of the thawing of frozen earth due to climate change (Schiermeier 2001; Blake 2010; Gillis 2011). In highlighting the variability and heterogeneity of frozen earth, humanists and social scientists extended the framework of cryology that Dobrowolski advocated and that Sumgin's critics embraced. Their cryohistories approached frozen earth as situated and three dimensional, focusing on local and planetary perspectives while critiquing the territorial gaze of the industrial project of conquering nature.

New histories of frozen earth foregrounded the political and social context in which knowledge about the phenomenon developed. They highlighted the historical processes of resource extraction, infrastructure construction, and military confrontation that motivated people's encounters with and investigations of frozen earth. In North America, these included the Klondike, Nome, and Fairbanks Gold Rushes and the building of the Canadian-Alaskan Military Highway during World War II (Nelson 2011; Cysewski 2013). They also included the creation of the town of Inuvik for supporting oil and gas development in the Mackenzie Delta, which transformed the lands and livelihoods of Inuit peoples (Stuhl 2016). In Eurasia, scholars emphasized the importance of the Gulag, the Soviet forced labor system, as a powerful sponsor of science and engineering (Khlevnyuk 2003; Siddiqi 2015). Administered by the Soviet interior ministry, which included the secret police, the

Gulag functioned as an instrument of not only political repression but also economic development. Although the Gulag operated throughout the USSR, its largest enterprises were located in the country's far northern and eastern regions, including ones underlain by frozen earth. In these lands, prisoners built railroads, mined coal and gold, and felled trees. To support these projects, knowledge about engineering in frozen earth environments was needed. A key site for researching frozen earth was Vorkuta, the center of a labor camp for mining coal in the Pechora river basin in the far north (Markova and Rodnyi 1998; Barenberg 2014). The science of frozen earth, scholars demonstrated, was intertwined with colonialism and state violence.

While exploring the external, social factors that have shaped frozen earth research, scholars shined fresh light on the phenomenon itself. A history of concepts approach revealed that *permafrost* constituted only one of multiple ways of understanding frozen earth and recovered other terms, including ones discussed in this chapter (Chu 2020). The different names given to frozen earth and changing motivations for studying it revealed the phenomenon as having many dimensions rather than a single essence. They suggested the need to pluralize rather than flatten knowledge about frozen earth and to incorporate local interactions and perspectives. As Arctic social scientists investigated local practices, they emphasized the relational, rather than fixed, materiality of frozen earth. The materiality of frozen earth depended on myriad local environmental variations, including human behaviors. Human activities and choices, including travel, settlement, construction, and design, altered the thermal conditions that influenced the formation and disappearance of frozen earth (Cho 2021). The materiality of frozen earth also depended on human perceptions of time and the rhythms of human lives. Taking temporality into consideration, frozen earth constituted a solid fluid, solid over shorter time scales but fluid over longer ones (Krause 2022). Social science scholarship therefore echoed the cryosphere framework's emphasis on the dynamism of frozen earth. But rather than focusing solely on physical substances and systems, it integrated human societies into analyses of nature.

Anthropologists have argued that appreciating the inherent dynamism of frozen earth requires recognizing alternative, non-Eurocentric ways of knowing. They have explored how communities that have lived for centuries in regions with frozen earth built livelihoods connected to the inherent variability of the phenomenon. Such communities included the Sakha and the Eveny of northeastern Siberia. Sakha historically herded cattle and cultivated hay, moving between summer pastures and winter homes. Eveny domesticated reindeer, historically as transport for nomadic hunting and more recently, since Soviet times, for producing meat (Vitebsky 2005). Both communities long understood the fluidity of frozen earth, depending on it for their sustenance and developing worldviews to adapt to it.

Rather than *permafrost*, for example, Sakha spoke of *alaas*, which recognized the impermanence of frozen earth. *Alaas* formed when ice-rich frozen earth thawed, creating a depression that filled with water to become a lake. The lake then dried or drained, and in its wake appeared grassland that Sakha used as pasture for their herds and meadow for haymaking. Sakha not only observed patterns in the changing landscape, but also actively encouraged them: to encourage the formation of *alaas*, especially its grassland, they selectively burned vegetation to thaw the underlying frozen earth and dug canals to drain the lake. They understood *alaas* to be a source of life and cultivated a spiritual relationship with the land that encompassed rituals of gratitude and reciprocity (Takakura 2010; Crate 2021).

Eveny also viewed changeability as a characteristic feature of the frozen earth landscape with both benefits and disadvantages. When frozen earth thawed, water from its melted ice replenished the rivers and streams that mobile Eveny used to navigate the land. Yet frozen earth could also block and divert the passage of water, leading to it unexpectedly erupting, flooding, and freezing over large areas and obstructing movement. Eveny observed that the lakes that formed from the thawing of ice-rich frozen earth shifted with strong winds, creating an uneven landscape of shallow depressions. The depressions resulted from the fact that, as pools of water moved, the energy contained in them promoted further thawing of underlying frozen earth. Although moving lakes contributed to further thawing, their movement enabled frozen earth to re-form in the places they vacated. The transitory quality of frozen earth supported the animist worldview of the Eveny in which nonhuman nature is also capable of sentience. For Eveny, the land constituted part of a web of interaction and dependence, continually responding to changes and triggering further changes in the environment. Survival meant accepting and adapting to change rather than expecting permanence (Ulturgasheva 2022).

Attention to the inherent variability of frozen earth has contributed to critiques of the concept of the Anthropocene. As geologists debated the stratigraphic requirements for officially recognizing the Anthropocene as a new epoch in earth history, others have debated the concept's political implications. For scholars in the humanities and social sciences, the Anthropocene concept valuably highlighted the enormity of human impacts on the planet, including but not limited to global warming. But it also obscured the political and social processes – especially capitalism, colonialism, and racism – that structured the choices and behaviors that led to those impacts (Haraway 2015; Barry and Maslin 2016; Mirzoeff 2018; Yusoff 2018). The Anthropocene concept, scholars argued, collapsed the human species into a universal mankind, homogenizing the unequal harms that have been done as well as the unequal responsibilities. By doing so, it encouraged reliance upon the same processes to solve the problems they created. From the point of view of this critical perspective, frozen earth, too, had been absorbed into a universalist, homogenizing pattern of thinking. Such thinking portrayed the thawing of frozen earth as novel, irreversible, comprehensive, and catastrophic, requiring salvation in the form of bold technological interventions. In actuality, the formation and disappearance of frozen earth were characterized by heterogeneity and discontinuity (Wrigley 2023). The discontinuities of frozen earth created dangers but also opened possibilities for survival rather than salvation.

5 Conclusion

To conclude this tour through the historiography of frozen earth, we might highlight three key claims. Firstly, human engagements with frozen earth invested the phenomenon with multiple essences and multiple names, of which *permafrost* was but one.

Nineteenth-century terms like *Boden-Eis*, *Eisboden*, *Eisfelsen*, and *Steineis* variously highlighted the substances of ice, soil, and rock that constituted frozen earth. Twentieth-century terms like "cryophilic rocks" and "cryolithozone" drew on the framework of the cryosphere advanced by Dobrowolski. Such a framework recognized dynamism as a characteristic feature of ice resulting from its constant exchanges of heat and energy with its surroundings. This embrace of the changeability of ice differed from the emphasis on stability suggested by *permafrost* and its definition. But specific historical circumstances enabled the idea of *permafrost* as ground that maintains a negative temperature for at least 2 years to gain acceptance. The historiography of frozen earth reveals both how *permafrost* became the dominant way of understanding frozen earth and how alternatives emerged, faded, and persisted.

Secondly, the historiography of frozen earth was shaped by colonialism and its legacies. Interest in the phenomenon was nurtured by connections between science and empire, including both external and internal colonization. States like the Russian Empire sponsored expeditions to Arctic and sub-Arctic regions in hopes of discovering and exploiting natural resources. Such expeditions generated incidental observations and descriptions of frozen earth. As exploration segued into colonization, industrialization, and militarization of peripheries, states sponsored systematic research into frozen earth. Empire- and state-building ambitions influenced ideas about and approaches to studying frozen earth, but their influence, like colonialism itself, had different facets. On one hand, they encouraged a utilitarian approach to frozen earth for the sake of ensuring stable engineering. This approach consolidated a view of frozen earth as an aggregate structure, ground. On the other hand, the process of expanding and asserting control over diverse and far-flung places gave rise to a "planetary consciousness" (Pratt 1992). It cultivated a habit of seeing connections in nature across geographical space and between local, regional, and global scales (Wulf 2015). This perspective nurtured a view of frozen earth as part of a cohesive planetary envelope defined by the presence and activity of ice. Although the utilitarian and earth systems approaches generated distinct views of frozen earth, both can trace their development to processes of colonialism.

The coloniality of the historiography of frozen earth sets up a final claim: it is necessary to recognize and critique how ideas about frozen earth have been shaped by colonialism and to recover and make room for other forms of understanding. Rather than expecting to escape from colonial legacies or purge them from the science of frozen earth, the goal is to foster critical awareness and distance. Such awareness and distance must extend to the term *permafrost* itself, which, despite its flaws, has persisted. Interrogating its implications and considering alternatives can clarify the interests, assumptions, and blind spots that people bring to frozen earth. Scholars must resist allowing a term and definition promulgated in the particular contexts of Soviet industrialization and US militarization to narrow the historiography of frozen earth. Historians, anthropologists, and geographers have begun to elucidate the multiple ontologies that humans have created for frozen earth as well as the heterogeneity and dynamism of the phenomenon itself. Exploring the richness of local interactions with frozen earth, past and present, may uncover narratives and metaphors that point to ways of living with frozen earth beyond aspirations of conquest and control.

References

Adams, Michael. 1808. Some account of a journey to the frozen-sea, and of the discovery of the remains of a mammoth. Trans. Benjamin Smith Barton. *The Philadelphia Medical and Physical Journal* 3: 120–137.

Baranov, I.Ia. 1933. Predvaritel'nye zamechaniia po voprosu o terminakh v nauke o merzlote. In *Na geologicheskom fronte Vostochnoi Sibiri*, 98–105. Moskva/Irkutsk: OGIZ.

Barenberg, Alan. 2014. *Gulag town, company town: Forced labor and its legacy in Vorkuta*. New Haven: Yale University Press.

Barnes, Lynn C. 1946. Permafrost: A challenge to engineers. *The Military Engineer* XXXVIII (243): 9–11.

Barry, Andrew, and Mark Maslin. 2016. The politics of the Anthropocene: A dialogue. *Geo: Geography and Environment* 3 (2): e00022.

Barry, R.G., J. Jania, and K. Birkenmajer. 2011. A.B. Dobrowolski, the first cryospheric scientist, and the subsequent development of cryospheric science. *History of Geo- and Space Science* 2: 75–79.

Beechey, Frederick William. 1831. *Narrative of a voyage to the Pacific and Beering's strait*. Vol. 1. 2 vols. London: Henry Colburn and Richard Bentley.

Ber, K.M. 2000. In *Materialy k poznaniiu netaiushchego pochvennogo l'da v Sibiri*, Sankt-Peterburg, 1842 g, ed. R.M. Kamenskii. Iakutsk: Izdatel'stvo Instituta merzlotovedeniia SO RAN.

Blake, Heidi. 2010. Climate change could be accelerated by 'methane time bomb.' *The Telegraph*, February 22. https://www.telegraph.co.uk/news/earth/environment/climatechange/7289698/Climate-change-could-be-accelerated-by-methane-time-bomb.html

Boev, Nikolai Nikolaevich. 1897. Izsliedovanie voprosa o vechnoi merzlote pochvy. *Biulleteni politekhnicheskago obshchestva, sostoiashchago pri Imperatorskom Tekhnicheskom Uchilishche* 6 (1–2): 14–25.

Bogdanov, N.S. 1912. *Vechnaia merzlota i sooruzheniia na nei*. Vol. LXXXIII, *Vysochaishe uchrezhdeniia, Osobaia Vyschaia Komissiia dlia vsestoronnogo issledovaniia zheleznodorozhnogo dela v Rossii*. St. Petersburg: Tipografiia t-va "Obshchestvennaia Pol'za".

Brown, Roger J.E. 1970. *Permafrost in Canada: Its influence on northern development*. Toronto: University of Toronto Press.

Bryan, Kirk. 1946a. Cryopedology—The study of frozen ground and intensive frost-action with suggestions on nomenclature. *American Journal of Science* 244: 622–642.

———. 1946b. Permanently frozen ground. *The Military Engineer* XXXVIII (246): 168.

Carey, Mark. 2010. *In the shadow of melting glaciers: Climate change and Andean Society*. New York: Oxford University Press.

Chakrabarty, Dipesh. Winter 2009. The climate of history: Four theses. *Critical Inquiry* 35 (2): 197–222.

Cho, Leena. 2021. Permafrost politics: Toward a relational materiality and design of Arctic ground. *Landscape Research* 46 (1): 25–35.

Chu, Pey-Yi. 2020. *The life of permafrost: A history of frozen earth in Russian and soviet science*. Toronto: University of Toronto Press.

Coen, Deborah. 2018. *Climate in motion: Science, empire, and the problem of scale*. Chicago: The University of Chicago Press.

Cohen, Claudine. 2002. *The fate of the mammoth: Fossils, myth, and history*. Trans. William Rodarmor. Chicago: The University of Chicago Press.

Crate, Susan. 2021. *Once upon the permafrost: Knowing culture and climate change in Siberia*. Tucson: The University of Arizona Press.

Crutzen, Paul J., and Eugene F. Stoermer. 2013. The 'Anthropocene'. In *The future of nature: Documents of global change*, 483–485. New Haven: Yale University Press.

Cysewski, Margaret Hope. 2013. *Initial permafrost engineering research in Alaska*. M.S., University of Alaska Fairbanks, Fairbanks.

Daum, Andreas W. 2019. German naturalists in the Pacific around 1800: Entanglement, autonomy, and a transnational culture of expertise. In *Explorations and entanglements: Germans in Pacific Worlds from the early modern period to World War I*, 79–102. New York: Berghahn Books.
Dietz, Albert G.H. 1963. Forward. In *Proceedings: Permafrost international conference, 11–15 November 1963 in Lafayette, Indiana*. Washington, DC: National Academy of Sciences and National Research Council.
Dobiński, Wojciech. 2006. Ice and environment: A terminological discussion. *Earth-Science Reviews* 79: 229–240.
———. 2011. Permafrost. *Earth-Science Reviews* 108: 158–169.
Dobrowolski, Antoni Bolesław. 1923. *Historja Naturalna Lodu z 340 rysinami w tekście*. Warszawa: Wydawnistwo Kasy Pomosy dla Osób Prasujących na Polu Naukowem Imienia D-ra J. Mianowskiego.
Elenevskii, V. 1937. Problemy stroitel'stva v raionakh merzloty. *Sotsialisticheskii transport* 8–9: 81–91.
French, Hugh, and Frederick E. Nelson. 2008. Introduction. In *Frozen in time: Permafrost and engineering problems*, ix–xxiv. Reston: American Society of Civil Engineers.
Gillis, Justin. 2011. Warming Arctic permafrost fuels climate change worries. *The New York Times*, December 16. http://www.nytimes.com/2011/12/17/science/earth/warming-arctic-permafrost-fuels-climate-change-worries.html
Gordin, Michael D. 2015. *Scientific babel: How science was done before and after global English*. Chicago: The University of Chicago Press.
Haraway, Donna. 2015. Anthropocene, capitalocene, plantationocene, chthulucene: Making kin. *Environmental Humanities* 6: 159–165.
———. 2016. *Staying with the trouble: Making kin in the chthulucene*. Durham: Duke University Press.
Hess, David. 1997. *Science studies: An advanced introduction*. New York: New York University Press.
Hulme, Mike. 2011. Reducing the future to climate: A story of climate determinism and reductionism. *Klima* 26 (1): 245–266.
Jones, H. Gerald. 2008. From commission to association: The transition of the International Commission on Snow and Ice (ICSI) to the International Association of Cryospheric Sciences (IACS). *Annals of Glaciology* 48: 1–5.
Josephson, Paul. 2014. *The conquest of the Russian Arctic*. Cambridge: Harvard University Press.
Kabakchi, Victor V., and Ronald R. Butters. 1989. Are permafrost and vernalization loan translations from Russian? *American Speech* 64 (3): 287–288.
Khlevnyuk, Oleg. 2003. The economy of the OGPU, NKVD, and MVD of the USSR, 1930–1953: The scale, structure, and trends of development. In *The economics of forced labor: The soviet gulag*, 43–66. Stanford: Hoover Institution Press.
Kojevnikov, Alexei. 1998. Rituals of Stalinist culture at work: Science and the games of intraparty democracy circa 1948. *Russian Review* 57 (1): 25–52.
Krause, Franz. 2022. The tempo of solid fluids: On river ice, permafrost, and other melting matter in the Mackenzie Delta. *Theory, Culture and Society* 39 (2): 31–52.
Kropotkin, P., and I.S. Poliakov. 1873. *Otchet ob Olekminsko-Vitimskoi ekspeditsii dlia otyskaniia skotoprogonnogo puti iz Nerchinskogo okruga v Olekminskii, snariazhennoi v 1866 godu pri sodestvii Sibirskogo otdela Geograficheskogo obshchestva*. Vol. 3, *Zapiski Imperatorskogo russkogo geograficheskogo obshchestva po obshchei geografii*. St. Petersburg: Tipografiia V. Bezobrazova i komp.
Krüger, Tobias. 2013. *Discovering the ice ages: International reception and consequences for a historical understanding of climate*. Trans. Ann M. Hentschel. Leiden: Brill.
L'vov, A.V. 1916. *Poiski i ispytaniia vodoistochnikov vodosnabzheniia na Zapadnoi chasti Amurskoi zhel. dor. v usloviiakh "vechnoi" merzloty pochvy (Letnii i zimnii rezhim rek, gruntovykh vod i usloviia pitaniia glubokikh vodonosnykh tolshch v raionakh sploshnogo rasprostraneniia "vechnoi" merzloty)*. Irkutsk: Tipo-litografiia P.I. Makushina i V.M. Posokhina.

Legget, R.F. 1954. Permafrost research. *Arctic* 153 (3 and 4): 153–158.
Markova, E.V., and A.N. Rodnyi. 1998. Nauka Vorkutlaga kak fenomen totalitarnogo gosudarstva. *Vestnik Instituta estestvoznaniia i tekhniki* 3: 60–77.
Marks, Steven G. 1991. *Road to power: The trans-Siberian railroad and the colonization of Asian Russia, 1850–1917*. Ithaca: Cornell University Press.
Martin-Nielsen, Janet. 2013. *Eismitte in the scientific imagination: Knowledge and politics at the center of Greenland*. New York: Palgrave Macmillan.
Matseevich, V. 1907. Sluchai vozvedeniia sooruzhenii na vehnoi merzlote. *Izvestiia obshchestva grazhdanskikh inzhenerov* 13 (1): 4–18.
McCannon, John. 2012. *A history of the Arctic: Nature, exploration and exploitation*. London: Reaktion Books.
McKay, John J. 2017. *Discovering the mammoth*. New York: Pegasus Books.
Meister, L.A., and N.I. Saltykov. 1958. *K istorii geokriologicheskikh issledovanii v SSSR*. Syktyvkar: Komi knizhnoe izdatel'stvo.
Meister, L.A., and P.F. Shvetsov. 1955. O nekotorykh terminakh v uchenii o zonakh merzlykh pochv i gornykh porod i ego meste sredi drugikh nauk. *Izvestiia Akademii nauk SSSR, seriia geograficheskaia* 1: 69–73.
Mirzoeff, Nicholas. 2018. It's not the Anthropocene, It's the white supremacy scene; or, the geological color line. In *After extinction*, 123–149. Minneapolis: University of Minnesota Press.
Muller, Siemon William. 1947. *Permafrost or permanently frozen ground and related engineering problems*. Ann Arbor: J.W. Edward, Inc.
Mushketov, I.V., ed. 1895. *Instruktsiia dlia izucheniia merzloty pochvy v Sibiri, sostavlena komissiei imperatorskogo russkogo geograficheskogo obshchestva*. Sanktpeterburg: Tipografiia Imperatorskoi Akademii Nauk.
"Nabliudeniia nad merzlotoi zemli v nerchinskom okruge 1836 goda, v mestakh naibolee neobitaemykh, pri rozyske zolotonosnykh rossypei issledovannkyh." *Gornyi zhurnal* Chast' II, no. Knizhka II. 1838. 115.
Nelson, Frederick E. 2011. 'America's glory road'…on ice: Permafrost and the development of the Alcan highway, 1942–1943. In *Engineering earth: The impacts of megaengineering projects*, 643–661. Dordrecht: Springer.
Obruchev, V.A. 1940. Predislovie. *Trudy komiteta po vechnoi merzlote* 9: 3–4.
———. 1947. Uspekhi merzlotovedeniia v SSSR. In *Iubileinyi sbornik: Posviashchennyi tridtsatiletiiu Velikoi Oktiabrskoi sotsialisticheskoi revoliutsii*, vol. 2, 217–237. Moskva: Izdatel'stvo Akademii Nauk SSSR.
Obruchev, V.A., and A.M. Chekotillo. 1943. Pamiati M.I. Sumgina. *Izvestiia Akademii Nauk SSSR, seriia geologicheskaia* 2: 28–30.
Parkhomenko, S.G. 1938. Merzlotovedenie kak uchenie o kriofil'nykh gornykh porodakh. *Trudy komiteta po vechnoi merzlote* 6: 177–194.
Pąssek, A.N. 1911. Mestnye usloviia, klimat i vechno-merzlye grunty Golovnogo Uchastka Zapadno Amurskoi zhel. dor. *Izvestiia sobraniia inzhenerov putei soobshcheniia* 3: 1–5.
Piper, Liza. 2009. *The industrial transformation of subarctic Canada*. Vancouver: UBC Press.
Pollock, Ethan. 2006. *Stalin and the soviet science wars*. Princeton: Princeton University Press.
Polynov, B. 1910. O 'vechnoi' merzlote i o formakh l'da i snega, perezhivaiushchikh leto, v Amurskoi oblasti. *Zemlevedenie* 17 (knizhka 3): 35–48.
Pörtner, H.-O., D.C. Roberts, E.S. Poloczanska, K. Mintenbeck, M. Tignor, A. Alegría, M. Craig, et al., eds. 2022. IPCC, 2022: Summary for policymakers. In *Climate change 2022: Impacts, adaptation, and vulnerability,* Contribution of working group II to the sixth assessment report of the intergovernmental panel on climate change, 3–33. Cambridge: Cambridge University Press.
Prasolov, L. 1911. O 'vechnoi' merzlote v stepnoi polose Zabaikal'ia. *Pochvovedenie* 13 (4): 3–57.
Pratt, Mary Louise. 1992. *Imperial eyes: Travel writing and transculturation*. London: Routledge.
Sarton, George. 1931. The discovery of the mammalian egg and the foundation of modern embryology, with a complete facsimile (No. XIII) of K.E. von Baer's fundamental memoir: De Ovi Mammalium et hominis Genesi (Leipzig, 1827). *Isis* 16 (2): 315–377.

Schiermeier, Quirin. 2001. Fears grow over melting permafrost. *Nature* 409 (6822): 751.

Schmidt, Friedrich. 1869. Vorläufige Mittheilungen Über Die Wissenschaftlichen Resultate Der Expedition Zur Aufsuchung Eines Angekündigten Mammuthcadavers, von Magister Fr. Schmidt (Lu Le 16 Avril 1868). *Bulletin de l'Académie Impériale Des Sciences de St.-Pétersbourg* 13: 97–130.

———. 1872. Wissenschaftliche Resultate der zur Aufsuschung eines angekündigten Mammuthcadavers von der Kaiserlichen Akademie der Wissenschaften an den Unteren Jenissei ausgesandten Expedition. *Mémoires de l'Académie Impériale des Sciences de St.-Pétersbourg* VIIe série, 18 (1): 1–168.

Shur, Yuri, Daniel Fortier, M. Torre Jorgenson, Mikhail Kanevskiy, Lutz Schirrmeister, Jens Strauss, Alexander Vasiliev, and Melissa Ward Jones. 2022. Yedoma permafrost genesis: Over 150 years of mystery and controversy. *Frontiers in Earth Science* 9: Article 757891.

Shvetsov, P.F. 1955. *Vvodnye glavy k osnovam geokriologii*. Vol. 1, *Materialy k osnovam ucheniia o merzlykh zonakh zemnoi kory*. Moskva: Izdatel'stvo Akademii Nauk SSSR.

Shvetsov, P.F., and B.N. Dostovalov, eds. 1959. *Osnovy geokriologii (merzlotovedeniia), chast' pervaia: obshchaia geokriologiia*. Moskva: Izdatel'stvo Akademii Nauk SSSR.

Siddiqi, Asif. 2015. Scientists and specialists in the Gulag: Life and death in Stalin's Sharashka. *Kritika: Explorations in Russian and Eurasian History* 16 (3): 557–588.

Sörlin, Sverker. 2015. Cryo-history: Narratives of ice and the emerging Arctic humanities. In *The new Arctic*, 327–340. Cham: Springer.

Steffen, Will, Paul J. Crutzen, and John R. McNeill. 2007. The Anthropocene: Are humans now overwhelming the great forces of nature? *Ambio: A Journal of the Human Environment* 36 (8): 614–621.

Steffen, Will, Jacques Grinevald, Paul Crutzen, and John McNeill. 2011. The Anthropocene: Conceptual and historical perspectives. *Philosophical Transactions of the Royal Society A* 369: 842–867.

Stuhl, Andrew. 2016. *Unfreezing the Arctic: Science, colonialism, and the transformation of Inuit lands*. Chicago: The University of Chicago Press.

Sukachev, V.N. 1911. K voprosu o vliianii merzloty na pochvu. *Izviestiia imperatorskoi Akademii Nauk* 5 (1): 51–60.

Sukhova, N.G., and E. Tammiksaar. 2015. *Aleksandr Fedorovich Middendorf: K dvukhsotletiiu so dnia rozhdeniia, izdanie vtoroe, pererabotannoe i dopolnennoe*. Sankt-Peterburg: Nestor-Istoriia.

Sumgin, M.I. 1927. *Vechnaia merzlota pochvy v predelakh SSSR*. 1st ed. Vladivostok: N.K.Z. Dal'ne vostochnaia Geofizicheskaia Observatoriia.

———. 1934. *Vechnaia merzlota*. Leningrad: Izd-vo AN SSSR.

———. 1935. Sovetskie oblasti vechnoi merzloty. *Pravda*, July 26.

———. 1937. *Vechnaia merzlota pochvy v predelakh SSSR*. 2nd ed. Moskva/Leningrad: Izdatel'stvo Akademii Nauk SSSR.

———. 1938. Neskol'ko zamechanii po povodu stat'i S.G. Parkhomenko, 'Merzlotovedenie kak uchenie o kriofil'nykh porodakh.'. *Trudy komiteta po vechnoi merzlote* 6: 195–201.

Sumgin, M.I., and B.N. Demchinskii. 1938. *Zavoevanie severa (v oblasti vechnoi merzloty)*. Moskva/Leningrad: Izd-vo Akademii nauk SSSR.

Sunderland, Willard. 2014. *The Baron's cloak: A history of the Russian empire in war and revolution*. Ithaca: Cornell University Press.

Takakura, Hiroki. 2010. The social and symbolic construction of Alaas landscapes in the Siberian Forest among the Sakha. *Acta Slavica Iaponica* 28: 51–69.

Tammiksaar, Erki. 2000. *Geograficheskie aspekty tvorchestva Karla Bera v 1830–1840 gg*. University of Tartu.

Tilesio. 1815. De Skeleto Mammonteo Sibirico Ad Maris Glacialis Littora Anno 1897, Effosso, Cui Praemissaie Elephantini Generis Specie Rum Distinctiones. *Mémoires de l'Académie Impériale Des Sciences de St.-Pétersbourg* V: 406–513.

Tolstikhin, N.I. 1935. Vechnaia merzlota ili merzlaia zona zemnoi kory. *Problemy Sovetskoi geologii* 5 (8): 765–769.

———. 1941. *Podzemnye vody merzloi zony litosfery.* Moskva/Leningrad: Gosudarstvennoe izdatel'stvo geologicheskoi literatury Komiteta po delam geologii pri SNK SSSR.
U.S. Geological Survey. 1943. *Permafrost or permanently frozen ground and related engineering problems*. Vol. 62, *Strategic engineering study.* Intelligence Branch Office, Chief of Engineers.
Ulturgasheva, Olga. 2022. 'Tears of the earth': Human-permafrost entanglements and science indigenous knowledge encounters in Northeast Siberia. In *Risky futures: Climate, geopolitics and local realities in the uncertain circumpolar north*, 58–88. New York: Bergahn Books.
van Everdingen, Robert O., ed. 1998. *Multi-language glossary of permafrost and related ground-ice terms*. Revised 2005. Ottawa: International Permafrost Association.
Vitebsky, Piers. 2005. *Reindeer people: Living with animals and spirits in Siberia*. London: HarperCollinsPublishers.
von Baer, K.E. 1866. Neue Auffindung eines vollständigen Mammuths, mit der Haut und den Weichtheilen, im Eisboden Sibiriens, in der Nähe der Bucht des Tas (Tazovskaia guba), von K.E. von Baer (Lu le 8 février 1866). *Bulletin de l'Académie Impériale des Sciences de St. Pétersbourg* 10: 230–296.
———. 2001. *Materialien zur Kenntniss des unvergänglichen Boden-Eises in Siberien: Unveröffentlichtes Typoskript von 1843 und der erste Dauerfrostbodenkunde*, ed. Erki Tammiksaar. Giessen: Universitätsbibliothek.
von Baer, K.E., H.R. Goeppert, Gr. von Helmersen, Al. Graf Keyserling, E. Lenz, A. Th. Middendorff, W. von Middendorff, Johannes Mueller, and Chr. Peters. 1848. *Dr. A. Th. v. Middendorff's Reise in den Äussersten Norden und Osten Sibiriens*. Band I, Theil I, *Einleitung, Meteorologische, Geothermische, Magnetische und Geognostische Beobachtungen, Fossile Hölzer, Mollusken und Fische*. St. Petersburg: Buchdruckerei der Kaiserlichen Akademie der Wissenshaften.
von Chamisso, Adelbert. 1821. *Bemerkungen und Ansichten*. Vol. 3, *Entdeckungs-Reise unternommen in den Jahren 1815–1818 auf kosten Sr. Erlaucht des Herrn Reichs-kanzlers Grafen Romanzoff auf dem Schiffe Rurick, unter dem Befehle des Lieutenants der Russisch-kaiserlichen Marine Otto von Kotzebue*. Weimar: Gebrüder Hoffmann.
von Kotzebue, Otto. 1821. *Entdeckungs-Reise in die Süd-See und nach der Berings-Strasse zur Erforschung einer nordöstlichen Durchfahrt, unternommen in den Jahren 1815, 1816, 1817, und 1818, auf kosten Sr. Erlaucht des herrn Reichs-kanzlers Grafen Romanzoff auf dem Schiffe Rurick, unter dem Befehle des Lieutenants der Russisch-kaiserlichen Marine Otto von Kotzebue, Erster Band*. Vol. 1. Weimar: Gebrüder Hoffmann.
von Middendorff, A.Th. 1867. *Dr. A. Th. v. Middendorff's Reise in den Äussersten Norden und Osten Sibiriens*. Band 4, Theil 1, *Einleitung, Geographie, Hydrographie, Orographie, Geognosie, Klima und Gewächse*. St. Petersburg: Buchdruckerei der Kaiserlichen Akademie der Wissenshaften.
———. *Dr. A. Th. v. Middendorff's Reise in den Äussersten Norden und Osten Sibiriens*. Band 4 Theil 2, *Die Thierwelt Sibiriens, Die Eingeborenen Sibiriens*. St. Petersburg: Buchdruckerei der Kaiserlichen Academie der Wissenschaften, 1875.
von Toll, Baron Eduard. 1895. Wissenschaftliche Resultate der von der Kaiserlichen Akademie der Wissenschaften zur Erforschung des Janalandes und der neusibirischen Inseln in den Jahren 1885 und 1886 ausgesandten Expedition, Abtheilung III: Die fossilen Eislager und ihre Beziehungen zu den Mammuthleichen. *Mémoires de l'Académie Impériale des Sciences de St.-Pétersbourg* 42 (13): 1–86.
Whelan, Heide W. 1999. *Adapting to modernity: Family, caste, and capitalism among the Baltic German nobility*. Köln: Böhlau Verlag.
Whyte, Kyle. 2020. Against crisis epistemology. In *Routledge handbook of critical indigenous studies*, 52–64. London: Routledge.
Woods, Rebecca. 2023. Telling time with mammoths: Frozen flesh and temporal arrangement in the science of the north since 1800. *Journal for the History of Knowledge* 4: 33–52.
Wright, Edmund A. 1986. *CRREL's first 25 years, 1961–1985*. https://usace.contentdm.oclc.org/digital/collection/p16021coll4/id/15

Wrigley, Charlotte. 2023. *Earth, ice, bone, blood: Permafrost and extinction in the Russian Arctic*. Minneapolis: University of Minnesota Press.
Wulf, Andrea. 2015. *The invention of nature: Alexander von Humboldt's new world*. New York: Vintage Books.
Yusoff, Kathryn. 2018. *A billion black Anthropocenes or none*. Minneapolis: The University of Minnesota Press.
Zukriegel, Josef. 1935. *Cryologia Maris*. Praha: The Geographical Institute of the Charles IV. University.

Part III

Institutions and Practices

Collections and Museums in the Historiography of the Earth Sciences

Irina Podgorny

Contents

Abstract

This chapter wants to show how museums and collections have been treated in recent historiographies of earth sciences as well as to examine the genesis, disappearance, and reemergence of museums and collections as spaces linked (or not) to knowledge production. It considers collections of minerals; rocks and fossils; and meteorites and gems; as well as inventories, books, and catalogues.

Keywords

Mineralogical collections · Geological museums · Lost museums

I. Podgorny (✉)
Archivo Histórico de la Facultad de Ciencias Naturales y Museo, Universidad Nacional de La Plata/CONICET, La Plata, Argentina
e-mail: ipodgo@isis.unlp.edu.ar

E. Aronova et al. (eds.), *Handbook of the Historiography of the Earth and Environmental Sciences*, Historiographies of Science,
https://doi.org/10.1007/978-3-031-40799-4_14

1 Introduction

In 1963, Claude C. Albritton (1913–1988), Hamilton Professor of Geology at Southern Methodist University, Dallas, Texas, published a seminal book, *The Fabric of Geology,* that paralleled his later, famous *The Abyss of Time* from 1980. It was devoted to the philosophy of geology and the production of geological thought and was one of the results of the dissatisfaction of many American geologists with the training that was being given to students of the earth sciences in the years immediately following World War II. Similar concerns had permeated the Society's debate over the years, and on the occasion of the celebration of its 75th anniversary in 1963, the councilors decided that the theme for the platinum jubilee meeting was going to be the philosophy of geology, a decision that resulted in the abovementioned book (Albritton 1963, v).

Consequently, the American Council of the Geological Society convened a committee to investigate the "state of the art" and to offer suggestions for its improvement. The logical structure – using the terminology of the philosophy of science of those days – of the earth sciences' disciplines was at stake (Albritton 1963, v).

The Fabric of Geology began with a toast to James Hutton and continued with several essays dealing with the problems of geological observation and the scale of geological time and space, one of Albritton's main themes of research throughout his life. Several chapters were centered on the question of the historic character of geology and whether or not this science had laws and theories of its own, another epistemological concern from those years. Plate tectonics were still far from the global agenda, but geologists were slowly moving toward the quantitative, to geochemistry and geophysics, and to the analysis of the chains of reasoning and inference (Podgorny 2005a).

The book devoted many pages to the tension between field observations and laboratory experiments. Fieldwork remained a crucial component of the discipline: in H. H. Read's words, "Any man looking out of any window sees a geological laboratory in constant and full-scale operation" (quoted by Albritton 1963, vi). Geologists, it was said, had the unusual opportunity to observe the results of complex natural experiments that involved larger masses and longer periods of time. However, the small-scale experiments of the laboratory could establish new facts and inspire innovative hypotheses (Greene 1998).

Not surprisingly, the terms "museum" and "collections" are almost absent from the 372 pages of the book. Each term appeared just twice as if they were irrelevant to the study of the Earth, and, therefore, there was no place for them in the geological thought from the 1960s. No explicit reason was given to this absence, but, essentially, they were perceived as institutional refuges for those who were still attached to a descriptive tradition that had been a necessary precursor to the language and practices of experimental, real science but which was now obsolete. Museums had lost the allure and their status as places of research and learning that they had acquired a century earlier.

In fact, after the Second World War it had been forgotten – or it was on an agenda of being forgotten – that there had been a time when the practice of geology could not have been imagined or described without "journeys of intense labor employed in seeing every collection and professor that could be heard of, and purchasing every map, book, and print that has been published relative to the earth sciences, or to the political economy of the countries" (Gordon 1894). These words used to describe the endeavors of the Oxford Professor of Geology William Buckland (1784–1856) are much more than a portrayal of the obsessions of a British individual. According to historians Martin S. J. Rudwick, Pietro Corsi, and James Secord, this picture can be applied to the labor of most of Buckland's contemporaries in England, the European continent or further afield.

Nineteenth-century geologists and anatomists – professionals and amateurs, provincial and metropolitan alike – expended much of their time and resources in the formation of collections. Collectors mobilized people, mules, boxes, bones, rocks, stones, gems, shells, books, maps, newspapers, and an incalculable amount of paper in their letters, labels, and packaging. Collectors dispatched and/or received objects by ship, train, and/or cart crossing and connecting the five continents. The scale of those movements relied upon freight and post and enabled the infinite possibilities of being inside a network for the exchange of data, specimens, and gifts. The span of a collection depended on the fortune or salary of the collector, on the institutional budget (if that was the case) as well as on the dimension of the network of correspondents and providers. The practice of collecting fossils, rocks, and stones created new objects especially designed to display or keep them, new skills, new professions, new economic activities, and new investigative tools linked to the provision of all that was needed for doing science and knowing the secrets of the Earth.

Having a collection often meant having the capacity to keep the objects in a space, to build and design special furniture and supports, to label, and to catalogue or record where the things were in order to manage their movements, as well as to teach and to learn how to read what these objects had to tell. Nineteenth-century public lectures or university classes were unthinkable without the collections in all their forms, still today a useful tradition if students want to distinguish a piece of pyrite from gold (Fig. 1).

The contrast with today cannot be sharper: Whereas the physical collection in the nineteenth century played a central role in the production and transmission of geological knowledge, a 100 years later they had vanished from the vocabulary and the practices of earth science practitioners, who, instead, had moved to a reliance on databases and the analysis of the data within them (see Aronova et al. 2017).

This chapter wants to show how museums and collections were treated by the historiography of earth sciences as well as to examine the coming into being, the vanishing and reemerging of museums and collections as spaces linked (or not) to knowledge production and artistic creation. It refers – without going into details – to the collections of minerals, rocks, fossils, meteorites, and gems, as well as inventories, books, catalogues, newspaper clips, and paper museums in the sense of Rudwick (2000).

A few notes for clarification are required. First,: in this chapter I use "earth sciences" to include paleontological collections, following the lines set by the History of Earth Sciences Society, although in some institutional settings paleontology is considered a branch of the life sciences.

Fig. 1 "Oro," photograph by Argentinean photographer Adriana Miranda, 2015. In fact, this is a sample of pyrite (fool's gold), used in this case to mock at the contemporary art market. (©Adriana Miranda, courtesy of the author)

Second, the chapter refers mostly to early nineteenth-century collections. Geology in its modern form emerged as a specific discipline late in the eighteenth-century, paleontology a couple of decades later. Comparative anatomy was the discipline and the institutional setting for the study of fossil vertebrates, and, therefore, the chapter uses "anatomists" to refer to people that collected fossils before the age of paleontology began. Botanists and zoologists, chemists and physicians, and men and women with no university education made collections, and it would be inaccurate to call them geologists or paleontologists.

Finally, the chapter does not follow a chronological line; it is structured around themes and reflections that reappear in the different sections.

2 Collections and Museums in Contemporary Literature: Reassessment and Neglect

In the 1980s, Martin J. S. Rudwick's work on the history of geology and paleontology focused attention on the practices involved in the amassing of collections in a space that was not always called a museum: Sometimes it was a private room, a professor's office, a cabinet, or a library. The diversity of the locations did not prevent the conception of collecting and collections as a practice and a space that, in concert with observations in the field, shaped the emergence of modern geology (Rudwick 1985).

Late twentieth-century historians of the earth sciences have shown that there was a continuum that connected the field, the library, the cabinet, university teaching, and the collections, and that there was not a clear divide between natural history and antiquarian practices (Rudwick 2008). They were connected by a visual language that had to be learned and transmitted by means of new visual devices such as geological maps and crystallographic models, which were combined with the use of topographic instruments and old miners' practices that relied also on taste, smell, and touch (Rudwick 1976; Guntau 1996).

Collecting rocks, minerals, and fossils implied being immersed in a network of correspondents, creating local, regional, and international alliances, a circuit where things could circulate, and be compared and studied (Rudwick 1997). Letters and drawings had been in use since the Renaissance as a means of "cheap" and two-dimensional representations of complex three-dimensional data. The expansion of the market and navigation routes to Australia and the Americas, the industrial revolution, and the use of fossil energy added speed and distance to the dynamics that had existed since the early modern period and allowed for the transportation of the objects themselves rather than just information about them. Collectors and collections emerged everywhere, and they usually knew of the existence of each other and tried to be connected either by personal communication, visits, or by proxy devices such as catalogues, drawings, and illustrated books. Collections in order to circulate beyond their actual form had to be translated into paper following certain conventions and knowing what to remark in order to make the object readable for all of those sharing the conventions. Far from new, this was common practice since nature was conceived as something one can possess in a room (Findlen 1994; Olmi 1997).

The collections were, thus, real and "virtual," three-dimensional, and in paper, objects, books, newspapers clips, and treatises,[1] but always conceived as a central research tool that required the organization of labor force and labor space, which very often interconnected the private and public, family life and professional activity. In France, England, Argentina, and Italy, wives, daughters and sons, brothers and sisters, cousins, in-laws, acquaintances, and relatives were mobilized in these kinds of small family enterprises, often structured against other family groups (Outram 1997; Podgorny 2021).

The description of French anatomist Georges Cuvier's working place in the Muséum National d'Histoire Naturelle was in that sense paradigmatic: as Eigen (1997) stressed, Cuvier's private study (or "laboratoire") was organized like a "public office." Secretaries, family members, collections, and plates were harnessed in mimicking the process of serial production. In contrast to the semblance of French order of Cuvier, William Buckland, who also worked with his wife and progeny, studied the collections either in the private space of their family kitchen and/or in his university room. Geologist Roderick Murchison (1792–1871), his colleague, has graphically described him sitting in the only empty chair, in his black gown, cleaning out a fossil bone from its matrix, and surrounded by rocks, shells, and bones in dire confusion. In Argentina, Florentino Ameghino (1853–1911) turned his house and bookstore in La Plata into one of the main centers of mammal paleontology from the late nineteenth century. There he not only kept and stored the Patagonian fossil collections made by his brother Carlos, but he also received his scientific guests, and wrote his extensive correspondence and voluminous publications (Podgorny 2005b)

[1] For the study of earthquakes by means of collecting newspapers clips, see Coen (2013).

(Fig. 2). His French wife, no doubt, helped in the editing part of his work and accompanied him on field excursions. Thus, the spaces where collections were located and studied differed greatly and combined the private, the familiar, the public, and the state-funded.

Making collections was an economic enterprise that required the investment of considerable sums of money for paying the salaries of the excavators and for dispensing recompenses among the local field collectors. As the cases of Mary Anning (1799–1847) in England and Auguste Bravard (1803–1861) in Massif Central in France and later in the Argentine pampas show, becoming a fossil supplier to the big collectors was an interesting option for those living all the year round close to the sites where fossils can be found (Podgorny 2001, 2020, 2022; Creese 2007).

With the possible exception of the Royal Cabinet of Madrid (est. 1771), where some of the objects that arrived were presents for the King, fossil and paleontological collections were mostly the result of private investments and endeavors, and as such part of the market of natural history that exploded in the nineteenth century. The private ownership of fossils would partially change in the 1900s when some countries (for example, in Argentina) included fossils (although not mineral collections), archaeological remains, and subsoil resources as the property of the state. Other countries, however, never considered fossils as such, and they remained as property that could be sold, purchased, and inherited. This is why many collections

Fig. 2 Florentino Ameghino's collections in his family house in La Plata as they were left at his death in 1911. (Photograph from Argentinean newspaper… (Archivo Histórico del Museo de La Plata))

became fragmented, as they were sold and lost after the death of the collector when, for example, the heirs did not inherit the obsessions of their parents or were not on the position to keep what absorbed their parents' time when they were alive. Keeping a geological and or/fossil collection alive required an ever increasing amount of storage room, resources for renting or purchasing furniture, shelves, and space or for purchasing a big house. Such was the case of the American millionaire Edward Drinker Cope (1840–1897), who used two of his Philadelphia houses for keeping the unopened boxes of the collections bought in Paris in 1878 (Podgorny 2021).

The problem of storage space was particularly an issue for the poor collectors and for the curation of fossil vertebrates, the mighty stars of the early nineteenth-century scientific world. The search for private and/or state patronage dates from those years, when George Cuvier, in his *éloge* to Louis-Jean-Marie Daubenton (1716–1800), his predecessor in the chair of Comparative Anatomy in Paris, pleaded for state support for the collections of the chair, an innovative approach that would see the emergence of the role of the state in funding both collections and their keepers (Outram 1978).

Nicolaas Rupke (1994) and Martin Rudwick (1985) have shown that in England this process happened later, as many of the gentlemanly geologists of those years were privately wealthy and did not work as employees. Rupke (1994, 12–13), for instance, argued that anatomist Richard Owen (1804–1892), in choosing museum work in the late 1830s, did not move into a ready-made institutional niche for scientific study. Owen, the "English Cuvier" who coined the term "dinosaurs," was the person in charge of describing the fossil Mammalia collected by Charles Darwin on his voyage on the Beagle. He spent most of his career as a museum curator and was the promoter of London's new Museum of Natural History in South Kensington that opened in 1881. For Rupke (1994, 12–13),

> both the concept and the architectural reality of museums as institutions of research, though at the time already well established in Paris, were still being developed in Britain. In fact, Owen himself was one of the main driving forces of the Victorian movement to provide accommodation for museum collections, to expand such collections, and to turn these to educational and research purpose.

That concept of the museum and its collections as the space for professional education happened also in Brazil and in the Parisian *École de mines*. In the latter institution, established in 1760 during the Ancien Régime, the mineralogical collections, comprising samples of raw and worked materials, were crucial for the training of prospective mining engineers who arrived from the Americas and other regions of the world to study in the French capital (Napolitani 2020).

In that context, learning how to manipulate minerals and fossils became crucial in a world where the exchange of specimens on an international scale helped in defining the global span of the new disciplines of geology and paleontology. Specimens were traded, accumulated in one place, cast, dispersed, sold, bought, and reshipped again to shape a universal vision of the history of the Earth and of the processes that defined the organization of crystals. Collections of minerals, models, and fossils became central to the modernization of the new disciplines and the reform of the language of natural history.

3 Absences, Forgetting, and Losses

The emergence and consolidation of the role of museums and collections as research and educational tools seems unquestionable. However, a review of some important general works published in English in the 1990s shows the same tendency observed in Albritton' *Fabric of Geology*.

For instance, in *Sciences of the Earth*, the monumental and useful encyclopedia of events, people, and phenomena edited in 1999 by earth science historian Gregory G. Good, there are no specific entries either for "museums" or for "collections" (Good 1998). In the very extensive analytical index to these almost 1000 pages, "museum" only appears once in relationship to "museums of natural history" (p. 664), namely, to the entry on "Paleontology in Australia," where the role of the exchange of fossil collections is explored and analyzed.

Martin Guntau, a refined German historian of mineralogy, contributed to this trend, placing the collections in a kind of prehistory of the discipline. Thus, his article from *The Cultures of Natural History* starts by saying:

> In the natural history museums of London, Berlin, Vienna, New York and many other cities, there are always substantial displays of mineral, rocks and fossils. These public exhibitions often date back to the eighteenth century and derive from still older collections of curiosities. Such cabinets were a common feature of courtly furnishing; they served the combined functions of display, entertainment and improvement. Gemstones and other visually appealing minerals and striking natural productions – together with countless works of art- fascinated the wealthy and powerful. (Guntau 1996, 211)

Guntau could be partially right regarding the origin of some of those specimens, but those collections, as already remarked, continued to be used or were constituted in a context that went far beyond the cultures of curiosity from the early modern period. This is particularly true for the mineralogical, anatomical, and "paleontological" collections organized in the *École Royale de Mines* and the *Muséum National d'Histoire Naturelle* in Paris (est. 1797), which, as will be discussed later, shaped the teaching collections of the Americas' new republics. Furthermore, the early nineteenth century had great expectations for the potentialities of the collections. Among others, Frank James (2018) has analyzed the formation of the British Royal Institution's mineral collection during the early nineteenth century and showed that, despite the initial enthusiasms and hopes entertained for it, geology and mineralogy did not then contribute to the process of industrialization, except in a negative manner. For him, applying scientific knowledge and method to practical problems was very difficult at the time and collections, therefore, were seen not as objects of fascination; quite the contrary, they were conceived as practical and useful tools. These were the years when the societies for the encouragement of arts, manufactures, and commerce in Europe and the Americas improved methods of extracting the ore, assaying, mining equipment, and crucibles and portable furnaces. Mineralogy and geology implied the encouragement of new collections and exhibitions.

Geological and mineralogical specimens and samples were everywhere, in part because of their connection to chemistry and the hopes of their use in modern industry. Alexandre Brongniart (1770–1847), the French chemist, mineralogist, geologist, paleontologist, and zoologist, famous for his collaboration with Georges Cuvier on a study of the geology of the region around Paris, was also the founder of the *Musée national de Céramique-Sèvres* (National Museum of Ceramics), having been director of the Sèvres Porcelain Factory from 1800 to his death in 1847. There he worked to improve the art of enameling in France, making Sèvres one of the leading factories in Europe (Millasseau 1997; Auffret 2014). The collection he organized in Sèvres, as well as his *Traité des arts céramiques*, was developed on an insistence that there was an intrinsic connection between geology and the expanding industry of European porcelain. That connection can also be observed in England where the instruments, developed for the pottery furnaces of Josiah Wedgwood (1730–1795), would go on to be applied in experiments to calculate the Earth's temperature (Chaldecott 1975). In the early nineteenth century, collections were in good shape, creating new products, and well-connected with the economic concerns of those years. In fact, one can argue that the emergence of paleontology and geological archaeoloy is rooted in the practice of recycling, namely, the excavation of recent and old garbage pits and the transformation of those deposits into scientific museum specimens (Podgorny 2016).

As for the historiography, the revision of the titles of the articles published in the biyearly journal *Earth Sciences History* exhibits a similar trend to that expressed before: collections and museum becoming marginal among the topics of this journal. The History of Earth Sciences Society started publishing its journal in 1982 to promote and publish historical work on all areas of the earth sciences: geology, geography, geophysics, oceanography, paleontology, meteorology, and climatology. The first issue included an article on an American fossil and mineral collection from the 1840s, but the reader has to wait until the second issue of 1986 for another note, in this case on the national gem collection at the Smithsonian Institution.

In its 40 years of existence, the journal has published ten articles whose titles refer to a specific collection and only five to museums, even when the journal encourages a variety of approaches to historical study that includes biography, history of ideas, social history, and histories of institutions, organizations, and techniques. The vast majority of the articles whose titles include "collections" dealt with fossils; some others – not included on the table – focused on the collector and a biographical approach. Only the two most recent articles on the role of museums in the exploration of the American West consider these institutions as a space of knowledge production (see Table 1 for the distribution of topic/publication year).

Some of the articles published by this journal analyzed "lost collections" or "lost legacies," such as Lardner Vanuxem's collection of minerals and fossils, "lost" since its sale in 1853–1854 to W. M. Stewart of Clarksville, Tennessee. However, about 5 percent of the original 6000-specimen mineral collection still exists at Southwestern University in Memphis. The fossil collection, on the other hand, contained some 16,000 specimens, most of which were from the Lower Paleozoic of New York State. W. M. Stewart organized the collection on the basis of geological periods. No

Table 1 Articles published in *Earth Sciences History* that contained "museum" or "collection/s" in the title. This survey does not include the few articles that referred to "collectors" or to the practice of collecting (https://meridian.allenpress.com/esh)

Year	Volume, issue	Title	Type of collection if not indicated in the title
1982	1, 1	The Vanuxem collection	Minerals and fossils
1986	5, 2	The Nation's gem collection	Gems
1992	11, 1	The Schlagintweit collections	Natural history
1994	13, 2	C. F. Hartt's contribution to Brazilian museums of natural history	
2008	27, 2	The Heringen collection	Publications, maps
2009	28, 2	A truly European forest: a historic lower Silesian Paleobotanical collection at the Museum of Natural History (Berlin)	
2015	34, 2	Reconstructions of a historic paleontological collection: diversity recreated	
2016	35, 2	C. W. Beehler's collection of vertebrate fossils: a lost legacy	
2018	37, 2	Negative geology: Humphry Davy and Forming The Royal Institution's Mineral Collection, 1803–1806	
2019	38, 2	(a) The 1477 Silver Find at the St. Georg Mine in Schneeberg and significant ore specimens in the Museum of Mineralogy and Geology in Dresden (b) The Catalog of the Mineralogical Collection of the Emperor Leopold II (1747–1792): collecting and learning in eighteenth-century Europe	
2020	39, 1	The importance of the Museum in Antebellum US Western territorial exploration: understanding the relevance of collecting fossils and their conservation	
2021	40, 1	More than just a rock collection. The meteorite collection of the Italian Geologist Teodoro Monticelli (1759–1845)	
2021	40, 2	The importance of the Museum in Antebellum US Western territorial exploration: part 2. The roles of Hayden and Meek in a paradigm shift in geologic and paleontologic studies	

trace of the Vanuxem fossil collection exists today, although it was still in use in 1923 by students at Southwestern Presbyterian University, in Clarksville, Tennessee. Only recently, historians have started reflecting upon the fact that collections and museums are very far from what the International Council of Museums currently defines as a museum:

> A museum is a non-profit, **permanent** institution in the service of society and its development, open to the public, which acquires, conserves, researches, communicates and exhibits the tangible and intangible heritage of humanity and its environment for the purposes of education, study and enjoyment. (ICOM 2007)

This is not what history shows. Collections (and museums) are the opposite of permanent: museums and collections are fragile, ephemeral, and subjected to the contingencies of history. As Italian art historian Adalgisa Lugli said in 1983:

> *Sans doute, n'y a-t-il pas de véhicule plus éphémère que la collection auquel on puisse confier une idée ou une image du monde. L'œuvre-collection est incroyablement périssable, corrodée par une fragilité presque pathétique, exposée comme elle l'est à la dispersion, au mouvement continuel qui déplace les objets, les soustrait à un ensemble pour les déposer dans un autre, ou les isoler définitivement* [...] *On pourrait la considérer comme le lieu de la non-existence, semblable au rivage de la mer, espace très particulier et insaisissable parce que perpétuellement bouleversé par le va-et-vient de la vague. Dans la collection, des mouvements contrastés et répétés modifient donc continuellement la physionomie de l'ensemble, lui interdisant d'atteindre à la stabilité de l'identification. Disons, tout au moins, que le moment définitif de la collection dure très peu de temps.* (Lugli 1998, 91)
>
> Without doubt, there is no more ephemeral vehicle than the collection, to which we can entrust an idea or an image of the world. The work-collection is incredibly perishable, corroded by an almost pathetic fragility, exposed as it is to dispersal, to the continual movement that displaces objects, removes them from one group to deposit them in another, or isolates them definitively [...] It could be considered as the place of non-existence, similar to the shore of the sea, a very particular and elusive space because it is perpetually disrupted by the coming and going of the wave. In the collection, therefore, repeated contrasting movements continually alter the physiognomy of the whole, preventing it from achieving the stability of identification. Let's say, at the very least, that the definitive moment of the collection lasts a very short time.

Even when this remarkable definition dates from the early 1980s, they were not not taken into consideration by the literature in English until recently (see Lubar et al. 2017; Jardine et al. 2019).[2]

This fate repeats itself in several American and British colleges, where the number of lost museums and collection is uncountable. It includes, among others, losses in small colleges such as Wofford Geological Museum and Collection (Spartanburg, South Carolina) and central institutions, such as Brown or Oxford University, where William Buckland's collection represents a paradigmatic case of neglect. During Buckland's lifetime, the collection was placed in Oxford's Clarendon Buildings. Buckland's wife Mary Morland had written or painted descriptive labels on all specimens that had come into her husband's possession. Buckland bequeathed the collection to the Vice-Chancellor of the University for the use of the professors of geology who might succeed him, with all the geological charts, sections, and engravings that might be in the Clarendon Building at the time of his death. John Phillips (1800–1874), the successor to his Chair, proposed that the collection should henceforth be known by the name of Bucklandean Museum. A new building was erected in about 1858, to which the collection was removed. "The subsequent history of the collection" – said his daughter – "is a melancholy record of reject. Owing to a variety of causes, a great part of this valuable bequest to the University remains in the same condition (and with perishing labels) in which it was removed from the Clarendon thirty-six years ago" (Gordon 1894, 76). After Phillips' death, the Buckland collection became "the collection in the cellars." Thus, a collection famous

[2] Some museographical proposals (see the Ethnographic museum of Neuchâtel https://www.men.ch/en/exhibitions/spotlight/the-impermanence-of-things/ and the several art exhibitions refer to the destruction of museums (Podgorny and Achim 2023)).

as the first of its kind in Europe lay neglected and lost its labels at the same pace that Buckland's name was being forgotten.

Whatever the chronology and the causes for this neglect, the result was a loss of meaning for the scientific agendas of successors. The stones and fossils were not the stable objects that they appeared to be. When Buckland was alive, the collection grew, was used, and generated new knowledge. Once he died, the collection as scientific instrument died with him and was turned into a historic object that occupied a space that other professors also wanted for placing their own books, instruments, and living collections. Scientific objects are transitory (Rheinberger 1997): Scientific museums and collections, as scientific tools that are part of a certain research agenda, share this character; they can disappear or turn into a mere historic curiosity. They can be revisited, of course, and reinterpreted, as has happened with several historic fossil collections where the specimens are named and renamed repeatedly. But despite this possibility, the neglected collections and the subsequent silence that certain historiography created around them speak of the artificiality of their existence, of the amount of care and money they needed to survive, and of the fact that this was not always available.

To conclude with the visibility of earth science history collections in recent literature, in 2020, *Colligo,* an independent journal devoted to the history of collections, published a special issue on the history of paleontological collections. Based upon papers given at a conference session entitled "How to build a paleontological collection: expeditions, excavations, exchanges," held at the fifth International Paleontological Congress in Paris, France, in July 2018, it discusses museums in Copenhagen, Bologna, Brazil, and Berlin as well as private collections from central France and Mauritius. The issue demonstrated that the landscape of nineteenth-century collections is certainly much wider, more nuanced, and more complex than the current historiography, centered on the metropolitan collections assembled by northern Europe's colonial powers, would suggest (Podgorny et al. 2020).

4 Collections and Museums in the Historiography from South America

Nearly two decades after the publication of *Sciences of the Earth* (Good 1998), the Archivo Histórico del Museo de La Plata (2016), a team with extensive research experience on the role of museums and collections and on the history of the training of scientists in general and geologists in particular, organized a *Diccionario Histórico de las ciencias de la Tierra en la Argentina* (2016). It was a collective work, much inspired by Good's monumental encyclopedia. However, it decided to include specific entries for "Museums" and "Mineralogical collections."

The first was dedicated to the history of local institutions. It described how the Minister Bernardino Rivadavia (1780–1845), one of the main promoters of the Buenos Aires public museum established in 1823, purchased from France a collection of 720 minerals and a group of scientific instruments for display in the cabinets of physics and natural history. The collections were devoted to the study of science and for teaching university students, reflecting the meaning that the word "museum" had in the Spanish dictionaries of the 1820s. For example, in the *Diccionario de la lengua castellana,* issued by the *Real Academia Española* in 1822 (pp. 544 and 554), "museo" had three definitions: first, the building devoted to the study of the sciences, the humanities, and the liberal arts; second, the place where several scientific curiosities, mathematical devices, and ancient medals were kept; and third, a coin collection, a "monetario," a term also applied to the shelves, drawers, and tables in which the series of coins and medals were properly housed.

The Buenos Aires museum, one can say, fulfilled this definition thanks to the coin and mineralogical collections, the devices bought for the physics cabinet, and its institutional foci. However, it was founded without either the allocation of a building or the shelves and drawers that defined a museum. The museum was more an ideal than a place; as such, it was conceived in theory as "público," emphasizing with this adjective that – even with no building – it belonged to the people and that it had been erected to serve all the citizens and interested people from other parts. It was only in 1826 that the museum was given a home – in the upper floors of the Santo Domingo Convent.

The acquisition of the French mineralogical collection can be seen as a device used to integrate the local samples of mineral into a universal order of minerals, namely, the mineralogical systems used in Paris. It can represent also what María Paola Rodríguez Prada, describing the early years of the National Museum from Bogotá, called a "global scientific geopolitics," namely, the hiring of French graduates and the expansion of the methods and systems used at the Paris School of Mines; this included, in particular, René Just Haüy's crystallographic system (Rodríguez Prada 2016). The French institutions were a very important model for the new republics, the Brazilian Empire (Lopes 2010, 2013), and also the USA, where American geologist Vanuxem held the chair of Chemistry and Mineralogy at South Carolina College (1792–1848) having graduated at the *École de Mines* in Paris in 1819. Vanuxem was well aware of the importance attached to museums and collections in France and spent some time in Albany arranging the state geological collection, out of which grew the New York State Museum. His private collection of mineral and geological specimens was considered at the time of his death to be "the largest, best arranged, and most valuable private collection in the country." Now, it is almost lost (Kemmerly 1982).

French mineralogical methods, collection, and classification systems expanded but were, however, far from being monolithic, universal, or adopted without resistance. Mineralogists trained in other traditions than Haüy's crystallography contested his terminology and method, not only in France and Europe but also in the new Republics. Thus, *La Abeja Argentina* from April 1822 – a kaleidoscope journal from Buenos Aires that included sciences, politics, and literature – translated "Analysis of the Red Silver Ore" and "P.S.," two articles published in 1822 in the *Edinburgh Philosophical Journal*, in which Haüy's crystallographic system was questioned. At stake was mineralogical taxonomy and related nomenclature, namely, forming names for objects in inorganic nature that could be accepted all over the world. This agenda was marked by considerable conflict; naming this "inorganic nature" with words signifying a quality, constituent, use, or locality proved to be very divisive, especially when mineralogists started naming minerals in a very generic way or after people.

When the Society for Physical Sciences and Mathematics of Buenos Aires recommended works to be used in the courses to be taught at the university, it included a variety of French works, notwithstanding the conflicting views of nature and of the history of the Earth that they expressed. "Ciencias, Lista de las obras que se pueden seguir para el curso completo de las ciencias físicas y matemáticas" (A list of the works to be used in the teaching of science) from *La Abeja Argentina* (3, 1822, pp. 105-107), included Alexandre Brongniart's *Traité élémentaire de minéralogie, avec des applications aux arts: ouvrage destiné à l'enseignement dans les lycées nationaux* (1807); Jean André Deluc's *Traité élémentaire de géologie* (1809, a defense of Mosaic Neptunism and a refutation of James Hutton and John Playfair); Georges Cuvier's famous *Le Règne Animal distribué d'après son organisation, pour servir de base à l'histoire naturelle des animaux et d'introduction à l'anatomie comparée* (4 volumes, 1817); and Charles-François Brisseau de Mirbel's *Éléments de physiologie végétale et de botanique* (1815), as well as André Marie Constant Duméril *Traité élémentaire d'histoire naturelle* (1803). In respect of mineralogy, geology, and natural sciences, it included Haüy's treatises (*Traité de minéralogie*, 5 vols, 1801, and *Traité élémentaire de physique,* 1803, 1806) but also his rival Jean-Claude Delamétherie's *Leçons de Géologie* (1816) et *de Minéralogie* (1811). Delamétherie was a contemporary of Haüy at Paris and, for a number of years, principal editor of the *Journal de Physique*. He gave offense to Haüy in some of his early publications by not crediting any of Haüy's work (Corsi 2020; Oldroyd 2009).

In a context and a discursive tradition that alluded to nature as a source of wealth and common welfare, an education in natural history and in the art of drawing and observing natural objects was seen as the fundamental ground of a new citizenship based on the universals of reason and humanity. However, many of the Latin American museums created in the 1820s were in fact linked to the possibility of surveying the resources of a country following the standards of the relatively new scientific disciplines, such as geology, and the new classification systems used for vertebrates, invertebrates, and plants alike. The classification of the mineralogical and natural resources was crucial to promote commerce and attract new investors.

Thus, the Museo Nacional established in 1854 in the city of Paraná, the then capital of the Argentine Confederation, received a series of minerals and animals from several provinces. This included, among others, a collection of quartzite plus six samples of gold, silver, copper, and nickel ore from the Province of La Rioja; a collection of minerals from the Province of Córdoba sent in by the provincial governor; samples of copper from Catamarca; and samples of limestone and shells, ocher, clays, fuller's earth, and fossil bones, all from the vicinity of the city of Paraná. The establishment of the Museo Nacional from Paraná is an episode connected to Argentina's first participation in a World Fair, as part of the propaganda campaign launched in the process of organizing the Argentine Confederation (Podgorny and Lopes 2008; Lopes and Podgorny 2000).

Starting in the second half of the nineteenth century, overseas products were displayed at periodical World Fairs held in various European capitals, where participating countries had to meet certain sets of rules regarding the selection and classification of their material. One was to form a local jury to evaluate the samples and determine whether they were truly representative of the country's wealth. Within this framework, the Museo was conceived of as the place that would collate examples of the potential wealth of the nation, and be the institutional backing for the items Argentina sent to the World Fairs. From there, the image of bountiful nature was spread as the central propaganda argument aimed at those who might be interested in Europe. Following the Great Exhibition of London in 1851, South American states were invited to participate in these events by sending in raw materials produced within their national territories. In late 1853, Argentina was invited to participate in Paris' 1855 *Exposition Universelle des produits de l'Agriculture, de l'Industrie et des Beaux-Arts*.

In August, just 1 month after the Museo Nacional had opened, the invitation to participate in the 1855 *Exposition Universelle* in Paris became the responsibility of its director. The provincial producers' response to this call managed to exceed by far the collections that the museum had previously been able to put together. The director wrote a descriptive report to accompany the exhibit, preceded by a study of the country's mines and mining industry, geographical notes, and an overview of the Confederation's political structure and current laws on mining. The report emphasized the superiority of Argentina's gold and silver mines over those of Australia and California, at least in terms of abundance. The Argentinean samples never returned to Argentina: Once the Universal Fair closed its doors, they were dispersed and donated to the French institutions that were willing to accept them (Podgorny and Lopes 2008).

Regarding the *Diccionario's* entry for "Mineralogical collections," Argentinean historian of science Susana V. García explored the role of mineral collections in the education of professional geologists. She explained how the nineteenth-century collections were sold and marketed by commercial houses intended for the teaching of mineralogy and the promotion of the associated classification systems. Whether in Paris, London, Rochester, or Buenos Aires, those houses prepared collections of mineral specimens classified following the system they were eager to expand through the museums, university chairs, and basic education schools. Another kind of collection sold by those commercial houses consisted of slides with transparent mineral cuts and sets of models of crystal shapes made of wood, most of them made or based on the French models. Today they are kept in the stores and displays of museums and university departments of mineralogy and crystallography, but they were sold originally to teach the external form of natural specimens. Their use was for identification and demonstration purposes, by scientists, teachers, and students from the late eighteenth century onward, a study that continued into the twentieth century.

The eye of professional geologists was not only trained by means of those collections: Collections taught them that these kinds of objects could be commercialized and could help to develop their local and international careers. In fact, what collections teach historians is that many geologists and natural history professors decided to partner with commercial providers for selling and organizing new teaching and museum collections. Collections as matters of exchange – in the sense of Harold Cook (2007) – played different roles: They were commercial objects, used for teaching, for the consolidation of classification systems and no less important for the building of international reputations.

The Capellini Museum in Bologna, Italy, still displays some of them, including the collection of Nummulite, a type of foraminifera, organized by the Austrian-Hungarian geologists Maximilian Hantken von Prudnik (1821–1893) and Eduard Sigismund von Madarász. The Nummulite collection from Bologna reveals how collections circulated connecting museums located in the capitals and also in the provinces. Collections today are used by historians to reconstruct the circles that met at the international exhibitions and congresses and related people and places (Lopes and Matos 2015). For instance, the Nummulite collection and its catalogue were exhibited in the universal exhibition in Paris in 1878, which was accompanied by an international geological congress that reunited, for instance, the American Edward Cope, the Italian Giovanni Capellini, the Italo-Argentine Florentino Ameghino, and Austrian-Hungarian Maximilian Hantken (Podgorny 2021).

5 Collection of Minerals as Source of Artistic Inspiration

The new disdain for the museum has been as international as was its nineteenth-century respect. One of the landmarks of this trend was the demolition late in the 1960s of half of the original building of the Geological Capellini Museum in Bologna (established in 1881) to use the land for erecting new laboratories. The old collections had to be moved to the remains of the old building which still keeps the historical showcases and collections but in a denser way that in its early days, creating structural problems given the extra weight the building has to carry. Undoubtedly, some geologists tried and succeeded in sending part of the collections to the cellars or to the garbage pit. However, at the same time the collections were being discarded from scientific use, now more concerned with the collection of data, those mineralogical collections became the source of inspiration for twentieth-century artists, thinkers, and architects dealing with shape, morphology, and structures (Tamborini 2022).

French mineralogist François Farges (2020) has dedicated part of his work to track the uses of minerals in the arts and applied industries. He has shown how minerals can be regarded from different point of views, including the scientific, technical, and artistic as well as their use and employment in jewelry. Collections are multidimensional, and the emphasis on the exhibits of wonder normally neglected that in the collection of natural samples there was always the potential for non-scientific use. In particular, minerals and stones, rocks, and fossils had been associated with mining and architecture. The columns and the walls from the Natural History Museum in Oxford or the Natural History Museum in Copenhagen are, in reality, samples of the local geological materials that could be used in architecture. The museum building in Oxford opened in 1860 and is an example of Victorian neo-Gothic architecture, a style strongly influenced by the ideas of nineteenth-century art critic John Ruskin. Around the court perimeter are 126 columns, each made from a different British decorative rock, labeled with the name of the stone and its source. As for Copenhaguen, the walls mimic fossils in their construction.

In the realm of literature, mineral collections inspired George Sand (1804–1876) who, in 1865, located her novel *Laura. A Journey into the Crystal* in the halls of a mineral collection that resembled that of the Parisian Muséum National d'Histoire Naturelle. In the plot, while working for his uncle, Alexis Hartz is introduced to a woman who shares his scientific interests, and in particular his fascination for crystals. She has discovered a way to enter this world, and together they travel the vast and glittering landscape. Similarly, Jules Verne (1828–1905) expressed his fascination with crystals in his *Voyage au centre de la Terre*, which shaped our vision of the giant crystals from Brazil that since the 1980s have been exhibited in the Mineralogy Gallery of the Muséum.

In the twentieth century, the crystal became the symbol of a new purity. American-German painter Lyonel Feininger (1871–1956) and German architect Walter Gropius (1883–1969), founder of the Bauhaus School, adopted the crystal as a principal motif. The cube that fascinated Haüy, and which was modeled to teach geometry, reappears in the works by French artists George Braque (1882–1963) and Marcel Duchamp (1887–1968). Around 1932, the Hungarian-French photographer Brassaï (1899–1984) took pictures of the minerals and crystals exhibited in the Muséum that were going to be published in the Surrealist-oriented magazine *Minotaure* by André Breton, who proclaimed that only these objects can explain what was real beauty. Adriana Miranda's photograph of the pyrite is an Argentine descendant of this tradition of working with the light of the minerals (see Fig. 1).

This relationship between art, writing, and mineral collections had not finished there. The French intellectual Roger Caillois (1913–1978), as it is well known, has accused André Breton and followers of having sunk into the veneration of the irrational. He proposed to join the two domains of "investigation and poetry," in order to establish a link between the weft of dreams and the chain of knowledge. Caillois had, already at the time of his rupture with Breton, begun the sketch of a subversive and revolutionary science, which "would not despise the obscure side of the Nature." His very personal vision of the natural sciences allowed Caillois to imagine a form of interdisciplinary knowledge in which invisible, subterranean relations had to be exhumed in order to allow the blooming of a new and more complex image of the universe. In that frame, Caillois published *L'écriture des pierres*, a late work from 1970 in which he applied his interpretative talents to the stones that he had collected for decades, gathering a 100 specimens preserved today in the Mineralogical Galleries of the Parisian *Muséum National d'histoire naturelle*. Caillois in 1970 "was struck by a curious property of the imagination, which is that it is not deterred when challenged to interpret or identify. I realized that men had never ceased to read the drawings on the stones and to recognize in them gods, landscapes, panoramas of ruined cities, dragons, unicorns, bulls, swordfish. Such a power to secrete metaphors is practically boundless, as seems irresistible the need that this faculty is not hardly asked to satisfy. It is not held back by verisimilitude: it is even often encouraged by the insufficiency or the ambiguity of the support" (Caillois 1970; Gioni 2014). For Caillois, nature revealed itself as a superfluous delight, as if moved by caprice. Art becomes a geological question, as vast as the world itself. Caillois' writing of the stones resulted from the crossroads of the visual order and the poetic order. In this sense, it was truly writing only made possible by the art of collecting.

This chapter concludes with reference to another rock collection, in this case, one made by German percussionist Alexej Gerassimez (1987–), who on May 19, 2022, presented at the Konzerthaus Berlin "Stonewave" a work for a percussionist's duo. I was inspired by a quotation attributed to J.W. Goethe and Erich Kästern: "Auch aus Steinen, die einem in den Weg gelegt werden, kann man etwas Schönes bauen "(Even from the stones that are put in your way, you can build something beautiful"). Gerassimez, in fact, extracted sound from his collection of stones. "As its core" – he said in the program of the performance – "I explore the sound and shape of rocks. From rather porous and capricious natures we can tell amazingly harmonious stories." May be, some of the stones he is using came from a lost collection, disguised as instruments of contemporary music.

6 Conclusion

Museums and collections generated and channeled a flow of data, natural specimens, and artifacts that, through their relationship with people, traveled to diverse places, and in a variety of directions. Collections connected places and people in most unexpected ways and generated new collaborations. Through their relationship with people, they traveled and connected the world from the nineteenth up to the twentieth century, creating networks that were not necessarily centralized around either European metropolises or their respective national museums.

In the analysis of collections, there is a historiographical problem linked to changing disciplinary boundaries and concerns and, therefore, the changing role that the same specimen could have played in different arenas and in different periods of history. For example, gems, rocks, fossils, and minerals – the objects of study of mineralogy and sedimentology – were collected and studied also in pharmacies and medical cabinets up until the eighteenth century; this is still reflected in some exhibitions such as those in the in the geological galleries at the Natural History Museum in Vienna. However, most medical mineral collections were transferred as curiosities to pharmacy museums, or discarded as mere rubbish from the past. Gem stores, on the other hand, continue to sell minerals using the teaching of old pharmacopeias but marketed as New Age goods (see Duffin et al. 2013).

Collections of minerals, rocks, and fossils, of course, did not start with geology. Are the mineral collections and lapidaries from ancient, medieval, and early modern times part of the history of the earth sciences? Some historians and museum curators would say yes, some would choose to treat them as prescientific, and some would discard them from the realm of natural history and science even when they sell well both on the Internet and in real life. This chapter, working in the instable borders of disciplines, aims at helping to answer these kinds of question, which still have implications for our contemporary order of things.

Returning to *The Fabric of Geology* with which this chapter began: The chapter by Arthur Hagner (1963) was devoted to the problem of scale. He suggested that a mountain could not be taken into the laboratory. Nineteenth-century geologists, however, did try, through their samples and series, to do just that; in many cases, their efforts have been lost, and they created collections that, even when dispersed, are inspiring today new human creations. Collecting is not only a catalyst, a good sample of how the changing institutions that house, or housed, them are linked to the unexpected uses to which scientific collections have been put – and, therefore, the changing fortunes and uses of museums and similar institutions.

Acknowledgments This chapter is a result of a RISE project SciCoMove (Scientific Collections on the Move: Provincial Museums, Archives, and Collecting Practices (1850–1950)) that has received funding from the European Union's Horizon 2020 research and innovation program under the Marie Skłodowska-Curie grant agreement No 101007579. Margaret Lopes, David Roe, Franck Laurent, and Marco Tamborini have read and commented on earlier drafts of this chapter.

References

Albritton, Claude, ed. 1963. *The fabric of geology*. Palo Alto: Addison-Wesley.

Archivo Histórico del Museo de La Plata. 2016. *Diccionario histórico de las Ciencias de la Tierra en la Argentina*. Rosario: Prohistoria Ediciones.

Aronova, Elena, Christine von Oertzen, and David Sepkoski. 2017. Historicizing big data. *Osiris* 32.

Auffret, Anne-Lise. 2014. *Le musée céramique de la manufacture de Sèvres: progrès techniques, enjeux artistiques et pratiques savantes en France (1800–1914)*. Ph.D. Dissertation. École doctorale Histoire de l'art, Paris.

Caillois, Roger. 1970. *L'écriture des pierres*. Genève: A. Skira.

Chaldecott, John. 1975. Josiah Wedgwood (1730–95) – Scientist. *The British Journal for the History of Science* 8 (1): 1–16.

Coen, Deborah. 2013. *The earthquake observers: Disaster science from Lisbon to Richter*. Chicago: The University of Chicago Press.

Cook, Harold. 2007. *Matters of exchange: Commerce, medicine, and science in the Dutch Golden age*. New Haven: Yale University Press.

Corsi, Pietro. 2020. A chair for two: Georges Cuvier and Jean-Claude Delamétherie at the Collège de France. In *Darwin au Collège de France*. Paris: Collège de France. http://books.openedition.org/cdf/7311.

Creese, M.R.S. 2007. Fossil hunters, a cave explorer and a rock analyst: Notes on some early women contributors to geology. In *The role of women in the history of geology*, ed. C.V. Burek and B. Higgs, 39–49. London: The Geological Society. https://doi.org/10.1144/SP281.3.

Duffin, Christopher, Richard Moody, and Christopher Gardner-Thorpe. 2013. *A history of geology and medicine*. London: The Geological Society.

Eigen, Edward. 1997. Overcoming first impressions: Georges Cuvier's types. *Journal of the History of Biology* 30 (2): 179–209. https://www.jstor.org/stable/4331431.

Farges, François. 2020. *Pierres précieuses*. Paris: Muséun National d'Histoire Naturelle.

Findlen, Paula. 1994. *Possessing nature: Museums, collecting, and scientific culture in early modern Italy*. Berkeley: University of California Press. https://public.ebookcentral.proquest.com/choice/publicfullrecord.aspx?p=848572.

Gioni, Massimiliano. 2014. In *La Lecture des pierres*, ed. Roger Caillois, 21–27. Paris: Atelier EXB.

Good, Greg, ed. 1998. *Sciences of the earth. An encyclopedia of events, people, and phenomena*. New York: Garland.

Gordon, Elizabeth. 1894. *The life and correspondence of William Buckland*. London: John Murray.
Greene, Mott. 1998. Alfred Wegener and the origin of lunar craters. *Earth Sciences History* 17 (2): 111–138. https://doi.org/10.17704/eshi.17.2.b342311n1w401vv7.
Guntau, Martin. 1996. The natural history of the earth. In *Cultures of natural history*, ed. Nicholas Jardine, James Secord, and Emma C. Spary, 211–229. Cambridge: Cambridge University Press.
Hagner, Arthur. 1963. Philosophical aspects of the geological sciences. In *The fabric of geology*, ed. C. Albritton, 233–241. Palo Alto: Addison-Wesley.
ICOM. 2007. Definition of "Museum", https://icom.museum/en/resources/standards-guidelines/museum-definition/
James, Frank. 2018. Negative geology: Humphry Davy and forming the Royal Institution's mineral collection, 1803–1806. *Earth Sciences History* 37 (2): 309–332.
Jardine, Boris, Emma Kowal, and Jenny Bangham. 2019. How collections end: Objects, meaning and loss in laboratories and museums. *BJHS Themes* 4: 1–27. https://doi.org/10.1017/bjt.2019.8.
Kemmerly, Phillip. 1982. The Vanuxem collection. *Earth Sciences History* 1 (1): 32–35. https://doi.org/10.17704/eshi.1.1.a527738u5q78m377.
Lopes, Maria Margaret. 2010. Compartir espacios, colgar ballenas y apoyar a las Universidades. Los museos de Historia natural en América Latina en el siglo XIX. In *El museo en escena: política y cultura en América Latina*, ed. Américo Castilla, 39–52. Buenos Aires: Paidós.
———. 2013. Minerales y fósiles para escudriñar el país, abarrotar las vitrinas y educar a la gente. In *Museos al detalle. Colecciones, antigüedades e historia natural, 1790–1870*, ed. Miruna Achim and I. Podgorny, 179–200. Rosario: Prohistoria.
Lopes, M.M., and Ana Matos. 2015. O que visitar em Paris durante a Exposição Universal de 1878: Um guia turístico para geólogos. *Revista Iberoamericana de Turismo* 5: 48–62.
Lopes, M.M., and I. Podgorny. 2000. The shaping of Latin American museums of natural history. *Osiris* 15: 108–118.
Lubar, Steven, Lukas Rieppel, Ann Daly, and Kathrinne Duffy. 2017. Lost Museums. *Museum History Journal* 10 (1): 1–14. https://doi.org/10.1080/19369816.2016.1259330.
Lugli, Adalgisa. 1998. *Naturalia et Mirabilia. Les cabinets de curiosités en Europe*. Paris: Adam Biro.
Millasseau, Sylvie. 1997. Brongniart as taxonomist and Museologist: The significance of the Musee Céramique at Sevres, the Sèvres porcelain manufactory. In *Alexandre Brongniart and the triumph of art and innovation, 1800–1847*, ed. Derek E. Ostergard. New Haven: Yale University Press.
Napolitani, Maddalena. 2020. *Les pierres de la Nation: les collections minéralogiques de l'École de Mines de Paris (1760–1825)*. Ph.D. Dissertation, Université Paris Sciences. https://www.theses.fr/2020UPSLE062
Oldroyd, David. 2009. Jean-André Deluc (1727–1817): An atheist's comparative view of the historiography. In *Geology and religion: A history of harmony and hostility*, ed. Martina Kölbl-Ebert, 7–15. London: The Geological Society.
Olmi, Giuseppe. 1997. *L'inventario del mondo: catalogazione della natura e luoghi del sapere nella prima età moderna*. Bologna: Il Mulino.
Outram, Dorinda. 1978. The language of natural power: The "Eloges" of Georges Cuvier and the public language of nineteenth century science. *History of Science* 16: 153–178. https://doi.org/10.1177/007327537801600301.
———. 1997. Institution scientifique ou champ de bataille pour les familles et les groupes d'influence ? In *Le Muséum au premier siècle de son histoire*, ed. Claude Blanckaert et al., 25–30. Paris: Éditions du Muséum National d'Histoire Naturelle.
Podgorny, Irina. 2001. El Camino De Los Fósiles: Las colecciones de mamíferos pampeanos en los museos franceses e ingleses del Siglo XIX. *Asclepio* 53: 2. https://doi.org/10.3989/asclepio.2001.v53.i2.161.
———. 2005a. La tierra en el laboratorio. las ciencias de la Tierra en el siglo XX. In *Filosofía de las ciencias naturales, sociales y matemáticas*, ed. Anna Estany, 129–162. Madrid: Profitós.

———. 2005b. Bones and devices in the constitution of paleontology in Argentina at the end of the nineteenth century. *Science in Context* 2: 249–283. https://doi.org/10.1017/S0269889705000475.

———. 2016. Recyclen. Vom Schrott der Ausrottung zur Ökonomie der (Sub-)Fossilien. In *Sammlungsökonomien*, ed. Ina Heumann and Nils Güttler, 23–46. Berlin: Kadmos.

———. 2020. La guerre, la paix et la querelle. Les sociétés paléontologiques d'Auvergne sous la Seconde Restauration. *Colligo* 3: 3. https://perma.cc/C6ZK-TQFQ.

———. 2021. *Florentino Ameghino y hermanos. Empresa Argentina de paleontología ilimitada.* Buenos Aires: Edhasa.

———. 2022. Una elefanta en el fondo del océano. Objetos fuera de lugar en la historia de la paleontología, con especial referencia a las Reliquias del Diluvio del Reverendo William Buckland (1784–1856). *Museologia & Interdisciplinaridade* 11: 280–294. https://periodicos.unb.br/index.php/museologia/article/view/40840.

Podgorny, Irina, and Miruna Achim. 2023. Les musées et les naufrages de l'histoire. *Communications* 113: 91–100.

Podgorny, I., and M.M. Lopes. 2008. *El desierto en una vitrina. Museo e historia natural en la Argentina, 1810-1899.* Mexico: Limusa.

Podgorny, Irina, Eric Buffetaut, and Maria M. Lopes. 2020. Paleontological collections in the making – An introduction to the special issue. *Colligo* 3: 3. https://perma.cc/L772-PP5T.

Rheinberger, Hans-Jörg. 1997. *Toward a history of epistemic things: Synthesizing proteins in the test tube.* Stanford: Stanford University Press.

Rodríguez Prada, María Paola. 2016. The creation of the National Museum of Colombia (1823–1830): A history of collections, collectors, and museums. *Museum history Journal* 9 (1): 29–44.

Rudwick, Martin J.S. 1976. The emergence of a visual language for geological science 1760–1840. *History of Science* 14: 149–195. https://doi.org/10.1177/007327537601400301.

———. 1985. *The great Devonian controversy: The shaping of scientific knowledge among gentlemanly specialists.* Chicago: The University of Chicago Press.

———. 1997. Georges Cuvier et la collecte d'alliés internationaux. In *Le Muséum au premier siècle de son histoire*, ed. Claude Blanckaert et al., 591–606. Paris: Éditions du Muséum National d'Histoire Naturelle.

———. 2000. Georges Cuvier's paper museum of fossil bones. *Archives of Natural History* 27 (1): 51–68. https://doi.org/10.3366/anh.2000.27.1.51.

———. 2008. *Worlds before Adam: The reconstruction of Geohistory in the age of reform.* Chicago: The University of Chicago Press.

Rupke, Nicolaas. 1994. *Richard Owen. Victorian naturalist.* New Haven: Yale University Press.

Tamborini, Marco. 2022. *The architecture of evolution. The science of form in twentieth-century evolutionary biology.* Pittsburgh: The University of Pittsburgh Press.

Expeditions and Fieldwork in Earth Sciences

Marianne Klemun

Contents

Abstract

What do journeys, expeditions, and fieldwork mean in the context of earth sciences? What status do they have, and what is their contribution to geology's historical development, and to the training of geologists? These are the main questions that this chapter seeks to answer. In the first section of the contribution, we will explore the phenomenon of travel and expeditions from a general point of view. Then we will analyze it on three levels. First, the practice of field research

M. Klemun (✉)
University of Vienna, Vienna, Austria
e-mail: marianne.klemun@univie.ac.at

E. Aronova et al. (eds.), *Handbook of the Historiography of the Earth and Environmental Sciences*, Historiographies of Science,
https://doi.org/10.1007/978-3-031-40799-4_15

will be examined on the basis of introductory books as a manifestation of the methodological development of geologists' targeted travel (questions and instructions), which reached its peak between 1750 and 1840. It was during this period that geology emerged as a discipline in its own right. Introductory books detailing the aims and procedures of fieldwork reflect the fact that field research was considered essential by contemporary scholars. Similarly, the geologist's hammer was invented in the eighteenth century and became the everyday tool of geologists in the field and ultimately the emblem of the profession.

The contribution identifies three thematic clusters that illustrate the heterogeneity of expeditions and fieldwork in the development of geology. First, both are essentially connected to practices. Second, we have to explore the question of space and place in relation to expeditions and fieldwork, and third, the interplay between practices and scientific discourse, or theory and fieldwork (observation), needs to be analyzed. All three perspectives help us understand what was specific to the science of geology in terms of fieldwork and expeditions that also involved other sciences.

Keywords

Practices · Epistemic virtues · Space and place · Geological hammer · Desmarest · Geikie · Lyell · Humboldt · Sausssure · Wegener

1 Introduction

In *Geological Sketches at Home and Abroad* (Geikie 1882), the Scottish geologist and historian of geology Sir Archibald Geikie (1835–1924), General Director of the Geological Survey of Great Britain from 1882, published a collection of essays that had previously appeared in several journals. He remembered his experiences during his geological journey to Wyoming: "after landing in New York my preparations for a journey to the Far West were completed... A project which had been little more than a dream for many years was now at last actually realised" (Geikie 1882: 180). He did not refrain from emphasizing in particular: "Let me briefly explain the project, that the purport of the journey, and of the following notes, may be understood. And first I would give the reader due warning that the object of the expedition was not sport or adventure, but science. My companion and I were not, indeed, wholly unarmed. To go without at least revolvers into these western wildernesses would, we were told, be sheer Folly" (Geikie 1882: 180).

Why did Geikie exclude sport and adventure commonly associated with expeditions, and why did he want to affirm the demarcation of both from doing science, when his narrative invoked elements that essentially accompanied the practice of outdoor experiences during expeditions and surveys conducted by his colleagues? A common feature is that expeditions were ventures based on a calculated risk. It is a key objective of any endeavor to minimize the risks from the very outset, and to be armed meant that the explorer calculated the danger appropriate to the situation. It could be argued that adventurism and sport were not among the "epistemic virtues" (Daston and Galison 2007) in Geikie's view of the ideal geologist. But we have to remember that fieldwork was essential to the making of the new earth sciences from 1760 to 1840, prior to Geikie's statement. Lorraine Daston and Peter Galison explain "epistemic virtues" as a manifestation of the scientific self that is grounded in the cultivation of a specific discipline or a group in terms of which it is understood as being essential to achieve knowledge in the first place. "Epistemic virtues" can apply to an academic discipline, but also to a "thought collective" (Fleck) within a discipline. In terms of adventure, expeditions and fieldwork tended to be associated with openness to the unexpected and the pursuit of realism. Furthermore, the convention of establishing credibility as an outdoor geologist was not important at this specific stage of Geikie's career. After the establishment of geological surveys in Europe, beginning with the British in 1835, these survey geologists were an increasingly more visible group in most western countries. They were trained through fieldwork, taught the next generation, traveled and did fieldwork, and felt themselves to be professionals, distinct from other geologists working in smaller numbers at universities and museums, who were more closely associated with amateurs (Oldroyd 1996: 329). It is well known, however, that the boundaries between professionals and amateurs were not fixed, but depended on the process of institutionalization throughout the nineteenth century and were therefore reciprocal. Nevertheless, they were productive and changed in different contexts and at different times (Gieryn 1999). To return to Geikie's quote: Geikie had been working his way through the British Geological Survey from 1855, and in 1871, he became the first Murchison Professor at the University of Edinburgh (1871–1881).

Following Ludwik Fleck's concept of "thought style" shared by a "thought collective" (Fleck 1935/1979), it can be argued that geological travelers prioritized practicality and everyday practices over a sense of adventure during their fieldwork. Epistemic virtues, as defined by Daston and Galison, can also be seen as dimensions shaping the thought collective. Thus the two concepts can used in conjunction with each other. One can also make the case, like Kohler, that field research has not been given due recognition, in historical scholarship, compared to laboratory experimentation in terms of status acquisition. This also influenced Geikie, who was interested in the history of geology. Expeditions were often seen by the public as purely adventurous explorations. This, however, was not relevant for a geologist affiliated with an institutionalized survey in terms of his or her self-presentation, despite the fact that geologists were participating in an increased number of expeditions. They were surveying territories outside Europe and were part of the exploration of previously unknown spaces (Bourguet 1997). And, even more boldly, by the end of the nineteenth century they focused on altitude, much like mountaineers, in order to identify the geology of the Himalayas or the Arctic (Jones 1875; Brescius 2019).

The ambiguity of Geikie's narratives raises many questions and presents a challenging constellation between two phenomena – expeditions and fieldwork – their relationship in different historical contexts, and the question of how earth sciences can be positioned in this framework. It must be noted that this constellation is far more intricate than the dichotomous narrative suggests at first glance. When

Peter Sloterdijk, in his philosophical considerations of globalism, describes a type of scientific journey, an expedition, as an "epistemological form of adventurism" (Sloterdijk 2006: 152), he is referring to a famous common phenomenon and an important connection between the spatial frame and the search for new knowledge that made expeditions so attractive to naturalists and the public. According to Sloterdijk, these global enterprises were accompanied by social and spatial exclusion and inclusion. What is often overlooked, however, are aspects that lead us to the scientific enterprise in the narrower sense: The 'as yet unknown' first had to be identified as such in the preparations for an expedition, enabling it to be integrated into exploratory questions. During the eighteenth and nineteenth centuries, scientific collectives and institutions with a claim to leadership, such as academies and scientific societies, institutionalized surveys, ultimately determined exploratory aims and spatial-cognitive definitions of these gaps in knowledge. The anchoring of the factual non-knowledge that mutually conditioned the scientific enterprise and expeditionary activity was of central importance. These preliminary considerations were incorporated into the instructions issued by the institutions and principals behind an expedition while the transformation from not knowing to knowing was similarly celebrated by the protagonists in narratives, newspapers, and books.

The following sections discuss three thematic clusters that illustrate the heterogeneity of expeditions and fieldwork in the development of earth sciences. First, both are essentially connected to practices. Second, we have to explore the question of space and place in relation to expeditions and fieldwork, and third, the interplay between practices and scientific discourse, or theory and fieldwork (observation), needs to be analyzed. All three perspectives help us understand what was specific to the science of geology in terms of fieldwork and expeditions that also involved other sciences.

2 Defining Scientific Expeditions

In standard histories of geology, which mainly focused on ideas (Geikie 1905; Gohau 1990; Oldroyd 1996), neither expeditions nor fieldwork were not explicitly included as a category or were not discussed together. Let us first look at expeditions from a general point of view. Much has been written about expeditions by historians from a national, post-colonial, or environmental perspective (Driver 2005; Stuhl 2016). From the seventeenth century, expeditions have played a significant role in acquiring knowledge about the world and bolstering imperialism (Miller and Reill 1996; Nielsen et al. 2012). As Roy MacLeod points out, expeditions and science became inextricably intertwined over the course of the nineteenth century (MacLeod 2009). They had their heyday as spectacular and heroic endeavors during the nineteenth and early twentieth centuries. Expeditions can be considered enterprises governed by metropolitan "centres of calculation" (Latour 1987). The traditional attention focused on the heroic European explorer has increasingly been replaced by critics of colonialism (MacKenzie 1990; Gascoigne 1998; Palladino and Worboys 1993) and by interest in organizational aspects (Anderson 2018). The focus has shifted from the performance of the individual to the connection among the participants. In this context, historians have given particular attention to communication with local people during the expedition. Traveling in connection with fieldwork is a cultural technique that is central to the practices of earth sciences. However, if we adopt a wider notion of expedition that also includes travel and mobility, we must consider the meaning of an expedition and how we can understand the role of expeditions in this wider sense for the development of geology. The term "expeditio" in Latin refers to a campaign, completion, or execution. In the German administrative bureaucratic context, the term "Expedieren" means dispatch, take care, complete. This refers to the meticulous scientific preparation for an expedition, which influenced both the journey and the expedition itself (Klemun 2012; Thomas 2015a, b). Instructions issued by the institutions and principals behind an expedition became a distinct genre (Despoix 2005). There is a connection between the formation of geology and the emergence of manuals providing specific guidance for practicing geologists during the eighteenth century. These guidebooks outlined the essential requirements for specific practices and interactions in nature (Vaccari 2007). In this respect, every journey that meets these requirements, especially in the sense of having a specific elaborated image of practices and a clear target, can be understood as an expedition. What can be classified as expeditions encompasses journeys of individual travelers, who embarked on short-term expeditions to well-known areas, to the great expeditions lasting several years with the aim of exploring previously undiscovered regions of the world or uncharted places in nature. In order to emphasize this diversity, the term "scientific expedition" could be used as an umbrella term. It defines a culturally and historically specific mission carried out by a group of individuals with specific work tasks, guided by "epistemic virtues" and with the aim of reducing the unknown as well as systematically acquiring, collecting, and documenting knowledge (Klemun and Spring 2016). "Thought collectives" (Fleck 1935/1980) played an important role in the planning of the expeditions and in the evaluation of the documentation.

3 Practices

3.1 Field and Laboratory

Let us first have a look at a concept suggested by Kuklick and Kohler which changed the understanding of field research on a general level (Kuklick and Kohler 1996; Kohler 2002). Kohler and Kuklick have valorized fieldwork as a practice by positing it in relation to laboratory research. The laboratory has long been privileged in the history of science. In contrast to the field, the laboratory is characterized by a culturally defined "placelessness," since, from the point of view of place as an analytical category, laboratories are independent of their surroundings. Their detachment from nature is marked by control, repeated intervention and domination, simplification, normalization and standardization, formalization and production. In contrast, the practice of fieldwork involves movement, observation, collection, description, and representation (Nielsen et al. 2012). Thus, the field implies, according to Kuklick and Kohler, an ideal-typical scenario of direct engagement with the environment.

Today, field research is no longer viewed as separate from laboratory work, but rather as an interaction between the two. However, the field has no direct influence because field researchers are exposed to a plethora of natural variables, which means overload and irritation. It would be naïve to assume that natural scientists engage with nature without prior theoretical considerations or preoccupations, particularly as observation is theory-guided and theory-laden (Duhem 1998). On the other hand, practices are "embedded in procedures, routines and knowledge, which guide without fully prescribing the way" (Bourguet et al. 2002). This is where Ludwik Fleck's concept of a "thought collective" and the influence of group pressure come into play. Historians of science could interpret this idea in relation to the interconnected practices of fieldwork, laboratory work, and the storage of instruments, materials, and bodies. As Latour has demonstrated, during a pedological-botanical excursion in the Amazon region, a cascade of epistemic contexts occurred in which neither pure field research nor pure laboratory work could succeed. He emphasized how the natural situation was shaped by the use of mapping instruments, recording, labeling, and a type of "proto-laboratory" that preceded actual field research (Latour 1999: 36).

In nineteenth-century geological fieldwork, the selection of topographical maps and their scales, as well as preparatory work, played an important pre-structuring role. We can understand this practice as a mode of the "proto-laboratory" that was a prerequisite for field research. The sorting and preparation of a collection of rocks at a cultural location formed a synopsis of the discoveries made in the field, and at the same time the original determinants of the natural objects were also present. This means that Latour favored using mutually complementary research strategies of laboratory and field research.

Klemun and Spring (2016) have addressed the question of the extent to which the laboratory was present in the field on expeditions in a very different way from the dichotomy between experiment and fieldwork. They have proposed understanding expeditions as experiments, serving as a heuristic tool that offers substantial benefits. Experiments provide a platform for testing new constellations in gaining knowledge, new practices, new organizational forms, identification, and objectives. In addition, they are also centers of negotiation, eliminating the dichotomy of center and periphery, and offering spaces for cooperation. We can consider circumnavigations of the globe and their ships as a laboratory and as a central reference point for mobile field research, particularly since in this floating laboratory the previously selected natural phenomena had already been prepared, identified, labeled, and thereby transformed into "epistemic objects" (Rheinberger 1997). So there is a close connection between field and laboratory research in museums (Kohler 2007: 1–17), surveys, and collections at universities, on ships, in tents or stations during and after the expedition. It is also methodologically essential that both in the laboratory and the field, the physical presence of the researcher contributes to scientific knowledge. As with all living beings, those conducting research in the field are inevitably influenced by the environment and have a reciprocal relationship with it. In geology this manifests itself in the fact that journeys, explorations, and surveys take place often in the summer – only in one particular season — while the work of synthesis takes place in the centers over the winter (Rudwick 1985: 40). To equip geologists for fieldwork, a particular form of compulsory clothing developed over the course of the nineteenth century, which also distinguished traveling geologists externally from teaching professors or members of an academy, where uniforms were common. In sum, it is important to understand the relationship between the laboratory and the field as central to the expedition phenomenon.

3.2 The Value of Travel, Instructions, and the Emergence of Fieldwork in Establishing Geology

The immediate link between geology and fieldwork is taken for granted by today's geologists in the classical setting. However, from a historical viewpoint, fieldwork was not always a given and had to be introduced and accepted epistemically among researchers as a means of knowledge production. This happened toward the end of the eighteenth century, when fieldwork-based earth science began to be established through a new respected practice: traveling. It is no coincidence that those researchers who were among the pioneers in conducting fieldwork in 'mineralogy' (at the time, they were not referred to as 'geologists' but rather as naturalists, 'oryctologists,' or 'geognosts' in German-speaking countries) and had an interest in gaining knowledge of the internal structures of rock formations also produced their own travel instructions.

In early modern times, travel already played a significant role in the professional life of many researchers. One example of this is Nicolaus Steno (1638–1686), who is readily acclaimed in accounts of the history of geology as "central to the birth of science" (Gohau 1990: 570). Steno traveled to Italy, France, and the Netherlands. In particular, he visited the famous mines in Banská Štiavnica, today in Slovakia. But this is not the important point in this context, as researchers have always traveled for a variety of reasons, including education, training, or curiosity. Only the explicit recognition of travel as a specific practical means of advancing geological knowledge is crucial for this transformation. This was accompanied by the epistemic appreciation of fieldwork (Rudwick 2005), which was reflected in the emergence of introductory books between 1760 and 1840. This new genre, whose rise can also be seen as related to the development of geology as a discipline (Vaccari 2007), was designed to assist natural history researchers in their travels and observations. Among them were John Woodward's *Brief Instructions for Making Observations* (1696, 1728), Carl von Linné's *Instructio Peregrinatoris* (1760), Horace-Bénédict de Saussure's *Agenda* (1796), de Motte Hacquet's Instructions for Being on the Way (1796: 222–246), Christian Carl André's *Anleitung zum Studium der Mineralogie* (1804), Cyperien Prosper Brard's *Manuel du Minéralogiste* (1805), Georg Gottlieb Pusch's *Geognostischer Katechismus* (1819), Ami Boué's *Guide du Géologue-Voyageur* (1835–1836), Henry De la Beche's *A Geological Manual* (1831), Karl

Caesar Leonhard's *Agenda Geognostica* (1838), and Charles Darwin's Geology as part of *A Manual of Scientific Inquiry* (1849). The titles of these introductory books reflect a shift from mere mountaineering expeditions to geognostic or geological ventures.

Initially, the focus was on the practice of collecting natural phenomena while at the same time ensuring no common features were overlooked in a certain area (as in Woodward's work). Soon, a growing number of proposals were made as to what criteria should be taken into account when observing the surface of the earth. Central to this was the characterization of the surface forms of mountains and valleys that had to be addressed by means of research questions.

Horace Bénédict de Saussure's *Agenda ou tableau général des Observations* (*Agenda or General Table of Observations*) was given particular importance, not only by his contemporaries but also by research publications written by historians of geology. This was partly due to the fact that it was very well received and translated into German and English (Carozzi 2000). The Scottish geologist Charles Lyell also advised his students to travel, travel, and keep on traveling, as the most important requirement for the work of a geologist (Lyell 1830–33: 202–215).

The development of fieldwork and the establishment of geology as an academic discipline over the course of the eighteenth and nineteenth centuries was accompanied by the emergence of guidebooks. Specific tools played a central role, which is discussed in more detail in the next section.

3.3 Instruments in the Field: The geologist's Hammer as Tool, Sign, Badge, and Emblem of the Geological Profession

Saussure considered the geologist's hammer the most important piece of equipment, soon overtaking the wedge, the chisel, and the tongs. The predecessor of the geologist's hammer was the mountaineer's or miner's hammer, from which the newly designed geologist's hammer was already distinct by the end of the eighteenth century, when its form was being perfected and discussed in instruction books (Klemun 2011). Saussure did not have himself portrayed in the seclusion of his studies, but outdoors with his working equipment and the hammer, which can be seen as an indicator of a new iconography of geology as a discipline to be practiced as fieldwork (Klemun 2000; Rudwick 2005). Previously, naturalists had preferred to be portrayed in their workrooms or, at best, surrounded by books in their libraries. Drawings of various forms of the hammer were published in instruction manuals as patterns or models from which a blacksmith could make a hammer on order. In fact, the literature of the first half of the nineteenth century is replete with evidence of great interest in, and constant demand for, the special design and production of hammers for geologists. In the course of the assimilation of this practical tool into scientific practice, the simple miner's hammer mutated into the geologist's hammer, with its special meaning associated with the specific profession and today widely used as a logo by scientific geological organizations (Klemun 2011). From the twentieth century, the hammer has been produced industrially. To this day, it serves as an identity marker for the geological profession; no other science has such a well-known tool, sign, and badge.

As early as 1980, the historian of science Derek J. de Solla Price criticized earlier studies of tools, pointing to the "blatant and naive insistence that scientific instruments arose as tools for measurement … . and testing hypothesis by experiment" (Price 1980: 75). Historians of science today understand tools as "mediators" (Meinel 2000: 11) and emphasize processes of mediation. The geologist's hammer is one of many instruments that is not suited for exact measurement. Since the emergence of photography as an important documentation tool in the second half of the nineteenth century, the hammer has been used to indicate the size of rocks and structures. It is also a special tool for the investigation of rock formations, employed in the field to break up rocky outcrops and obtain manageable fragments with new fracture surfaces. It is also used in the search for fossils. The famous petrographer Hans Höfer von Heimhalt (1843–1924) gave a succinct and unambiguous explanation of the use of the hammer in his *Instructions on Geological Observation, Mapping and Profiling*, published repeatedly from 1915 onward and originally intended for his students at the Mining Academy in Leoben (Styria, Austria). As Höfer explained, "the hammer is not only used to break and crush the rock formation but also to format it: this consists of bringing the rock samples into a rectangular shape and achieving at least one fresh fracture line" (Höfer 1915: 17). The high esteem in which the hammer was held as a sign of fieldwork was devalued when theory as such regained value among contemporaries. Over the course of the nineteenth century, the long-term devaluation of the hammer in the explorer's arsenal inevitably followed the process of professionalization and the increasing importance of theoretical questions in geology. For example, when James Geikie, brother of Archibald Geikie, professor in Edinburgh and later president of the Geological Society of London from 1890, pointed out "the failure of geologists to furnish an adequate theory" of the question of the causes of cosmic changes in climate, he spoke condescendingly of "our hammer-bearers" (Geikie 1874: 110), suggesting that many practitioners and fieldworkers had a narrow empirical focus and lacked theoretical imagination (Klemun 2011).

Many surveys equipped their geologists with geologist's hammers bearing the institution's logo. In the sense of Ludwik Fleck's "Denkkollektiv" (Fleck 1935/1980), this equipment underlined collective solidarity. The geologist's hammer is a striking symbol of fieldwork, but it is also important to consider the historical context in which it appeared and the meaning that contemporaries ascribed to it.

3.4 The Increasing Importance of Field Practice: The Life of a Protagonist

The increasing importance of field research to geological practice during the eighteenth century was also evident in the generation of naturalists who increasingly identified themselves as geologists (Rudwick 2005). Let us take a look at one protagonist who represents the period in which geology came into being and in which fieldwork gradually became more important. The life of Nicolas Desmarest (1725–1815) is a prime example of this change, as Ken Taylor's source-critical historical work demonstrates. Initially, Desmarest had educated himself exclusively by reading, with some emphasis on texts of antiquarian or historical scholarship—which he viewed as providing evidence regarding changes on the Earth's surface—and of physical geography, particularly bearing on worldwide patterns in the configuration of mountain ranges. (Taylor 1969: 342). In the next phase of his self-fashioning, he began to collect mineral specimens for utilitarian purposes and participated in the "promenades savantes" organized by Bernard de Jussieu (1699–1777) in the countryside around Paris (Taylor 2001: 59). Botanists had established a special tradition of fieldwork decades earlier (Allen 1976; Larsen 1996). Field experience came about through the development, learning and refining of skills, and there was a growing awareness among this generation of geologists "regarding the cultivation of appropriate empirical methods of investigation, symbolized by (but not confined to) the act of geological fieldwork" (Taylor 2001: 59). The year-long journey to Italy in 1765–1766 as the companion of the Duc Louis-Alexandre de la Rochefoucauld can be seen as a further manifestation of the increasing intensification of fieldwork in Desmarest's life. Travels to Italy were also important for other geologists of his time: the establishment of contacts, direct discussions with Italian experts in the field, and the confrontation with a different and geologically diverse country were the main features. This experience led to the realization that "distinctions among successive formations" (Taylor 1995: 103) and agents that destroyed them were essential. A further increase in the importance of fieldwork came with the desire to map the Auvergne geologically, after the naturalist Jean-Etienne Guettard had recognized the volcanic character of the region. Desmarest "made the additional find that the abundant prismatic basalts, whose configurations had long been considered indicative of an aqueous formation, were associated with the congealed lavas and must be volcanic as well" (Taylor 1994: 130). The collaboration with ingénieur-géographe François Pasumot and Dailley lasted longer than Desmarest had anticipated. The map set new standards for the interpretation of the terrain and precise graphic representation and positioning of different rocks. For leveling, Desmarest was able to use a newly manufactured graphometer with telescopic sights, made by Cavinet. Desmarest had a portable barometer for leveling (Taylor 1994: 133). Maps were developed as the most important "visual language" (Rudwick 1976) in geology, particularly as an aid in the process of fieldwork. In the eighteenth century, however, they still had no relation to the much richer verbal description of data that characterized the work of the most highly esteemed geological empiricists such as Horace Bénédict de Saussure, Peter Simon Pallas, Déodat Gratet de Dolomieu, and Belsazar de la Motte Hacquet (Taylor 1985). Demarest's biographical development as a geologist was used to show how field research became increasingly important, determining the praxis of geologists in the eighteenth century. Travel played a decisive role in this. It was only in the nineteenth century that topographic maps became the point of departure into which the stratigraphers and geologists of the national surveys integrated the results of their fieldwork, but what role did observation play in this context?

3.5 Observation and Mapping: Fieldwork and Stratigraphy

There is arguably no better practice than observation as a precondition for gaining knowledge that is so obvious, fundamental, omnipresent, and at the same time so ambiguous. Every observation, of whatever kind, is a central component of active questioning. What is questioned depends on the geologist's collective group which is always part of a "Denkstil" (Fleck 1935) determined by shared epistemic values. And questioning is directly linked to documentation. This includes the most varied forms of recording, the "little tools of knowledge" (Becker and Clark 2001), such as lists and labels as well as journals, notebooks, diaries, sketches, pictures, and instruments. Drawings became necessary as mimetic instruments to record observations. New ways of recording observations were constantly being discovered in geology as the most important component, and for this reason the productivity of observation is essential to fieldwork. Here we may follow Daston's apt description: "As a practice, observation is an engine of discovery and a bulwark of evidence" (Daston and Lunbeck 2011: 7). Often, observation is not only the means, but also the end, and "a learned reflection as a distinct form for knowledge" (Daston and Lunbeck 2011: 12) with its own standards and conditions. In the protagonists' self-representations, observing and collecting appear predominantly as inseparable characteristics of the practices used during fieldwork. Observation dominates perception, the activation of the senses, the selection of phenomena, and the orientation of questioning and assessment. The registering process during the observation procedure created and developed new terms of classification. In the early nineteenth century, stratigraphy emerged. Secord has emphasized that "for an understanding of the practical activity of geological classification we must follow them [the geologists who were mostly stratigraphers] from the rooms of the Society into the field" (Secord 1986: 25). In their baggage, British geologists carried the introductory books, the hammer, a notebook, and instruments such as the barometer and compass. In order to give themselves a preliminary overview of a district, they first had to find an ideal mountain or other elevation from which they could determine in advance either the parallel lines of the traverses for their sections or at right angles to the outcrops of the strata. With this initial assessment in mind, the next steps in the route were planned, following the dip and strike of the strata encountered. Secord has reconstructed this procedure in detail on the basis of De la Beche's contemporary manual (1831, esp. 598–606), and for this reason we shall follow him:

> The angle of dip, usually estimated by sight, measured the maximum inclination of tilted strata from the horizontal; the point of the compass towards which the rock beds inclined was known as the direction of dip. The strike of the strata, always at right angles to the direction of dip, represented the line of intersection of the tilted plane of the strata with the horizontal. Along the line of his traverse the geologist also noted the lithological characters; thicknesses; and fossil contents of each outcrop, The resulting traverse (also known as horizontal) sections united isolated exposures and observations into a connected slice of the succession, cutting across the countryside like a trench. Such sections were best exposed along stream beds and costal cliffs, two of the favourite haunts of the Victorian geologist. In addition to

> these traverse sections, geologists very occasionally traced out an individual stratum along the entire length of its outcrop. Mapped in this way, a bed with distinctive lithology could serve as a marker horizon, with its appearance in the separate traverse sections allowing them to be coordinated into a general picture of regional geological structure. (Secord 1986: 25/26)

This practice became standard in the nineteenth century and perhaps not only in Britain but also in Switzerland, France, and the Austrian Monarchy, as well as in the non-European colonies.

Bourguet et al. have argued in general that there is a close relationship between mapping and observation. "Both depend on the ordering and co-ordering of places of performance and dialogue, of accounting and recounting" (Bourguet et al. 2002: 8). Whereas maps had previously been based on the exegesis of ancient geographical descriptions and travel accounts, from the seventeenth, and especially from the eighteenth century on, the replacement of narratives by instrumentally acquired data changed the role of the use of instruments, barometers to determine height, and compasses to determine direction. The prerequisite was that these instruments should be standardized and decoupled from the place of their original production, in order to be generally usable in field research anywhere (Richard 1996). Bouguet et al. understand these processes as "delocalizing or decontextualizing knowledge as a consequence of the relationship between instruments, travel and science" (Bourguet et al. 2002: 9).

Even though observation is a very difficult variable to determine, it is a crucial aspect of field research in the context of expeditions. This bridging function will be analyzed in more detail in the next section.

4 Places and Spaces

4.1 Extension of Space Via Expedition, Intension Via Fieldwork

Expeditions and field science are fundamentally connected to space and place (Ophir and Shapin 1991). It is about here and there, the contrast between the familiar and the unfamiliar, and, as Geikie put it, "home and abroad" (Geikie 1882). While expeditions seem to expand space, leading to the further spaces in the sense of expanding space, fieldwork goes deep into a place in the sense of deep concentration and observation of places.

At the end of the eighteenth century, some sites were not only discovered by locals and foreigners, but also held a special fascination for hybrid visitors, such as naturalists, mining officials, and travelers. Think of the Giant's Causeway on the coast of Antrim (now Northern Ireland) or the Campi Phlegraei in Italy (Rudwick 2005). A typical example is Auvergne, which attracted almost 80 naturalists, mining experts and geologists between 1751 and 1800, who visited the volcanoes and wrote about their experiences (Taylor 2007). There were many reasons to visit the site: to discover extinct volcanoes in the heart of France with one's own eyes, to map the volcanic terrain, and to interpret the recognition of the association of columnar basalt with the products of volcanic action in geohistorical terms. The Auvergne became a "geological mecca," "a fixed element of an international canon for the informed geologist" (Taylor 2007: 74). It was the most prominent place when geology was in the process of developing "a place-specificity" and a dazzling example of the argument that local experience, combined with unfamiliar features, had significance in developmental theories, if we consider the most prominent eighteenth-century controversies between neptunistic, vulcanistic, and plutonistic concepts (Fritscher 1990; Guntau 2009).

On many expeditions, however, the short duration of voyage meant that it was not possible to carry out this fieldwork in a way that adequately met the explorer's high standards. This was the case with Bonaparte's Egyptian Expedition (1798–1801), when the mineralogist Déodat Gratet de Dolomieu, frustrated by the obstacles that prevented him from properly carrying out his observations of the volcanoes and mountains he expected to find, withdrew his participation prematurely in 1799. This expedition was unprecedented, not only for the number of its participants (60, including 40 polytechnicians), but also for the 35,000 military men who took part. Previous academic expeditions around the world had only included individual scientists and artists, such as expeditions leaded by Bougainville, Cook, La Pérouse, and Malaspina. The Egyptian Expedition, however, had a dual purpose: a scientific and a military one (Herald 1992). Initially, an institute was moved to Cairo, and the particular field explorations around Cairo and in the south were closely linked to the army and its operations. All departments, including the military, were directed toward a common object: space, territory, and its material appropriation. At this point we might ask, "to what extent does science depend on the place where it is practised?" Bourguet, who has studied this expedition in great detail, gives the following explanation: "the framework of an unequal, asymmetric discourse emerges from the contrast that the French established between the science they practiced in Egypt and the science they knew in the metropolis. The former – and, science in general – practised outside Europe in colonized lands or countries recently opened to Western influence – was shaped by the obstacles which, from the exterior, hindered its progress. In other words, for the French scientists, the construction of knowledge in Egypt was a result of the social, cultural and material conditions of its production and it shared the limitation of the colonial context. Science in the metropolis was liberated from the bounds of its context" (Bourguet 2002: 100). The success of the expedition, however, was due in no small part to the rich collections of drawings, the measurements of monuments and landscapes, and the precise topographical descriptions made with compass and theodolite. For reliable distance measurements, they relied on the step of a camel – a "veritable animal pendulum" (Bourguet 2002: 105). Even more lasting was the monumental publication under the auspices of the government, in which the maps had to remain secret until 1818.

4.2 Expeditions Building on Previous Enterprises and Pointing to Future Ones

If one argues for the importance of an expedition in terms of its results, the 29 volumes of Alexander von Humboldt's (Beck 1959–1961; Daum 2019) travels to the Americas, with its panorama of culture, nature, and politics, would probably rank first among many examples. It was very different from another narrative of Humboldt's travels, based on his Russian-Siberian journey of 1829 and concentrating on climate, terrestrial magnetism, and geology. In contrast to the work on the Americas, there was a clear division of labor, with additional members of the expedition involved. Humboldt had given up his previous financial and political independence and, after 4 years of preparation, embarked on the expedition initiated by Nicholas I.

He was accompanied by the biologist Christian Gottfried Ehrenberg, the mineralogist Gustav Rose, his own right-hand man Johann Seifert, the Russian mining official Dmitri Stepanovich [Stepanovič], Menschenin [Mern'šenin] a cook, two servants, and a courier. They traveled in four coaches, with 12,000 horses at their disposal. In the New World, Humboldt had covered some 8,000 km in 5 years, while in the Russian Empire the group traveled more than 18,000 km in 8 months (Suckow 2001: 253). What was important was the high speed, which placed an enormous burden on the field research. A clear manifestation of the density of the scientific work and the incredibly high speed of travel are the diaries, which even in their handwriting give clear evidence of the cramped and incomplete style of reporting (Kraft 2018: 64). While Humboldt initially showed little enthusiasm for the barren landscape, which he perceived as "unnatural," he was all the more fascinated by the visits to the mines around Yekaterinburg.

It was only through the deliberate suppression of political and cultural phenomena in Humboldt's *Fragmens de géologie et de climatologie asiatiques* [*Fragments of Asian Geology and Climatology*] (1831) that there was a focus on the analysis of the Asian mountain ranges and their geological origins and on the connection between the earth's surface, temperature, and climate variation. Humboldt was extensively preoccupied with attempts to explain the special geological problem of various subsidences in the Caspian Sea. A quarter of the pages in this work were annotations that Humboldt owed to correspondents such as Alexander Kasimowitsch Kasem-Beck (Kazembek), Anton von Klostermann, Heinrich Lenz, François Désiré Roulin, and Julius Klaproth, all of whom knew the area well. It was not by chance that Humboldt had to defend himself against the criticism of the Berlin geographer Carl Ritter that he had written his books with the help of others. More important was his three-volume work on his trip to Russia, entitled *Asie centrale. Recherches sur les chaines de montagnes et de la climatologie comparée* (2 Vol. 1843), on which Humboldt had worked for more than 10 years. He used the time particularly to gather information from colleagues in the relevant fields. Humboldt considered the work to be so important for geological research "that it throws the entire ancient mountain geography of Asia into confusion," as he wrote to his publisher Johann Georg Cotta in 1849 (Humboldt and Cotta 2009: 360).

Recent scholarship has also seen the importance of this work in the fact that his magnetic research brought about a worldwide expansion of special observatories and further expeditions (Kraft 2018). Humboldt's travels and travel writings are a particular example of how each expedition was not only a singular event but also a process: expeditions built on previous enterprises and observations of contemporary travelers, and pointed to future ventures. At the same time, each had its own rules, depending on circumstances.

Geographically, Humboldt did not enter new territories. His intensive preparation included reading all the reports of his predecessors, such as those of the "Great Nordic Expedition" or the Kamchatka Expedition (1733–1744) on behalf of the Russian state, as well as overland expeditions sent out by the Petersburg Academy of Sciences, especially under the direction of the natural scientist Peter Simon Pallas, and also Heinrich Julius Klaproth's expeditions to the Caucasus and Georgia as far as the borders of China, as well as those of many other authors.

Humboldt considered Pallas to be a multi-talent, but his theses on the formation of mountains in Central Asia were "caught up in the hypotheses of a dogmatic and imaginative geology that was prevalent at that time" (Humboldt 1849, vol. I.: 93). Humboldt's expedition is a striking example of the characteristic of every expedition that each built on previous ones and gave rise to new ones. However, this only worked if the expedition was documented, a phenomenon that will be analyzed in more detail in the next section.

4.3 Expeditions and Documentation in the Nineteenth Century

The reliability of documentation was a key factor for all expeditions, regardless of their differences. Few circumnavigations have attracted as much attention to the question of the relationship between documentation and expedition as that of the HMS Beagle (1831–1836) with Charles Darwin on board. On the one hand, this significance is due to the rich source of Darwin's material that resulted from his complex modes of documentation. On the other hand, it can also be attributed to the work and painstaking research of the historian Sandra Herbert. Darwin saw himself as a geologist before becoming a biologist on the basis of his theory of selection (published in 1859). Darwin's recording system during the expedition was multi-faceted: it consisted of field notebooks, of four specimen notebooks, the "geological notes proper, generally organized according to locality and running to 1383 pages, as opposed to 368 pages for the zoological notes" (Herbert 2005). The latter had to demonstrate his own system of recording: one page was always left blank, reserved for additional notes. The records were also interspersed with numbers referring to the specimen, and there were also synthetic essays that Darwin wrote toward the end of his journey, reflecting his observations. At the same time, the notebooks were by no means dedicated solely to his observations, but were also places where he formulated general questions and arguments. In addition, Darwin developed a second argument, entitled "reflecting reading my geological notes," which emphasized the attitude of reflection in a particular way. He thus juxtaposed observation and reflection.

Larger frameworks were dominant, less determined by a particular site. This ensured a step-by-step approach, which can also be seen as an intermediate step toward a later publication.

Darwin was inspired by Humboldt, from whom he not only adopted the "habit of viewing the Globe as great whole" (Herbert 2005: 15) but also the ideal of simplicity and its style. According to Herbert, Darwin belonged, "as writer of his travel narrative to the past," but as a geologist, theoretically combining his considerations, he pointed to the future. For Darwin did not work within the paradigm of stratigraphy, but was guided by the search for a geology determined by other components: "namely, understanding the vertical motions of earth's crust, elevation and subsidence": "Part of Darwin's scheme for a simple geology has survived close to the form in which he imagined it: his theory of structure and distribution of coral reefs, which emphasized the role of subsidence in the creation of coral islands, most strikingly atolls. Coral Reefs held a special place in Darwin's estimation, and, alone of three parts of his geological trilogy from the voyage, he rewrote it substantially when it appeared as second edition in 1874" (Herbert 2005: 356). The records of the expeditions were not only the starting point for Darwin's multiple revisions for publication but remained a principle throughout the nineteenth century.

4.4 Empirical Versus Speculative, Observation (Field Science) Versus Geological Theory, Local Knowledge Versus Universal Knowledge

To better understand knowledge production and acquisition in geology, we need to explore the interplay between practices (fieldwork) and scientific discourses (theory). The misleading historical notion, and common narrative today, that observation and system (theory) were antithetical in the eighteenth century goes back to Charles Lyell's historical introduction in his *Principles of Geology* (1837). Of the well-known empirical geologist Horace-Bénédict de Saussure, Lyell wrote that he "employed the chief portion of his time in studying the structure of the Alps and Jura, and he provided valuable data for those who followed him. He did not pretend to deduce any general system from his numerous and interesting observations; and the few theoretical opinions which escaped from him, seem, like those of Pallas, to have been chiefly derived from the cosmological speculations of preceding writers" (Lyell 1838, Vol 1: 80). On the one hand, traditional cosmological theory was seen as an obstacle to all progress, while observation seemed to pave the way for generalization only slowly. The division between two ways of doing geology, the empirical and the speculative, was also emphasized by Karl Alfred Zittel in his influential *History of Geology and Palaeontology* in 1898 (Tamborini 2015a, b). Under the headings "Hypothesis of the Earth's Origin and History, and Beginning of Geological Observation" Zittel argued that "the keen interest in minerals and fossils and the flourishing condition of the mining industry gradually attracted the attention of scientific men to the investigation of the earth itself. Two methods of research, the empirical and the speculative, developed alongside one another. The one had for its immediate aim the determination of facts, and in its further outlook, the possible construction of some suitable theory; the other contented itself with a minimum of observation, accepted the risks of error, and set about explaining the past and the present from the subjective standpoint. The latter method naturally attained no higher results than the geogenetic fantasies of classical antiquity. And it certainly could never have gathered sufficient energy to roll aside the mass of philosophical and doctrinal tradition that blocked the path of progress" (Zittel 1901: 23).

For a long time, the history of the development of geology was also understood as a simple displacement of the theory of the Earth by field geology. Frank Dawson Adams (1938) spoke of fables, of which he mentioned Buffon's *Theory and Epoques* as the best example: "the close of this long period of imaginative effort" (Adams 1938; Taylor 1992). The polar opposition between speculative theory of little value and good empirical work by naturalists such as Simon Pallas, Horace B. de Saussure, Dolomieu, and Mitchel was also taken for granted by Gillispie: "The science of geology represents a coming together of lore from the ancient practice of mineralogy with speculation about the origin of the earth, seventeenth and eighteenth century cosmogonies which have in them more of science fiction than of science" (Gillispie 1960: 291–292).

It was necessary for historians of geology to paint a picture of a more complicated and ambiguous relationship between theory and factual knowledge, rather than a black and white narrative. Rachel Laudan has rejected such a simple contrast between "speculators" and "empiricists": "All the geologists in the eighteenth and nineteenth centuries expected theory or 'system' to play a role in geology. And they all demanded that theory should be warranted by evidence. The point at issue was not whether to opt for theory or for fact gathering. Rather, it was relationship between theory and facts" (Laudan 1987: 8). In contrast to the old dichotomy, Kenneth Taylor has argued that the old tradition of cosmology was an integral part of the making of the new geology in the second half of the century. Based on close readings of Desmarest's sources, Taylor has demonstrated that Nicolas Desmarest's observational preoccupations were derived from theories of the Earth written by scholars such as Louis Bourguet, Nicolas Antonie Boulanger, and Georges Louis Buffon. They emphasized natural "regularities" in what they often referred to as "dispositional" regularities in the configuration of geological features. The marginalization of the extraordinary and irregular in observation was shared by many scholars. Jean-Louis Giraud Soulavie, for example, wrote in his *Natural History of Southern France* that the naturalist must follow nature in the most general and widest operations. In Desmarest's case, this meant that he had already "settled on a clear position about how observations must be treated: their significance depended upon their integration into a framework of generalization – observations must be used to find general relations, which in turn offer some prospect of yielding explanatory principles" (Taylor 2001).

In the nineteenth century, field geologists were still in the majority, given the large number of surveys responsible for mapping of national territories. The work of Rudwick, Oldroyd, and Secord valued mapping as both a theoretical and an observational activity. Le Grand has suggested that field geologists, working in specific geographical regions on which they focused their research, were particularly attached to a kind of "localism" (Grand 1986). This was often framed nationally or according to a "thought collective" (Fleck 1935). Questioning Le Grand's term, Oreskes suggested that it would be better to speak of "epistemological affinity" –

meaning that "geologists would weigh the available observations differently, based on their affinities" (Oreskes 53).

Naomi Oreskes, who has studied in depth the decades-long rejection in the United States of the continental drift theory proposed by Alfred Wegener (1880–1930) in 1912, considered field research as one factor among many. She has identified three research traditions: field mapping, geodetic surveying, and atomic physics. Uniformitarianism had a chance in the debate because Wegener allowed the field evidence to be interpreted logically (Oreskes 1999: 314). As field research lost some of its authority during the twentieth century, so did the specific theories attached to it.

However, many geologists continued to work in the field because of the authenticity it provided, while in the twentieth century at the latest, the decline of fieldwork was heralded by laboratory geologists. As the precision and accuracy of laboratory research was now sufficient for orientation, the value shifted more toward laboratory research. New instruments, which, according to Galison, produced new data in physics were far less decisive in geology: "They rejected well-established evidence-based traditions in favour of a new tradition whose reliability could have been considered as suspect as the theory it was used it to support" (Oreskes 1999: 311).

Field research, previously the unquestioned backbone of geology, had lost its dominance. But what remained was the habitus, the clothing. Even geologists who worked purely theoretically and stayed away from the field adapted the clothing that had been introduced to the habitus of field geologists. The difference to physicists can still be seen today at world congresses of geology.

5 Conclusion

In order to analyze expeditions and situate their specific meaning within the framework of earth sciences, this chapter has first proposed to start from an expanded concept of expedition, in contrast to earlier approaches, and to avoid a dualistic view of exploration and science. While adventure had long been associated exclusively with exploration, this study does not exclude it per se. At best, it had no place in contemporaries' "epistemic virtues," although adventure was of course part of every voyage.

By broadening the concept of expedition, this chapter has discussed traveling and fieldwork practices associated with the establishment of geology as a discipline. It has also taken into account the ambiguity of the expedition phenomenon, an undertaking involving from a single individual up to several thousand people. It is therefore possible to build a new and more consistent historiographical bridge between fieldwork and expedition. The reference to Daston's "epistemic virtues" and Fleck's "thought collective" provides further methodological arguments in favor of a broader concept of expedition and thus for a new emphasis on the heterogeneity of the expedition phenomenon.

For expeditions represented a space of negotiation in which both different "epistemic virtues" and group pressure, in the sense of Fleck's "thought collective," played a crucial role in the preparations prior to departure and during the processing of findings. The chapter has also suggested that rather than making a general and idealized distinction between fieldwork and the laboratory, it is more fruitful to follow Bruno Latour and see the two as interrelated or complementary. Stratigraphic research in the earth sciences, for example, is a prime example of how intensive preparation preceded and followed the actual work in the field.

The value placed on travel and the development of fieldwork is shown by the emergence of travel guides. They are indicative of the specific profile of the earth sciences from the mid-eighteenth century. It is no coincidence that the geologist's hammer was declared the most important tool, distinguishing it from the traditional miner's hammer. It takes us back to the period between 1750 and 1840, when geology was established as an independent science.

There is no question that field research must be regarded as a major preoccupation in the practice of geologists. The process of establishing geology took place at the same time as it was being promoted in instruction books. With its acceptance as an integral part of journeys and expeditions, field research gained recognition and became the axiomatic essence of geology. But how did fieldwork become integrated into the traditional habits of the study room over the course of the eighteenth century? This process happened gradually, as has been shown using the example of one geologist's biography. The starting point for most geologists was self-study by reading books. The transition from the study room to the field took place in the context of mineralogical collecting activities. This was followed by a period of fieldwork with like-minded people. Finally, as a bridge between the study room and the field, mapping became a central goal of the geologist.

The close relationship between mapping and documentation can be seen as specific to geology. Both are involved in a mutual process of fixation and reference. The use of measuring instruments led to a close relationship between travel and research. Its importance is demonstrated by the combination of the significance of observation, documentation during a journey, and relevance to the formation of knowledge, always in close interaction with theoretical concepts. Journeys and expeditions, in turn, led to an expansion of territorial knowledge, initially focusing on Europe, then on the colonial world and beyond. Field research, however, centered more intensively on a particular site. Large-scale expeditions, such as Napoleon's in Egypt, could also pose problems characteristic of the tension between expedition, movement within a particular territory and field research on site. No other field of knowledge has such a clear symbol of its professional basis as geology, which gave fieldwork a high status by means of the geologist's hammer and which, by no means coincidentally, still uses the geologist's hammer as a batch and symbol today, despite its loss of importance. And yet there is another specificity in the earth sciences. This is the close dependence on places and spaces, which has been referred to as "place specificity" of the earth sciences.

One site could become so famous that it was visited as a “geological mecca” by every geologist who could travel there. Expeditions also evoked successor enterprises. In any case, they had to be well documented, like those famous ones of Humboldt and Darwin. The practice of documentation consisted of various forms of recording, which fascinated both later expeditioners and the historians of science who have increasingly studied these practices in recent decades.

The relationship between empiricism and speculation, between fieldwork and theory, has long been understood as antithetical. In contrast, this chapter has argued that it is important not to choose one over the other, but discuss the relationship between them. Mapping, for example, has been assessed by historians such as Secord and Rudwick as both a theoretical and an observational activity. Geologists assessed observation according to their affinities and different scientific traditions or thought collectives. In the rejection of Wegener’s drift theory, fieldwork and observation were only one factor among many. In the twentieth century, however, fieldwork lost its status and the theories associated with it.

References

Adams, Frank Dawson. 1938. *The birth and development of the geological sciences*. Baltimore: Williams and Wilkins.

Allen, David Elliston. 1976/1994. *The Naturalist in Britain*. Princeton: Princeton University Press.

Anderson, Katharine. 2018. Natural history and the scientific voyage. In *Worlds of natural history*, ed. H.A. Curry et al., 304–318. Cambridge: Cambridge University Press.

André, Christian Carl. 1804. *Anleitung zum Studium der Mineralogie, für Anfänger*. Vienna: Camesianische Buchhandlung.

Beck, Hanno. 1959/1961. *Alexander von Humboldt. Eine wissenschaftliche Biographie*, 3 Vol. Wiesbaden: Franz Steiner Verlag.

Becker, Peter, and William Clark, eds. 2001. *Little tools of knowledge: Historical essays on academic and bureaucratic practices*. Ann Arbor: The University Michigan Press.

Boué, Ami. 1835–1836. *Guide du Géologue-Voyageur, sur le Modèle de l’Agenda Geognostica de M. Leonhard*, 2 Vols. Paris: F. G. Levrault.

Bourguet, Marie-Noëlle. 1997. Le collecte du monde: voyage et histoire naturelle (fin XVIIème siècle –débuit XIXème siècle). In *Le Muséum ou premier siècle de son histoire*, ed. Claude Blanckaert, Claudine Cohen, Pietro Corsi, and Jean-Louis Fischer, 163–196. Paris: Éditions du Muséum d’Histoire Naturelle.

———. 2002. Science and memory: The stakes of the expedition to Egypt. In *Taking liberties. Problems of a new order from the French revolution to Napoleon*, ed. Howard G. Brown et al., 92–109. Manchester/New York: Manchester University Press.

Bourguet, Marie-Noëlle, et al. 2002. Introduction. In *Instruments, travel and science*, ed. Marie-Noëlle Bourguet et al., 1–19. London/New York: Routledge.

Brard, Cyprien Prosper. 1805. *Manuel du Minéralogiste et du Géologue Voyageur*. Paris: Maradan.

Brescius, Moritz von. 2019. *German science in the age of empire: Enterprise, opportunity and the Schlagintweit brothers*. Cambridge: Cambridge University Press.

Carozzi, Albert V. 2000. *Horace-Bénédict de Saussure (1740–1799): Un pionnier des sciences de la terre*. Geneva: Editions Slatkine.

Darwin, Charles. 1849. Geology. In *Manual of scientific inquiry; prepared for the use of her majesty' navy: And adapted for Travellers in general*, ed. John F.W. Herschel, 156–195. London: John Murray.

Daston, Lorraine, and Peter Galison. 2007. *Objectivity*. New York: Zone Books.

Daston Lorraine and Elizabeth Lunbeck 2011. Introduction: Observation observed. In *Histories of scientific observation*, eds. Lorraine Daston and Elizabeth Lunbeck, 1–9. Chicago: The University of Chicago Press.
Daum, Andreas W. 2019. *Alexander von Humboldt*. München: H. C. Beck.
De la Beche, Henry T. 1831. *A Geological Manual*. London: Treuttel and Würtz.
de Saussure, Horace-Bénédict. 1796. Agenda, ou tableau général des Observations et des Recherches dont les résultats doivent servir de base à la théorie de la terre. *Journal des Mines* 4: 1–70.
Despoix, Philippe. 2005. *Le monde mesuré: Dispositifs de l'exploration à l'âge des Lumierès*. Geneva: Librairie Droz.
Driver, Felix. 2005. Distance and disturbance: Travel, exploration and knowledge in the nineteenth century. *Transactions of the Royal Historical Society* 14: 73–92.
Duhem, Pierre. 1998. *Ziel und Struktur der physikalischen Theorien*. Hamburg: Meiner Verlag.
Fleck, Ludwik. 1935. *Entstehung und Entwicklung einer wissenschaftlichen Tatsache. Einführung in die Lehre von Denkstil und Denkkollektiv*. Basel: Bruno Schwabe. Second edition: Frankfurt am Main: Suhrkamp 1980.
Fritscher, Bernhard. 1990. Die Verwissenschaftlichung der Geologie: Zur Bedeutung phänomenologischer und konstruktiver Erfahrungsbegriffe im Vulkanismusstreit. *Sudhoffs Archiv* 74 (1): 22–44.
Gascoigne, John. 1998. *Science in the service of Empire*. Cambridge: Cambridge University Press.
Geikie, Archibald. 1882. *Geological sketches at home and abroad*. New York: Macmillan and Co.
———. 1905. *The founders of geology*. London: Macmillan and Co.
Geikie, James. 1874. *The great ice age, and its relation to the antiquity of man*. New York: D. Appleton and Company.
Gieryn, Thomas F. 1999. *Cultural boundaries of science: Credibility on the line*. Chicago: University Press of Chicago Press.
Gillispie, Charles C. 1960. *The edge of objectivity: An essay in the history of scientific ideas*. Princeton: Princeton University Press.
Gohau, Gabriel. 1990. *A history of Geology* (revised and translated by Albert V. Carozzi and Marguerite Carozzi). New Brunswick/London: Rutgers University Press.
Guntau, Martin. 2009. The rise of geology as a science in Germany around 1800. In *Geological Society, Special Publications* 317, 163–177.
Hacquet, Belsazar de Motte. 1790–1796. *Neueste physikalisch-politische Reisen,* 4 Vol. Nuremberg: Raspe.
Herald, Christopher. 1992. *Bonaparte in Egypt*. London: Hamish Hamilton.
Herbert, Sandra. 2005. *Charles Darwin. Geologist*. Ithaca/New York: Cornell University Press.
Höfer von Heimhalt, Hans. 1915. *Anleitung zum geologischen Beobachten, Kartieren und Profilieren*. Braunschweig: Viehweg & Sohn.
Humboldt, Alexander von. 1849. *Ansichten der Natur mit wissenschaftlichen Erläuterungen,* 2 vol. Stuttgart und Tübingen: Cotta.
Humboldt, Alexander von and Johann Georg Cotta. 2009. *Briefwechsel*, ed. Ulrike Leitner under cooperation of Eberhard Knobloch. Berlin: Akademie Verlag.
Jones, Rupert T. 1875. *Manual of the natural history, geology, and physics of Greenland and the neighbouring regions; prepared for the use of the arctic expedition of 1875, under the direction of the arctic committee of the Royal Society*. London: Majesty's Office.
Klemun, Marianne. 2000. … .. *mit Madame Sonne konferieren. Die Großglockner-Expeditionen 1799 und 1800*. Klagenfurt: Das Kärntner Landesarchiv.
———. 2011. The Geologist's Hammer – 'fossil' tool, equipment, instrument and/or badge? In *Seeing and measuring, constructing and judging: Instruments in the history of the earth sciences*, ed. Ana Carneiro and Marianne Klemun, *Centaurus, An International Journal of the History of Science and its Cultural Aspects*, Vol. 53, 86–101.
———. 2012. Verwaltete Wissenschaft – Instruktionen und Forschungsreisen. In *Ordnung durch Tinte und Feder? Genese und Wirkung von Instruktionen im zeitlichen Längsschnitt vom Mittelalter bis zum 19. Jahrhundert*, ed. Anita Hipfinger et al., 391–414. Vienna: Veröffentlichungen des Instituts für Österreichische Geschichtsforschung.

Klemun, Marianne, and Ulrike Spring, eds. 2016. *Expeditions as experiments*. London/New York/ Melbourne: Palgrave & Macmillan.

Kohler, Robert F. 2002. *Landscapes and Labscapes: Exploring the lab-field border in biology*. Chicago/London: The University of Chicago Press.

Kohler, Robert E. 2007. Finders, keepers: Collecting sciences and collecting practice. *History of Science* 45: 1–27.

Kraft, Tobias. 2018. Das Russisch-Sibirische Reisewerk. In *Alexander von Humboldt-Handbuch. Leben – Werk – Wirkung*, ed. Ottmar Ette Ottmar, 59–72. Stuttgart: J.B Metzler und Springer-Verlag.

Kuklick, Henrika, and Robert E. Kohler. 1996. Introduction. In *Science in the field, Osiris 11*, ed. Kuklick Henrika and E. Kohler Robert, 1–14. Chicago/London: The University of Chicago Press.

Larsen, Anne. 1996. Equipment for the field. In *Cultures of natural history*, ed. N. Jardine, J.A. Secord, et al., 358–377. Cambridge: University Press.

Latour, Bruno. 1987. *Science in action: How to follow scientists and engineers through society*. Cambridge, MA: Harvard University Press.

———. 1999. *Pandora's hope: Essay on the reality of science studies*. Cambridge, MA: Harvard University Press.

Laudan, Rachel. 1987. *From mineralogy to geology: The foundations of science, 1650–1830*. Chicago/London: University of Chicago Press.

Le, Grand, Homer Eugene. 1986. Specialities, problems and localism: The reception of continental drift in Australia, 1920–1940. *Earth Sciences History* 5: 84–95.

Leonhard, Karl Caesar. 1838. *Agenda Geognostica. Hülfsbuch für reisende Gebirgsforscher und Leitfaden zu Vorträgen über angewandte Geognosie*. 2nd ed. Heidelberg: Mohr.

Linné, Carl. 1760. *Instructio Peregrinatoris quam sub praesidio D.D. Car. Linnaei Proposuit. Amoenitates Academicae*, 298–313. Holmiae: L. Salvii.

Lyell, Charles. 1830–1833. *Principles of Geology: Being an attempt to explain the former changes of the earth's surface, by reference to causes now in operation,* 3 vols. London: John Murray.

———. 1837. *Principles of geology: Being an inquiry how far the former changes of the earth's surface are referable to causes now in operation*. 4 Vol., Vol. 1. London: John Murray.

MacKenzie, John M. 1990. *Imperialism and the natural world*. Manchester: Manchester University Press.

MacLeod, Roy. 2009. Discovery and exploration. In *Cambridge history of science, vol, 6. The modern biological and earth sciences*, ed. Peter J. Bowler and John V. Pickstone, 34–59. Cambridge: Cambridge University Press.

Meinel, Christoph. 2000. Vorwort. In *Instrument – Experiment. Historische Studien*, ed. Christoph Meinel, 9–12. Berlin/Diepholz: Verlag für Geschichte er Naturwissenschaften und Technik.

Miller, David Philip, and Peter Hanns Reill, eds. 1996. *Visions of empire: Voyages, botany, and representations of nature*. Cambridge: Cambridge University Press.

Nielsen, Kristian H., Michael Harbsmeier, and Christopher J. Ries. 2012. Studying scientists and scholars in the field: An introduction. In *Scientists and scholars in the field: Studies in the history of fieldwork and expeditions*, ed. Kristian H. Nielsen, Michael Harbsmeier, and Christopher J. Ries, 9–12. Aarhus: Aarhus University Press.

Oldroyd, David R. 1996. *Thinking about the earth. A history of ideas in geology*. Cambridge, MA: Harvard University Press.

Ophir, Adi, and Steven Shapin. 1991. The Place of Knowledge: A Methodological Survey. *Science in Context* 4: 3–21.

Oreskes, Naomi. 1999. *The rejection of continental drift: Theory and method in American earth science*. New York/Oxford: Oxford University Press.

Palladino, Paolo and Michael Worboys. 1993. Science and Imperialism. *Isis* 84: 91–102.

Price, Derek de Solla. 1980. Philosophical mechanism and mechanical philosophy. Some notes toward a philosophy of scientific instruments. *Annali dell'Instituto e Museo di Storia della scienza di Firenze* 5(1):75–85.

Pusch, Georg Gottlieb. 1819. *Geognostischer Katechismus oder Anweisung zum praktischen Geognosten für angesehene Bergleute und Geognosten*. Freiberg: Craz und Gerlach.
Rheinberger, Hans-Jörg. 1997. *Toward a history of epistemic things*. Palo Alto: Stanford University Press.
Rudwick, Martin J.S. 1976. The emergence of a visual language for geological science 1760–1840. *History of Science* 14: 149–195.
———. 1985. *The great Devonian controversy: The shaping of scientific knowledge among gentlemanly specialists*. Chicago: Chicago University Press.
———. 2005. *Bursting the limits of time. The reconstruction of Geohistory in the age of revolution*. Chicago/London: University of Chicago Press.
Secord, James A. 1986. *Controversy in Victorian geology. The Cambrian-Silurian dispute*. Princeton: Princeton University Press.
Sloterdijk, Peter. 2006. *Im Weltinnenraum des Kapitals. Für eine philosophische Theorie der Globalisierung*. Frankfurt am Main: Suhrkamp.
Sorrenson, Richard. 1996. The ship as a scientific instrument in the eighteenth century. In *Science in the field, Osiris 11*, ed. Henrika Kuklick Henrica and Robert E. Kohler, 221–236. Chicago/London: The University of Chicago Press.
Stuhl, Andrew. 2016. *Unfreezing the Arctic: Science, colonialism, and the transformation of Inuit lands*. Chicago/London: The University of Chicago Press.
Suckow, Christian. 2001. Alexander von Humboldt und Russland. In *Alexander von Humboldt – Aufbruch in die Moderne*, ed. Ottmar Ette, 247–264. Berlin: Akademie Verlag.
Tamborini, Marco. 2015a. Die Wurzeln der Ideographischen Paläontologie: Karl Alfred von Zittels Praxis und sein Begriff des Fossils. *NTM Zeitschrift für Geschichte der Wissenschaften, Technik und Medizin* 23: 117–142. https://doi.org/10.1007/s00048-015-0130-3. 2016.
———. 2015b. 'If the Americans can do it, so can we': How dinosaur bones shaped German paleontology. *History of Science* 54 (3): 225–256.
Taylor, Kenneth L. 1969. Nicolas Desmarest and geology in the eighteenth century. In *Toward a history of geology*, ed. Cecil J. Schneer, 339–356. Cambridge, MA: The M.I.T. Press.
———. 1985. Early geoscience mapping, 1700–1830. *Proceedings of the Geoscience Information Society* 15: 15–49.
———. 1992. The historical rehabilitation of theories of the earth. *The Compass* 69: 334–345.
———. 1994. New light on geological mapping in Auvergne during eighteenth century: The Pasumot-Desmarest collaboration. *Revue d'Histoire des Sciences* 47: 129–136.
———. 1995. Nicolas Desmarest and Italian geology. In *Rocks, fossils and history*, ed. Gaetano Giglia, Carlo Maccagni, and Nicoletta Morello, 95–109. Florence: Edizioni Festina Lente.
———. 2001. The beginnings of a geological naturalist: Desmarest, the printed word, and nature. *Earth Sciences History* 20: 44–61.
———. 2007. Geological travellers in Auvergne, 1751–1800. In *Four centuries of geological travel: The search for knowledge on foot, bicycle, sledge and camel*, ed. Patrick N. Wyse Jackson, 73–93. London: Geological Society of London.
Thomas, Martin, ed. 2015a. *Expedition into empire: Exploratory journeys and the making of the modern world*. New York: Routledge.
———. 2015b. What is an expedition? An introduction. In *Expedition into empire: Exploratory journeys and the making of the modern world*, ed. Martin Thomas, 1–15. New York: Routledge.
Vaccari, Ezio. 2007. "The organized traveler: Scientific instructions for geological travels in Italy and Europe during the eighteenth and nineteenth centuries". In *Four centuries of geological travel: The search for knowledge on foot, bicycle, sledge and camel*, Patrick N. Wyse Jackson, 7–17. London: Geological Society of London.
Woodward, John. 1696. *Brief instructions for making observations in all parts of the world as also for collecting, preserving, and sending over natural things [...] presented in the Royal Society*. London: Wilkin.
Zittel, Karl Alfred von. 1901. *History of geology and Palaeontology to the end of the nineteenth century* (trans: Ogilvie-Gordon Maria, M.). London: Walter Scott.

Earth Science and Extractive Capitalism in the Age of Empire

Lukas Rieppel

Contents

Abstract

This chapter situates the practice of geology within the history of European imperialism, arguing that the earth sciences are deeply bound up with the political economy of extractive capitalism. On a mundane level, the practice of geology primarily consists of extracting specimen collections from far-flung parts of the world to make broader claims about the history, morphology, and physical structure of the earth. In addition, the earth sciences also developed in tandem with the imperial project of resource extraction. In what follows, I draw out these connections by offering a schematic history of geology that links the practice of science to the political economy of extractive capitalism.

Keywords

Extractive capitalism · Temporalities of capitalism · Indigenous dispossession · Othniel Charles Marsh · Georges Cuvier · Red Cloud · Resource extraction · Geological mapping · Colonial expansion · bureaucracy.

L. Rieppel (✉)
Brown University, Providence, RI, USA
e-mail: lukas_rieppel@brown.edu

E. Aronova et al. (eds.), *Handbook of the Historiography of the Earth and Environmental Sciences*, Historiographies of Science,
https://doi.org/10.1007/978-3-031-40799-4_16

1 Introduction

This chapter brings the history of geology into conversation with economic history to suggest that the earth sciences are deeply bound up with the imperial project of resource extraction. The core thesis I wish to defend is that geology ought to be understood as an extractive science. By this, I mean two things. First, on a mundane level, the practice of geology primarily consists of extracting samples – rocks and minerals, fossils, larger-scale distributional patterns, and so on – from far-flung parts of the world to make broader claims about the history, morphology, and physical structure of the earth. Second, and more profound, is the fact that the earth sciences developed in tandem with the imperial project of extractive capitalism. To substantiate those claims, I offer a schematic history of geology that highlights the theme of extraction by linking the practice of science to the project of imperial statecraft and the political economy of capitalism.

To be sure, this is hardly the first attempt to connect the earth sciences to mining and other forms of extractive industry. Classic work by Roy Porter (Porter 1977), Ted Porter (Porter 1981), Stephen Turner (Turner 1987), and William Goetzmann (Goetzmann 1978), as well as more recent contributions by James Secord (Secord 2018, 1986), Paul Lucier (Lucier 2008, 2018), Shellen Wu (Wu 2015), and Simon Knell (Knell 2000, 2012), all place the history of geology squarely within an economic context. At the same time, another strand in the literature primarily focuses on intellectual history, sometimes going so far as to argue that theoretical debates in geology and paleontology are completely divorced from politics and economics (Parsons 2001). Martin Rudwick's ambitious syntheses are particularly notable for their remarkable breadth and depth (Rudwick 2005, 2008, 2014), but the earlier work of George Merrill (Merrill 1964) and Mott Greene (Greene 1982) remain instructive as well. Perhaps inspired by widespread anxieties about the so-called "Anthropocene," however, recent years have seen a flurry of new publications that explicitly integrate the intellectual history of the earth sciences with the cultural history of capitalism and imperialism. These include a spate of new books about dinosaurs (Brinkman 2010; Nieuwland 2019; Noble 2016; Rieppel 2019), mammals (Manias 2023), humans (Milam 2019), earthquakes (Valencius 2015), and extinction (Sepkoski 2020), among many others. More recently, still, historians have called for a new earth history to be written from the South (Bashford et al. 2023), especially focusing on the "modern" history of the supercontinent Gondwanaland (Bashford et al. 2021; Chakrabarti 2020).

This chapter brings that literature into conversation with a lively debate about the emergence of "extractivism" as a distinct mode of capital accumulation in modern Latin America. While Sandro Mezzadra and Brett Neilson offer a fairly neutral description of economic extraction as "the forced removal of raw materials and life forms from the earth's surface, depths, and biosphere" (2017, 188), most scholars building on the foundational work of Eduardo Gudynas (2015, 2021) stress the high levels of exploitation and violence that often obtain in extractive economies, whose lack of attention to social equality and ecological reciprocity makes them fundamentally unsustainable. Thus, Macarena Gomez-Barris argues that extractive capitalism "engages in thefts, borrowings, and forced removals, violently reorganizing social life as well as the land by thieving resources from Indigenous and Afro-descendant territories" (Gómez-Barris 2017, xvii), while a recent review in *The Journal of Peasant Studies* describes the complex assemblage of "practices, mentalities, and power differentials" that structure "socio-ecologically destructive modes of organizing life through subjugation, depletion, and non-reciprocity" in extractive contexts (Chagnon et al. 2022, 4).

In what follows, then, I want to use the conceptual vocabulary developed by Latin American political economists to highlight some of the specific ways in which the earth sciences contributed to, and benefitted from, the violent and exploitative history of extractive capitalism. In that respect, my project resembles Kathryn Yusoff's provocative "geophilosophy of the Anthropocene" (2017), which seeks to expose how the earth sciences "naturalize (and thus neutralize) the theft of extraction" by framing "geosocial 'events'" like climate change as a purely geological process (Yusoff 2018, 12–13). However, rather than focus on the discursive realm, as Yussof primarily does, I want to interrogate the material entanglements between the intellectual history of geology and the economic history of resource extraction. To that end, I also draw on critical insights from political ecologists like Andreas Malm and Tim Mitchell, who have documented how the transition to a carbon economy allowed for greater centralization and control over production processes, yielding increased profits and power for economic elites (Malm 2016; Mitchell 2011). Finally, because the history of extractive capitalism cannot be understood without interrogating its role in imperial statecraft, this chapter also engages with recent work that documents how the earth sciences contributed to the history of settler colonialism (Wolfe 2006) by rendering the rich mineral resources of conquered lands available for exploitation (Lester 2023a, b). Supplemented by the work of Indigenous STS scholars like Zoe Todd (Todd Forthcoming), Paulette Steeves (Steeves 2021), and Vine Deloria Jr. (Deloria 1995), the chapter's overarching goal is thus to frame the development of the earth sciences within the context of what Michael John Witgen evocatively describes as "the political economy of plunder" (Witgen 2022).

2 The Deep Time Horizons of Modern Capitalism

One way to begin fleshing out the argument sketched out above is to consider what William Sewell memorably described as the "temporalities" of modern capitalism (2008). Following Marx, social and cultural historians of the economy often describe the capitalist mode of production as a temporally extended process that centers on self-perpetuating cycles of accumulation in which "the past devours the future," as Thomas Piketty famously put it (Piketty 2014). For example, Jonathan Levy describes capital as a "pecuniary process of forward-looking valuation," (2017, 483) whereas the sociologist Jens Beckert argues that capitalism engenders a future-oriented cognitive disposition. According to Beckert, "the evolution of capitalism was accompanied by fundamental changes in the temporal orientations of actors," who came to regard "the future as open, containing opportunities to be seized and risks to be calculated" (2016, 22; see also Muniesa 2017). In other words: Because money only becomes capital once it is invested in the means of production, the political economy of capitalism orients itself around a speculative future that has been framed as a lucrative space wherein profits are found. So far, so good. What this narrative leaves unaddressed, however, is the coincident invention of deep time. Precisely at the same moment that saw economic actors begin to devise formal tools like discounting curves to project themselves forward in time (Deringer 2018), practitioners of what came to be called the "earth sciences" also developed sophisticated techniques to imagine themselves back into the furthest recesses of prehistory by reconstructing a lost world from the material remains left behind (Rudwick 2005).

The new sciences of deep time helped to extend the temporal horizons of capitalist accumulation in more ways than one. In addition to expanding our imaginative capacity to travel through time, the earth sciences also contributed to the creation of a new energy regime whose commercial and geopolitical impact can hardly be overstated. Although the metabolic capacity of human and animal bodies to perform work did not become any less valuable during the industrial revolution, the invention of increasingly efficient steam engines incentivized the exploitation of fossil fuels on a scale that was simply unimaginable before (Smith and Norton Wise 1989; Odum 1970; Rosen 2010). Insofar as it entailed harnessing millions of years' worth of metabolic labor performed by long-deceased microorganism to power industrial technologies, the energy revolution of the nineteenth century constituted a form of exploitation across vast chasms of time – the future also devours the past! – creating a temporal rift in the carbon cycle that contributed directly to global climate change (Foster 2000; Clark and York 2005; Beldo 2017). Finally, the transition to steam power also allowed for an unprecedented centralization of the industrial production process, further concentrating political power in the hands of entrenched economic elites and all but ensuring that popular opposition to the implementation of so-called "labor-saving" technologies would not get very far (Bailey 1998; Hobsbawm 1952; Malm 2016; Thompson 1963).

According to Kenneth Pomeranz (2000), the productive capacity that geologists helped to unleash by documenting the location and abundance of fossil fuels played a decisive role in bringing about the Great Divergence, a dramatic difference in economic wealth and political power that came to characterize the relationship between Europe and other parts of the world during the long nineteenth century. Of course, the global inequities that emerged at this time were not simply created by burning coal. The systematic application of physical violence was decisive as well, meaning that colonial dispossession and coerced labor were central to the creation of European hegemony (Beckert 2014; Liu 2020). The combustion of fossil fuels and the violent expropriation of economic resources went hand in hand, mutually reinforcing each other to create the foundation of Europe's immense military, political, and economic might (Barak 2020). By creating the epistemic foundation that allowed European states to extract immense thermodynamic power from fossil fuels, the earth sciences thus played a central role in what Marx described as "primitive" (or *ursprüngliche*) accumulation, the violent extraction of resources whose devastating socio-ecological consequences are slowly unfolding today (Daggett 2019; Marx 1872; Moore 2015; Nixon 2011).

The production and circulation of geological maps offers an especially informative trace of the history sketched out thus far, linking the intellectual labor of scientists with the extraction of mineral wealth in far-flung parts of the globe. Some of the earliest systematic attempts to represent the distribution of valuable resources cartographically took place during the mid-eighteenth century. In 1746, for example, a French naturalist named Jean-Etienne Guettard published a *Carte Mineralogique* of mineral deposits in France, England, and Wales. By the end of the eighteenth century, over a dozen similar maps had been produced, but they were still restricted to locating various rock types and mineral resources on a two-dimensional plane. That changed at the turn of the nineteenth century, as naturalists and surveyors began to experiment with new techniques to represent the three-dimensional structure of geological formations in section. An especially notable example was produced by the orphaned son of a village blacksmith from Oxfordshire named William Smith, who figured out how to use the fossil remains of extinct life forms to ascertain the relative age of various rock types while he worked as a "Mineral Surveyor" for the wealthy owners of recently enclosed land. This gave Smith a considerable advantage over other mining consultants in that it allowed him to make stratigraphic inferences when called on to speculate whether a particular locality contained enough coal to warrant the expense of sinking a shaft. By the start of the nineteenth century, it also gave him the idea of producing a comprehensive geological map of the British Isles. When it was finally published in 1815, the result measured over 8 feet by 6 feet in total. But what made it especially important was Smith's use of watercolors to indicate the different layers or "strata" of rocks. In addition, clever use of shading allowed readers to surmise the stratigraphic superposition of different rock types, which made it possible to predict where a particular stratum was likely to continue away from a surface outcrop. Thus, although it was both esthetically beautiful and conceptually ambitious, the primary purpose of Smith's map was practical: to aid in the efficient extraction and accumulation of valuable resources (Sharpe 2015; Torrens 2016; Macfarlane et al. 2020) (Fig. 1).

Scientific representations of "vertical" space also functioned as tools of imperial expansion, helping to turn far-flung territories into zones of economic extraction. As the geographer Bruce Braun explains, geological maps did far more than just make otherwise invisible resources legible from above. By "bringing into view the 'verticality' of the state's territory," Braun contends, practitioners of the earth sciences helped to incentivize colonization, enabling recently conquered territories "to be drawn into global circuits of extractive capital" (Braun 2000, 13, 24). For example, the abundant coal deposits whose location Smith indicated in dark gray and black on his map directly fueled the stupendous growth of the British Empire during the nineteenth century, whose insatiable demand for mineral resources increased dramatically as it grew larger in size and power. As

Fig. 1 William Smith, *A Delineation of the Strata of England and Wales with Part of Scotland, Exhibiting the Collieries and Mines, the Marshes and Fen Lands Originally Overflowed by the Sea, and the Varieties of Soil According to the Variations in the Substrata*, 1815. (Courtesy of the Natural History Museum, London)

the historian On Barak has shown, Britain's armada of steam-powered warships not only reinforced the empire's naval supremacy, but it also touched off a race to identify overseas coal deposits to supply the complex network of off-shore bases and refueling depots being constructed across the globe (Barak 2020). As a result, the British dispatched a small army of geologists to render the world's mineral resources visible to colonial administrators and available for economic extraction.

The legions of survey geologists who were charged with mapping the Empire's mineral resources often participated in celebrating new scientific discoveries as well. An especially notable case in point occurred when the British East India Company ordered a colonial engineer named Captain Proby Cautley to begin excavating a section of the famed Doab Canal, which led him to unearth abundant coal and lignite deposits in a borderland region located between modern India and Pakistan. Cautley's work eventually prompted the Scottish physician Hugh Falconer, who had just taken up a post as the assistant surgeon for the EIC in Bengal, to begin exploring the so-called "Siwalik Hills," where he found rich trove of prehistoric fossils. Relying on manual labor from local people who had long used "Bijili ki har" or "lightning bones" as traditional medicine, Falconer and Cautley disinterred literally tons of vertebrate fossils that caused a sensation throughout the empire. Among many others, this included a giant tortoise named *Colossochelys atlas* that was widely associated with Hindu traditional knowledge about turtles carrying the world on their back (Chakrabarti 2020; Nair 2005). In other words, what might at first sight appear to have been purely intellectual achievements by scientists without any practical applications were often connected to the imperial history of extractive capitalism.

Of course, not every geologist was directly engaged in mineral extraction and colonial administration. Still, an argument can be made that the very idea of "earth science" rests on the epistemology of empire. For the earth, itself, to be conceptualized as a unified object of knowledge with a single, coherent history, European scientists had to develop techniques that would allow them to cast their gaze over – and into – the entire globe. Such an ambition only took shape once scientists began traveling to distant parts of the world, usually as part of state-sponsored projects to promote imperial expansion along both a vertical and horizontal dimension (Pratt 1992; Anthony 2020). Moreover, the synoptic desire to map the earth's interior structure and catalogue its most valuable resources would not have been possible without a sophisticated bureaucracy that organized the creation of huge specimen collections housed in well-funded institutions of knowledge production. Examples could be adduced almost indefinitely, ranging from the global network of botanical gardens that sought to acclimatize exotic plants in the name of imperial self-sufficiency (Drayton 2000; Endersby 2008; Schiebinger and Swan 2005; Williams 2021) to the construction of so-called "universal" or "encyclopedic museums" in metropolitan centers of learning like Paris and London (Delbourgo 2017; Hicks 2021; Ratcliff 2016; Sheets-Pyenson 1988). Regardless of the specificities that characterized each of these institutions, their core mission was remarkably similar: to amass large and supposedly comprehensive collections of specimens in a single, circumscribed space, making it possible to gain a synoptic overview of the world's vast, complex, and abundant natural, cultural, and economic resources (Latour 1986; Delbourgo 2017; Rieppel 2016).

An especially fascinating case in point is the Royal Cabinet of Madrid, which was inaugurated in 1776 after Pedro Franca Dávila persuaded King Carlos III to establish a natural history museum in the Spanish capital. The very same year, Carlos III issued a "Royal Order" instructing local governors throughout the colonial administration to "gather any curious pieces that they may find in their districts" and send them back to the Iberian metropole. In this way, it was hoped that the Royal Cabinet would amass a complete collection of "whatever Nature produces in his Majesty's vast Domains." Among many things, this resulted in seven boxes containing the fossil remains of a ponderous quadruped that were found on the estate of Francisco Aparicio near the Luján River being shipped from Buenos Aires to Madrid in 1788. When a series of engravings commissioned by Juan Bautista Bru from the Royal Cabinet made their way to Paris, the French anatomist Georges Cuvier compared them to his specimen collection at the *Jardin des Plantes* and concluded the massive bones must have belonged to an extinct relative of the modern ground sloth that he christened *Megatherium*. Eventually, Cuvier went so far as to argue that, given only a single bone, his comparative method allowed paleontologists to reconstruct the entire skeleton of previously unknown organisms with uncanny accuracy, so long as they had a suitably comprehensive collection at their disposal (Dawson 2016; Podgorny 2019; Pimentel 2017).

Cuvier's daring feat of informed speculation now ranks among the most celebrated events in the founding mythology of vertebrate paleontology, allowing him to proclaim that he had become "a new species of antiquarian" who knew "how to burst the limits of time and … recover the history of the world" from a careful comparison of its fragmentary remains (Cuvier 1812; Rudwick 1997, 183, 185). In other words, earth scientists such as Cuvier explicitly claimed that extracting specimens from distant parts of the world and accessioning them to metropolitan institutions like the *Jardin des Plantes* in Paris, where they could perform synoptic comparisons, would finally allow scientists to begin piecing the earth's one true creation story together. Hence, as the extraction and accumulation of specimens like the *Megatherium* fossil plainly shows, practitioners of the earth sciences contributed to the imperial hegemony of powerful states in more ways than one. In addition to being involved in the extraction of valuable resources like coal, their more self-consciously intellectual pursuits were part and parcel of the imperial project as well, making it possible for them to claim epistemic mastery over the entire globe, its interior structure, as well as its deep history.

3 Extractive Capitalism in the United States

So far, this chapter has offered an overview of the material and conceptual entanglement between the earth sciences and European imperialism. In what follows, I want to travel across the Atlantic and take a somewhat closer look at the United

States during the long nineteenth century. In part, this choice is informed by my scholarly background and expertise. However, historiographical considerations are at play as well. The territorial reach of the United States increased dramatically during this period, which resulted in violent conflict with many Indigenous Nations. Moreover, white settlers who colonized the North American interior primarily moved west in response to economic incentives, often involving mineral extraction. Imperial expansion and economic extraction also fueled industrialization. As a result, the US economy grew at a stupendous pace during the decades after the Civil War, so much so that by the eve of World War I its industrial output exceeded that of France, Britain, and Germany combined. Taken together, these factors helped make the earth sciences a favorite pursuit among upwardly mobile elites. In turn, that helped to ensure practitioners of the earth sciences would enjoy lavish funding from both the US Federal Government and wealthy philanthropists, whose industrial firms benefited so much from the period's booming extractive economy.

When British colonists began to settle the eastern coast of North America, they were especially interested in its abundant resources. Hence, it was not long before they began cataloguing the riches one could find in the proverbial "New World." Hoping that it might encourage others to "helpe forward the Plantation," for example, William Wood issued a "true, lively, and experimentall description" of Massachusetts Bay in 1634. According to Wood, New England featured the "best ground and sweetest Climate." It also contained abundant "timber" that "grows straight, and tall," as well as meadows so full of wild strawberries that "one may gather half a bushel in a forenoon"(Wood 1634, 3, 7, 8, 9, 12, 13). Wood's text helped to establish a whole genre of similar boosterish catalogues, including Thomas Jefferson's famous *Notes on the State of Virginia* (which praised the state's coal deposits as being of "very excellent quality" (Jefferson 1787, 42)). A third example was by the French nobleman Constantin François Volney, whose (1803) study of North America's climate and soil included a detailed map that William Maclure went on to use as the basis for the first geological map of the United States. (Like both other texts discussed in this paragraph, Volney's book also included a lengthy discussion of North America's Indigenous people.) Although it was not as detailed or sophisticated as Smith's map of England, Maclure nonetheless hoped that his efforts would yield "many practical advantages," especially in the "search for coal, gypsum, salt, limestone" and "all kinds of metallic veins and repositories" (Maclure 1818, 4).

Around the time that Maclure published his geological map, the US Federal Government dispatched several armed expeditions to the North American interior. By far the most well-known example was the Corps of Discovery, a special unit of the US Army led jointly by Captain Meriwether Lewis and Second Lieutenant William Clark that Jefferson sent to explore the Louisiana purchase. These expeditions were complex enterprises whose brief was to document the natural resources one could find across the continent, to form trading relationships with Native Nations, and to lay claim to vast swaths of territory that, in practice, remained under Indigenous control. In effect, the United States sought to use survey documents to project epistemic power over Indigenous lands, using two-dimensional representations of the continent's physical geography to assert sovereignty from afar. But survey documents had a more practical purpose as well. By publicizing the location, extent, and abundance of the continent's valuable resources for economic extraction, they also incentivized white settlers to physically colonize large swaths of territory wherein the United States had only nominal power. To that end, the first decades of the nineteenth century witnessed a concerted effort to map the entirety of North America (Goetzmann 1978; Short 2001; Schwartz 1980). Besides the US Army Corps of Topographical Engineers, much of this work was done by geological surveys, which were initially organized on a state-by-state basis. In addition, the General Land Office, which functioned as something akin to a federal real estate agent and eventually evolved into the Bureau of Land Management, also hired geologists to map the lands that it offered for sale (Ablavsky 2020; Hubbard 2009; Rohrbough 1968).

These efforts increased dramatically during the mid-to-late nineteenth century, especially after the California gold rush of 1848 touched off a mining boom in the North American interior (Paul 1963). The 1859 discovery of immense silver deposits near Virginia City, for example, drew hordes of prospectors to what is now western Nevada. They were joined by wealthy speculators whose capital investments turned the Comstock Lode into one of the world's largest mining districts, producing upward of $325 million (over $8 billion in today's currency) in silver over the next several decades (Lucier 2018). The frenzy only increased after the Civil War, especially once a transcontinental railroad was completed in 1868. Without ready access to a reliable means of transporting freight, it did not pay to work even the most lucrative lode. Knowing that an extensive transportation infrastructure was essential to the development of a thriving extractive economy, the US Federal Government invested enormous resources to make these capital-intensive projects attractive to private corporations like the Union Pacific. To that end, it authorized numerous surveys to explore the feasibility of various routes. The United States also helped finance construction directly by issuing government bonds and paying for public bailouts when railroad companies threatened to declare bankruptcy (which they frequently did). Finally, the Federal Government offered generous land grants along proposed railway lines, essentially inducing private developers to fund railway construction with mineral-rich lands dispossessed from Indigenous people (White 2011).

Before the overland route was even completed, the United States Congress dispatched several additional surveys to locate and publicize the vast tracts of mineral lands that were about to be made accessible to extractive industry. The first was led by Clarence King, whose brief was to map the land adjacent to the fortieth parallel (along which the railroad was being built). At the same time, Ferdinand V. Hayden led a team of surveyors into Nebraska, Wyoming, and Colorado. Then, in 1871, two more surveys were sent out from Washington: The Geological and Geographical Survey of the Rocky Mountain Region, led by John Wesley Powell, and the Geographical Surveys West of the One Hundredth Meridian, led by the Army Corps of Engineers Lieutenant George M. Wheeler. Their disparate efforts were consolidated in 1879, when Congress passed legislation to form the US Geological Survey (USGS) under the direction of King, whose exploration of the fortieth parallel was praised as an especially successful example of how federal investment could underwrite the creation of a thriving extractive economy. To that end, King decided that one of the new survey's first orders of business would involve a thorough investigation of the country's most lucrative mining districts (which included Eureka and Comstock, as well as Leadville, Colorado), hoping they might serve as a model that could be replicated elsewhere as well (Rabbitt 1979).

The Federal Government also supported basic research without any direct ties to the mineral industry. Long before the US Geological Survey had been established, for example, the United States began building an encyclopedic museum in Washington DC. The impetus that led to its creation was a philanthropic bequest by the English polymath James Smithson, who donated 11 boxes of gold sovereigns toward the foundation of "an Establishment for the increase & diffusion of knowledge among men" (Goode 1897, 20). By the time that its first physical building (the "Castle") was completed in 1855, the Smithsonian Institution had already amassed a sizable collection, including thousands of specimens the United States Exploring Expedition brought back from the Pacific. The Smithsonian also acquired an immense collection of fossils and minerals, especially after being designated as the official repository for US Government Surveys. Moreover, under the direction of John Wesley Powell (who would go on to lead the USGS), the Smithsonian also embarked on a program to "salvage" the material culture of countless Indigenous people. This effort was informed by a widely shared but erroneous belief that Indigenous cultures were naturally doomed to extinction, which scientists such as Powell believed was a tragic but inevitable consequence of evolutionary "progress" (Dippie 1991; Redman 2021). As a result, the Smithsonian quickly amassed an enormous collection of plants, animals, fossils, and minerals, as well as ethnographic objects and human remains (Hinsley 1994; Oehser 1970).

The Smithsonian's collections resembled a geological map in that both aimed to provide a totalizing vision of North America and its abundant resources. Maps and specimen collections both also functioned as an expression of bureaucratic authority, engendering the illusion that scientists and government officials could gain mastery over vast areas of geographic space, its biological and cultural diversity, as well as its mineralogical wealth, all without having to leave the comfort and security of the nation's capital. As Fig. 2 clearly indicates, however, museum curators soon found themselves overwhelmed by the sheer volume of material under their care. If the task of organizing an enormous specimen collection exceeded the organizational capacity of Smithsonian scientists, the United States found itself even less capable of governing the vast territories to which it lay claim. Indeed, the Federal Government's inability to control unruly settlers often resulted in violence. During the 1860s, for example, throngs of mineral prospectors began traveling along the Bozeman Trail on their way to the Montana Gold Rush, violating a treaty the United States had negotiated with Plains Indians Tribes in 1851. Tensions especially flared over their habit of harassing the buffalo, respected kin as well as an indispensable source of protein to many tribes in the region. When the US military intervened with shocking brutality, the Cheyenne, Lakota, and Arapaho forged a powerful alliance led by the Lakota headman Mahpiya Luta (Red Cloud). Between 1866 and 1868, the Lakota and their allies won many pitched battles in what came to be known among the

Fig. 2 Government scientists attempting to organize specimens for the Bureau of American Ethnology at the Smithsonian Institution. (Image NHB-3680, Smithsonian Institution Archives, Record Unit 95, Box 32, Folder 19)

American public as Red Cloud's War, forcing the United States to negotiate a new treaty at Fort Laramie during the summer of 1868. But it was not long before the insatiable desire for mineral wealth led white settlers to violate the terms of the 1868 treaty as well. This resulted in renewed conflicts and eventually culminated in the brutal massacre of some 300 Lakotas (more than two-thirds of whom were women and children) at Wounded Knee (Estes 2019, 89–131; Ostler 2004).

The career of Othniel Charles Marsh weaves some of these disparate threads into a single narrative, demonstrating how the scientific study of seemingly arcane topics intersected with the imperial expansion of American capitalism. In addition, Marsh also ties the territorial dispossession of Indigenous people to the extraction of specimens from Native lands. Finally, Marsh had the good fortune of being the nephew of George Peabody, one of nineteenth-century America's wealthiest business tycoons who endowed a new museum of natural history to secure Marsh a professorship at his alma mater, Yale University. Thus, Marsh brings many of this chapter's elements into relation, showing how the practice of imperial statecraft, the political economy of modern capitalism, the institutional history of museums and other archival institutions devoted to amassing huge specimen collections, and the intellectual pursuits of earth scientists were all entwined, mutually reinforcing each other during the long nineteenth century.

Let us follow the money: George Peabody established himself as a leading financier during the mid-nineteenth century, principally by marketing railroad bonds to European investors in London. After his retirement in 1864, he passed on his firm to Junius Spencer Morgan, so he could devote all his attention to charitable causes. In addition to founding numerous libraries and institutes – including a music conservatory that is now part of Johns Hopkins University in Baltimore – he donated large sums toward the creation of several museums, such as the Peabody Museum of Archeology and Ethnology at Harvard University. At the request of his nephew, Peabody also donated the funds to establish a natural history museum at Yale. He did so on the condition that Yale appoint Marsh as the museum's first curator of paleontology and grant him an honorary professorship in the college. Within just a few years, Marsh had leveraged his position at Yale – as well as the $100,000 inheritance he received after his uncle's death in 1869 – to become one of the most widely known and respected geologists in the United States. Thus, when Powell replaced King as Director of the US Geological Survey in 1882, he put Marsh in charge of its Division of Vertebrate Paleontology. Since the USGS was one of the most well-endowed federal agencies at the time, Marsh now had access to almost limitless funds. As a result, he enjoyed a huge advantage in the race to uncover and name the abundant prehistoric fossils that one could find in the North American interior. In turn, Marsh's success in the field allowed him to compile a series of lavishly illustrated monograph publications, each of which claimed to be last word about a particular group of extinct organisms. Taken together, Marsh's extensive collection of fossils coupled with his impressive monograph publications aimed to provide a comprehensive account of North America's deep history, illustrating how present life forms had evolved from their extinct predecessors over vast swaths of geological time (Schuchert and LeVene 1940; Jaffe 2000).

Marsh was a shrewd social climber, and he began making his name as an intrepid geologist through a series of expeditions he organized almost immediately after arriving at Yale. While attending a scientific meeting in Chicago during the summer of 1868, he accepted a complimentary trip on the Union Pacific Railroad to Omaha. "As we passed westward," Marsh recalled, "I felt sure that entombed in the soft sandy clays ... there must be remains of many strange animals new to science" (Schuchert and LeVene 1940, 97). As a result, Marsh decided to lead a group of Yale undergraduates into the North American fossil fields during several seasons starting in 1870. These excursions would prove very successful, yielding many invaluable specimens that Marsh eagerly accessioned to his growing collection at Yale. Among the most notable was a series of toothed birds that Marsh used to establish a phylogenetic link between extinct dinosaurs and modern birds, and which Charles Darwin described as "the best support to the theory of evolution, which has appeared within the last 20 years."[1] But these expeditions were more than just a scientific endeavor. They also provided experiences that Marsh used to fashion himself into the masculine figure of an intrepid explorer who helped conquer the western frontier.

[1] Darwin to Marsh, 31 August 1880, O. C. Marsh Correspondence (MS 343), Archives of the Yale Peabody Museum, New Haven, CT.

Throughout the rest of his life, Marsh developed a reputation for recounting interminable adventure stories to anyone who would listen. These invariably included a confrontation with hostile Indigenous people, necessitating an official escort from the United States army (which detailed William "Buffalo Bill" Cody to accompany the Yale expedition during its first field season). Truth be told, however, the situation was far more complex than Marsh's grandiose tales of geological adventure let on. In fact, he was absolutely dependent on the extensive knowledge of Native guides, who had long taken a keen interest in the fossilized remains of prehistoric creatures that littered their ancestral hunting grounds (Dussias 1996; Mayor 2005). Relying on Indigenous knowledge, Marsh removed literally tons of valuable fossils from recently conquered tribal lands, accessioned them to the so-called permanent collection of the Peabody Museum, and built a reputation as the foremost expert on North America's deep history (Bradley 2014). Once they arrived at the museum, these fossils were inserted into a speculative theory of evolutionary progress wherein the extinction of supposedly "primitive races" was believed to open the ecological space for more "advanced" lineages to evolve (Manias 2023; Rieppel 2020), thus helping to supply an ideological foundation for the removal and dispossession of Native people from their ancestral lands (Dippie 1991; Qureshi 2013) (Fig. 3).

Although Marsh always insisted that he was only interested in science, Native people had good reason to be wary of this strange foreigner who seemed to exhibit such an unhealthy obsession with fossils. They understood perfectly well the extent to which the "pure" science of geology was entangled with mineral extraction. And Marsh was no exception. For example, when Colonel George Armstrong Custer mounted a military expedition into the Black Hills during the summer of 1874, Marsh was invited along. Although he was busy conducting his own fieldwork that summer, Marsh made sure to send a promising student – George Bird Grinnell – in his stead. Custer's expedition was in direct violation of the 1868 Treaty of Fort Laramie, and it was designed to engineer a gold rush that would force the US Federal Government to open the Black Hills to mineral prospectors (Ostler 2010). Thus, when Marsh approached Red Cloud and asked for permission to collect fossils in the White River Badlands later that very same summer, it is easy to understand why the Lakota headman was more than a little suspicious. After several days of tense negotiations, Red Cloud only allowed Marsh to proceed under armed escort if he agreed to bring Red Cloud's complaints about the corruption of local Indian Agents, who routinely embezzled funds the United States had promised his people, to Washington. This agreement allowed Marsh to enter the Badlands and collect specimens for his museum back in New Haven. As these fossils were transported east on the recently constructed transcontinental railroad, they were stripped of their Indigenous meanings and repurposed as evidence for a new, teleological narrative about the history of life on earth (Ruse 1996). Thus, the extraction of specimens worked as a form of temporal dispossession, all the more so since they were subsequently used to rewrite the deep history of lands being colonized by the United States (Schuller 2016).

Fig. 3 Othniel Charles Marsh posing with the Oglala Lakota headman Red Cloud while holding a sacred pipe, 1883. (National Portrait Gallery, Smithsonian Institution)

In addition to the prominent role that earth scientists played in the extraction of epistemic and economic resources, they were also deeply entangled in a process that I have elsewhere described as cultural resource extraction (Rieppel 2019). This was a result of their close connection to some of the period's most famous philanthropists, of whom George Peabody was a prominent early example. During the decades that followed his endowment of the Yale natural history museum, Peabody's conspicuous generosity came to be widely admired and imitated among subsequent generations of wealthy elites. The latter were especially keen to be seen as magnanimous benefactors of culture during a period of widespread class conflict. The industrial juggernaut might have been booming at the end of the nineteenth century, but American capitalism was in a state of crisis as well. As the social theorist Henry George famously wrote, "It is as though an immense wedge were being forced, not underneath society, but through society. Those who are above the point of separation are elevated, but those who are below it are crushed down" (George 1879, 9). This sentiment was especially widespread among working people, which touched off a wave of labor unrest that could be remarkably violent and bloody, often resulting in pitched battles on the streets of industrial cities such as Chicago, Philadelphia, and New York. Anxious that incendiary labor leaders might succeed in bringing the industrial economy to its knees, wealthy elites responded with highly visible acts of public munificence that were designed to demonstrate that modern capitalism worked for the good of all mankind, not just the wealthy few (Carnegie 1900).

To make a convincing claim that, through philanthropy, industrial capitalism could generate public goods in addition to profits, it did not suffice for wealthy elites to build monuments to their own vanity. It was absolutely essential for their philanthropic bequests to attract the attention, admiration, and approval of diverse audiences, including working class people. Thus, in addition to constructing grand concert halls for the performance of symphonic music and magnificent art galleries in which to display classical statuary and paintings by European masters, they also invested in more popular sites of public amusement. At the same time, however, rich benefactors of high-brow cultural institutions remained equally anxious to ensure their philanthropic bequests also help them perform their social distinction (Bourdieu 1986; DiMaggio 1982). For that reason, they did not want to be seen as engaging in mere showmanship, treading carefully to avoid the impression of pandering to popular tastes. As luck would have it, they discovered an institution that met all these complex and seemingly contradictory demands. Natural history museums were already developing a hybrid institutional mission that mixed public education with popular amusement and the furtherance of original scientific research (Rader and Cain 2014). Hence, it did not take long for these institutions to become a favored target of philanthropic largesse, which resulted in the creation of several new civic museums of natural history at the turn of the twentieth century. In addition to the American Museum in New York, this included the Carnegie Museum in Pittsburgh and the Field Museum in Chicago, all of which engaged in a fierce competition over who could attract a larger and more socially diverse crowd of visitors into their public galleries (Brinkman 2010) (Fig. 4).

Of all the specimens on display in philanthropic museums of natural history, none drew in so many visitors as the spectacular fossils of prehistoric monsters that paleontologists such as Marsh unearthed in the American interior. As the late nineteenth century turned into the early twentieth century, the earth sciences therefore garnered a disproportionate amount of financial support. The menagerie of extinct life forms that geologists disinterred were not just immensely popular, it also garnered a great deal of social and intellectual prestige. Indeed, geology and paleontology were widely regarded as the most notable and successful branches of American science during this period. Finally, their close ties to the mineral industry offered a welcome opportunity to transform what might otherwise have been a purely utilitarian enterprise into a noble scientific pursuit. For wealthy elites whose vast fortune was often a direct product of economic extraction, philanthropic support of the earth sciences thus offered an especially promising means to distance themselves from the rude and uncultivated aspects of commercial culture. Finally, philanthropists such as Andrew Carnegie often invoked popular theories of evolutionary progress to naturalize the political history of capitalism, claiming that one and the same "developmental" process explained how prehistoric fossils gave rise to

Fig. 4 A massive plant-eating sauropod dinosaur from the North American interior on display in the Hall of Fossil Reptiles, American Museum of Natural History, 1921. (Image ID: 38715)

more advanced organisms and agrarian societies begat the political economy of modern capitalism (Rieppel 2020). Thus, wealthy capitalists who had grown rich from the exploitation of North America's abundant natural resources discovered a way to extract social legitimacy from the deep past as well.

4 Conclusion

By way of conclusion, I want to offer a final reflection on how the earth sciences helped shape the temporalities of modern capitalism. As the historian Martin Rudwick has shown, the geological concept of deep time is based on the once-revolutionary idea that human history and geological history take place on radically different timescales (Rudwick 2005, 2008). During the eighteenth and early nineteenth century, European naturalists such as Georges Cuvier theorized that our planet is unfathomably old, existing on a temporal register that is almost incomprehensible to a comparatively short-lived species such as ourselves. Cuvier also argued that Earth history is episodic, linear, and punctuated by catastrophic upheavals that he likened to revolutions, claiming that "nature has also had its civil wars, and the surface of the globe has been upset by successive revolutions and various catastrophes" (Cuvier 1812; Rudwick 1997, 186). Although subsequent geologists, Charles Lyell most famously among them, argued in favor of a more "uniformitarian" vision of the deep past, the scientific community has by and large come to embrace the idea that catastrophes like meteor impacts and volcanic eruptions were a decisive factor in the mass extinction events that are used as convenient markers to divide earth history into discrete geological epochs (Sepkoski 2020). Geologists thus came to regard our planet as dumbfoundingly old and the deep past as almost incomprehensible alien. In turn, this created a radical discontinuity between human history and earth history, implying that deep time constituted nature's own temporality and rendering prehistory as an extraterritorial space that is completely divorced from human culture and politics (Chakrabarti 2020; Rieppel and Chang 2023).

The temporal chasm opened by the invention of deep time created a moral caesura as well, marking an ethical rift between geological events and human affairs. In a fascinating account of the profound effect this had on the historical imagination, Lydia Barnett observes that "[o]nce the planet's timescale deepened into the prehuman past, there was no longer a clearly defined role for humanity in earth's history, either as agents of geological change or as the victims or beneficiaries of its aftermath." In this way, she continues, the enlightenment science of modern geology "paved the way for Romantic conceptions of nature as wild, alien, and completely elusive of the human" (Barnett 2019, 194). There is a deep irony here. While its inventors understood the immense scale of the prehistory as a means to express awe for the stupendous power of nature, they inadvertently created the moral foundations for its exploitation as well. As I have tried to suggest throughout this chapter, the connection between these dynamics is far from coincidental. If extractive capitalism is characterized by lack of reciprocity, it requires an intellectual foundation to justify exploitation. Here too, the earth sciences did important work. By casting the creation of so-called "natural resources" into a temporal realm that is utterly divorced from our own, the invention of deep time created the conditions for thinking the earth as unworthy of reciprocal care.

As humans began grappling with the reality that extractive capitalism is fundamentally unsustainable, earth scientists proposed that our species has entered a new epoch in which we have been elevated to the level of a "geological force" (Crutzen and Stoermer 2000; Steffen et al. 2007; Chakrabarty 2009). What dystopian narratives about the so-called "Anthropocene" often leave unanswered, however, is the question of how we might enter into more nurturing relationships with what Zoe Todd describes as our "fossil kin" (Todd Forthcoming). And doing that will require a temporal order fundamentally at odds with the imperial project of extractive capitalism. Instead of a linear time marked by repeated catastrophes which produce the sorts of "golden spikes" that the Anthropocene discourse requires, the sustainable relationship that critical theorists like Todd ask us to cultivate with the earth is more likely to emerge from an embrace of what the philosopher Kyle Whyte describes as "spiraling time." This is an "intergenerational time," Whyte explains, wherein the needs of one's ancestors and descendants are equally present with the demands of the now (Whyte 2018). Thus, whereas geologists such as Cuvier and Marsh embraced a teleological narrative wherein extinction was required for evolutionary advancement, Whyte's concept of spiraling time engenders a radically different temporal disposition, creating a sense of inter-generational responsibility that invites us to understand the past, present, and future as inextricably inter-linked and interdependent.

Acknowledgments I would especially like to thank Staffan Müller-Wille, Xan Chacko, Lawrence Bradley, David Sepkoski, and Jarrod Hore for illuminating conversation and helpful feedback while writing this chapter.

References

Ablavsky, Gregory. 2020. *Federal ground: Governing property and violence in the first U.S. Territories*. Oxford University Press.
Anthony, Patrick. 2020. Mines, mountains, and the making of a vertical consciousness in Germany ca. 1800. *Centaurus* 62 (4): 612–630. https://doi.org/10.1111/1600-0498.12337.
Bailey, Brian J. 1998. *The luddite rebellion*. New York: University Press.
Barak, On. 2020. *Powering empire: How coal made the Middle East and sparked global carbonization*. Berkeley: University of California Press.
Barnett, Lydia. 2019. *After the flood: Imagining the global environment in early modern Europe*. Johns Hopkins University Press.
Bashford, Alison, Pratik Chakrabarti, and Jarrod Hore. 2021. Towards a modern history of Gondwanaland. *Journal of the British Academy* 9 (s6): 5–26.
Bashford, Alison, Emily M. Kern, Adam Bobbette, and Dipesh Chakrabarty, eds. 2023. *New earth histories: Geo-cosmologies and the making of the modern world*. Chicago: University of Chicago Press.
Beckert, Sven. 2014. *Empire of cotton: A global history*. New York: Alfred A. Knopf.
Beckert, Jens. 2016. *Imagined futures: Fictional expectations and capitalist dynamics*. Cambridge: Harvard University Press.
Beldo, Les. 2017. Metabolic labor: Broiler chickens and the exploitation of vitality. *Environmental Humanities* 9 (1): 108–128.
Bourdieu, Pierre. 1986. *Distinction: A social critique of the judgment of taste*. London: Routledge & Kegan Paul.
Bradley, Lawrence. 2014. *Dinosaurs and Indians: Paleontology resource dispossession from Sioux lands*. Denver: Outskirts Press.
Braun, Bruce. 2000. Producing vertical territory: Geology and governmentality in late Victorian Canada. *Ecumene* 7 (1): 7–46.
Brinkman, Paul. 2010. *The second Jurassic dinosaur rush: Museums and paleontology in America at the turn of the twentieth century*. Chicago: University of Chicago Press.
Carnegie, Andrew. 1900. *The gospel of wealth, and other timely essays*. New York: Century.
Chagnon, Christopher W., Francesco Durante, Barry K. Gills, Sophia E. Hagolani-Albov, Saana Hokkanen, Sohvi M.J. Kangasluoma, Heidi Konttinen, et al. 2022. From extractivism to global extractivism: The evolution of an organizing concept. *The Journal of Peasant Studies* 0 (0): 1–33.
Chakrabarti, Pratik. 2020. *Inscriptions of nature: Geology and the naturalization of antiquity*. Baltimore: Johns Hopkins University Press.
Chakrabarty, Dipesh. 2009. The climate of history: Four theses. *Critical Inquiry* 35: 197–222.
Clark, Brett, and Richard York. 2005. Carbon metabolism: Global capitalism, climate change, and the biospheric rift. *Theory and Society* 34 (4): 391–428.

Crutzen, Paul, and Eugene Stoermer. 2000. The 'Anthropocene'. *IGBP Newsletter* 41: 17–18.
Cuvier, Georges. 1812. *Recherches Sur Les Ossemens Fossiles de Quadrupèdes*. Vol. 4 vols. Paris: Deterville.
Daggett, Cara New. 2019. *The birth of energy: Fossil fuels, thermodynamics, and the politics of work. Elements*. Durham: Duke University Press.
Dawson, Gowan. 2016. *Show me the bone*. Chicago: University of Chicago Press.
Delbourgo, James. 2017. *Collecting the world: Hans Sloane and the origins of the British museum*. Cambridge: The Belknap Press of Harvard University Press.
Deloria, Vine. 1995. *Red earth, White lies: Native Americans and the myth of scientific fact*. New York: Scribner.
Deringer, William. 2018. *Compound interest corrected*. Ed. Lukas Rieppel, Eugenia Lean, and William Deringer. Osiris 33.
DiMaggio, Paul. 1982. Cultural entrepreneurship in nineteenth-century Boston. *Media, Culture, and Society* 4: 33–50.
Dippie, Brian W. 1991. *The vanishing American: White attitudes and U.S. Indian Policy*. Lawrence: University Press of Kansas.
Drayton, Richard. 2000. *Nature's government: Science, Imperial Britain, and the "improvement" of the world*. New Haven: Yale University Press.
Dussias, Allison. 1996. Science, sovereignty, and the sacred text: Paleontological resources and native American rights. *Maryland Law Review* 55: 84–159.
Endersby, Jim. 2008. *Imperial nature: Joseph hooker and the practices of Victorian science*. Chicago: University of Chicago Press.
Estes, Nick. 2019. *Our history is the future: Standing rock versus the Dakota access pipeline, and the long tradition of indigenous resistance*. London/New York: Verso.
Foster, John Bellamy. 2000. *Marx's ecology: Materialism and nature*. New York: Monthly Review Press.
George, Henry. 1879. *Progress and poverty: An inquiry into the cause of industrial depressions and of increase of want with increase of wealth*. National Single Tax League Publishers.
Goetzmann, William H. 1978. *Exploration and empire: The explorer and the scientist in the winning of the American West*. New York: Norton.
Gómez-Barris, Macarena. 2017. *The extractive zone: Social ecologies and decolonial perspectives. Dissident acts*. Durham: Duke University Press.
Goode, George Brown, ed. 1897. *The Smithsonian Institution, 1846–1896: The history of its first half century*. Washington, DC: Government Printing Office.
Greene, Mott T. 1982. *Geology in the nineteenth century: Changing views of a changing world*. Ithaca: Cornell University Press.
Gudynas, Eduardo. 2015. *Extractivismos: ecología, economía y política de un modo de entender el desarrollo y la naturaleza*. Cochabamba: CEDIB, Centro de Documentación e Información Bolivia.
———. 2021. *Extractivisms: Politics, economy and ecology*, Critical development studies. Vol. 5. Fernwood.
Hicks, Dan. 2021. *The brutish museums: The Benin bronzes, colonial violence and cultural restitution*. London: Pluto Press.
Hinsley, Curtis M. 1994. *The Smithsonian and the American Indian: Making a moral anthropology in Victoria America*. Washington, DC: Smithsonian.
Hobsbawm, E.J. 1952. The machine breakers. *Past & Present* 1 (1): 57–70.
Hubbard, Bill. 2009. *American boundaries: The nation, the states, the rectangular survey*. Chicago: University of Chicago Press.
Jaffe, Mark. 2000. *The gilded dinosaur: The fossil war between E.D. Cope and O.C. Marsh and the rise of American science*. New York: Crown.
Jefferson, Thomas. 1787. *Notes on the state of Virginia*. London: Printed for John Stockdale.
Knell, Simon J. 2000. *The culture of English geology, 1815–1851: A science revealed through its collecting*. Aldershot: Ashgate.

———. 2012. *The great fossil enigma: The search for the conodont animal*. Bloomington: Indiana University Press.
Latour, Bruno. 1986. Visualization and cognition: Thinking with eyes and hands. *Knowledge and Society* 6: 1–40.
Lester, Gustave. 2023a. Land, fur, and copper: The Union of settler colonialism and industrial capitalism in the great lakes region, 1815–1842. *Early American Studies: An Interdisciplinary Journal* 21 (1): 122–165.
———. 2023b. *Mineral lands, mineral empire: Mapping the raw materials of US industrial capitalism, 1780–1880*. PhD Thesis, Harvard University, Cambridge.
Levy, Jonathan. 2017. Capital as process and the history of capitalism. *Business History Review* 91 (3): 483–510.
Liu, Andrew. 2020. *Tea war: A history of capitalism in China and India*. New Haven: Yale University Press.
Lucier, Paul. 2008. *Scientists & swindlers: Consulting on coal and oil in America, 1820–1890*. Baltimore: Johns Hopkins University Press.
———. 2018. Comstock capitalism: The law, the lode, and the science. *Osiris* 33 (1): 210–231.
Macfarlane, Robert, Duglas Palmer, John Henry, Peter Wigley, Tom Sharpe, Jill Darrell, Diana Clements, John Mather, and Dave Williams. 2020. *Strata: William Smith's geological maps*. Chicago: The University of Chicago Press.
Maclure, William. 1818. Observations on the geology of the United States of North America; with remarks on the probable effects that may be produced by the decomposition of the different classes of rocks on the nature and fertility of soils. *Transactions of the American Philosophical Society* 1: 1–91.
Malm, Andreas. 2016. *Fossil capital: The rise of steam-power and the roots of global warming*. New York: Verso.
Manias, Chris. 2023. *The age of mammals: Nature, development & paleontology in the long nineteenth century*. Pittsburgh: University of Pittsburgh Press.
Marx, Karl. 1872. *Das Kapital: Kritik Der Politischen Oekonomie*. Hamburg: O. Meissner.
Mayor, Adrienne. 2005. *Fossil legends of the first Americans*. Princeton University Press.
Merrill, George P. 1964. *The first one hundred years of American geology*. New York: Hafner Pub. Co.
Mezzadra, Sandro, and Brett Neilson. 2017. On the multiple frontiers of extraction: Excavating contemporary capitalism. *Cultural Studies* 31 (2–3): 185–204.
Milam, Erika Lorraine. 2019. *Creatures of cain: The hunt for human nature in cold war America*. Princeton: University Press.
Mitchell, Timothy. 2011. *Carbon democracy: Political power in the age of oil*. New York: Verso.
Moore, Jason. 2015. *Capitalism in the web of life: Ecology and the accumulation of capital*. London: Verso, Verso Books.
Muniesa, Fabian, ed. 2017. *Capitalization: A cultural guide*. Paris: Mines Paris tech.
Nair, Savithri Preetha. 2005. 'Eyes and no eyes': Siwalik fossil collecting and the crafting of Indian palaeontology (1830–1847). *Science in Context* 18 (3): 359–392.
Nieuwland, Ilja. 2019. *American dinosaur abroad: A cultural history of Carnegie's plaster diplodocus*. Pittsburgh: University of Pittsburgh Press.
Nixon, Rob. 2011. *Slow violence and the environmentalism of the poor*. Cambridge, MA: Harvard University Press.
Noble, Brian. 2016. *Articulating dinosaurs: A political anthropology*. Toronto: University of Toronto Press.
Odum, Howard T. 1970. *Environment, power, and society*. New York: Wiley.
Oehser, Paul H. 1970. *The Smithsonian Institution*. New York: Praeger.
Ostler, Jeffrey. 2004. *The plains Sioux and U.S. colonialism from Lewis and Clark to wounded knee*. Cambridge: Cambridge University Press.
———. 2010. *The Lakotas and the Black Hills: The struggle for sacred ground*. New York: Viking Penguin.

Parsons, Keith. 2001. *Drawing out Leviathan: Dinosaurs and the science wars*. Bloomington: Indiana University Press.
Paul, Rodman. 1963. *Mining frontiers of the far west, 1848–1880*. New York: Holt, Rinehart and Winston.
Piketty, Thomas. 2014. *Capital in the twenty-first century*. Trans. Arthur Goldhammer. Cambridge: The Belknap Press of Harvard University Press.
Pimentel, Juan. 2017. *The rhinoceros and the megatherium: An essay in natural history*. Trans. Peter Mason. Cambridge, MA: Harvard University Press.
Podgorny, Irina. 2019. Bureaucracy, instructions, and paperwork – The gathering of data about the three Kingdoms of nature in the Americas, 1770–1815. *Nuevo Mundo Mundos Nuevos. Nouveaux Mondes Mondes Nouveaux – Novo Mundo Mundos Novos – New World New Worlds*, February.
Pomeranz, Kenneth. 2000. *The great divergence: China, Europe, and the making of the modern world economy*. Princeton: Princeton University Press.
Porter, Roy. 1977. *The making of geology: Earth science in Britain, 1660–1815*. Cambridge: Cambridge University Press.
Porter, Theodore. 1981. The promotion of mining and the advancement of science: The chemical revolution of mineralogy. *Annals of Science* 38 (5): 543–570.
Pratt, Mary Louise. 1992. *Imperial eyes: Travel writing and transculturation*. London: Routledge.
Qureshi, Sadiah. 2013. Dying Americans: Race extinction and conservation in the New World. *Proceedings of the British Academy* 187 (May 30, 2013): 269–288.
Rabbitt, Mary C. 1979. *Minerals, lands, and geology for the common defense and general welfare*. Vol. 4 vols. Washington, DC: U.S. Govt. Printing Office.
Rader, Karen, and Victoria Cain. 2014. *Life on display*. Chicago: University Of Chicago Press.
Ratcliff, Jessica. 2016. The East India Company, the Company's museum, and the political economy of natural history in the early nineteenth century. *Isis* 107 (3): 495–517.
Redman, Samuel J. 2021. *Prophets and ghosts: The story of salvage anthropology*. Cambridge: Harvard University Press.
Rieppel, Lukas. 2016. Museums and botanical gardens. In *A companion to the history of science*, ed. Bernard Lightman, 238–251. New York: Wiley-Blackwell.
———. 2019. *Assembling the dinosaur: Fossil hunters, tycoons, and the making of a spectacle*. Cambridge: Harvard University Press.
———. 2020. How dinosaurs became tyrants of the prehistoric. *Environmental History* 25 (4): 774–787.
Rieppel, Lukas, and Yu-chi Chang. 2023. Locating the central Asiatic expedition. *Isis* 14 (4).
Rohrbough, Malcolm J. 1968. *The land office business: The settlement and administration of American Public Lands, 1789–1837*. New York: Oxford University Press.
Rosen, William. 2010. *The most powerful idea in the world: A story of steam, industry, and invention*. 1st ed. New York: Random House.
Rudwick, Martin J. 1997. *Georges Cuvier, fossil bones, and geological catastrophes: New translations & interpretations of the primary texts*. Chicago: University of Chicago Press.
———. 2005. *Bursting the limits of time: The reconstruction of geohistory in the age of revolution*. Chicago: University of Chicago Press.
———. 2008. *Worlds before Adam: The reconstruction of geohistory in the age of reform*. Chicago: University of Chicago Press.
———. 2014. *Earth's deep history*. Chicago: University of Chicago Press.
Ruse, Michael. 1996. *Monad to man: The concept of Progress in evolutionary biology*. Cambridge: Harvard University Press.
Schiebinger, Londa, and Claudia Swan, eds. 2005. *Colonial botany: Science, commerce, and politics in the early modern world*. Philadelphia: University of Pennsylvania Press.
Schuchert, Charles, and Clara Mae LeVene. 1940. *O.C. Marsh, pioneer in paleontology*. New Haven: Yale University Press.

Schuller, Kyla. 2016. The fossil and the photograph: red cloud, prehistoric media, and dispossession in perpetuity. *Configurations* 24 (2): 229–261.
Schwartz, Seymour I. 1980. *The mapping of America*. New York: H. N. Abrams.
Secord, James. 1986. *Controversy in Victorian geology: The Cambrian-Silurian dispute*. Princeton: Princeton University Press.
———. 2018. Global geology and the tectonics of empire. In *Worlds of natural history*, ed. Helen Curry, Nicholas Jardine, James Secord, and Emma Spary. Cambridge: Cambridge University Press.
Sepkoski, David. 2020. *Catastrophic thinking: Extinction and the value of diversity from Darwin to the Anthropocene*. Chicago: University of Chicago Press.
Sewell, William H. 2008. The temporalities of capitalism. *Socio-Economic Review* 6 (3): 517–537.
Sharpe, Tom. 2015. The birth of the geological map. *Science* 347 (6219): 230–232.
Sheets-Pyenson, Susan. 1988. *Cathedrals of science: The development of colonial natural history museums during the late nineteenth century*. Vol. 25, 279. Montreal: McGill-Queen's University Press.
Short, John R. 2001. *Representing the Republic: Mapping the United States, 1600–1900. Picturing history*. London: Reaktion.
Smith, Crosbie, and M. Norton Wise. 1989. *Energy and empire: A biographical study of Lord Kelvin*. Cambridge/New York: Cambridge University Press.
Steeves, Paulette F.C. 2021. *The indigenous paleolithic of the western hemisphere*. Lincoln: University of Nebraska Press.
Steffen, Will, Paul Crutzen, and John McNeill. 2007. The Anthropocene: Are humans now overwhelming the great forces of nature? *Ambio* 36 (8): 614–621.
Thompson, E.P. 1963. *The making of the English working class*. New York: Vintage Books.
Todd, Zoe. Forthcoming. Fossil fuels and fossil kin: An environmental kin study of weaponised fossil kin and Alberta's so-called 'energy resources heritage.' *Antipode*.
Torrens, Hugh. 2016. William Smith (1769–1839): His struggles as a consultant, in both geology and engineering, to simultaneously earn a living and finance his scientific projects, to 1820. *Earth Sciences History* 35 (1): 1–46.
Turner, Stephen P. 1987. The survey in nineteenth-century American geology: The evolution of a form of patronage. *Minerva* 25 (3): 282–330.
Valencius, Conevery Bolton. 2015. *The lost history of the new Madrid earthquakes*. Chicago: University of Chicago Press. https://press.uchicago.edu/ucp/books/book/chicago/L/bo15987819.html.
Volney, Constantine-Francois. 1803. *Tableau du climat et du sol des États-Unis d'Amérique: suivi d'éclaircissemens sur la Floride, sur la colonie française au Scioto, sur quelques colonies canadiennes, et sur les sauvages: enrichi de quatre planches gravées, dont deux cartes géographiques et une coupe figurée de la chûte de Niagara*. Paris: Courcier/Dentu.
White, Richard. 2011. *Railroaded: The transcontinentals and the making of modern America*. New York: Norton.
Whyte, Kyle P. 2018. Indigenous science (fiction) for the Anthropocene: Ancestral dystopias and fantasies of climate change crises. *Environment and Planning E: Nature and Space* 1 (1–2): 224–242.
Williams, J'Nese. 2021. Plantation botany: Slavery and the infrastructure of government science in the St. Vincent botanic garden, 1765–1820's. *Berichte zur Wissenschaftsgeschichte* 44 (2): 137–158.
Witgen, Michael John. 2022. *Seeing red: Indigenous land, American expansion, and the political economy of plunder in North America*. Chapel Hill: University of North Carolina Press.
Wolfe, Patrick. 2006. Settler colonialism and the elimination of the native. *Journal of Genocide Research* 8 (4): 387–409. https://doi.org/10.1080/14623520601056240.
Wood, William. 1634. *New England's prospect: A true, lively, and experimental description of that part of America, commonly called New England*. Printed at London: by Tho. Cotes, for Iohn

Bellamie, and are to be sold at his shop, at the three golden Lyons in Corne-hill, neere the Royall Exchange.
Wu, Shellen Xiao. 2015. *Empires of coal: Fueling China's entry into the modern world order, 1860–1920*. Stanford: Stanford University Press.
Yusoff, Kathryn. 2017. Geosocial strata. *Theory, Culture & Society* 34 (2–3): 105–127.
———. 2018. *A billion black Anthropocenes or none*. Minneapolis: University Of Minnesota Press.

Mining and the Formation of Modern Geology

Sebastian Felten

Contents

Abstract

Mining mineral resources and knowing the Earth are historically entwined activities. This connex of practical and theoretical interest in the Earth has long been noted by historians of science. This historiographical chapter uses Vaccari and Morello's survey as a starting point. It revisits the idea that an eighteenth-century conjuncture of mining and scholarship brought about modern geology, tracing some strands of this narrative into the period of institutional entrenchment and global diffusion in the nineteenth century. It surveys canonical and recent works in the history of geology. Section 2 situates the interaction of mining and geology in a broader debate on practical and theoretical knowledge in the Scientific and Industrial Revolution. Sections 3, 4, 5, and 6 describe four ways in which historians have found mining (including other forms of earthwork like digging canals or quarrying) to have shaped the modern earth sciences. Section 7 points to a new avenue for research about the interface between mining and science: the impact of specifically *bureaucratic* knowledge production on modern geology

S. Felten (✉)
Department of History, University of Vienna, Wien, Austria
e-mail: sebastian.felten@univie.ac.at

E. Aronova et al. (eds.), *Handbook of the Historiography of the Earth and Environmental Sciences*, Historiographies of Science,
https://doi.org/10.1007/978-3-031-40799-4_17

during its formation. The Conclusion asks how the interaction between mining and science was shaped, or even constituted, by social mechanisms of inclusion and exclusion based on class/status, gender, and race.

Keywords

Bureaucratic knowledge · Circulation of knowledge · Experts · Gender · Labor · Practical knowledge · Race

1 Introduction

Mining mineral resources and knowing the Earth are historically entwined activities. This connex of practical and theoretical interest in the Earth has long been noted by historians of science. In a survey written 25 years ago, Ezio Vaccari and Nicoletta Morello proposed the following narrative: Mining has informed knowledge about the Earth and stones since the antiquity. This flow of information became stronger in the sixteenth century, but weakened in the seventeenth century. Over the course of the eighteenth century, mining was vital for molding traditional Earth knowledge into modern science, while mining practice received little useful knowledge in return from scholars. During the nineteenth century, geology became institutionalized at universities, especially after 1850. This changed the flow of information: geology now provided important indications for mining activities, "above all for the exploitation of the mines, which is now referred to as applied geology" (Vaccari and Morello (1998, quotation on p. 592. See also Vaccari 2000; Eyles 1969; Guntau 1978).

This framework is embedded in a larger narrative about the emergence and spread of geology as a modern science. Central works in the history of geology are focused on a European story of discipline formation that took place between 1780 and 1830 across Britain, Germany, France, and Italy (Porter 1977; Guntau 1984b; Laudan 1987; Oldroyd 1996; Rudwick 2005). For the nineteenth century, narratives highlight how geology (and related disciplines such as mineralogy, petrology, and paleontology) became established at academies of science, universities, and government offices across Europe, its colonies as well non-colonial countries in South America and East Asia (Mayer et al. 2017). Geological surveys, often driven by the search for useful minerals, play an important part in this narrative of transforming local cultures of knowing the Earth into an array of global geosciences (Lucier 2009; Westermann 2015).

This historiographical chapter uses Vaccari and Morello's synthesis as a starting point. It revisits the idea that an eighteenth-century conjuncture of mining and scholarship brought about modern geology, tracing some strands of this narrative into the period of institutional entrenchment and global diffusion in the nineteenth century. The following sections survey canonical and recent works in the history of geology. Section 2 situates the interaction of mining and geology in a broader debate on practical and theoretical knowledge in the Scientific and Industrial Revolution.

Sections 3, 4, 5, and 6 describe four ways in which historians have found mining (including other forms of earthwork like digging canals or quarrying) have shaped the modern earth sciences. Section 7 points to a new avenue for research about the interface between mining and science: the impact of specifically *bureaucratic* knowledge production on modern geology during its formation. The Conclusion asks how the interaction between mining and science was shaped, or even constituted, by social mechanisms of inclusion and exclusion based on class/status, gender, and race.

2 Broader Historiographical Context: Scholars and Craftsmen in Early Modern Mines

Historians of science and technology have long placed European mining into master narratives of the Scientific and Industrial Revolution, putting an emphasis on innovation in the mechanics of draining, ventilating and hauling, and the chemistry of refining (Weber 1976; Porter 1981). The debate around innovation has revolved around an old question: how were theoretical and practical knowledge linked, despite the great social distance between intellectuals and practitioners? This problem was formulated early by Edgar Zilsel.[1] He pointed out that late medieval and early modern "university-scholars and humanists despise the uneducated lower classes," but that "Real science is born when, with the progress of technology, the experimental method of the craftsmen overcomes the prejudice against manual work and is adopted by rationally trained university-scholars." Key for this marriage of theory and practice were "superior craftsmen" such as the "artist-engineers" who built fortresses and canals and who needed more theoretical knowledge than other craftsmen. One of Zilsel's case studies was the natural philosopher William Gilbert who interacted with navigators to work out the magnetism of the Earth – an intriguing intellectual puzzle that had great practical importance for the seaborne trade of early modern England (Zilsel 1941).

This strand of scholarship has argued that collaboration between "the scholar and the craftsman"[2] was key, both for the new experimental science of the sixteenth and seventeenth centuries, and for industrialization in the eighteenth and nineteenth centuries (Smith 2004; Jones 2008; Jacob 2014). This collaboration was pronounced in sites like arsenals, engineering project, and manufactories. Pamela Long argued especially for the early, post-medieval period that the lack of formal training for engineers that mines could function as "trading zones" for theoretical and practical knowledge, and that this opportunity closed after the eighteenth century with increased professionalism (Long 2011, 2015). Moving on in time, Ursula Klein draws on case studies from eighteenth- and nineteenth-century Prussian (and Saxon)

[1] Zilsel (1939). The following quotations are on pp. 4–5.

[2] This polemical phrase was coined by Hall (1959), in response to Merton (1938) and subsequent work. For the history of this debate, see Klein (2016a).

mining to develop the concept of "hybrid experts," that is, actors with multiple training (scholarly, technical, administrative) who were able to bridge the practice/ theory divide and lay the ground for the industrialization of "late-developing" economies like Prussia/Germany (Klein 2015, 2016b, 2017, 2020). Working in mines required constant engagement of water, rock, and air, which prompted mining experts to advance natural sciences from metallurgy to the study of gases and, most important for this handbook, geology. Increased attention to technical and labor processes in the mines and refineries gave rise to hybrid technosciences such as "mining science," which was now teachable via textbooks and university lectures.

Historians of the formative years of modern geology understand mines in this way, even if they do not use Klein's term. They note different forms, or intensities, of engagement between scholarship and mining practice. Using eighteenth-century Italy as a case, Vaccari distinguishes two modes of "using mines" for scholarship (Vaccari 2000, p. 179). One mode was exemplified by two Paduan professors of medicine, Antonio Vallisneri (1661–1730) and Bernardino Ramazzini (1633–1714). From 1704 to 1711, Vallisneri travelled through the northern Apennines to gather and describe fossils, plants, and animals. He visited mines and caves to collect data on rock strata and underground water, which revealed the "anatomy of mountains" to him. This experience, he claimed, helped him solve the question of the origins of springs (Luzzini 2013). Ramazzini, likewise, was present during the drilling of artesian wells near the city of Modena. Observations made at these occasions allowed him to work out a stratigraphical sequence of the Po River Valley and to argue that the reservoir feeding the wells was located underneath the Apennines. Ramazzini is also known as a pioneer of occupational medicine. Like the medical reformer Paracelsus (Theophrastus von Hohenheim, 1493–1541) before him, he visited mines to describe miners' diseases (Rosen 1943; Paracelsus 2013). In short, scholars like Vallisneri and Ramazzini used mines as sites of empirical observation. They occasionally visited them during field trips to collect data and specimens which would help them to resolve their scholarly questions.

A more common way of "using mines," however, was that the scholar's livelihood became enmeshed with mining practice. This allowed for more meaningful interactions between mining and science and makes it difficult to mark off clear boundaries between these two activities.[3] Georg Bauer/Georgius Agricola (1494–1555), the best-known "earth scientist" of the European Renaissance, had a personal stake in mining. He was a physician in, and mayor of, Chemnitz, a town benefitting from the nearby silver boom, and he also acted as court historian of the dukes of Saxony, who drew income and prestige from mining (Agricola 1912, 1955–1997). Vannoccio Biringuccio (1480–1539), another "founding father" of the earth sciences, supervised iron mines near Siena and throughout his checkered career directed work in related sites such as mints and foundries (Bernardoni 2012; Biringucci and Smith 1942). Vaccari's examples from eighteenth-century Italy include the scholar Giovanni Arduino (1714–1795), who surveyed and directed

[3] On "boundary work" between science and the management of mining, see Felten (2018).

mines in Tuscany, the Apennines, Veneto, and Lombardy, and the scholar Benedetto Spirito Nicolis di Robilant (1724–1801), who was an inspector of the Piedmont mines (Vaccari 1993, 2000, pp. 172–178). Many important contributions during the formative years of modern geology came from officials working for Central European mining administrations, most notably Abraham Gottlob Werner, who began his career as the accountant of an iron works in Lower Silesia and ended it training officials in the Saxon mining town of Freiberg (1749–1817) (Albrecht and Ladwig 2002; Guntau 1984a). Many, if not all, geological surveys that have mapped the Earth's surface since the eighteenth century involved mining experts and/or had clear utilitarian goals of unlocking a country's mineral resources (Kozák et al. 2016; Mayer et al. 2017).

In short, the interaction between mining and earth science during the formative years of geology is evident both on a biographical and institutional level. The focus here is on the epistemological consequences that this enmeshment might have had: How did mining practice shape scientific ways of knowing the Earth? The following sections look at four dimensions that are commonly associated with modern science as a form of knowledge: that it is based on empirical data, interprets this data with the help of analytical tools and technical terms, that it is institutionalized, and that it is a collective and transnational endeavor.

3 Experiencing Mines: Empirical Data

Scholars have used mines to gather empirical data through on-site observation, the collection of specimens, and by talking to local informants. Mines allowed scholars to observe otherwise inaccessible phenomena. The visits of mines by Italian, German, British, and French scholars can therefore be situated in a broader history of scientific travel, expeditions, and peripatetic scholarship that intensified in early modern Europe (Harris 2006; Klemun and Spring 2016; chapter ► "Expeditions and Fieldwork in Earth Sciences"). Yet the experience of mines as sites of labor had specific impacts on how scholars thought about the Earth.

A very specific case of such an impact would be the experience of wet galleries on theories of metal generation. Vannoccio Biringuccio noted in 1540 that, "because of water above and water below I was soaked as if I had passed through a heavy rain, but this did not surprise me since I had always understood that water was the primary and peculiar companion of minerals, indeed that it was perhaps the very reason for the generation of their substance."[4] Water was, in fact, a constant companion of sixteenth-century mining, when complex hydro-powered machines were used to drain ever-deeper shafts (Ludwig and Schmidtchen 1992; Agricola 1912). The prevailing theory at that time was that metals were formed through "exhalations" and the combination of "sulfur" and "mercury" (which were principles rather than the modern chemical elements). However, scholars with the experience of abundant

[4]Biringuccio, Pirotechnia (ed. Smith), 20. Luzzini (2020).

water in mines found an aqueous origin of metals more plausible (Norris 2006, 2007). Johannes Mathesius (1504–1565), Lutheran preacher of the mining town of St. Joachimsthal/Jáchymov, reported on the belief among miners that metals formed from a sort of cold slime (*gur*), which may reflect observations of natural mineral decay specific to the mines in that region (Norris 2014; Alfonso-Goldfarb and Ferraz 2013; Smith 2017). In the eighteenth century, the mineralogist and mining official Heinrich von Trebra marshalled the first-hand experience of violent water that tortured workers and soaked his own clothes to counter the idea that fire was the primary causal agent of mineral change and to confer credibility on the *gur* theory of mineral growth.[5]

A very general case of an impact of earthwork on earth science would be the visual impression of cut walls – the "section" that shows different layers of rocks and soils – and stratigraphy (Eyles 1969, esp. 167). Taking Britain as an example, early practitioners of stratigraphy were mining surveyors who claimed expertise to indicate coal beds from the relief of a landscape or trial borings. "However dubious much of his 'expertise', the logic of a mines expert's situation ultimately forced him to address himself to relations of strata to the Earth at large. And, of course, the operations of digging shafts perpendicularly through the strata, and extracting minerals laterally along them, also gave immediate practical experience of the geometrical parameters of stratification" (Porter 1977, p. 60). Canal-digging was a similarly important site of experiencing strata as it cut open landscapes for many miles, forced workers and engineers to understand how the different layers of the ground interact with water, and gave them ample opportunity to find fossils (ibid., p. 138).

A well-known case of a practitioner who developed stratigraphic thinking from experiencing earthwork was the British surveyor William Smith (1769–1839). Working on an estate in the Somerset coalfield in the 1790s, he experienced hard layered Secondary rocks in shafts and adits and learned from miners that coal beds lay in the same sequence wherever they were found. The digging of the Somerset Coal Canal allowed him to extend his observations across the county and include data on soft clays, sandstones, and limestones. By travelling to canal projects in the north of England, he extended his observations once again and mapped outcrops across England with an innovative method using graded bands of color. By around 1806, he was able to show that the concept of characteristic fossils was valid across much of England and Wales, and that they could be used to make fine distinctions between different formations. Only a fraction of his work was published in print, but his ideas circulated among provincial agricultural improvers and eventually reached metropolitan scientists. Overall, such interaction between earthwork and earth science "was a decisive element for the development of lithostratigraphy, especially in Germany and most of Central Europe, in Sweden, in Great Britain, in Russia, and also in Italy" (Vaccari 2000, p. 166).

[5]Discussed in Felten (2018, p. 14, note 60). Another eighteenth-century case is described by Luzzini (2011).

4 Making Sense of Mines: Analytical Tools and Technical Terms

So far, this survey has established that mines were a prototypical "trading zone" between practical and scholarly knowledge, and that the experience of being underground prompted specific scholarly ideas such as wet mineral generation and a layered structure of the Earth. However, the epistemic impact of mining on earth science was more profound than just providing sites for empirical observation. As Martin Rudwick argued in a seminal paper, geology developed a specific "visual language" of diagrams and maps "which could complement verbal descriptions and theories by communicating observations and ideas that could not be expressed in words" (Rudwick 1976; see also Kusukawa 2016). This visual language had several sources, but one particularly important one was mining: "Mining provided geognosy [a precursor of geology] not only with empirical data on the dimension of depth in the earth's crust, but also—far more importantly—a distinctive way of thinking and even of seeing. Anyone involved in the mining industry, from ordinary miners right up the social scale to those who managed and administered mines, worked in a three-dimensional world of rock structures" (Rudwick 2005, p. 84)[6].

Depth was visualized through vertical sections of one or several mines. This was a widespread management practice existing since at least the sixteenth century, but rarely appeared in print for reasons of secrecy (Weiß 1997; Neubert and Stein 1958). Scholars such as John Strachey (1671–1743) in England, Giovanni Arduino (1714–1795) in Italy, or Georg Christian Füchsel (1722–1773) in Germany adopted and adapted these techniques in their notes and publications (Fuller 1992; Vaccari 2000, p. 174; Rudwick 2005, pp. 89 and 124). Sections thus became "the most characteristic graphical tool in geognosy" (Rudwick 1976, p. 87), where it aided three-dimensional thinking and facilitated the communication of arguments about stratigraphy. The Prussian naturalist Alexander von Humboldt (1769–1859), who was trained as a mining official, used vertical mapping techniques during his American expeditions, "representing entire countries as one would a mine" (Anthony 2018, quotation on p. 48, 2020). By 1830, sections were used to make correlations between rock formations in different regions and to demonstrate the lateral variation in their thickness. In this way, mining sections were at "the origin of the correlation diagrams that later became such an indispensable part of the visual language of stratigraphical geology" (Rudwick 1976, p. 168).

Mining also provided actual language, in the form of technical terms or concepts, that proved useful for geological thinking. "Many terms came from a double tradition: some of them came from the mining workers' language – that is to say, from national languages and dialects; other terms came from the scientific literature of the previous centuries and resulted from linguistic choices made by individual authors, who often adopted Greek and Roman words" (Vaccari and Morello 1998,

[6]Geognosy was a science of classifying rock formations practiced mostly in Germany. See Klemun (2015).

pp. 591–592). In German geognosy, the object of study was "Gebirge," literally mountains, which in mining language was a generic term for rock surrounding ores, even if these rocks were found in a valley; similar terms such as "montagnes" and "monti" were used throughout Europe Rudwick (2005, p. 86).[7] The classification of rock layers by scholars such as Johann Gottlob Lehmann (1719–1767), in turn, drew on vocabulary current among miners of the time, such as *Stinkstein* (malodorous stone), *Zechstein* (mine stone), or *rothe tode Lager*, which as Old Red Sandstone "entered the geologist's regular vocabulary."[8] The development of a special language in German – used in mining regions across Central and East Central Europe due to the migration of experts – can be traced through different genres of text.[9] Some of the early evidence (ca. thirteenth century) comes from codifications of mining law. Technical handbooks, starting with Rülein von Calw's *Bergbüchlein* (ca. 1500), normally contained a glossary of terms that were deemed too esoteric for a broader readership. Georg Agricola's *De re metallica* (1556) collected many vernacular German terms and, since classical Latin was not precise enough, rendered many of them in neologisms. Standalone dictionaries in German were printed as early as the seventeenth century, a tradition that peaked in Heinrich Veith's *Deutsches Bergwörterbuch* (1871). The use of special terminology provided precision to describe and manage complex labor processes, especially when it was combined with technical images as in *De re metallica* or the Schwazer Bergbuch (Kessler-Slotta 1994; Lefèvre 2010). But it arguably also provided a way to signal expertise or at least a connection to mining. As Ilpo Piirainen notes, in this tradition "a miner is not only someone who does physical work in mining, but everyone who is productively involved in mining, from miners working underground to overseers and mining officials, mining judges and scientists. Therefore, one cannot assume a homogeneous group when it comes to text production and reception in mining" (Piirainen 1998). This explains why specific terms, including those for rocks and mountains, were used simultaneously in worksites, bureaus, and the academy. It is therefore not surprising that German mineralogists drew heavily on vernacular terms, some of which gained international currency, not least through the spread from Saxony of Gottlob Abraham Werner's ideas.

Several historians emphasized that geology borrowed its fundamental concepts of matter, and practices to determine its properties, from metallurgy and assaying. Sally Newcomb begins her investigation of laboratory practice in geology with Renaissance alchemy, as both mineralogy and mining used its "considerable wealth of knowledge about the behaviors of natural materials" (Newcomb 2009). The testing and refining of ores that was practiced in mining regions contributed to geology by strengthening the experimental tradition in mineralogy. The chemical laboratory, which shared equipment such as furnaces with refining sites, "proved to be fundamental in determining parameters of rock and mineral behavior that didn't so much

[7]Art. "Gebirge" in Veith (1871).

[8]Oldroyd (1996), quotation on p. 79. See also Engelhardt (1992), and Norris (2014).

[9]The following overview is largely based on Piirainen (1999).

explain their origin, but gave insight into the possible as well as the impossible" (ibid., p. 1) David Oldroyd similarly draws a direct line from mining practice via analytical chemistry to modern geologists' fundamental ideas about matter: "to all practical purposes, the assayers deployed the idea of elements as simple substances, occurring as the last terms of analysis, long before Antoine-Laurent Lavoisier defined elements thus" (Oldroyd 1996, p. 68; see also Porter 1981; Fors 2014). In sum, then, mining not only provided empirical data to the emerging earth sciences but also some of the conceptual tools and terms to make sense of this data.

5 Managing Mines: Institutions

The realization that mining was important for the emergence of modern earth science in eighteenth- and nineteenth-century Europe forced historians to take into account the social conditions, even if they believed like Rachel Laudan that, "the social history of science must play second fiddle to the intellectual history of science because only when we have traced the development of scientists' ideas do we know what needs to be explained – namely, the causes of those ideas, the conditions that conduce to their success" (Laudan 1987, p. 18). Martin Guntau, writing in socialist East Germany, started from a materialist position stressing the material needs of the mining industries, but his conclusions are similar to Laudan who gives methodological primacy to intellectual history (Guntau 1984b). Laudan devoted a considerable part of her widely cited book to describe state-directed mining in German lands as the social context in which the "Becher-Stahl school of mineralogy and cosmogony" flourished in the eighteenth century. This tradition, associated with the scholar and projector Johann Joachim Becher (1635–1682) and the physician Georg Ernst Stahl (1659–1734), used models inspired by alchemy/chemistry to conceptualize changes of the Earth.

States like the Electorate of Saxony or Prussia-Brandenburg managed industrial enterprises such as mining-refining complexes and porcelain manufactories. These enterprises "provided bureaucratic positions for mineralogists" (Laudan 1987, p. 49), such as the eminent chemist and physician Johann Friedrich Henckel (1678–1744). Mineralogists were employed as teachers in the new mining academies in Freiberg (Saxony, today Germany) and Schemnitz (Lower Hungary, today Slovakia), which inspired similar schools in Prussia and France. "Institutions were set up to supply bureaucrats for state industries, and the graduates of these institutions could count on employment" (ibid., p. 52). A number of earth scientists thus had a stable income and the leisure to engage in research, even if their employers expected them to focus on utilitarian aspects that would aid the management of mines. Laudan names four specific areas in which institutionalization had a discernible impact on earth science: mineralogists became more interested in rocks rather than minerals; their description of the primordial ocean that supposedly deposited these rocks became more refined; they improved their ability to distinguish minerals, especially earths, by chemical means; and they improved their understanding of heat and water as causal agents (ibid., p. 56).

The protagonist of Laudan's narrative is the mineralogist, mining official, and teacher at the mining academy in Freiberg Abraham Gottlob Werner, whose "theory marks a pivotal point in the development of geology. The succeeding generation looked to Werner and to his 'school of Freiberg' as the source of most of their geological theories" (Laudan 1987, p. 87). He was steeped in the tradition of chemical cosmogony, and considered rock as depositions from water, in analogy to laboratory observation, and thus as unique historical entities. He visited mines and used inspection logs as evidence, but abstracted from them to point out correlation of specific strata sequences in different regions. During and after Werner's lifetime, his theory of rock formation percolated through a network of peers and students. Through this process of "radiation," Wernerians came to control geological teaching institutions, textbooks, and journals in the first half of the nineteenth century (ibid., pp. 102–109). A tradition emerging in the mining bureaucracy of Saxony thus "formulated the conceptual foundations of geology and dominated its intellectual and institutional development in the period between 1780 and 1830" (ibid., p. 223).

Ursula Klein's recent exploration of scientific culture in Prussia (and Saxony) provides important context to Laudan's story as she embedded scholar-officials like Werner in a broader pursuit of "useful knowledge" in service of German states. The protagonist of her narrative is Alexander von Humboldt, whose training as a civil servant at the Freiberg Mining Academy was formative for his personality as scientist. "There he also learned to write detailed reports about travels and completed tasks which carry the imprint of reports on mining inspections and official business, but also had similarities to geological-mineralogical travel reports" (Klein 2015). When, at the age of 22, he was put in charge of all mining and refining in the Prussia territory of Franconia, he wrote to his friend Johann Carl Freiesleben: "I will now live entirely devoted to practical mining and mineralogy. [...] I am giddy with joy" (ibid., p. 154). Humboldt immediately used his access to the underground for botanical and chemical studies. His investigations into underground airs resulted not only in learned treatises but also in the invention of a safety miner's lamp and a breathing mask.

Humboldt and Werner are only the most prominent of a group of men who had different careers and personalities but who together embodied a historically specific persona: that of the "hybrid scientific-technical expert in a state ambience" (ibid., p. 9). This was a Continental-European equivalent to the "savant fabricant" that Peter Jones has identified as important for English industrialization. This figure had, according to Klein, a "key social function that was equally instrumental in the onset of industrialisation, the institutionalisation of the useful sciences and the differentiation of the exact, analytical sciences" (ibid., pp. 9–10). They differed from both "scholastic university scholars" and "tradition-bound craftsmen" (Klein 2016b, p. 90), but also from the figure of the apolitical "scientist" constructed by Western historians of science during the Cold War. Humboldt and Werner must thus be seen as part of "an alliance of technically skilled and scientifically educated civil servants, technicians and naturalists." Their basic consensus was that trade and manufactory had to be promoted by the state, through technical improvements and training of technicians and naturalists (Klein 2015, p. 8). The Freiberg Mining

Academy, where hybrid experts were trained and hybrid bodies of knowledge were taught and researched, had a "hermaphroditic position" (Klein 2016b, p. 16). The experts who advanced these useful sciences were in turn "bridge builders" between the "world of work" and "that of reading and writing" (ibid., p. 92). According to Klein, they were "social figures that are difficult for us to grasp today. From today's perspective, they are historical transitional figures, as they combined functions that are now distributed among several specialist groups" (Klein 2015, p. 131).

Saxony and Prussia are not the only cases in which mining bureaucracies (rather than mining per se) fostered transfer and co-production of Earth-related knowledge. Hjalmar Fors and Jacob Orrje showed that mining was organized along similar lines in Sweden, creating a similar social context in which technical experts were trained and employed by a powerful *Bergskollegium* (Fors and Orrje 2019; Fors 2014, 2015). Emanuel Swedenborg (1688–1772) and Axel Cronstedt (1722–1765) are examples of mining officials who also made important contributions to mineralogy. For France, Isabelle Laboulais has shown that mining was established as an administrative field in the eighteenth century via the activities of the *bureaux du contrôle générale*, the *intendance des mines*, and the *école royale des mines*. The *maison des mines*, a mining school set up in Paris after the revolution and comparable to the academies in Freiberg and Schemnitz, institutionalized a common professional culture of mining engineers who used its archive, library, and mineral collection as knowledge tools. The survival of pre-revolutionary institutions in this form was, as Rudwick argued, "important for the sciences of the earth" in France (Rudwick 2005, p. 325). It was given the task of coordinating the exploitation of mineral resources and issued a journal. This professional culture diffused via education, journal, and publications. "This house of learning, which has close links with the neighbouring offices responsible for the administration of mines, must be the heart of 'mining science', i.e. the place where all the knowledge required to develop mineral resources is mobilised" (Laboulais 2013). An increasingly sophisticated bureaucracy formed a nexus between mining practice and earth science also in Italy, Spain, and Spanish America, as well as Russia (Lang 2008; Escamilla-González and Morelos-Rodríguez 2020).[10]

6 Travelling Between Mines: Transnational Expert Communities

The emerging system of mining managed, or fostered, by state institutions facilitated mobility of scientists and knowledge transfer between local contexts: "European states had constructed a system in which a number of scientists [...] were financially secure and had sufficient leisure to contemplate theoretical as well as practical matters. Scholars travelled to the leading mining academies, and the international exchange of ideas became easier" (Laudan 1987, p. 55). Because of dynastic

[10]For Russia, see chapter ▶ "Mining and Metallurgy in Early Imperial Russia" by Graber.

alliances, many Central European polities extended beyond the bounds of the Holy Roman Empire. For example, Saxony and Poland-Lithuania were at some point ruled by a monarch of the Wettin dynasty, whereas Brunswick-Wolfenbüttel and Britain were united through the Welfs. Many Central European territories, especially those in the Habsburg Empire, had complicated cultural and linguistic dynamics due to migration. In Lower Hungarian mining towns, for example, the German language was associated with high-status migrants attracted through medieval colonization schemes (Slotta and Labuda 1997). All of these territories had a global colonial history that was sometimes obvious, as in case of the Habsburg dynasty chartering the Ostend Company to trade with India, sometimes more oblique, as in the case of the dukes of small Saxe-Gotha acquiring Asian, African, and American objects for their cabinet of curiosity (Meisterle 2020; Collet 2007).

The transregional and transcontinental movement of knowledge through people and objects was particularly pronounced in mining. Mining existed in specific locations, with communities that formed a specific socio-economic niche in which unusual types of knowledge, such as underground surveying or analytical metallurgy flourished (Bartels 2010). These interstitial communities had arguably more in common with each other than with their surrounding societies, sharing practical problems, legal arrangements, and a specialized language. There is evidence for transregional careers, espionage, and co-constitution of knowledge between the Harz Mountains and the Ore Mountains; between these areas and Lower Hungary as well as Sweden; and between Central Europe and Mexico and the Andes (Hillegeist and Ließmann 2001; Fettweis and Hamann 1989; Fors 2015; Escamilla-González and Morelos-Rodríguez 2020; Guerrero 2017). The mining academies in Schemnitz/Banská Štiavnica and Freiberg (both founded in the mid-eighteenth century) formalized and increased this movement as they drew students from Europe, its empires, liberated South America as well as Japan and the USA (Schleiff and Konečný 2013). The educational model of the mining academies with its successful merging of theory and practice, in turn, proved a transregional success "which was confirmed by its rapid diffusion throughout Europe and overseas" (Brianta 2000, quotation on p. 279; Brianta 2007). They served as "centers for geological research and education" of an increasingly transnational scientific network (Vaccari 1998, 2009).

The intellectual and geographical journey of the Brazilian mineralogist and mining official Manuel Ferreira da Camara (1764–1835) can illustrate how knowledge was circulating, or co-produced, within this network (Figueiroa and da Silva 2000). Born in Minas Gerais into a family of gold mine owners, he was sent to Portugal to study law and philosophy, which included science classes. Shortly after graduation in 1788, he became a member of the Portuguese Royal Academy of Science. He was tasked by the crown to visit important mining districts across Europe and take classes in mineralogy, chemistry, metallurgy, and other "mining arts" at various mining schools, especially the Freiberg Academy where he became a student of Werner's. He returned to Portugal in 1789, and in 1800 to Brazil. He was made mining intendant of Minas Gerais where he tried, unsuccessfully, to set up a school similar to Freiberg. His plans to reform mining by replacing enslaved labor with improved technology (for efficiency, not humanitarian reasons) were equally thwarted by resistance by local mine owners. Camara can serve as an example of "enlightened Portuguese and Brazilians who believed that improvements in mining and agriculture would bring economic development." He showed a distinctively "Brazilian tendency to appropriate, amalgamate, and employ the various scientific traditions," which goes against historians' "linear diffusionist models of the spread of scientific knowledge" (ibid., p. 188).

7 New Questions: Geology as Bureaucratic Knowledge

The importance of the state ambience for the flourishing of hybrid experts brings up interesting question for the history of earth science. Did earth science borrow anything specific from mining *bureaucracies* (rather than mining tout court) during its formative period in early modern Europe? In Klein's narrative, administration appears as a medium of change. Even if many reform impulses in Prussia did not yield many visible results, "they were, however, fundamental in that they affected the value catalogue and behavioural repertoire of the civil service, and gradually changed it" (2015, p. 305). The "scientific ethos of open communication" (2015, p. 80 and 2016b, p. 151) and careful recordkeeping, Klein argues, defined the virtues of a new type of civil servant. As Klein shows, ideals and practices seem to have been imported from Saxony to Prussia through the pivotal figure of Anton von Heynitz (1725–1802). Would it not be plausible that the reform efforts in Prussia and Saxony not only brought a scientific ethos into the administration, but that civil servant ideals also entered the sciences? A closer investigation of bureaucratic knowledge production could support Klein's assessment that the interpretation of a one-sided "scientification" (2016b, p. 12) of practical production contexts is insufficient to understand the dynamic of eighteenth-century "useful knowledge."

Central European mining administrations embraced experimentation, openness, and precision already in the seventeenth century because they were increasingly committed to the idea of a "common good" and had to provide accountability to mine owners (Jobst and Schellhas 1994; Bartels 1992). Central European regions responded in similar ways to the structural crisis that beset many mining regions after 1550: princely officials became more involved in the management of ailing mines (Fessner and Bartels 2012). The resulting institutional structure is called "principle of direction" in the literature, meaning that officials actively directed mines, in contradistinction to a government that only supervised companies making their own managerial decisions, as favored by nineteenth-century liberals (Lück 2015). Central Europe shared this political economy with Scandinavia, where a specialized mining administration was also developed. It stood in a certain contrast to Spain and Spanish America, Italy, and France where the sovereign and its administration were more content to simply enforce legal institutions and collect taxes (Fors 2015; Gliech 2013; Clair 2009; Gauché 2003).

Electoral Saxony, which is central in many histories of geology due to the Freiberg Mining Academy and Abraham Gottlob Werner, is considered a textbook case of a mining administration that pushed investors out of an active entrepreneurial role (Schirmer 2009). Yet investors had certain information rights, which resulted in an ideal of good recordkeeping and conduits for information (Felten 2020b). A topically named draft for a new mining code by the noble mining official Abraham von Schönberg, the *Berg-Information* of 1693, was an attempt, among other things, to disclose the procedures of the administration to foreign investors: "for my own sake, so that I can have sufficient knowledge of the work of one of the officials and servants under me"; for the mining officials "to have necessary information, so that no one can excuse themselves with ignorance"; and for the owners of the mines "to awaken in them a greater desire for mining" (von Schönberg 1693, unpaginated preface).

Bureaucratic procedures of knowledge production and controlled circulation shaped the scientific practice of the hybrid expert. These inspections were highly formalized procedures, usually preceded by careful study of files and maps, the results of which were in turn recorded in a regulated manner and filed away for later inspection. In other words, officials like Werner and Humboldt had procedural knowledge: they had learned sophisticated procedures for obtaining and condensing information, which Klein discusses in part as "Heinitz's conception of mining science" (Klein 2016b, pp. 145–152). What is interesting here is that mining administrations had long been working on the core problem of "useful science" suggested by Klein: How to abstract from individual cases for planning purposes so that the plan works in individual cases? Administrations offered a very special "epistemic setting" (Brendecke 2009) that was not only exhausted by the fact that civil servants were confronted with complex natural and social situations, but that also provided them with the procedural means to respond adequately to these situation – for example, through the above-mentioned detailed reports that Werner required from his students and that Humboldt learned to write in Freiberg.

Recent work has investigated bureaucracies as knowledge-producing contexts, whose history runs in parallel and is often intertwined with that of science (Felten and von Oertzen 2020; Raphael 2020; Westermann 2011). The expansion of a dedicated bureaucracy in many Central European territories, one could argue, transformed "mining" on a conceptual level. The hierarchical structure favored the consideration of systemic connections without losing sight on the material needs of individual mines (Felten 2018). This might be significant for the history of geology, as Deborah Coen's study of Habsburg climate science suggests: both the ideology and the administrative structure of the Habsburg Empire were conducive to the emergence of climate science that is both attuned to local and global scales (Coen 2018). In contrast, Renee Raphael case study of silver refining in Potosí would be a counterpoint to the idea that bureaucracy supported science. She shows through detailed textual analysis that administrative reports were not faithful record of practice, as which they are often used by historians of science and technology to recuperate Iberian science. Administrators emphasized procedural aspects and were less interested in recording technological details, which means that historian need to pay more "attention to the process of bureaucratic knowledge production when interpreting the written traces of the scientific and technical culture of the early modern Iberian world" (Raphael 2020).

The primary contrasting case, however, to the continental story of close institutional links between mining industry and science is Britain. Gold and silver were, as in continental lands, claimed by the monarch but they were not very plentiful, and a high-profile lawsuit in the sixteenth century ruled that all other metals belonged to the landowners (Ash 2001). This means that there was no incentive for the state to build a mining bureaucracy similar to that in Sweden or the German lands. As Laudan argued, "Both the nature and the relative value of British mineral resources encouraged the isolation of British mining engineers and mineralogists from continental practice" (Laudan 1987, p. 55). There were no mining schools or government employment. Yet, geology "is exceptionally dependent upon deep social and institutional foundations," as Roy Porter argued, which means that it had to be a different "social base" (Porter 1977, p. 7). Universities, museums, "invisible colleges" of like-minded scholars and discussions at the Royal Society (founded 1660) gave some coherence to the study of the Earth, though the period lacked a grand ideas (ibid., pp. 1–22, 53–54). The lack of a state ambience led to a disconnect between geology and industry: "Very few who possessed intimate professional knowledge of the Earth made significant contributions to the development of geological concepts or theory — whereas those who did owed little directly to personal contact with surveyors, miners, or with mining. The leading early nineteenth-century geologists were only marginally involved in the practical deployment of their knowledge." While industrialization created "unprecedented needs for knowledge of the Earth, and potential capacity for supporting practical geology experts, the very laissez-faire capitalist structure of that industrializing process was simultaneously militating against creating these experts and that practical science" (Porter 1977, p. 132; Porter 1973).

An institutional interface between mining and science was only created through the rise of specialist societies. Societies were important for making geologists a cohesive group, most importantly the *Geological Society of London* (1807) which helped to create cohesion and group identity among the gentlemen amateurs who made up its ranks (Porter 1977, p. 147). Some of regional societies had links to mining industries, such as *The Literary and Philosophical Society of Newcastle-upon-Tyne* (1793) or the *Royal Geological Society of Cornwall* (1814). "That mining interests underlay the societies — not dilettante gentlemen — ensured that the interest in the Earth would be geological — focusing on strata — not cosmogonical or natural historical" (Porter 1977, p. 135). In Scotland, improvement societies served as important vehicles that "put agricultural advancement, mineral discovery and exploitation, quarrying and coal-mining high on the agenda" (ibid., p. 151).

Over the course of the nineteenth century, the contrast between Britain and continental Europe became less significant as geology became bureaucratized in its own right, through university chairs and, importantly, geological surveys. This was a process that took place across Europe, and indeed the world.[11] The Habsburg

[11] Oldroyd (1996, pp. 124–127) (Table 5.2 "Bureaucratic structures for the geological mapping of the world"). Many case studies on geological surveys from Lithuania to Argentina are collected in Mayer et al. (2017).

Empire is an instructive case that has been well-documented by the work of Marianne Klemun. As in Saxony, mining was "fundamental to the development of geology" (Klemun 2017, quotation on p. 264) here as it provided scholars with employment and access to empirical material. Pivotal figures like Ignaz von Born (1742–1791) show how the bureaucratization and centralization of the Empire created a structure for scientific work. A high-ranking mining official in Schemnitz, he reorganized the princely cabinet of curiosity in Vienna into a scientific collection (Klemun 2004). When he carried out mineralogical fieldwork in Temeswar, Transylvania, and Hungary, the state administration offered resources for him to pursue his travels. He made observations directly in the mines that feed into larger scientific arguments, but interests were clearly shaped by his work as mining administrator (Klemun 2007). The foundation of the Mining Academy in Schemnitz/Selmecbánya/Banská Štiavnica (ca. 1763) helped to formalize Earth-related knowledge as teachable curriculum long before chairs for earth sciences were established at the university in 1862.

Surveys of mineral resources have been carried out by mining officials since the eighteenth century, but in the early nineteenth century, there were new impulses by local associations and societies. "Every region of the Habsburg Monarchy sought to have a geognostical map" (Klemun 2017, p. 266). These mapping initiatives were eventually consolidated and superseded by the empire-wide geological survey carried out by a central institution, the *Geologische Reichsanstalt* (1849), which was the first continental equivalent to the English geological survey founded in 1835 (Klemun 2020). The survey mapped the Empire's vast and diverse space in less than 20 years, between 1848 and 1867. As in other large scientific projects of the time, this led to bureaucratic approach to data collection and processing (Daston 2017b). To carry out the survey, the *Reichsanstalt* established a network of correspondents who send objects and data, and many rounds of fieldwork, which in turn, involved "countless administrative acts and procedures" in the form of structured note-taking (Klemun 2014; see also Westermann 2011).

8 Conclusion

As this survey has shown, the impact of mining on modern geosciences was profound. Famous earth scientists but also the "broad army of workers bestowing abundant data from diverse terrains" (Porter 1977, p. 6) were through their livelihood involved with mining. Conceptual and material tools to make sense of the underground were imported from mining contexts into emerging geoscientific disciplines. Fundamental methods and models such as stratigraphy or the modern theory of matter were shaped by the mining experience. Yet entwining the history of mining and geology has a critical potential that is not yet fully realized. Perhaps more so than other aspects of the earth science history, mining can bring to the fore how contributions to earth science have been highlighted or hidden through social mechanisms of inclusion and exclusion.

The research discussed in the previous sections shows that around mining emerged stable conduits for empirical data, conceptual tools, and procedures of knowledge production between "theory" and "practice" of engaging with the Earth. It provided scientific endeavors with institutional backing and offered individuals remarkable geographical and social mobility. This was true in particular for the formative years of modern geology between 1750 and 1850. Less clear from existing literature is how these conduits were conditioned, or even constituted, by intersections of status/class, gender, and race. Mining, which often took place in hierarchical, gender-divided, and colonized spaces, is a useful focus to investigate the question of inclusion and exclusion in earth science history.

In a trailblazing recent article, Lydia Barnett is committed to recuperate hidden contributions by "earth workers" but makes the tactical choice to analyze scientists' practices of inclusion and exclusion. She follows historians of science who have sought to recover "invisible laborers whose social marginalization by virtue of their race, class, or gender kept them hidden in the shadows for far too long, leading to a skewed perception of premodern science, especially European and Euro-American science, as far more socially homogenous than it actually was" (Barnett 2019, quotations on pp. 246–247). Earth workers fall into that category of knowers as they were "laborers of very low social rank whose work brought them into regular contact with the earth's surface or interior, such as miners, quarrymen, ditch-diggers, and fieldworkers." Her method is to read printed works of eighteenth-century natural history, written by elite men, for traces of earth workers who made contributions as fossil finders in Italy, France, Switzerland, Britain, Ireland, Germany, Spain, and colonial America. As these sources are very muted on their lives, "it is both necessary and generative to acknowledge the limits of what these traces can tell us." Barnett therefore shifts the attention of her analysis to the "norms of elite masculinity that shaped the emerging earth and field sciences in the Enlightenment, seeking to reconstruct the complex attitudes of gentlemen naturalists toward the lower-ranking people whose labor they both devalued and depended on." Gentlemen scientists almost never named low-class fossil finders and often used the passive voice and other grammatical constructions to hide the contribution of laborers in their text. Sometimes they highlighted them in "an anonymous acknowledgment, in which an author briefly referenced an individual or group as discoverers of a fossil specimen without naming them or disclosing any other individuating detail beyond social rank and occupation" (ibid., pp. 249–250). By closely reading scientific texts for such traces, it is possible to recover some of earth workers' contribution to historical earth science. Combining this corpus with material and oral culture may lead to further insights into "the rich vernacular knowledges of the earth that existed beyond the gentlemanly worlds of eighteenth-century natural history" (ibid., p. 274).

Barnett's work is thought-provoking because it directly confronts assumptions about knowledge hierarchies that are implicit and explicit in existing narratives about early modern earth science. The inclusion of workers into the standard narrative of modern science has been an aim expressed in the "scholar and craftsman" debate, though this normally meant high-status (white, European, male) workers. Edgar Zilsel, who taught science classes in an adult education center in a working-class district in Vienna, highlighted cases in which scholars learnt from "superior craftsmen" (see above, Sect. 1). Pamela Long and Pamela Smith have continued in this vein but likewise work on artisans and technical experts who were unusually eloquent in the written word. As Long pointed out, the "trading zones" between science and practice "engaged artisanally trained individuals" who followed somewhat fluid careers and were not locked into a traditional craft. Such artisanally trained individuals were "engineer/architects, not shoemakers. They engaged in the new techniques of etchers and engravers, not the traditional craft of spinners. They were surveyors, not farmers" (Long 2015, p. 842). The history of the earth sciences follows this general trend. The eloquent and well-connected surveyor William Smith is prominent in the standard account, while the knowledge of workers sometimes acknowledged but often as a hypothetical contribution to the widening stream of published science, or in a summary and anonymous fashion.[12]

As Barnett showed in great details, the expansion beyond the circle of "superior craftsmen" and eloquent experts is difficult because accessible archives (both administrative and scholarly) contain mostly materials written by social elites or adopt their perspective. Historians have therefore used different strategies to recuperate earth knowledge by non-elite actors. John Norris turned to an unusual text type to understand early modern geochemistry, a collection of printed sermons by Johannes Mathesius (1504–1565), preacher in the Bohemian mining town of Joachimsthal/Jáchymov. Norris assumes that this text not only expressed learned ideas in a way that was appealing to Mathesius' laboring congregation but that he "learned and obtained specimens directly from Joachimstal miners" (Norris 2014, p. 39). It is unclear from the source alone who exactly they were. Mathesius calls them "deeply learned, experienced, and skilful miners," which is reminiscent of Zilsel's "superior artisans," but does cite an unnamed "foreman of the St. Lorenz mine in Abertam, about 10 kilometers from Joachimstal," i.e., a low-ranking official with regular working hours underground. A close reading of Mathesius' text thus offers a "valuable glimpse into the early modern interpretation of Earth processes," and specifically "the ways in which the sulfur-mercurius theory was applied to observations within the mines and how at least some miners viewed the composition and generation of the ore veins in which they toiled" (ibid., pp. 48 and 36; see also Dym 2006).

Moving on to the eighteenth century, Patrick Anthony seeks to excavate workers' ecological thinking through a close reading of mining lore and records from Alexander von Humboldt's mining school in Prussian Franconia (Anthony 2021b; see also Anthony 2021a). While this school had the explicit aim of expelling "superstition," its aim of installing regular and sustainable mineral extraction seems to have

[12] See for example: "Miners and quarrymen must long have known that some rocks were stratified and that individual strata varied greatly in character. Such knowledge is likely to have spread among interested persons, though it may not always have found its way into print. It would have been used in the search for economic minerals, such as coal" (Eyles 1969, quotation on p. 167).

borrowed from the region's "mining mythology." Johann Georg Spörl's "faint paper trail" in the school's records is used to reconstruct the mental world of foremen around 1800. The article's primary access to working miners' mentality, however, are collections of mining folklore which "offers a rare, if highly mediated, impression of labourers' own understandings of nature and its exploitation." Analyzing stories of Rübezahl, a mining spirit who punishes greed and safeguards the Earth, is thus an attempt to "bring 'vernacular knowledges' to bear on histories of science and environment" (Anthony 2021b, p. 588).[13] In particular, Anthony argues that "What administrators codified as Raubbau [unsustainable extraction] resonated also through centuries of mining myth." However, the source base for this exploration remains problematic as most stories were written down only in the nineteenth century by the workers' social superiors, and their oral circulation and original content before codification can only be assumed.[14] "Vernacular culture was contested terrain within the mining industry. Songs not only strengthened solidarity amongst the miners, but also served the commercial ends of their lords by 'reproducing labour-power'" (Anthony 2021b, p. 602).

An article by Sebastian Felten is resonant with Barnett's work in the sense that it investigates processes that both valued and marginalized low-status knowers. It is focused on Werner not as a geognost-mineralogist but as a teacher of mining officials who directed labor in the hierarchically structured Saxon mining industry (Felten 2020c). Werner's writing on *Gestein* (rock) invoked workers as knowledgeable informers. He distinguished five "different kinds of hewers' work," each kind defined by a specific tool. He then uses these five kinds of work to construct a 5-point scale to measure the solidity of rock. In this case, rock was not an epistemic object in the emerging earth sciences, that studied fossils, minerals, and the formation of mountains, but in an early attempt at the scientific management of mines. Scholars like Werner transformed workers' knowledge into a "science" (Wissenschaft) of mining work. As Felten argues, current models of early modern knowledge production engaging both scholars and practitioners, such as the trading zone and hybridization, do not adequately explain this transformation, which may warrant complementing these models with the notion of extraction. Scholar-officials extracted knowledge from workers, amalgamated it with other knowledge (such as mineralogy), and then re-introduced it to the workforce, not in the form of new theories, but as instructions for how to perform labor. Unlike in the trading zone and hybridization models, this process tended to reinforce, rather than weaken, the distinctions between different types of knowledge and different groups of knowers.

Gender intersected with questions of class/status and should be a more explicit category of analysis in the history of earth sciences. This is particularly salient for the focus of this review: If mining was as important to the emergence of modern geology as the reviewed work suggest, then this implies that most women were excluded from contributing to geology through this route. Women (in Europe) had limited

[13] See, in a similar vein, Asmussen (2020).

[14] Compare Felten (2020a).

access to mines due to a gendered division of labor that became stricter since the sixteenth century, and that was enshrined in laws against female and child labor in the nineteenth century (Vanja 1987, 1993). Mining administrations did not hire women to work as mining inspectors. In other words, the contact zones between science and mining in early modern Europe that were laid out in the previous sections were not easily accessible to women. Only exceptional high-status women visited mines and wrote about their experience in the early modern period, a well-known example being the mineralogist and mining expert Martine de Bertereau (ca. 1600–1642) who travelled with her husband (Chermette 1985; Vanja 1989). In the nineteenth century, women married to geologists made important, not always acknowledged contributions to the work of their husbands, often by providing technical drawings (Kölbl-Ebert 2012). Some exceptional and high-class women were able to be recognized for their geological work, like Etheldred Benett (1776–1845) who built an important fossil collection (some of which collected with the help of a quarry foreman) and published her research (Creese and Creese 1994). Mary Ann Anning (1799–1847), a carpenter's daughter who collected fossils for income and out of scientific curiosity, was extremely talented at finding and extracting fossils. She acquired considerable knowledge and corresponded with many important geologists of the time (often through their wives), but published little and was not always acknowledged when her spectacular finds furthered the careers of geologists like William Conybeare (1787–1857) (ibid.; Torrens 1995).

Racialization was third dimension that shaped access to mines, to channels of scientific communication, and by consequence to leaving a trace in the textual archives on which most historical research still relies. In Andean and Mexican mines of the colonial period, indigenous peoples worked in the mines as wage laborers or because they were drafted through *mita* (tribute) systems (Bakewell 1971, 1984). A recent study by Allison Bigelow on Spanish-American metallurgy reconstructed "the intellectual contributions of people whose ways of knowing have gone underacknowledged in the colonial scientific historiography, such as mining women, Afro-Latin artisans, and Indigenous experts" (Bigelow 2020). Quechua-speaking workers adapted the Mexican patio technique of silver refining to the different ores and ambient temperatures of the Andean highlands. The Spanish priest and scholar Álvaro Alonso Barba (1569–1662) described this process by using the Castilian word "concho" for the Quechua "qqunchu" in his 1640 book on metallurgy (ibid., p. 255). Through this encoding, a faint trace of the knowledge of Quechua-speaking workers circulated across Europe in numerous translations and reprints.

The complex and well-documented case of Kwasi Boakye (1827–1904) shows how the career of even a high-status, formally educated, and eloquent engineer-scientist was eventually derailed because of race-based prejudice (Blakely 2001). Boakye was sent to the Netherlands as part of a troop recruitment contract that his father king Kwaku Dua I of Ashanti made with King Willem I of the Netherlands in 1837. He was educated under the sponsorship of the Colonial Ministry, learnt among other things German and science, and became friends with the crown prince William III. He studied at the Mining Academy in Delft from 1843 and then, in 1848, in Freiberg. Surprisingly, he claims to have never experienced prejudice during his sojourn in Saxony (Arnold 1996). Instead, he was given warm welcome as his aristocratic and emissary status was acknowledged by city elites. After graduation from Delft a year later, he entered government service to work as a mining engineer in the Dutch East Indies. He thus partook in the international professional culture of mining engineers described by Brianta and Laboulais, and was highly mobile. However, when he arrived in the Dutch East Indies, he was denied promotion due to overt racism by colonial administrators in Holland and Batavia, in spite of his excellent education and publications on local coal deposits (Boachi 1856a, b). "While his engineering career was a notable achievement in itself, it was scarred by bitterness and by complaints against racial discrimination and systematic limitation of his level of achievement" (Blakely 2001, p. 257). He resigned from civil service and unsuccessfully tried to establish a plantation. He was a corresponding member of the Society of Civil Engineers in 1900 and also kept a sporadic correspondence with his teacher at Freiberg, the notable geologist Bernhard von Cotta (1808–1879).

In sum, then, the connex of mining and earth science was constituted by distinctions between status/class, race, and gender. Making this explicit bring to the fore the problem of archives that historians of earth science use. The social mechanisms at work that determine access to mining and scientific communication elicit questions about the boundaries of "science" – a broader discussion taking place in the discipline – and what kind of research should be included in this historiography (Mulsow and Daston 2019; Daston 2017a). Some may agree with Martin Rudwick that the core business of historians of science must be to analyze the "specific concrete claims to reliable knowledge were formulated, argued over, and consolidated or rejected," and that this is best done by focusing on the activities of the scientific elite, which after all is "not one based primarily on birth, wealth, or social class, but on hard intellectual work" (Rudwick 2005, p. 4). But as the work cited in this concluding section shows, the opportunities to perform intellectual work, and be recognized for it, were not equally distributed during the formative years of modern geology. Future standard works might therefore have to find ways of acknowledging hidden and failed work, or at least account for mechanisms of inclusion and exclusion.

Acknowledgments Funded by the European Union (ERC, SCARCE, Grant number 101076422). Views and opinions expressed are however those of the author(s) only and do not necessarily reflect those of the European Union or the European Research Council Executive Agency. Neither the European Union nor the granting authority can be held responsible for them.

Funded by the European Union

European Research Council
Established by the European Commission

References

Agricola, Georgius. 1912. *De Re Metallica* (trans: Clark Hoover, Herbert, and Hoover, Lou Henry). London: The Mining Magazine.

———. 1955–1997. In *Ausgewählte Werke*, ed. Hans Prescher, 10 vols. with two supplements. Berlin: Deutscher Verlag der Wissenschaften.

Albrecht, Helmuth, and Roland Ladwig, eds. 2002. *Abraham Gottlob Werner and the foundation of the geological sciences*. Freiberg: Technische Universität Bergakademie.

Alfonso-Goldfarb, Ana Maria, and Marcia H.M. Ferraz. 2013. Gur, Ghur, Guhr or Bur? The quest for a metalliferous prime matter in early modern times. *British Journal for the History of Science* 46 (1): 23–37.

Anthony, Patrick. 2018. Mining as the working world of Alexander von Humboldt's plant geography and vertical cartography. *Isis* 109 (1): 28–55. https://doi.org/10.1086/697061.

———. 2020. Mines, mountains, and the making of a vertical consciousness in Germany ca. 1800. *Centaurus* 62 (4): 612–630. https://doi.org/10.1111/1600-0498.12337.

———. 2021a. Introduction to 'working at the margins: Labor and the politics of participation in natural history, 1700–1830'. *Berichte zur Wissenschaftsgeschichte* 44 (2): 115–136. https://doi.org/10.1002/bewi.202000034.

———. 2021b. Labour, folklore, and environmental politics in German mining around 1800. *The Historical Journal* 64 (3): 583–605. https://doi.org/10.1017/S0018246X20000588.

Arnold, Anne-Sophie. 1996. Fremde Heimat – heimatliche Fremde: Kwasi Boakye in Deutschland. In *Fremde Erfahrungen: Asiaten und Afrikaner in Deutschland, Österreich und in der Schweiz bis 1945*, ed. Gerhard Höpp, 29–32. Berlin: Das Arabische Buch.

Ash, Eric H. 2001. Queen v. Northumberland, and the control of technical expertise. *History of Science* 39 (2): 215–240. https://doi.org/10.1177/007327530103900204.

Asmussen, Tina. 2020. Spirited metals and the oeconomy of resources in early modern European mining. *Earth Sciences History: Journal of the History of the Earth Sciences Society* 39: 371–388.

Bakewell, Peter J. 1971. *Silver mining and society in colonial Mexico, Zacatecas 1546–1700*. Cambridge: Cambridge University Press.

———. 1984. *Miners of the Red Mountain: Indian labor in Potosí, 1545–1650*. Albuquerque: University of New Mexico Press.

Barnett, Lydia. 2019. Showing and hiding: The flickering visibility of earth workers in the Archives of earth science. *History of Science* 58 (3): 245–274. https://doi.org/10.1177/0073275319874982.

Bartels, Christoph. 1992. *Vom frühneuzeitlichen Montangewerbe zur Bergbauindustrie: Erzbergbau im Oberharz, 1635–1866*. Deutsches Bergbau-Museum: Bochum.

———. 2010. The production of silver, copper, and lead in the Harz Mountains from late medieval times to the onset of industrialization. In *Materials and expertise in early modern Europe: Between market and laboratory*, ed. Ursula Klein and E.C. Spary, 71–100. Chicago: University of Chicago Press.

Bernardoni, Andrea. 2012. *La conoscenza del fare: Ingegneria, arte, scienza nel De la pirotechnia di Vannoccio Biringuccio*. Roma: L'Erma di Bretschneider.

Bigelow, Allison Margaret. 2020. *Mining language: Racial thinking, indigenous knowledge, and colonial metallurgy in the early modern Iberian world*, 12. Chapel Hill: University of North Carolina Press.

Biringuccio, Vannoccio, and Cyril Stanley Smith. 1942. *The Pirotechnia of Vannoccio Biringuccio* (trans: Gnudi, Martha Teach). Seeley W. Mudd Series. New York: American Institute of Mining and Metallurgical Engineers.

Blakely, Allison. 2001. *Blacks in the Dutch world: The evolution of racial imagery in a modern*, 256–259. Bloomington: Indiana University Press.

Boachi, Aquasie. 1856a. Onderzoek naar de kolen, gevonden langs het strand der Meeuwenbaai. *Natuurkundig tijdschrift voor Nederlandsch Indië* 9: 49–52.

———. 1856b. Onderzoek naar het aanwezen van steenkolen in het terrein aan de Tjiletoekbaai, residentie Preanger-regentschappen. *Natuurkundig tijdschrift voor Nederlandsch Indië* 11: 461–464.

Brendecke, Arndt. 2009. *Imperium und Empirie: Funktionen des Wissens in der spanischen Kolonialherrschaft*. Böhlau: Cologne.

Brianta, Donata. 2000. Education and training in the mining industry, 1750–1860: European models and the Italian case. *Annals of Science* 57 (3): 267–300.

———. 2007. *Europa mineraria: Circolazione delle élites e trasferimento tecnologico, secoli XVIII-XIX*. Milano: FrancoAngeli.

Chermette, Alexis. 1985. Martine de Bertereau (1590–1643), une femme ingénieur des mines au XVIIe siècle. *Publications de la Société Linnéenne de Lyon* 54 (10): 69–80.

Clair, Eduardo Flores. 2009. El Colegio de Minería: Una institución ilustrada en el siglo XVIII novohispano. *Estudios de Historia Novohispana* 20: 33–65. https://doi.org/10.22201/iih.24486922e.1999.020.3492.

Coen, Deborah R. 2018. *Climate in motion: Science, empire, and the problem of scale*. Chicago: University of Chicago Press.

Collet, Dominik. 2007. *Die Welt in der Stube: Begegnungen mit Außereuropa in Kunstkammern der Frühen Neuzeit*. Göttingen: Vandenhoeck & Ruprecht.

Creese, Mary R.S., and Thomas M. Creese. 1994. British women who contributed to research in the geological sciences in the nineteenth century. *The British Journal for the History of Science* 27 (1): 23–54.

Daston, Lorraine. 2017a. The history of science and the history of knowledge. *KNOW: A Journal on the Formation of Knowledge* 1 (1): 131–154. https://doi.org/10.1086/691678.

———. 2017b. The immortal archive: Nineteenth-century science imagines the future. In *Science in the archives: Pasts, presents, futures*, ed. Lorraine Daston, 159–182. Chicago: University of Chicago Press.

Dym, Warren. 2006. Mineral fumes and mining spirits: Popular beliefs in the Sarepta of Johann Mathesius (1504–1565). *Reformation & Renaissance Review* 8 (2): 161–185. https://doi.org/10.1558/rrr.v8i2.160.

Escamilla-González, Francisco Omar, and Lucero Morelos-Rodríguez. 2020. Bringing Werner's teachings to the New World: Andrés Manuel Del Río and the chair of mineralogy in the School of Mines of Mexico (1795–1805). *Earth Sciences History* 39: 246–261.

Eyles, V.A. 1969. The extent of geological knowledge in the eighteenth century, and the methods by which it was diffused. In *Towards a history of geology: Proceedings of the New Hampshire inter-disciplinary conference on the history of geology, September 7–12, 1967*, ed. Cecil J. Schneer, 159–183. Cambridge, MA: MIT Press.

Felten, Sebastian. 2018. The history of science and the history of bureaucratic knowledge: Saxon mining, circa 1770. *History of Science* 56 (4): 403–431. https://doi.org/10.1177/0073275318792451.

———. 2020a. Mining culture, labour, and the state in early modern Saxony. *Renaissance Studies* 34 (1): 119–148. https://doi.org/10.1111/rest.12583.

———. 2020b. Sustainable gains: Dutch investment and bureaucratic rationality in eighteenth-century Saxon mines. *Journal for the History of Knowledge* 1 (1): 14. https://doi.org/10.5334/jhk.19.

———. 2020c. Wie fest ist das Gestein? Extraktion von Arbeiterwissen im Bergbau des 18. Jahrhunderts. *WerkstattGeschichte* 81: 15–35.

Felten, Sebastian, and Christine von Oertzen. 2020. Bureaucracy as knowledge. *Journal for the History of Knowledge* 1 (1): 8. https://doi.org/10.5334/jhk.18.

Fessner, Michael, and Christoph Bartels. 2012. Von der Krise am Ende des 16. Jahrhunderts zum deuschen Bergbau im Zeitalter des Merkantilismus. In *Geschichte des deutschen Bergbaus*, Der Alteuropäische Bergbau: Von den Anfängen bis zur Mitte des 18. Jahrhunderts, ed. Christoph Bartels and Rainer Slotta, vol. 1, 453–590. Münster: Aschendorff.

Fettweis, Günter B., and Günther Hamann. 1989. *Über Ignaz von Born und die Societät der Bergbaukunde*. Wien: Verlag der Österreichischen Akademie der Wissenschaften.

Figueiroa, Silvia, and Clarete da Silva. 2000. Enlightened mineralogists: Mining knowledge in colonial Brazil, 1750–1825. *Osiris* 15 (1): 174–189.
Fors, Hjalmar. 2014. Elements in the melting pot: Merging chemistry, assaying, and natural history, Ca. 1730–60. *Osiris* 29: 230–244.
———. 2015. *The limits of matter: Chemistry, mining, and enlightenment*. Chicago: Chicago University Press.
Fors, Hjalmar, and Jacob Orrje. 2019. Describing the world and shaping the self: Knowledge-gathering, mobility and spatial control at the Swedish Bureau of Mines. In *Transnational cultures of expertise: Circulating state-related knowledge in the 18th and 19th centuries*, vol. 36. Berlin/Boston: De Gruyter Oldenbourg. https://doi.org/10.1515/9783110553734-007.
Fuller, J.G.C.M. 1992. The invention and first use of stratigraphic cross-sections by John Strachey, F.R.S., (1671–1743). *Archives of Natural History* 19 (1): 69–90. https://doi.org/10.3366/anh.1992.19.1.69.
Gauché, Sylvain. 2003. La formation du droit minier français: L'exemple du Dauphine (XIVe-début XIXe siècle). *Revue historique de droit français et étranger* 81 (1): 75–89.
Gliech, Oliver. 2013. Der mexikanisch/neuspanische Minensektor und die internationale Silberwirtschaft (1750–1810). In *Staat, Bergbau und Bergakademie: Montanexperten im 18. und frühen 19. Jahrhundert*, ed. Hartmut Schleiff and Peter Konečný, 309–335. Stuttgart: Franz Steiner.
Guerrero, Saúl. 2017. *Silver by fire, silver by mercury: A chemical history of silver refining in New Spain and Mexico, 16th to 19th centuries*. Leiden: Brill.
Guntau, Martin. 1978. Bergbauproduktion und geologische Erkenntnis in der Geschichte. In *ICOHTEC: Internationales Symposium zur Geschichte des Bergbaus und Hüttenwesens: Vorträge, 2 vols*, ed. Eberhard Wächtler and Gisela Ruth Engewald, vol. 1, 91–100. Freiberg: Bergakademie Freiberg.
———. 1984a. *Abraham Gottlob Werner*. Leipzig: Teubner.
———. 1984b. *Die Genesis der Geologie als Wissenschaft: Studie zu den kognitiven Prozessen und gesellschaftlichen Bedingungen bei der Herausbildung der Geologie als naturwissenschaftliche Disziplin an der Wende vom 18. zum 19. Jahrhundert*. Berlin: Akademie-Verlag.
Hall, Rupert. 1959. The scholar and the craftsman in the scientific revolution. In *Critical problems in the history of science*, ed. Marshall Clagett, 3–23. Madison: University of Wisconsin Press.
Harris, Steven J. 2006. Networks of travel, correspondence and exchange. In *The Cambridge history of science*, Early modern science, ed. Katharine Park and Lorraine Daston, vol. 3, 341–362. Cambridge: Cambridge University Press.
Hillegeist, Hans-Heinrich, and Wilfried Ließmann, eds. 2001. *Technologietransfer und Auswanderungen im Umfeld des Harzer Montanwesens*. Berlin: Lukas.
Jacob, Margaret C. 2014. *The first knowledge economy: Human capital and the European economy, 1750–1850*. Cambridge: Cambridge University Press.
Jobst, Wolfgang, and Walter Schellhas. 1994. *Abraham von Schönberg, Leben und Werk: Die Wiederbelebung des erzgebirgischen Bergbaus nach dem Dreißigjährigen Krieg durch Oberberghauptmann Abraham von Schönberg*. Leipzig: Deutscher Verlag für Grundstoffindustrie.
Jones, Peter. 2008. *Industrial enlightenment: Science, technology and culture in Birmingham and the West Midlands 1760–1820*. Manchester: Manchester University Press.
Kessler-Slotta, Elisabeth. 1994. Die Illustrationen in Agricolas 'De re metallica': Eine Wertung aus kunsthistorischer Sicht. *Der Anschnitt* 46 (2–3): 55–67.
Klein, Ursula. 2015. *Humboldts Preußen: Wissenschaft und Technik im Aufbruch*, 70. Darmstadt: Wissenschaftliche Buchgesellschaft.
———. 2016a. Kuhn in the Cold War. In *Shifting paradigms: Thomas S. Kuhn and the history of science*, ed. Alexander Blum et al., 115–121. Berlin: Max Planck Institute for the History of Science.
———. 2016b. *Nützliches Wissen: Die Erfindung der Technikwissenschaften*. Göttingen: Wallstein.

———. 2017. Hybrid experts. In *The structures of practical knowledge*, ed. Matteo Valleriani, 287–306. Cham: Springer. https://doi.org/10.1007/978-3-319-45671-3_11.
———. 2020. *Technoscience in history: Prussia, 1750–1850*. Cambridge, MA: MIT Press.
Klemun, Marianne. 2004. The Royal Natural History Collection in Vienna (18th century): From possessing minerals as a private treasure towards territorial ambitions as consciousness. *Scripta Geologica. Special Issue* 4: 193–199.
———. 2007. Writing, 'inscription' and fact: Eighteenth century mineralogical books based on travels in the Habsburg regions, the Carpathian Mountains. In *Four centuries of geological travel: The search for knowledge on foot, bicycle, sledge and camel*, ed. Patrick Wyse Jackson, 49–61. London: Geological Society.
———. 2014. Administering science: The paper form of scientific practice and geological fieldwork. *Earth Sciences History* 33 (2): 279–293.
———. 2015. Geognosie versus Geologie: Nationale Denkstile und kulturelle Praktiken bezüglich Raum und Zeit im Widerstreit. *Berichte zur Wissenschaftsgeschichte* 38: 227–242. https://doi.org/10.1002/bewi.201501728.
———. 2017. Spaces and places: An historical overview of the development of geology in Austria (Habsburg monarchy) in the eighteenth and nineteenth centuries. In *History of geoscience: Celebrating 50 years of INHIGEO*, ed. Wolf Mayer et al., 263–270. London: The Geological Society.
———. 2020. *Wissenschaft als Kommunikation in der Metropole Wien: Die Tagebücher Franz von Hauers der Jahre 1860–1868*. Wien: Böhlau.
Klemun, Marianne, and Ulrike Spring, eds. 2016. *Expeditions as experiments: Practising observation and documentation*. London: Palgrave Macmillan.
Kölbl-Ebert, Martina. 2012. Sketching rocks and landscape: Drawing as a female accomplishment in the Service of Geology. *Earth Sciences History* 31 (2): 270–286.
Kozák, Jan T., et al. 2016. *Early geological maps of Europe: Central Europe 1750 to 1840*. Cham: Springer International.
Kusukawa, Sachiko. 2016. Classics from this journal: Martin Rudwick's 'the emergence of a visual language for geological science 1760–1840,' History of Science, Xiv: 3, 1976, Pp. 149–95. *History of Science* 54 (1): 98–104. https://doi.org/10.1177/0073275315624927.
Laboulais, Isabelle. 2013. *La maison des mines: La genèse d'un corps révolutionnaire d'un corps d'ingénieurs civils*, 14. Rennes: PU Rennes.
Lang, Heinrich. 2008. Technologisches Wissen und Transfer. Die Scuola für metallurgische Wissenschaften des Spirito Benedetto Nicolis di Robilant in Savoyen-Piemont in der Mitte des 18. Jahrhunderts. *Technikgeschichte* 75 (1): 33–50. https://doi.org/10.5771/0040-117x-2008-1-33.
Laudan, Rachel. 1987. *From mineralogy to geology: The foundations of a science, 1650–1830*. Chicago: University of Chicago Press.
Lefèvre, Wolfgang. 2010. *Picturing the world of mining in the Renaissance: The Schwazer Bergbuch (1556)*. Berlin: Max-Planck-Institut für Wissenschaftsgeschichte.
Long, Pamela O. 2011. *Artisan/practitioners and the rise of the new sciences, 1400–1600*. Corvallis: Oregon State University Press.
———. 2015. Trading zones in early modern Europe. *Isis* 106 (4): 840–847.
Lucier, Paul. 2009. Geological industries. In *The Cambridge history of science*, ed. Peter J. Bowler and John V. Pickstone, vol. 6, 108–125. Cambridge University Press. https://doi.org/10.1017/CHOL9780521572019.008.
Lück, Heiner. 2015. Die Entwicklung des deutschen Bergrechts und der Bergbaudirektion bis zum Allgemeinen (preußischen) Berggesetz 1865. In *Geschichte des deutschen Bergbaus*, Salze, Erze und Kohlen. Der Aufbruch in die Moderne, ed. Wolfhard Weber, vol. 2, 111–216. Münster: Aschendorf.
Ludwig, Karl-Heinz, and Volker Schmidtchen. 1992. *Metalle und Macht: 1000 bis 1600*, Propyläen-Technikgeschichte. Vol. 2. Berlin: Propyläen-Verl.

Luzzini, Francesco. 2011. Matrices, not seeds: Vallisneri's research on mines. Between empiricism and philosophy. In *History of research in mineral resources*, ed. José Eugenio Ortiz, Octavio Puche, and Isabel Rábano. Madrid: Instituto Geológico y Minero de España.

———. 2013. *Il miracolo inutile: Antonio Vallisneri e le scienze della Terra in Europa tra XVII e XVIII secolo*. Firenze: Leo S. Olschki.

———. 2020. 'Per la vertù propria dell'acque.' Vene minerali e sorgenti nella Pirotechnia di Vannoccio Biringuccio (1540). *Acque Sotterranee* 9: 65–67.

Mayer, Wolf, et al., eds. 2017. *History of geoscience: Celebrating 50 years of INHIGEO*. London: The Geological Society.

Meisterle, Stefan. 2020. Kaiserlich Ostindien: Die indischen Niederlassungen Karls VI. In *Herrschaft und Repräsentation in der Habsburgermonarchie (1700–1740)*, ed. Sandra Hertel and Stefan Seitschek, 269–282. Berlin/Boston: De Gruyter.

Merton, Robert K. 1938. Science, technology and society in seventeenth century England. *Osiris* 4: 360–632.

Mulsow, Martin, and Lorraine Daston. 2019. History of knowledge. In *Debating new approaches to history*, ed. Marek Tamm and Peter Burke, 159–187. London: Bloomsbury Academic.

Neubert, Karl, and Walther Stein. 1958. *Plan- und Risskunde*. 2nd ed. Bergakademie Freiberg: Freiberg.

Newcomb, Sally. 2009. *The world in a crucible: Laboratory practice and geological theory at the beginning of geology*, 3. Boulder: Geological Society of America.

Norris, John A. 2006. The mineral exhalation theory of metallogenesis in pre-modern mineral science. *Ambix* 53 (1): 43–65.

———. 2007. Early theories of aqueous mineral genesis in the sixteenth century. *Ambix* 54 (1): 69–86.

———. 2014. Auß Quecksilber und Schwefel Rein: Johann Mathesius (1504–65) and Sulfur-Mercurius in the silver mines of Joachimstal. *Osiris* 29 (1): 35–48.

Oldroyd, D.R. 1996. *Thinking about the earth: A history of ideas in geology*, 76–79. London: Athlone.

Paracelsus. 2013. In *Von der Bergsucht und anderen Bergkrankheiten (De morbis fossorum metallicorum)*, ed. Irmgard Müller. Berlin: Springer.

Piirainen, Ilpo Tapani. 1998. Die Fachsprache des Bergbaus. In *Fachsprachen/languages for special purposes: Ein internationales Handbuch zur Fachsprachenforschung und Terminologiewissenschaft/an international handbook of special-language and terminology research*, ed. Lothar Hoffmann, Hartwig Kalverkämper, and Herbert Ernst Wiegand, vol. 1, 1092–1098. Berlin/New York: De Gruyter.

———. 1999. Die Fachlexikographie des Bergbaus: eine Übersicht. In *Fachsprachen/languages for special purposes: Ein internationales Handbuch zur Fachsprachenforschung und Terminologiewissenschaft/an international handbook of special-language and terminology research*, ed. Lothar Hoffmann, Hartwig Kalverkämper, and Herbert Ernst Wiegand, vol. 2, 1930–1937. Berlin/New York: De Gruyter.

Porter, Roy. 1973. The industrial revolution and the rise of the science of geology. In *Changing perspectives in the history of science: Essays in honour of Joseph Needham*, ed. Mikuláš Teich, Robert M. Young, and Joseph Needham, 320–343. London: Heinemann Educational.

———. 1977. *The making of geology: Earth science in Britain 1660–1815*. Cambridge: Cambridge University Press.

Porter, Theodore M. 1981. The promotion of mining and the advancement of science: The chemical revolution of mineralogy. *Annals of Science* 38 (5): 543–570. https://doi.org/10.1080/00033798100200371.

Raphael, Renée. 2020. In pursuit of 'useful' knowledge: Documenting technical innovation in sixteenth-century Potosí. *Journal for the History of Knowledge* 1 (1): 11. https://doi.org/10.5334/jhk.16.

Rosen, George. 1943. *The history of miners' diseases: A medical and social interpretation*. New York: Schuman.

Rudwick, Martin J.S. 1976. The emergence of a visual language for geological science 1760–1840. *History of Science* 14 (3): 149–195. https://doi.org/10.1177/007327537601400301.

———. 2005. *Bursting the limits of time: The reconstruction of geohistory in the age of revolution*. Chicago: The University of Chicago Press.

Schirmer, Uwe. 2009. Entwicklung, Ausbreitung und Verfestigung des Direktionsprinzips in der kursächsischen Montanwirtschaft (1470–1624). In *Wirtschaftslenkende Montanverwaltung – Fürstlicher Unternehmer – Merkantilismus: Zusammenhänge zwischen der Ausbildung einer fachkompetenten Beamtenschaft und der staatlichen Geld- und Wirtschaftspolitik in der Frühen Neuzeit*, ed. Angelika Westermann, 145–158. Husum: Matthiesen.

Schleiff, Hartmut, and Peter Konečný, eds. 2013. *Staat, Bergbau und Bergakademie: Montanexperten im 18. und frühen 19. Jahrhundert*. Stuttgart: Franz Steiner.

Slotta, Rainer, and Joseph Labuda, eds. 1997. *"Bei diesem Schein kehrt Segen ein": Gold, Silber und Kupfer aus dem slowakischen Erzgebirge*. Deutsches Bergbau-Museum: Bochum.

Smith, Pamela H. 2004. *The body of the artisan: Art and experience in the scientific revolution*. Chicago: University of Chicago Press.

———. 2017. The codification of vernacular theories of metallic generation in sixteenth-century European mining and metalworking. In *The structures of practical knowledge*, ed. Matteo Valleriani, 371–392. Cham: Springer International Publishing.

Torrens, Hugh. 1995. Presidential address: Mary Anning (1799–1847) of Lyme; 'the greatest fossilist the world ever knew.' *The British Journal for the History of Science* 28 (3): 257–284.

Vaccari, Ezio. 1993. *Giovanni Arduino (1714–1795): Il contributo di uno scienziato veneto al dibattito settecentesco sulle scienze della terra*. Firenze: Leo S. Olschki

———. 1998. Mining academies. In *Sciences of the earth: An encyclopedia of events, people, and phenomena*, ed. Gregory A. Good, vol. 2, 585–589. New York/London: Garland.

———. 2000. Mining and knowledge of the earth in eighteenth-century Italy. *Annals of Science* 57 (2): 163–180. https://doi.org/10.1080/000337900296236.

———. 2009. Mining academies as centers of geological research and education in Europe between the 18th and 19th centuries. *De Re Metallica* 13: 35–41.

Vaccari, Ezio, and Nicoletta Morello. 1998. Mining and knowledge of the earth. In *Sciences of the earth: An encyclopedia of events, people, and phenomena*, ed. Gregory A. Good, vol. 2, 589–593. New York/London: Garland.

Vanja, Christina. 1987. Bergarbeiterinnen: Zur Geschichte der Frauenarbeit im Bergbau, Hütten- und Salinenwesen seit dem späten Mittelalter. Teil I: Spätes Mittelalter und Frühe Neuzeit. *Der Anschnitt* 39 (1): 2–15.

———. 1989. Von der Bergingenieurin zur dichtenden Bergarbeiterin: Wissenschaftliche und literarische Beschäftigung von Frauen mit dem Bergbau der Neuzeit. In *Frauen und Bergbau: Zeugnisse aus fünf Jahrhunderten*, ed. Werner Kroker, 84–93. Bochum: Selbstverlag des Deutschen Bergbau-Museums.

———. 1993. Mining women in early modern European society. In *The workplace before the factory: Artisans and proletarians, 1500–1800*, ed. Thomas Max Safley, 100–117. Ithaca: Cornell University Press. https://doi.org/10.7591/9781501737732-007.

Veith, Heinrich. 1871. *Deutsches Bergwörterbuch*. Breslau: Wilhelm Gottlieb Korn.

von Engelhardt, Wolf. 1992. Die Namen der Steine: Zur Sprache in deutschen geowissenschaftlichen Texten des ausgehenden 18. und beginnenden 19. Jahrhunderts. In *Language and earth: Elective affinities between the emerging sciences of linguistics and geology*, ed. Bernd Naumann, Frans Plank, and Gottfried Hofbauer, 161–192. Amsterdam: John Benjamins.

von Schönberg, Abraham. 1693. *Ausführliche Berg-Information*. Leipzig: David Fleischer.

Weber, Wolfhard. 1976. *Innovationen im frühindustriellen deutschen Bergbau und Hüttenwesen: Friedrich Anton von Heynitz*. Göttingen: Vandenhoeck & Ruprecht.

Weiß, Alfred. 1997. Grubenrisse: Zu Unrecht wenig beachtete montanhistorische Quellen. *Berichte der Geologischen Bundesanstalt* 41: 261–266.

Westermann, Andrea. 2011. Geologiegeschichte als Verwaltungsgeschichte: Stabilisierungseffekte zwischen Amtshandeln und Forschungshandeln bei Hans Conrad Escher (1767–1826). *Traverse: Zeitschrift für Geschichte* 18 (2): 57–74.

———. 2015. Geology and world politics: Mineral resource appraisals as tools of geopolitical calculation, 1919–1939. *Historical Social Research* 40 (2): 151–173. https://doi.org/10.12759/hsr.40.2015.2.151-173.

Zilsel, Edgar. 1939. The social roots of science. In *The social origins of modern science*, ed. Diederick Raven and Wolfgang Krohn, 3–6. Dordrecht: Kluwer. (Reprinted 2000).

———. 1941. The origins of William Gilbert's scientific method. *Journal of the History of Ideas* 2 (1): 1–32. https://doi.org/10.2307/2707279.

Data, Aesthetics, and Visualizations of Deep Time

Marco Tamborini

Contents

Abstract

This chapter accomplishes a phenomenology of deep time visualizations. It examines the power and limits of a series of visual devices used in paleontology and geology to access, and eventually work with the earth's deep past. First, I discuss how paleontologists visualize and sort data in the field; second, I examine the practices used to illustrate and validate knowledge about extinct animals; third, I explore what function visualizations play in supporting the transition between data collection and the possible explanation of global biological phenomena such as the estimation of diversity through geological time; fourth, I inquire into the role of the computer; and fifth, I discuss the recent intersection between paleontology and different kinds of new technologies such as augmented and visual reality and robotics, both to shed light on some aspects of the past and to generate new research questions. In conclusion, I reflect on the intersection between earth science visual cultures and knowledge production. In particular, I point out the function of knowledge circulation and in between scientists in the production of visual devices as well as the importance of

M. Tamborini (✉)
Institut für Philosophie, Technical University of Darmstadt, Darmstadt, Hessen, Germany
e-mail: marco.tamborini@tu-darmstadt.de

E. Aronova et al. (eds.), *Handbook of the Historiography of the Earth and Environmental Sciences*, Historiographies of Science,
https://doi.org/10.1007/978-3-031-40799-4_18

aesthetics for cooperative research projects and knowledge production in the earth sciences.

Keywords

Fossil · Data · Graphs · Technology · Media

1 Introduction

Earth's geological past extends far beyond the human capacity for comprehension. Deep time dilates natural history to the extent that human reasoning seems to lose its power; human's theories of the earth reach their "darkest point [...] we start falling into the abysses of time. There, genius' insight seems not to be enough and, due to lack of observations, it is powerless to guide us forward" (Comte de Buffon 1778, p. 44). While this vast temporal dimension clearly attracts scientists, it also traps them in a void and threatens the production of clear thought and relevant theories. Visualizations are one of the primary strategies used by earth scientists to improve their understanding and make deep time manageable.

Earth sciences are indeed extremely visual disciplines. It is quite difficult to reconstruct the history of, for instance, paleontology or geology without considering all the visual devices and technologies used to picture and investigate earth's deep past. Indeed, and quite paradoxically, paleontologists sought to develop a series of visual and technological devices to make visible something inaccessible: earth's deep time. Quoting Steve Shapin and Simon Schaffer's seminal book *Leviathan and the Air Pomp* (1985), science historian Martin Rudwick affirmed that "any scene from deep time embodies a fundamental problem: it must make visible what is really invisible. It must give us the illusion that we are witnesses to a scene that we cannot really see; more precisely, it must make us 'virtual witnesses' to a scene that vanished long before there were any human beings to see it" (Rudwick 1985, 1). He used Shapin and Schaffer's term "virtual witness" to emphasize that the process "makes real" that which is otherwise impossible to observe (ibid., 255).

Following Rudwick's insights, in this chapter, I examine the power and limits of a series of visual devices used mainly in paleontology and geology to make visible, access, and eventually work with earth's deep past. Accomplishing a phenomenology of visualizations used in paleontological knowledge production,[1] I will analyze the functions of these visualizations as essential means for working with the past. By so doing, I will defend the argument that visualizations help validate earth science as a field of scientific knowledge production.

[1] My phenomenology of deep time visualizations does not claim to be complete and conclusive. On the contrary, it provides the starting point from which other elements of the visual culture of the earth sciences could be analyzed. For example, my phenomenology needs to be integrated with the analysis of visualizations used to study human history as well as phylogenetic taxonomy (Shepherd 2015; Ebach 2015; Carter 2007; Bigg 2015; Podgorny 2003; Schäffner 1999).

The focus on the epistemic powers and limits of visualizations within earth science practices is relatively recent. Although Rudwick, as a consequence of his work as a paleontologist and morphologist during the 1960s,[2] called for a solid investigation of the visual language of geology as early as 1978, paleontological visualization began to be used as a strategy starting in the twenty-first century, following the broader material and visual turn in science study. From a variety of perspectives, scholars called attention to the different powers of visual media used to explore the earth's (and human) past.

Before engaging with the phenomenology of deep-time visualizations, one key epistemic issue of earth science disciplines should be mentioned. Earth sciences are peculiar disciplines since they are located between historical and natural disciplines. For instance, one of the most representative earth sciences, paleontology, finds itself at the intersection of history, biology, and the natural sciences. This unique place within the system of science is due in part to the status of its data. The fossil record is always imperfect and incomplete. Only a tiny number of organisms and events that have shaped earth's deep history have been preserved as fossilized objects. Hence, before using the record of the past for producing paleontological knowledge, paleontologists should analyze whether record of the past is biased. To do so, the fossil record is treated statistically and taken as a bulk to avoid possible biases and misleading interpretations of data.

Having mentioned the main epistemic peculiarity of paleontology, in the following section, I first focus on how paleontologists visualize and order field data; second, I look at the practices used to illustrate and validate knowledge of extinct animals; third, I explore the role visualization plays in supporting the transition between the collection of data and possible explanation of global phenomena; fourth, I investigate the role of computer; and fifth, I move to the recent collision of paleontology and new technologies to make visible some aspects of the past and generate new research questions. In the conclusion, I will reflect on the intersection between visual cultures and knowledge production.

2 Visualizing Field Data

In this section, I focus on the first set of visual devices used to work with fossilized and incomplete data gathered in the field to generate knowledge about the past. Visual practices were adopted during an emblematic paleontological expedition between 1989 and 1913: the Tendaguru expedition. Over the course of 4 years, scientists from the Berlin Museum für Naturkunde [Berlin Museum of Natural History] excavated more than 225 tons of fossils in present-day Tanzania and transported them to Berlin. Among these were the bones of the *Brachiosaurus brancai*, which would later become the largest erected dinosaur in the world, and which dominates the Dinosaur Exhibition Hall in Berlin to this day. The challenge

[2] For Rudwick's morphological analysis, see (Rudwick 2017, 2018; Tamborini 2022a).

faced by paleontologists and locals was both quantitative and quantitative: How could they deal with such a bulk of data, make sense of it, and exhibit the previously classified specimens in Berlin?

First, paleontologists advanced three kinds of considerations. Indeed, philosopher Darek Turner noted that in deciding what to collect and what to leave on the floor, scientists always ask whether, first, the specimen might be relevant to someone's ongoing research; second, whether the specimen is well preserved, complete, or beautiful; and third, how much effort is required to unearth it and prepare it (Turner 2019). Second, as Tamborini described, pictorial representations produced through drawings and other so-called paper-based techniques were central to making sense of incomplete and imperfect fossils (Tamborini 2020). First, fossils were presented through simple means such as paper and pencil. Second, figures drawn on paper were cut and pasted together to present a potentially unified and coherent image of the fossilized organism. This first set of basic steps helped to tentatively visualize and categorize the excavated finds.[3]

Although illustrations played a central role in visualizing deep-time organisms and guiding field work, they were not sufficient either to display a complex extinct organism in a museum hall or to present it as a coherent deep-time scene (sensu Rudwick) in textbooks or scientific papers. In other words, although field illustrations and sketches of fossils provide the semantic space of deep time, they still lack a syntax that brings the illustrations together. A different kind of knowledge and medium contributes to this task: bureaucratic knowledge and its technologies.

The transfer of knowledge between bureaucracy and natural history occurred continuously throughout the eighteenth and nineteenth centuries. Paleontologists, for example, have used tables and quantitative practices, which have already served an established function in the bureaucracy to reconstruct the history of nature (see next section). Natural history field work is characterized by the same transfer of knowledge. As Tamborini noted regarding the Tendaguru expedition, paleontologists on the field behave like bureaucrats. They had to meticulously classify, tabulate, and list the findings and link them to the particular spots on stratigraphic maps previously sketched. In this practice, the metadata assigned in the tables, lists, and book records determine how to re-read the fossils once they're transported into the artificial space of a laboratory. In other words, they provide a syntax for reading the data decoupled from the field.

The starting point of this practice is the identification of the earth layers with a natural archive (Sepkoski 2017). In the earth's natural archive, the fossils are

[3]As geologist Haakon Fossen wrote "Geology is, by nature, a field-based branch of science, and making on-site sketches is the best way to approach an outcrop. Why? Because field sketching sharpens your senses and turns you into a much better observer. It forces you to focus, to make important decisions regarding crucial structures and features, such as cross-cutting relations, sequence boundaries, fault geometry, layer continuity and grain size variations. It makes you discover important details that otherwise might go unnoticed. And it makes the locality stick to your mind" (Fossen 2013).

preserved together with other important information about, for instance, the geological formation and possible ecological events which might be indicated in that particular earth and temporal stratum. In the field, the paleontologist can read the temporal and environmental features that make up the differences and similarities between different fossil groups. They take notes on every possible element, drawing on the stratigraphic profile of the landscape and earth layers, that might help them to decode all the information once they are in the lab. As a result, the fossils become documents to be read, written, and carefully kept: "Fossils are not stamps or rarities, but they are, in a sense, documents from long ago periods of our earth" (Fraas 1910, 4).

Practiced since the eighteenth century, this bureaucratic method to visualize and keep record of the past was institutionalized during the nineteenth century and reached its peak during the paleontological expeditions of the mid-nineteenth and early twentieth centuries (Rieppel 2019; Nieuwland 2019). As stated in various paleontological and geological guides and textbooks, "When traveling or collecting in different horizons, one should also become accustomed to conscientiously attach to each find a label noting the locality and geological horizon" (Fraas 1910, 8). Labels are the first step toward establishing the grammar useful to coherently visualize the past.

In addition to the proper labeling of excavated fossils and the appropriate selection and preservation of metadata, lengthier journal entries are needed to explain the necessary brief notes written about the circumstances of the find. In diaries, paleontologists cataloged what was excavated at each site so that the earth's archive could be fully preserved and transported into a different context, such as museum's laboratory, with minimal error. In addition, these diaries, which in the emblematic case of the Tendaguru expedition resembled a double-entry bookkeeping system, had to clearly establish "which fragments or parts belong together. The drawing of sketches or photographic records with the numbering of parts should not be omitted in the process" (Stromer 1920, 10).

Paper technologies and knowledge from bureaucratic disciplines, as well as background biological theories, were therefore essential to work, identify, and transport data and supporting information from the field into the lab.

3 Illustrations and Validation of Extinct Animals

Like in all biological and taxonomic disciplines, illustrations play a crucial role in paleontology. After fossils have been excavated, transported, and finally reassembled, they are described and classified in painstaking detail. By accomplishing these tasks, these specimens can be eventually shared with the broader scientific community through specific publications. This practice relies on a mixture of words and images. Fossil illustrations are published in compendia, textbooks, scientific articles, or even sent to colleagues as paper placeholders of the material fossil record. In effect, scientific illustrations represent a piece of the record of the past to an audience that cannot directly perceive it; the original record is mostly incomplete, imperfect, and impossible to transport (see the Rudwick's account of Cuvier's paper museum of fossil bones (Rudwick 2000)).

Driven first primarily by paper and pencil and a strong cooperation between artists and scientists, this practice and resulting visualizations have received a new impetus and rhetoric with the advent of technologies such as the electron microscope and 3D scanners. These tools enable paleontologists to perceive the so-called microcosmos, thus enabling the scientists to perceive microstructures otherwise invisible and non-representable.

Advertised to the broader public as the conquering of the invisible world, the integration of different kinds of technologies within paleontological research has had three main consequences. First, it re-established the gap between the scientist and the technician. This gap had narrowed to some degree through the drawing techniques learned by the paleontologists themselves during their course of study, when paleontologists learn the motto "to see is to draw" (see Heumann et al. 2018). As philosopher Adrian Currie noted, "Sketching geological formations is important for training students to see like geologists: picking out pertinent details, learning how to represent them; making judgements about what is of epistemic import in geological formations" (p. 21). The same can be said for archeology and other earth sciences. This kind of formation represents an exercise in tacit knowledge. Sociologist Caitlin Wylie pointed indeed out that "distinguishing coveted fossilized bone from useless rock requires the tacit knowledge of expert visual judgment and manual skill to reveal the fossil by destroying the rock (i.e., the process of fossil preparation). Because of this, the same starting rock can result in a variety of different specimens" (Wylie 2019, 24). However, because of the difficulty of using highly specialized new technological tools, such as the electron microscope or 3D-scanners, scientists are now not generally able to produce images independently. They therefore collaborate with technicians to produce and visualize data. This has produced new social orders, as Wylie pointly illustrated.[4]

Second, through the visualization of invisible and highly aesthetic images and structures (such as the inner composition of a bone, sand dollars (Fig. 1), etc.), the use of these technologies set the preconditions for a multi- and interdisciplinary approach to the morphological study of organisms. Earth scientists, architects, and engineers started collaborating more than ever to understand the principles responsible for the structures which were made visible. A classic example is the transfer of the morphological features of sand dollars from morphology into architectural manufacturing.

Third, these highly technological practices have pushed the use of images beyond their purely representational character and toward a presentational use (see sections below). Images have been used not merely to bring the past back to light, but rather to create a different planform upon which to investigate the past.

[4] For more on the role of technicians in producing and shaping paleontological knowledge, see, for instance, (Wylie 2009, 2015, 2018, 2021).

Fig. 1 Skeleton of sand dollars used for designing research pavilion. Sand dollars' morphology paved the way for cooperative enterprises – https://www.trr141.de/index.php/research-areas-2/a07/

Besides presenting the past, images of fossilized organisms were used to validate the past (as Rudwick posed), to rhetorically promote paleontology to obtain financial support for expeditions and other activities, and in popular books and museum exhibitions to make visible the quite remote dimension of deep time to a broader audience.

4 From Local Data to Global Phenomena: Lists, Graphs, and Diagrams

The visual culture of earth sciences is marked by a polyvalent use of lists, graphs, and charts. As described in Sect. 1, scientists must perform a quasi-bureaucratic task of compiling fossils in the field to make sense of the bulk of incomplete and imperfect data unearthed. The extensive use of lists is also responsible for the transition from the collection and analysis of local data, for example all sorts of data gathered during an expedition, to the generation and analysis of global phenomena, such as phenomena of mass extinctions in the case of paleobiology and stratigraphy.

German geologist and paleontologist Albert Oppel (1831–1865), the founder of the modern zone concept, used lists as epistemic tools for collecting and comparing stratigraphic data on a global scale.[5] First, he collected all the species found in one specific earth layer and described the layer as the Early Jurassic (*Unterer Lias*).

[5] This practice was also well-established in bio-geography and eventually culminated into *Tabellenstatistik* and Humboldtian arithmetic (Ebach 2015; Schäffner 1999; Klein 2015; te Heesen 2005).

Second, to facilitate further analysis and comparison of this set of fossils, Oppel applied the name of the fossil most frequently found in the area to the entire zone. For instance, one zone of the Early Jurassic was named *Ammonite planorbis* after the massive presence of this fossil in this spatial and temporal dimension.[6] This operation was conducted through the morphological analyses of the fossils gathered, thus using the visual devices described in Sect. 2. Third, Oppel focused on the description of the geographical dispersion of these faunas by listing the places in Germany, France, and England where these fossils have been found. Comparing and bringing together different kinds of lists, Oppel identified the possible patterns of geographical dispersion of the species in Europe.

In the visual culture of earth sciences, lists also serve a function. They can be used to prompt the narration of deep time phenomena. Indeed, they can be seen as raw materials for what has been called data crunching. German paleontologist Heinrich Georg Bronn (1800–1862) came up with a complex system of lists to catalog the fossil record. He named it *Index palaeontologicus* (Sepkoski and Tamborini 2018; Tamborini 2015; Sepkoski 2012b). Successively, he took the previously listed fossils and tabulated them according to their number and place of discovery. Furthermore, he applied a quantitative approach to data proper to bureaucratic sciences to natural history. He used a series of pseudo-quantitative practices called statics to generate numerical patterns concerning the deep past. He successfully transferred another visual culture into the study of earth deep past: he took up the practices used in human history to visualize human past and implemented them into paleontology to narrate the natural history of data (Figs. 2 and 3).

Bronn's practice and set of visualizations became quite prolific in twentieth and twenty-first-centuries paleontology. For instance, J. John "Jack" Sepkoski Jr's *A Compendium of Fossil Marine Families* (1982) lists approximately 3500 marine family fossils according to their first and last occurrence. The times of the first and last occurrence generates "81 stratigraphic intervals ranging from 1.6 to about 20 m.y. long" (Sepkoski Jr. 1994, 133). This *Compendium* was a model of data. It was namely a tool provisioning information for complex operations on its entries. It was very similar to Bronn's *Index palaeontologicus* and other nineteenth-century enterprises. In fact, Sepkoski used his *Compendium* to study "mass extinction, evolutionary rates, and patterns of diversification" (ibid., 140) in the same was that Bronn used his *Index* to study the laws of diversification of the organisms in the past.

5 Computing the Past

Visualizations generated by data simulations provide another means of dealing with deep time. One classical example is the MBL Model, which originated at a meeting at the Marine Biological Laboratory (MBL) in Woods Hole, Massachusetts. The

[6] It is important to stress again that it is not the presence of a *Leitfossil* which characterizes a zone, but the entire fauna.

X. PELECYPODA, II. DIMYA, B. HOMOMYA, I. INTEGRIPALLIATA. 289

Benennungen.	Weltgegend.	a b c d e f g	h i k l	m n o p	q r f	s t u v w x	y z
Pleurodon Wood 2							.0
miliaris Dsh. *sp.*						t	
ovalis Wood						u	
Solenella So. 0							.2
β. Myophorieae.							
? **Disteira** Eichw. 1							.0
triangularis Eichw.	E²	b					
† **Schizodus** King. 11							.0
? Devonicus Vern.		c					
axiniformis Vern.		d					
sulcatus So. *sp.*		e					
parvus Vern.		f					
pusillus Vern.		f					
rotundatus Vern.		f					
undatus Vern.		f					
minimus Vern.		g					
† parallelus King		g					
Rossicus Vern.	S²	g					
† truncatus King		g					
Myophoria Br. 14							.0
Blainvillei Kli.			h				
? inaequicostata Kli.			h				
lineata Mü.			h				
ornata Mü.			h				
cardissoides Alb.			i k				
curvirostris (Br.) Alb.			i k				
laevigata Alb.			i k l				
vulgaris (Br.) Alb.			i k ?				
Goldfussi Alb.			k				
orbicularis Br.			k				
ovata Br.			k				
pes-anseris Br.			k				
simplex Schlth. *sp.*			k				
Whateleyae Bu. *sp.*			k				
Cryptina Boué 1.							.0
= ?Lyriodon =							
‡ Raibelana Boué			?	?			
Lyriodon (So.) Br. 87							.1
(Trigonia Lk., *non* Aubl.)							
* *Simplices.*							
Gaytani Kli.			h				
** *Harpiformes.*							
harpa Gf. *sp.*			h				
*** *Scaphoides Ag.*							
pulchellus Ag. *sp.*				m			
navis Br.				n¹			
rostrum Ag. *sp.*				o			
scapha Ag. *sp.*					q		
Robineauinus d'O. *sp.*					q		

19 (18[illegible])

Fig. 2 Bronn's list used to catalog the record of the past. See, Heinrich Georg Bronn, Untersuchungen Über Die Entwickelungs-Gesetze der organischen Welt Während der Bildungs-Zeit unserer Erd-Oberfläche (Stuttgart: Schweizerbart'sche Ver- lagsbuchhandlung, 1858)

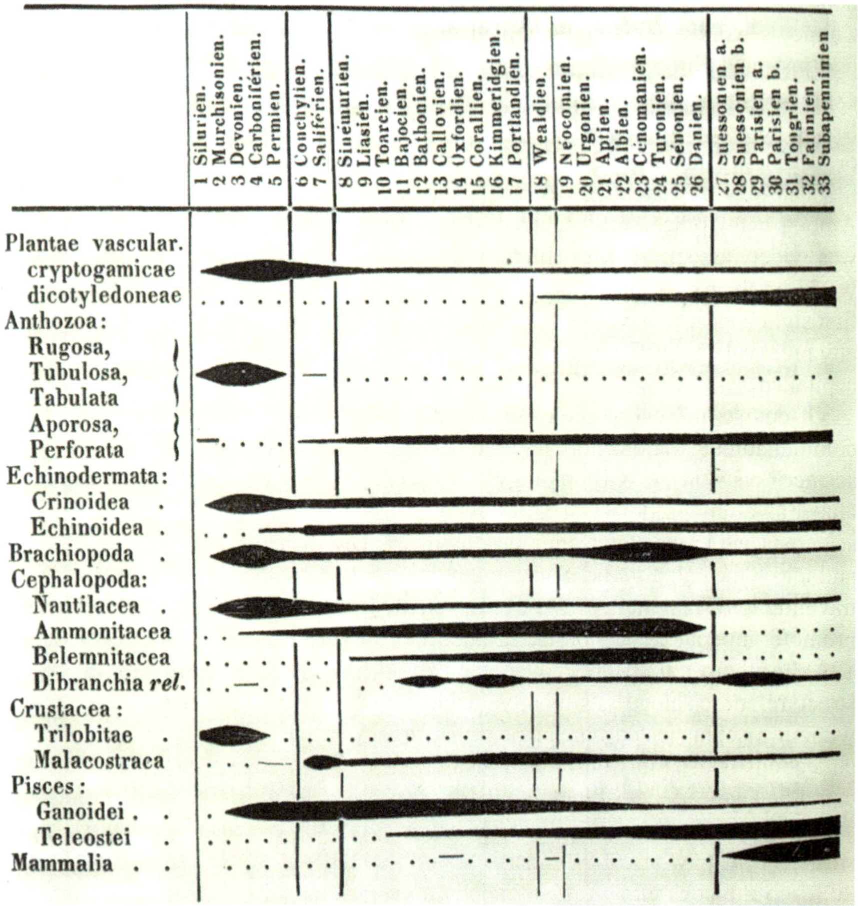

Fig. 3 Bronn's graph used to narrate earth's deep history. See, Heinrich Georg Bronn, Untersuchungen über sie Entwickelungs-Gesetze der organischen Welt während der Bildungs-Zeit unserer Erd-Oberfläche (Stuttgart: Schweizerbart'sche Ver- lagsbuchhandlung, 1858)

resulting computer program simulates the possible fate of a lineage. It can either die out or evolve, with or without speciation (Sepkoski 2016). As historian David Sepkoski commented, this model "was an application of a randomization process known as a Monte Carlo simulation. The computer was used to randomly draw numbers to determine outcomes with prespecified possibilities, much as a dealer might randomly draw cards from a deck and arrange the outcomes into hands. At the end of the run, the program output the results graphically in the form of a branching phylogenetic tree" (Sepkoski 2012a, 2228). Computer-based simulations present possible and highly contingent historical scenarios thus offering working hypotheses regarding evolutionary processes. They offer possibilities for intervention in a deeper temporal dimension that would otherwise have remained inaccessible.

A similar approach can be found in the study of the morphology of extinct organisms. American paleontologist David Raup pioneered this field of investigation.

In the early 1960s, following D'Arcy Thompson's analogical analyses of form development, Raup identified four parameters responsible for the shells of coiled gastropods and used a "digital computer with automatic plotting equipment [...] to make graphical reconstructions of a shell from any given values of the four parameters" (Raup 1962, 150). He then presented a hypothetical snail form simulated according to four previously identified parameters. In this case, form resulted from growing processes. In addition, he conceived of a virtual space (the so-called morphospace) in which all theoretically possible shapes could be simulated starting from the parameters responsible for their formation. The particularity of a morphospace lies in the way the forms were visualized. As Raup put it, these "can be combined to define a 'four dimensional' space which contains most of the theoretically possible shell forms. When the geometries of naturally occurring species are plotted in this space, it becomes evident that it is not evenly filled. Evolution has favored some regions while leaving others essentially empty. In the empty regions, we are presumably dealing with forms which are geometrically possible but biologically impossible or functionally inefficient" (Raup 1965) (Fig. 4).

In 1969, Raup co-authored another seminal paper in which he combined his computational ability to simulate theoretically possible shape processes, i.e., Raup's use of morphospace, with the morphological viewpoint adopted by German

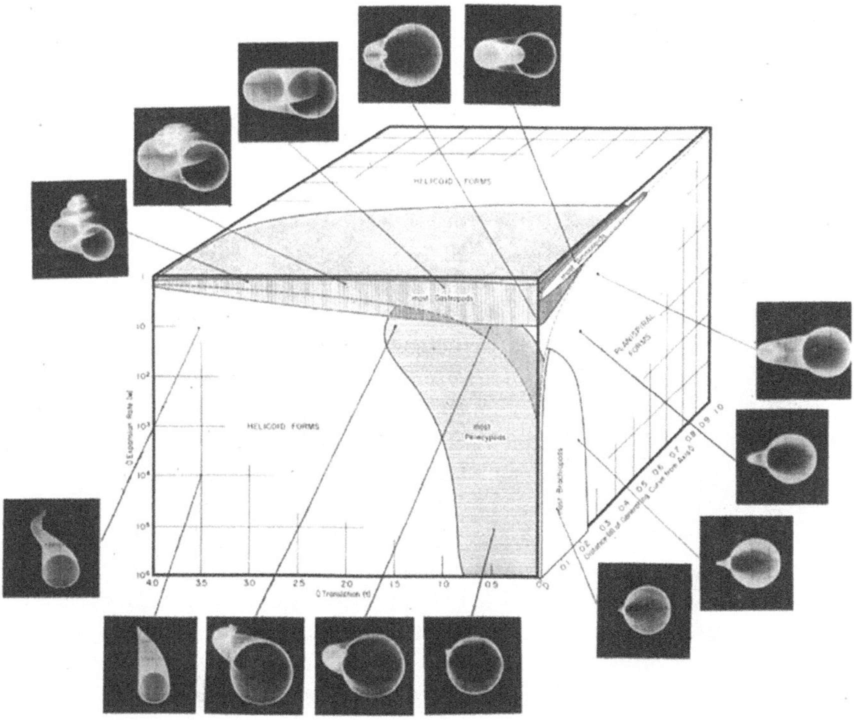

Fig. 4 Raup's morphospace of possible shell forms. See, Raup, David M. "Geometric Analysis of Shell Coiling: General Problems." Journal of Paleontology 40 (1967): 1178–90.

paleontologist Adolf Seilacher. Their idea was that "meander patterns produced by ancient sediment feeders can be simulated by a digital computer with x-y plotter output" (Raup and Seilacher 1969, 994). Assuming a hypothetical animal, the program could sense its immediate surroundings and convert the resulting information into behavioral instructions. However, the interesting part of their computer-based design was that "changing behavioral controls from one run to the next causes more pronounced differences in the model, which are comparable to the differences actually observed between species and genera" (ibid.). Raup and Seilacher concluded that "many more species models could be simulated with little or no program processing. This would lead to a better understanding of the change in parameters required to transform one variant species into another and would also have some effect on the classification of these trace fossils" (ibid.). Thus, shape transformation resulted from changing parameters in the morphogenetic program. This, in turn, could be studied through computer simulation (see also Tamborini 2022a, b).

As for the use of images described in Sect. 2, in this case, visualizations of the past do not merely reproduce something that happened in the past nor make visible different aspects of a possible *explanandum*. Instead, they present new kind of phenomena, thus extending paleontologists' access to deep time. Lorraine Daston and Peter Galison defined this activity as follows: presentation is "no longer necessarily focused on copying what already exists – and instead becomes part of a coming-into-existence" (Daston and Galison 2007, 383). By computing the past, paleontologists were able to create new phenomena, in effect "making real a scene or a process" (sensu Rudwick) not otherwise visible.

6 Virtual Reality, CT Scans, and Robots

In this section, I will take a step forward and explore how technology and different use of visualizations are essential not only for creating a possible access to deep time (as in the cases analyzed above), but also for generating new research questions. Following the architect Kostas Terzidis, I will examine the transition from the computerization of paleontology to computer-based paleontology. According to Terzidis, the computerization of a discipline is characterized by the insertion, manipulation, or storage on a computer system processes already conceived in terms of the project (Terzidis 2004, 2006). Computerized research allows scientists to ask new questions and possibilities that would not have been possible with classical technologies. To examine this transition, I will now turn to the characteristics of recent virtual paleontology.

As explained at the beginning of this chapter, fossils are imperfect and incomplete. They are also extremely fragile and difficult to separate from the rock in which they were fossilized. Some more or less effective techniques have been developed to free fossils from their rocks. The classic approach is to use tools to physically remove the fossil. Alternatively, chemical preparations can be used to try to remove the fossil from the rock, or the fossil can be chemically dissolved to release a cast that can in turn be filled. In all these cases, however, there is a risk of fracture or deformation of the fossil and, above all, the internal morphology of the inaccessible remains. The insertion of recent technology into the field, called virtual paleontology, provides new solutions to these problems. Today, fragile fossil material is displayed on the computer screen after being scanned. These new methods make it possible to experience the microcosm of the fossil and view it from all directions. In addition, they make it possible to perform a range of activities and investigations that would not otherwise be possible in a more traditional paleontological practice. For example, the scientist can non-invasively zoom in and out and rotate objects directly on their computer screen. The three-dimensional virtual visualizations of the material constitute the basis for descriptive publications and for further experimental and quantitative approaches to the evolutionary dynamics of organisms in deep time. In addition, the data generated by 3D scans and visualizations can be easily manipulated, reproduced, and shared with other scientists. Although there are still many problems associated with sharing fossils between institutions,[7] Cunningham and colleagues enthusiastically explain that "digital datasets have been touted as a panacea for the problems of limited access to fossil specimens. In principle, they can be shared online to make them available to the entire community" (Cunningham et al. 2014, 353). Digitization has the potential to bring the "openness enjoyed by other biological sciences" to paleontology" (ibid.).

Through digitization, fossil structures can also be made visible that could not otherwise be accessed without damaging the original. This technology-driven virtual approach "is not just a solution for problematic material, but a powerful new set of techniques for re-viewing fossils preserved in three dimensions" (Garwood et al. 2010, 96–97). 3D modeling of fossils, therefore, allows for technical access and intervention on the workings of individual structures of fossilized creatures. As Cunningham and colleagues assert, "Computer-assisted visualization and analysis of fossils has revolutionized the study of extinct organisms" (Cunningham et al. 2014, 347).

Furthermore, augmented reality is also being used to visualize the fossil record on the computer screen and merge it with other specimens. In this case, "a smartphone, a tablet, or any device with a camera and internet connection would suffice to grant access to the actual 3D morphology of the specimens and their tiniest features" (Bartolini Lucenti et al. 2020, 11; Bartolini-Lucenti et al. 2021) (Fig. 5).

To analyze how virtual paleontology is actually changing the way the geological and evolutionary past can be both accessed and studied, I will focus on two case studies. The first took place at the Naturkunde Museum Berlin. In recent years, the skeleton of *Branchiosaurus Brancaii* has been studied using virtual paleontology techniques. In a study published in 2020, paleontologists focused in particular on the morphology and musculature of the tail. The classical method of inferring the muscle-skeletal structure of an extinct organism is based on a comparison of this structure with that of evolutionarily related animals. In the case of dinosaurs, for

[7] On the problem of accessibly of (big) data see, for instance, (Strasser 2019; Aronova et al. 2017; Davies et al. 2017; Leonelli 2016).

Fig. 5 QR code and augmented reality marker used by the scientists to reproduce the comparison of teeth morphologies of C. *etruscus* from Upper Valdarno (blue), *Canis* from Dmanisi (red), and extant C. *lupus* (grayish). See Bartolini Lucenti et al. (2020, 10)

example, paleontologists have used the morphology of birds and crocodiles as a possible comparison. The paleontologists of the Museum für Naturkunde in Berlin accepted this classical method of comparison and implemented it in the context of virtual paleontology. They digitally reconstructed the tail of this sauropod using "photogrammetric 3D digitization and 3D modeling tools in combination with information provided by dissections of extant crocodilians (*Alligator mississippiensis*) [...] and an Extant Phylogenetic Bracket approach [...] comparing the anatomy of the caudal vertebrae and muscles of Giraffatitan with that of extant crocodilians" (Díez Díaz et al. 2020, 2). Following these analyses, the paleontologists were able to determine that *Branchiosaurus Brancaii* had an extremely powerful tail that played a functional role in the body: It helped "in its stabilization, and propulsion, but also as counterweight for the presacral part of the body" (ibid., 14). In addition, they determined that the total weight of the tail was about 250 kg. These results are extremely significant because the musculature of an extinct organism is never fossilized; since the muscles are made of soft material, they are not preservable.

By using 3D photography and various modeling techniques, paleontologists are able to gain access to something that is not a given. They used these results as possible platforms to experiment and work with deep time. Then, by examining the fossil virtually, new research questions were asked about the posture, strength, and biomechanical properties of the real fossil. In other words, reality (the material specimen) and virtuality (what can be visualized through different technologies) merged and supported each other. The resulting "methodology allows a better-constrained reconstruction of muscle volumes and masses in extinct taxa, and thus force and weight distributions throughout the tail, than non-volumetric approaches" (ibid., 1).

Another emblematic case of the dissolution of boundaries between biology and technology is the use of virtual methods to overcome the lack of fossilized information. As explained above, fossils are inherently imperfect and incomplete. In addition, the process of fossilization can cause severe deformation. Using virtual paleontology techniques, paleontologists examined the severely deformed fossil of *Equus stenonis*, one of the most widely distributed horse fossils of the *European Pleistocene*. Fossil horses are often deformed in the skull due to their long and thin morphology and the presence of multiple pneumatic sinuses that occur inside the skull. Paleontologists introduced "a new virtual reconstruction protocol, termed Target Deformation, which takes advantage of recent progress in digital restoration of fossil specimens [...] digital alignment of disarticulated portions [...] and 3D thin plate spline transformation (tps3d) [...] to provide the virtual reconstruction of badly deformed, partially incomplete cranial material by using target fossil remains of the same species" (Cirilli et al. 2020, 2). Target deformation aims to eliminate asymmetric changes due to taphonomic processes by applying a set of corresponding bilateral reference points.

Like in the case of the *Brachiosaurus brancai* muscle reconstruction, paleontologists in this study used technology in addition to fossil remains of other organisms to merge the virtual with the real. In this case, two fragmentary but well-preserved skulls of *Equus stenonis* from Olivola, Italy, and *Dmanisi*, Georgia, were used. Paleontologists showed that the 3D virtual reconstruction protocol used was "capable of virtually restoring a severely crushed specimen by using partially complete skull specimens, such as the fragmentary *E. stenonis* skulls from Olivola and Dmanisi" (ibid., p. 9).

One last element should be briefly mentioned here. Presently, the visual culture of earth science is merging with an engineering approach to morphology. As a result, paleontologists are teaming up with engineers to design robots that resemble extinct organisms. This robotic fabrication of extinct forms allows scientists to indirectly experiment and manipulate the deep past. The rationale of this practice is to create robots able to generate further biological questions that wouldn't have been possible otherwise. In this process, the robotic and the digital merge in the materialization of the morphology of extinct organisms. This integration has been recently labeled the "twenty-first-century material turn in the study of form" (Tamborini 2021, 2022b).

7 Conclusion: Visual Cultures and the Dynamics of Knowledge Production

Paleontology, geology, and other earth sciences address "configurations that [can] not be adequately conveyed by words or mathematical symbols alone" (Rudwick 1967, 151). Within this arena, different kinds of visualizations support earth scientists' reasoning, in effect enabling scientists to work with deep time. In this process, science, theory, and technology are fused together to produce knowledge. With the imperfect and incomplete record of the past as a point of departure, scientists have developed a set of practices to both represent and present deep time. Technology has allowed for researchers to overcome the imperfection and incompleteness of the record of the past. As a result, a working version of the past is produced which allows scientists to think with pictures.

At its core, the visual culture of paleontology is based on circulation of skills, practices, and technologies between different disciplines (see also Tamborini 2022a, c). For example, several circulations of practices took place between bureaucratic sciences and paleontology. The same can be said for the exchange of knowledge between paleontology and human history as well as for paleontology and engineering sciences. This has two main implications. First, paleontology, and broadly earth sciences, can be seen as methodological "omnivorous disciplines" (Currie 2015). They take up and implement every possible set of practices, visualizations, and tools to make real and visualize something otherwise non-perceptible. Second, as Rudwick (1967), Hentschel (2014), Sepkoski and Tamborini (2018) have aptly indicated, the transition of knowledge is made possible by scholars who were between different disciplines and therefore could pre-adapt and translate a particular visual culture, say the bureaucratic one, into another discipline such as paleontology or geology. This transition of media and visual practices characterized not only the development of earth sciences, but it also was a common issue in other disciplines.

Second, as my phenomenology of visualizations has indicated, every medium and visual tool has its own specific epistemic and ontological power. The images of extinct organisms used for taxonomic scopes have a quite different appeal, function, and epistemic power from the form simulated and visualized through a morphospace. The former represents the past, the latter presents a possible platform for creating the conditions for the manifestation of deep-time phenomena. The same can be said for the different modalities of working with data variously visualized and diversely visible. As seen, there is a wide range of epistemic possibilities when working with data previously listed, tabulated, or put into a graphical representation: they convey their representational meaning in quite different and sometimes complementary ways.[8]

However, although every visualization conveys its meaning differently, diverse visual devices should be put into relation to efficiently work with the past. One peculiarity of the earth sciences is that both the fossil record and its representation can be considered to be both data for further works and the sort of phenomena the scientists seek to explain (the so-called explanandum). In paleontology, data, such as fossilized bones, a displayed dinosaur, or the diversity curve generated through a database (such as the phanerozoic diversity curves generated by Heinrich Bronn or similar computer-based enterprises produced in the contemporaneity), are data used for understanding broader biological phenomena (such as the broader patterns and mechanisms of adaptability or biodiversity). At the same time, this data is what paleontologists seek to explain – they try to figure out why an organism had this particular form or why the data show a logarithmic instead of an exponential curve.

[8] As Hentschel noted, the question is "less *whether* images have a logic of their own than *how* they convey their meaning" (Hentschel 2014, 326).

To handle this dual function and the related epistemic and practical issues, scientists adopt a continuous flow of representations in which the visualized object becomes the starting point for generating new phenomena and visualizations, which in turn become the objects to be explained.

Third, knowledge production and visualizations are tied with the use of different technologies – ranging from pen and paper, through computers, to robots and virtual and augmented reality. Therefore, to understand how knowledge is produced and visualized in earth sciences, the productive intersection between technology, background theory, and images should be taken into account. Earth sciences' visual culture cannot be fully understood if studied in isolation. Rather, image production should always be put into the broader technological and theoretical framework adopted by the specific earth science disciplines.

Last, by working on the different modalities of visualizing the past, paleontologists and earth scientists often produce images which have a strong aesthetic appeal. The structures and beauty of fossilized structures have been used as guiding model to design buildings or as the inspiration for paintings. Geological and stratigraphic maps have inspired literary and poetic works. What is worth emphasizing from this is twofold. First, the aesthetics of the past have paved the way for many interdisciplinary and joint scientific projects. These were set up to understand how technical form could be best represented and built emulating the form-function-beauty complex of deep time organisms. In turn, paleontologists took up from these joint projects a nuanced engineering perspective toward organism construction. This has enabled paleontologists to develop new practices and vocabulary to visualize the past. The role of aesthetics as a binding element between different practitioners and scientists from diverse disciplines is a further element of the visual culture in earth science.

The aesthetics of the past fosters also its possible epistemic investigation. The aforementioned Tendaguru expedition provides one emblematic example for the intersection between aesthetics and knowledge production. Once the giant bones were transported to Berlin Museum of Natural History, they were vertically displayed in a museum main exhibition room, evoking all the majesty and beauty of the columns of an ancient temple. The director of the Museum used this to start a successful crowdfunding campaign to support the expedition and broadly paleontological work. Hence, as Turner rightly noted, "Collection decisions are heavily (though not exclusively) guided by aesthetic norms and goals. Some of the reasons not to collect a *Hadrosaur* fossil, [...], are distinctively aesthetic" (Turner 2019, 13). This point implies what Turner called a rejection of the bias toward the epistemic.[9] He did so "by showing that paleontological research is a form of aesthetic engagement with fossils and with landscapes" (ibid., 1). Particularly, he argued that "aesthetic engagement and historical investigation are mutually facilitating" (ibid., 29).

[9]Philosopher Adrian Currie called for an analysis of the role of aesthetics in paleontological knowledge production. See Currie (2020). Broadely on aesthetics and science see for instance (Ivanova 2021; Breitenbach 2013; Tamborini 2022a, b; Hentschel 2014).

Hence, and to conclude, the aesthetics of deep time should be also considered as relevant components when working toward an understanding of the role and function of different media and technologies in producing and sustaining deep-time knowledge production. As a result, earth sciences' visual culture intersects with the history of aesthetics, design, technology, and biological sciences. To fully comprehend this visual culture, a transdisciplinary effort is required.

References

Aronova, Elena, Christine von Oertzen, and David Sepkoski. 2017. Introduction: Historicizing big data. *Osiris* 32 (1): 1–17.

Bartolini Lucenti, Saverio, Maia Bukhsianidze, Bienvenido Martínez-Navarro, and David Lordkipanidze. 2020. The wolf from Dmanisi and augmented reality: Review, implications, and opportunities. *Frontiers in Earth Science* 8: 131.

Bartolini-Lucenti, Saverio, Joan Madurell-Malapeira, Bienvenido Martínez-Navarro, Paul Palmqvist, David Lordkipanidze, and Lorenzo Rook. 2021. The early hunting dog from Dmanisi with comments on the social behaviour in Canidae and hominins. *Scientific Reports* 11 (1): 1–10.

Bigg, Charlotte. 2015. Travelling scientist, circulating images and the making of the modern scientific journal: Norman Lockyer's visual communication of astrophysics in nature. *Nuncius* 30 (3): 675–698.

Breitenbach, Angela. 2013. V – Aesthetics in science: A Kantian proposal. *Proceedings of the Aristotelian Society* 113: 83–100.

Comte de Buffon, Georges-Louis Leclerc. 1778. "Les époques de la nature." In Histoire naturelle. Paris: Imprimerie royale.

Carter, R.M. 2007. Stratigraphy into the 21st century – Carter. *Stratigraphy* 4: 187–193.

Cirilli, Omar, Marina Melchionna, Carmela Serio, Raymond L. Bernor, Maia Bukhsianidze, David Lordkipanidze, Lorenzo Rook, Antonio Profico, and Pasquale Raia. 2020. Target deformation of the Equus stenonis holotype skull: A virtual reconstruction. *Frontiers in Earth Science* 8: 247.

Cunningham, John A., Imran A. Rahman, Stephan Lautenschlager, Emily J. Rayfield, and Philip C.J. Donoghue. 2014. A virtual world of paleontology. *Trends in Ecology & Evolution* 29 (6): 347–357.

Currie, Adrian. 2015. Marsupial lions and methodological omnivory: Function, success and reconstruction in paleobiology. *Biology and Philosophy* 30: 187–209.

Currie, Adrian Mitchell. 2020. Epistemic engagement, aesthetic value & scientific practice. *The British Journal for the Philosophy of Science* 74: 714802.

Daston, Lorraine, and Peter Galison. 2007. *Objectivity*. New York: Zone Books.

Davies, Thomas G., Imran A. Rahman, Stephan Lautenschlager, John A. Cunningham, Robert J. Asher, Paul M. Barrett, Karl T. Bates, Stefan Bengtson, Roger B.J. Benson, and Doug M. Boyer. 2017. Open data and digital morphology. *Proceedings of the Royal Society B: Biological Sciences* 284 (1852): 20170194.

Díez Díaz, Verónica, Oliver E. Demuth, Daniela Schwarz, and Heinrich Mallison. 2020. The tail of the late Jurassic sauropod Giraffatitan Brancai: Digital reconstruction of its epaxial and hypaxial musculature, and implications for tail biomechanics. *Frontiers in Earth Science* 8: 160.

Ebach, Malte Christian. 2015. *Origins of biogeography. The role of biological classification in early plant and animal geography*. Dordrecht/Heidelberg/New York/London: Springer.

Fossen, Haaokon. 2013. Do you sketch in the field? Structural geology blog. https://structuralgeo.wordpress.com/2013/11/20/do-you-sketch-in-the-field/

Fraas, Eberhard. 1910. *Der Petrefaktensammler. Ein Leitfaden Zum Sammeln Und Bestimmen Der Versteinerungen Deutschlads*. Stuttgart: Lutz.

Garwood, Russell J., Imran A. Rahman, and Mark D. Sutton. 2010. From clergymen to computers – The advent of virtual palaeontology. *Geology Today* 26 (3): 96–100.

te Heesen, Anke. 2005. Accounting for the natural world. Double-entry bookkeeping in the field. In *Colonial botany: Science, commerce, and politics in the early modern world*, ed. Londa Schiebinger and Claudia Swan. Philadelphia: University of Pennsylvania Press.

Hentschel, Klaus. 2014. *Visual cultures in science and technology. A comparative history*. Oxford: Oxford University Press.

Heumann, Ina, Holger Stoecker, Marco Tamborini, and Mareike Vennen. 2018. *Dinosaurierfragmente: Zur Geschichte Der Tendaguru-Expedition Und Ihrer Objekte, 1906–2017*. Göttingen: Wallstein Verlag.

Ivanova, Milena. 2021. The aesthetics of scientific experiments. *Philosophy Compass* 16 (3): e12730.

Klein, Ursula. 2015. *Humboldts Preußen. Wissenschaft Und Technik Im Aufbruch*. Darmstadt: Wissenschaftliche Buchgesellschaft.

Leonelli, Sabina. 2016. *Data-centric biology: A philosophical study*. Chicago: University of Chicago Press.

Nieuwland, Ilja. 2019. *American dinosaur abroad: A cultural history of Carnegie's plaster diplodocus*. Pittsburgh: University of Pittsburgh Press.

Podgorny, Irina. 2003. Medien Der Archäologie. In *Archiv Für Mediengeschichte – Medien Der Antike*, ed. Lorenz Engell, Bernhard Siegert, and Joseph Vogl. Weimar: Universitätsverlag Weimar.

Raup, David M. 1962. Computer as aid in describing form in gastropod shells. *Science* 138 (3537): 150–152.

Raup, David M., and Arnold Michelson. 1965. Theoretical morphology of the coiled shell. Science 147 (3663): 1294–1295.

Raup, David M., and Adolf Seilacher. 1969. Fossil foraging behavior: Computer simulation. *Science* 166 (3908): 994–995.

Rieppel, Lukas. 2019. *Assembling the dinosaur*. Cambridge: Harvard University Press.

Rudwick, Martin J.S. 1967. The emergence of a visual language for geological science, 1760–1840. *History of Science* 14 (3): 149–195.

———. 1985. *Scenes from deep time: Early pictorial representations of the prehistoric world*. Chicago: The University of Chicago Press.

———. 2000. Georges Cuvier's paper museum of fossil bones. *Archives of Natural History* 27 (1): 51–68.

———. 2017. The fate of the method of 'paradigms' in paleobiology. *Journal of the History of Biology* 51 (3): 479–533.

———. 2018. Functional morphology in paleobiology: Origins of the method of 'paradigms'. *Journal of the History of Biology* 51 (1): 135–178.

Schäffner, Wolfgang. 1999. Verwaltung Der Kultur. Alexander von Humboldts Medien (1799–1834). In *Interkultularität Zwischen Inszenierung Und Archiv*, ed. Stefan Rieger, Schamma Schahadata, and Manfred Weinberg. Tübingen: Narr.

Sepkoski, David. 2012a. *Rereading the fossil record: The growth of paleobiology as an evolutionary discipline*. Chicago: University of Chicago Press.

———. 2012b. Towards "a natural history of data": Evolving practices and epistemologies of data in paleontology, 1800–2000. *Journal of the History of Biology* 46: 401–444.

———. 2016. Simulations, metaphors, and historicity in Stephen Jay Gould's 'view of life'. *Studies in History and Philosophy of Biological and Biomedical Sciences* 58: 73–81.

———. 2017. The earth as archive. In *Archiving sciences: Pasts, presents, futures*, ed. Lorraine Daston. Chicago: The University of Chicago Press.

Sepkoski, David, and Marco Tamborini. 2018. 'An image of science': Cameralism, statistics, and the visual language of natural history in the nineteenth century. *Historical Studies in the Natural Sciences* 48 (1): 56–109. https://doi.org/10.1525/HSNS.2018.48.1.56.

Sepkoski, J. John, Jr. 1994. What I did with my research career. In *Mass–extinction debates: How science works in a crisis*, ed. William Glen. Stanford: Stanford University Press.

Shepherd, Nick. 2015. *The mirror in the ground: Archeology, photography and the making of a disciplinary archive*. Cape Town: Jonathan Ball Publishers and the Centre for Curating the Archive.

Strasser, Bruno J. 2019. *Collecting experiments: Making big data biology*. Chicago: University of Chicago Press.
Stromer, Karl Heinrich Ernst Freiherr von Reichenbach. 1920. *Paläozoologisches Praktkum*. Berlin: Gebrüder Borntraeger.
Tamborini, Marco. 2015. Paleontology and Darwin's theory of evolution: The subversive role of statistics at the end of the 19th century. *Journal of the History of Biology* 48 (4): 575–612.
———. 2020. Technoscientific approaches to deep time. *Studies in History and Philosophy of Science Part A* 79 (1): 57–67. https://doi.org/10.1016/j.shpsa.2019.03.002.
———. 2021. The material turn in the study of form: From bio-inspired robots to robotics-inspired morphology. *Perspectives on Science* 29 (5): 643–665. https://doi.org/10.1162/posc_a_00388.
———. 2022a. *The architecture of evolution: The science of form in twentieth-century evolutionary biology*. Pittsburgh: University of Pittsburgh Press. ISBN: 9780822947356
———. 2022b. *Entgrenzung: Zur Biologisierung der Technik und der Technisierung der Biologie*. Hamburg: Meiner.
———. 2022c. The circulation of morphological knowledge: Understanding 'form' across disciplines in the twentieth and twenty-first centuries. *Isis*, 113, 4: 747–766. https://doi.org/10.1086/722439
———. ed. 2023. *Die Ästhetik der Technowissenschaften des 21. Jahrhunderts*. Darmstadt: wbg Verlag.
Terzidis, K. 2004. *Expressive form: A conceptual approach to computational design*. London: Taylor & Francis.
———. 2006. *Algorithmic architecture*. London: Architectural Press.
Turner, Derek D. 2019. *Paleoaesthetics and the practice of paleontology*. Cambridge: Cambridge University Press.
Wylie, Caitlin Donahue. 2009. Preparation in action: Paleontological skill and the role of the fossil preparator, 3–12. Citeseer.
———. 2015. 'The artist's piece is already in the stone': Constructing creativity in paleontology laboratories. *Social Studies of Science* 45 (1): 31–55.
———. 2018. Trust in technicians in paleontology laboratories. *Science, Technology, & Human Values* 43 (2): 324–348.
———. 2019. Overcoming the underdetermination of specimens. *Biology and Philosophy* 34 (2): 1–18.
———. 2021. *Preparing dinosaurs: The work behind the scenes*. Cambridge: MIT Press.

Part IV

Perspectives

Indigenous Knowledge and Perspectives

Stephen Bocking

Contents

Abstract

Indigenous knowledge, once dismissed as mere folklore, is now widely recognized as an essential dimension of global environmental knowledge. Indigenous people, once excluded, now participate across a range of environmental affairs. Understanding how and why this has occurred requires attention to a complex history of scientists and others constructing ideas about Indigenous knowledge. A variety of scholars, including historians of science, environmental historians, and political ecologists have examined this history, identifying the factors that have influenced expert, public, and institutional perceptions of Indigenous knowledge. These include various colonial and postcolonial contexts, ideas about development, changes in the natural environment, disciplinary perspectives (such as those of anthropology), and shifting views of human-environment relations. Indigenous peoples – as knowledge producers, brokers, and intermediaries – have been crucial to these evolving perceptions, by asserting that their knowledge can be a means of achieving change in both knowledge and politics. The Arctic provides a

S. Bocking (✉)
Trent University, Peterborough, ON, Canada
e-mail: sbocking@trentu.ca

E. Aronova et al. (eds.), *Handbook of the Historiography of the Earth and Environmental Sciences*, Historiographies of Science,
https://doi.org/10.1007/978-3-031-40799-4_20

distinctive setting in which the historical construction of Indigenous knowledge can be examined in more detail.

Keywords

Anthropology · Arctic · Colonial sciences · Ecology · Identity · Indigenous knowledge · Indigenous people · Inuit · Local knowledge · Political ecology · Postcolonial studies · Racial sciences

1 Introduction

Indigenous knowledge has become an important dimension of environmental affairs. Inuit insights on Arctic sea ice in a warming climate and the roles of Indigenous peoples as stewards of biodiversity are among the many ways in which the knowledge of people with deep ties to place has emerged as essential to understanding and conserving the environment. Indigenous knowledge has also become a major theme in scientific work, contributing insights into ecosystems and human-nature relations, while challenging scientists' epistemological, ontological, and political assumptions. And it has gained a presence in studies of the history of science, as evident in, for example, histories of colonial and postcolonial science (see chapter ► "History of the Earth Sciences from the South" by Jarrod Hore) and of studies of global environmental change.

The significance of Indigenous knowledge raises many questions relevant to the historiography of the earth and environmental sciences. Why is this knowledge, once widely dismissed as mere stories or folklore, now invoked in scientific and environmental affairs, including by authoritative institutions such as the Intergovernmental Panel on Climate Change and the Convention on Biodiversity? How does its status relate to the evolving political contexts of environmental affairs? What does this status tell us about the intersections of knowledge and identity?

These and other aspects of Indigenous knowledge are relevant to a range of disciplines. Its historiography therefore encompasses not just the work of historians but anthropologists, political ecologists, and scholars in other fields. Insights from imperial and postcolonial studies, science and technology studies, the study of racial and cultural identities, and community-based knowledge systems are also relevant. Work across these fields presents opportunities to enhance our understanding of the historical geography of race, nature, and knowledge.

One challenge that Indigenous knowledge presents to historians is that of terminology. A variety of terms are used when discussing this topic, such as "vernacular knowledge," "local knowledge," and "traditional ecological knowledge." These terms must be used with care, as they reflect diverse and often contested theoretical, practical, and political commitments. So does "Indigenous epistemology" – a term that some Indigenous scholars reject, since it implies that the nature of knowledge can be considered in isolation from its political and social contexts. The signifier "Indigenous" itself is marked by its colonial origins. Geographic distinctions are also important: in settler societies such as North America, "Indigenous knowledge" refers to First Peoples' knowledge and associated ethical relations both today and in the past; in South Asia, it refers to precolonial forms of knowledge; historians of Latin America view it as one among many forms of knowledge that form a landscape of "creole science"; and in Europe, the term can refer to historical knowledge of local nature, distinct from knowledge of more exotic places (Cooper 2007; McCook 2013; Menon 2022).

Indigenous knowledge presents other challenges to the historiography of the earth and environmental sciences. Indigenous peoples themselves have challenged how scientists use evidence to adjudicate knowledge claims; they have critiqued the assumption that knowledge can exist in an objective space unsullied by politics; and they have insisted that researchers acknowledge the epistemic violence that has so often accompanied knowledge production in Indigenous contexts, and hence the ethical responsibilities that accompany knowledge production today. Historians working in Indigenous contexts must therefore acknowledge these challenges and revise their practices and purposes accordingly.

Focusing primarily on debates in the twentieth and early twenty-first centuries, this chapter examines the attitudes and interpretations of Indigenous knowledge formed by observers, usually situated outside Indigenous communities, including scholars of environmental knowledge in a variety of institutional and political contexts. The chapter begins with overviews of the erasure, reconsideration, and resurgence of Indigenous knowledge in colonial and postcolonial contexts. It then acknowledges the necessity of geographic specificity through a focus on the historiography of Arctic Indigenous knowledge.

2 Erasure

Ignorance of Indigenous knowledge was once widespread across empires and states. This ignorance was not just an absence of knowledge but a condition actively produced (Proctor and Schiebinger 2008). Science – as a source of both empirical information and authority – has been a consistent presence in the histories of imperialism and state formation: contributing to the imposition of order and rationality, the pursuit of efficiency and "improvement," and the formation of modern societies, while justifying erasure of other forms of knowledge, including that held by Indigenous people (Drayton 2000). Science has been coproduced with political authority: guiding and justifying both physical transformations (for example, building dams, designating game parks, and relocating people to reserves and reservations), and the discursive construction of landscapes as either managed systems of production or pristine wilderness – displacing thereby both people and their knowledge (Neumann 1998; Nixon 2011, 150–174). While this unfolded across diverse geographic and historical contexts, a common element has been the production of difference: distinguishing standardized, universal science from "local" ways of knowing. These distinctions have been inherently intersectional, produced amidst the assertion of differences based on gender, ethnicity, class, and location (Radcliffe 2017).

These physical and discursive transformations imply opportunities to link the history of science and environmental history. In diverse ways, perceptions of Indigenous knowledge have been linked to changing natural and social landscapes. In the late nineteenth and early twentieth centuries, environmental and social transformation – such as drought, disease, and war in East Africa, or the near-elimination of bison and displacement by settlers in the North American Plains – disrupted Indigenous societies, including their relations with local environments, encouraging the assumption that "traditional" precolonial knowledge and ways of life had been rendered irrelevant. Assumptions regarding Indigenous knowledge have also been embedded in interpretations of the history of local landscapes. From Madagascar to Southeast Asia to Australia, and across diverse ways of life, of pastoralism, shifting cultivation, or managing fire, narratives of environmental crisis were justified by the uncritical transfer of theories of a balanced, primeval nature. These narratives then justified intervention to "save" local people from their own "irrational" behavior (Fairhead and Leach 1996; Kull 2000; Munro 2020; Oba 2020; O'Brien 2002). Changing environmental conditions reinforced these perceptions. Drought in the 1930s encouraged the view that local people were overgrazing arid land, encouraging desertification: a diagnosis rooted in the belief that their knowledge was unable to adjust to changing conditions (Davis 2016). In the postcolonial era, these rationales for intervention were readily translated into the imperatives of international development: the formation of modern economies and societies through science and technology imparted urgency to the erasure of Indigenous knowledge (Scott 1998).

3 Recovery

In recent decades, historians have reconsidered this history of erasure, describing relations between Indigenous and scientific knowledge that include not just displacement but exchange, the agency of local knowledge holders, and the formation of hybrid forms of knowledge. One theme has been the roles of Indigenous peoples in exploration: dismantling narratives of heroic travelers by describing their reliance on local people for guidance, means of travel, and often survival. Colonial maps were the product of efforts to make places knowable by drawing on local knowledge, enabling, alongside collecting and other field practices, the colonial appropriation of Indigenous cultures into global systems of territorial and taxonomic classification (Bryan 2009). Attention to Indigenous peoples' roles as knowledge brokers and intermediaries has also provided the basis for recovering their agency in the history of field and museum sciences (Goss 2021). Historical studies of fieldwork have emphasized the roles of lay participants (including Indigenous peoples) as field assistants or local informants, forming distinctions between cosmopolitan and residential ways of knowing (Kohler 2011; Vetter 2011). "Working" landscapes such as agriculture formed particular contexts for these interactions, with the production of knowledge involving a variety of local actors (McCook 2011). While scientists often included little information about the Indigenous sources of their samples and observations, close examination of the knowledge they produced can illuminate their interactions with and reliance on local informants. The emerging image, therefore, is of a "middle ground" (as Richard White once described it), or "contact zones" of exchange and accommodation, where knowledge was transacted, with scientists paying close attention to local conditions and knowledge, and local people instructing and challenging imperial ideologies and scientific agendas (Anderson 2009; Hodge 2007; White 1991).

These knowledge relations were commonly the products of local negotiation. There was no consistent "colonial mind" but instead a complex interplay between local places and wider networks of knowledge (Chambers and Gillespie 2001). Less-developed colonial scientific capacities tended to encourage more interest in Indigenous knowledge, with their uneven distribution influencing debates about what counted as science, and hence where its boundaries would be drawn (Tilley 2010). Cultural attitudes also mattered: as Londa Schiebinger has explained, knowledge of the use of the peacock flower as an abortifacient failed to travel from the Caribbean to Europe for a variety of reasons, including attitudes regarding gender and female bodies, economic interests, and the shifting professional identity of medical practice (Schiebinger 2008). The relations that scientists formed with Indigenous knowledge were shaped not just by their imperial roles and by universalizing discourses of modernity and reason, but by their own institutions, disciplines, and environmental circumstances (Radcliffe 2017). Diverse scientific perspectives on Indigenous knowledge imply that the colonial and postcolonial state should not be viewed as a monolith, pursuing a coordinated agenda of domination guided by experts. Instead, scientific expertise might not only support but also critique colonial power structures.

Postcolonial studies of science have been complemented by work in science and technology studies. Both present a view of science, including assertions of its universal authority, as a situated practice: an amalgam of practical skills, technical devices, theory, and social strategies, tied to political, social, and institutional contexts. This implies study of the sites in which knowledge is produced and applied, how it travels between these sites, and how different forms of knowledge interact. According to this perspective, knowledge does not compel acceptance merely because it corresponds to the "truth"; rather, acceptance must be understood in terms of the interpretation of knowledge claims within particular cognitive or institutional frameworks (Anderson and Adams 2008). Close examination of controversies has also been important, by demonstrating how knowledge is the product of negotiation between scientists, their patrons, and other interests (Martin Rudwick's *The Great Devonian Controversy* is considered an early classic among "controversy studies"; Rudwick 1985).

Boundary work – how scientists assert distinctions between their work and other ways of knowing – has been essential to these relations between science and Indigenous knowledge (Gieryn 1999). Those seeking to dismiss Indigenous knowledge have done so by excluding it from science, often by describing it as anecdotal, inaccurate, or mere spirituality. In contrast, scientists seeking to draw on Indigenous knowledge have often emphasized those aspects, such as taxonomic classifications, that can be most readily removed from its context and understood in scientific terms. In effect, they redraw the boundary of science to include those aspects of Indigenous knowledge that have meaning within it, while excluding those that would not contribute to their scientific purposes, such as its social, cultural, and racial dimensions.

The historical authority of Indigenous knowledge is therefore tied to both the history of efforts to assert that science is distinctive, and the many practical links between these forms of knowledge. Historians must therefore pay close attention to the interactions and negotiations across the shifting and porous boundaries between science and other forms of knowledge. The image is of "Indigenous" and "scientific" as fractured and indistinct categories, with varying degrees of similarity, difference, and overlap, often combined in processes of knowledge formation, and often distinguished not on the basis of epistemological content but their political implications (Agrawal 1995).

Political ecologists have provided important perspectives on this nexus of situated history of science and Indigenous knowledge (Goldman et al. 2011). Motivated by a concern for the relations between knowledge and the exercise of power, they have examined, among other topics, the formation of environmental subjects in relation to the state and other forms of power, and the relations between discourses of environmental crisis, Indigenous knowledge, and critical perspectives on assumptions of equilibrium. Influential works such as Donna Haraway's *Staying with the Trouble* (2016) and Anna Tsing's, *The Mushroom at the End of the World* (2015) have challenged the notion of a dichotomy between nature and culture, asserting views compatible with Indigenous perspectives on relations between humans, other species, and physical and ethical worlds.

The disciplinary structure of science has been essential to historical perceptions of Indigenous knowledge: shaping how scientists think, and what practices, evidence, and ideas are included or excluded from science. Since the 1930s, a few colonial scientists in Africa, particularly in health, ecological, and agricultural disciplines, began to look at how people actually used and understood their environments. Although paternalistic and grounded in faith in superiority of science, they challenged assumptions of Indigenous ignorance and irrationality, instead describing knowledge of medicinal plants and ecological conditions, and the adaptation of farming activities to local soils and climates (Adams 2003; Tilley 2011). Each of these disciplines guided scientists' evaluations of the relevance and credibility of local knowledge.

Since its formation as a distinct discipline, anthropology has had a distinctive role in shaping perceptions of Indigenous knowledge. As European empires expanded, researchers followed, studying the ways of knowing of colonized peoples. Consistent with their focus on cultures that were thought to be closer to nature, anthropologists had a particular interest in how they related to their environment, including, as Bronislaw Malinowski described in the case of the Trobriand islanders, the "primitive knowledge" relevant to immediate concerns such as agriculture, healing, and navigation (Orr et al. 2015; Tilley 2010). By the 1950s, anthropologists were developing a variety of perspectives on Indigenous environmental relations. In forming cultural ecology as a new research field, Julian Steward and his colleagues emphasized the adaptive links between culture and environment. Ethnoscience (the study of conceptions of the world held by a people or culture), ethnoecology, and human ecology also became active research fields. As articulated by Harold Conklin in 1954, the "ethnoecological approach" compelled attention to how people understand their local environment. "Cognitive anthropology" suggested that instead of fitting cultures within an ethnocentric narrative of primitive to modern, anthropologists could understand them on their own terms, with Indigenous knowledge part of comprehensive, holistic perspectives, embedded in systems of resource use, choice, and decision-making. These trends led many anthropologists to view Indigenous knowledge as part of the defining subject matter of their discipline (Orr et al. 2015; Sillitoe 1998).

Anthropological perspectives on Indigenous knowledge have often been contentious, raising questions regarding, for example, the relative roles of material, social, and symbolic explanations of human-environment relations; the ontological status of social relations between humans and other species; and the extent to which anthropologists have reframed, simplified, or misrepresented Indigenous knowledge to fit within their discipline (Orr et al. 2015; Pualani Louis 2007). Anthropologists have also assumed a novel and often contested political status, through their role as mediators between Indigenous people and more powerful actors. This shift has implied a move from the apolitical study of local human-environment relations to consideration of politics and power, influenced by the work of critical scholars such as Arturo Escobar (1995). There are, therefore, a variety of ways in which historians might consider the relations between anthropology and Indigenous knowledge: in terms of the conceptual development of the discipline, the formation of research methods and practices (such as participant observation), or the evolving political implications of anthropology.

Study of these relations between science and Indigenous knowledge has been accompanied by reconsideration of the global diffusion of science, displaced by ideas of mobility, circulation, and transformation that challenge distinctions between center and periphery (Anderson 2018). The situated nature of science also challenges the distinction between local and global: what might be considered "global" (such as scientific perspectives) reflects the influence of local circumstances, while the "local" (such as Indigenous knowledge) gains its meaning at least in part in relation to global factors. Related issues include the implications of "decolonizing" knowledge; allocation of access and control over resources; intellectual property rights; and the challenges of communicating between science and oral knowledge (Hayden 2004). Various works have reframed the "globalization" of science as a contingent and heterogeneous assemblage, effectively "indigenizing" science in the contact zones shared by different forms of knowledge (Anderson 2020; Delbourgo 2019; Sim 2021). However, these contact zones must include not just the specific places where newcomers and Indigenous people interact, but all the sites where these relations are negotiated, such as the museums where field specimens are classified, and the scientific societies that serve as gatekeepers for disciplinary knowledge (Schiebinger 2011).

Anthropologists and historical ecologists have demonstrated that "pristine" habitats such as the Amazon Basin and North America before European settlement in fact exhibit long histories of manipulation by Indigenous peoples. In doing so, they have made visible not only the physical traces of these activities but also the knowledge that guided these activities (Denevan 2011; Raffles and WinklerPrins 2003). These conclusions continue to challenge biodiversity conservation and its concern for a nature defined in terms of the absence of humans.

Throughout this history of encounters between different forms of knowledge, ideas about race have been a continuing presence. Shifting imperial geographies – as seen, for example, in the British Empire – encouraged novel ideas about the relations between race and environments, with the capacity to develop technologies in response to local diseases interpreted in terms of racial differences, thereby linking knowledge, adaptation, and racial identity (Benson 2020, 48–77). Elements of critical race theory, such as the concept of "whiteness," provide one way of thinking about these histories of knowledge and racial thought, including racialized distinctions between science and Indigenous knowledge. Whiteness was not just about skin color but about how colonial officials, scientists, and doctors evaluated a range of characteristics and associated identities in terms of racial theories. As an identity defined as racially "unmarked" and thus the normal against which other identities are evaluated, whiteness became both an interpretative framework and a set of power relations (Garner 2007; Kobayashi 2003). It also maintained an analogical relationship with science: both asserted their status as unmarked and objective – and thus, as defined in terms of absence, whether of racial identity or of bias. Both also exercised through social relationships the power that accompanies this status, in ways that are time- and place-specific. Today, science, by asserting an absence of racial identity, continues to tacitly affirm its whiteness, while Indigenous knowledge is accorded an identity defined in terms of both race and place.

Historical perceptions of Indigenous knowledge can therefore be understood in terms of the situated construction of the authority of knowledge: the roles of places, practices, and tools, of professional and disciplinary communities and institutions, and of power in the making of what becomes defined as truth. The concept of Indigenous knowledge itself – including the tendency to collect and examine it in forms detached from its social and cultural contexts – reflects the limits implied by how scientists recognize evidence, authority, and objects of knowledge (Cameron et al. 2014). Reframing the history of science as the history of knowledge is one response: acknowledging that science is only one among many ways of knowing the world (Beattie and Morgan 2021). However, this reframing must also acknowledge the continuing influence of the assumptions and political commitments that underpin the distinctive authority of science.

4 Reconsidering Contemporary Indigenous Knowledge

Historical analyses of Indigenous knowledge exist in dialogue with studies by scientists, scholars, and activists of the contemporary status of Indigenous knowledge. The latter have often been framed in empirical terms: evaluating the accuracy of Indigenous knowledge in relation to reality as defined by science. A related strategy has been to assert its practical benefits – as seen, for example, in the conservation of traditional crop varieties, or the species and habitats of Indigenous-managed territories. This perspective received prominent notice at the 1992 Earth Summit, when the contribution of Indigenous knowledge to biodiversity conservation was highlighted. These analyses frame Indigenous knowledge in terms of a pragmatic realism: its benefits confirm its correspondence with empirical reality. Analyses of intellectual property issues raised by the application of Indigenous knowledge to bioprospecting are grounded in a similar perspective: useful species and knowledge of them are both viewed as distinctively valuable resources (Hayden 2004; Osseo-Asare 2013).

Realist perspectives on Indigenous knowledge have also drawn on ideas from ecological science. Ecologists have reconsidered assumptions of equilibrium; instead, change and disturbance have been recognized as intrinsic to ecological systems even in the absence of human interventions. A related development has been the formation by C. S. Holling and colleagues of the concepts of adaptive management and ecological resilience. These concepts imply that scientific knowledge is necessarily incomplete; that "surprises" are always possible; that resilience not control should be the goal; and that management must allow for continuous learning. Some scientists see in these concepts a ratification of Indigenous knowledge, particularly in how it exhibits adaptation to nonequilibrium conditions – more so than resource management regimes founded on control. In effect, Indigenous knowledge has been conceived as a way of working with, rather than against, nature. Numerous scholars have described how systems of local resource management are often highly adapted to more variable environments. For example, in sub-Saharan Africa, mobile pastoralism is more suited to rangelands exhibiting variations in rainfall and other conditions than are resource management systems grounded in assumptions (often imported from temperate regions) of equilibrium (Davis 2016). A similar argument has been made regarding Indigenous use of fire to manage ecosystems (Kull 2000). Overall, Indigenous knowledge has gained authority by according with, and even anticipating, innovations in ecological theory, as well as with doubts regarding the capacity of science to impose control over nature.

While persuasive in public, political, and scientific contexts, these pragmatic perspectives are only a starting point for understanding the contemporary status of Indigenous knowledge. In the postcolonial era, development agencies became of central significance to Indigenous knowledge: evaluating, extracting, and applying it in practical and regulatory contexts across a variety of scales, from local institutions to multilateral organizations such as the World Bank (Brokensha et al. 1980). This reflected several factors: the influence of anthropological research; recognition of the practical value of this knowledge in the context of agriculture, conservation, and forestry (encouraging the formation of, for example, agroforestry to codify longstanding practices of mixed agriculture); advocacy on behalf of Indigenous peoples; and alternative framings of the relations between states and peoples, including communities and the subaltern. By the late 1970s, critiques of development itself were emerging, including the assumption of a single path through science to modernity. Nevertheless, in development practice, Indigenous knowledge remained framed in terms of science: evaluated for accuracy in relation to scientific knowledge, defined as specifically local knowledge to distinguish it from "universal" science, translated into scientific terms (as in agroforestry, for example) to be viable in the modern world, and adhering to a firm distinction between nature and culture (Agrawal 2002). This framing gained prominence through the concept of sustainable development: while the World Commission of Environment and Development's report, *Our Common Future* (1987), expressed sympathy for Indigenous peoples, it also continued a decades-long tradition of presenting their knowledge as an endangered historical relic that had only survived through isolation from the modern world (O'Brien 2002). The critique of this perspective that O'Brien references parallels Kim TallBear's analysis of efforts to present Indigenous peoples' identities and knowledge as historical relics that can be validated (and therefore circumscribed) through genetic science (TallBear 2013).

This reconstruction of Indigenous knowledge in terms of science and science-based development became evident in its codification as traditional ecological knowledge (TEK). The concept of TEK epitomized the extraction of Indigenous knowledge from its cultural context, to be commodified into a form compatible with environmental administration and resource management (Nelson 2005). TEK exemplifies how these categories must be interpreted not as self-evidently distinct forms of knowledge but in terms of specific institutional, political, and social contexts.

There is also a parallel history of public discourse regarding Indigenous knowledge. In America, notions of native "wisdom" and the "ecological Indian" as an authentic alternative to industrial society were invoked by Ernest Thompson Seton, Henry David Thoreau, John Muir, and other writers, becoming fixtures of popular culture – eventually invoked in the era of environmentalism through the image of the "crying Indian": an icon conducive to mass persuasion but disengaged from Indigenous realities (Dunaway 2015). In defense of the forests and rights of Indigenous peoples such as the Penan, environmentalists reframed their knowledge in terms compatible with a Western romantic tradition (Brosius 1997). Critiques of science (including the notion of the history of science as a story of progress) and modern society became projected onto Indigenous knowledge, now designated as a source of new environmental attitudes (Krech 1999; Smithers 2015). Viewed as timeless and outside modernity, Indigenous knowledge became defined in oppositional terms that eschewed any attempt to engage with Indigenous people and their histories on their own terms (Nadasdy 2005).

5 Resurgence

Indigenous people have been essential to the formation of Indigenous knowledge as a research area, negotiating its relationship with science and dominant institutions. They have done so in conventional research contexts, including universities, but also in communities and Indigenous territories, state and international agencies, and the wider public sphere. In many instances, these contexts have formed sites of struggle over both material environments and contrasting ways of knowing and living in places. Indigenous peoples' assertions and interpretations of their knowledge extend the historiographic agenda of understanding knowledge production in diverse environments, but they also challenge academic history by questioning the boundaries between knowledge and its political implications and between knowledge and other aspects of society, including the ethical responsibilities that link humans, other species, and the wider environment. (There are important parallels between this perspective and work that has identified the racial and colonial violence embedded in the sciences of the Anthropocene, including geology [Yusoff 2018].) The interest therefore is in moving beyond description and analysis, towards political action. This work can be framed in terms of regional environmental histories, including accounts of struggles over landscapes and environmental features and the consequences of dams, pipelines, and other energy facilities, resource developments, and military activities. However, environmental historians have been slow to incorporate these dimensions of Indigenous knowledge into their discipline: by one account, scholarship in American environmental history has tended to assume that Indigenous people have not been significant actors after 1900 (Rosier 2013).

One way in which Indigenous people have asserted these perspectives on knowledge and its boundaries has been by insisting that research adhere to ethical relations with communities: that it respect Indigenous points of view, incorporate ethical practices, and contribute to community well-being. Accordingly, Indigenous communities and agencies have formed protocols and structures of review and accountability to facilitate respectful and ethical relationships between researchers and communities, founded on collaboration rather than extraction of knowledge, and an acknowledgment of the political character of research (Kovach 2009; Latulippe and Klenk 2020; Liboiron 2021; Smith 2021).

However, ethical research principles – like the notion of knowledge itself as something distinct from daily life – also reflect an accommodation to science-based epistemological principles. Similarly, Indigenous peoples' efforts to assert claims to land and resources must still adhere to colonial requirements, including demonstration that their stewardship of land and animals meets the standards of science-based environmental management. Participatory mapping techniques that draw on local knowledge to assert territorial authority are similarly embedded in state-based frameworks of territorial authority. In practice, Indigenous knowledge has often only been recognized when it has been backed by access to or control over land – adhering to rather than challenging conventional standards of authority.

Beyond its assertions in specific contexts and communities, many Indigenous authors have presented more general perspectives on the meanings and politics of Indigenous knowledge. In the North American context, these have included, among others, Vine Deloria, Deborah McGregor, Winona LaDuke, and Leanne Simpson. They have stressed that Indigenous knowledge is about more than epistemology: it implies ethical commitments and forms of action, as well as alternative ontologies to those asserted by "Western" territorial authority and by science (such as social and ethical relations between humans and other species) (Liboiron 2021; Whyte 2017).

In the context of this chapter, the significance of these authors is that they challenge conventional assumptions about academic research (including the history of science): how evidence should be interpreted; that this research is, or even can be, politically neutral (instead, it expresses the values of the settler colonial state); narratives of the progressive formation and diffusion of scientific knowledge (rather, this knowledge has been founded on the epistemic violence inherent in the marginalization of Indigenous perspectives); and the colonial construction of the signifier "Indigenous" itself (Parrenas 2020). To decolonize both science and history, it is therefore necessary to affirm the autonomy of other knowledge traditions and resist their extraction and inclusion within dominant ways of knowing (Radcliffe 2017b). This implies a critical perspective on postcolonial historians' analyses of exchanges between science and Indigenous knowledge and the formation of hybrid knowledge: this, it is argued, downplays the exploitation of Indigenous people and their knowledge throughout the history of science. Instead, they stress the incommensurability of Indigenous and scientific knowledge, including differences in what counts as evidence, explanation, and authorship, ways in which knowledge relates to communities, and perceptions of time and history (Tuck and Yang 2012). One expression of this has been efforts to "decolonize" or "indigenize" the university, to provide space for indigenous ways of knowing on their own terms (Latulippe and Klenk 2020).

The global extension of Indigeneity (the expression of distinctive Indigenous identities) – evident in, for example, the United Nations Statement on Indigenous Rights, and the increasing presence of Indigenous peoples in international conservation affairs – has its own implications for the historiography of environmental science. Like Indigenous knowledge, indigeneity is neither imposed nor an autonomous product but the outcome of interactions between how a people view themselves and their contexts. Although in part the product of globalization, Indigeneity is also geographically specific. Key aspects of the concept, including a tendency towards essentialism and an unambiguous demarcation between Indigenous and non-Indigenous identities, are rooted in settler societies, particularly North America. In contrast, Indigeneity carries different meanings in, for example, Southeast Asia or in Africa (Mavhunga 2018; Parrenas 2020; Sim 2021). Thus, the historiography of Indigenous knowledge as an expression of indigeneity must encompass how it has circulated across diverse geographical, political, and cultural contexts.

6 Histories of Arctic Indigenous Knowledge

Situated in specific places and communities, Indigenous knowledge should be studied accordingly. The Canadian Arctic presents distinctive opportunities to explore its situated nature, including its historical relations with scientific knowledge. These relations have exhibited in distinctive ways many of the issues discussed in this chapter; they therefore illustrate the value of studying the history of Indigenous knowledge in regions with unfamiliar, intemperate environments that embody a challenge to assumptions of scientific expansion and domination. Knowledge in the Canadian Arctic has been shaped by both local circumstances, including the distinctive environment and political, economic, and institutional conditions, and by factors that are not specific to the region, including transnational scientific disciplines, the global resource economy, and relations between nations. The history of Indigenous knowledge in the Canadian Arctic is also relevant to other issues, such as how climate change and the transition to the Anthropocene involve not just transformation of the physical environment but social, ethical, and human rights concerns.

For centuries, the Canadian Arctic has been understood in terms of the meanings and values of powers located elsewhere: as imperial territory, a place of economic opportunity, or a hoped-for passage to more profitable regions. The twentieth-century history of Arctic knowledge has been marked by the intersection of colonial and scientific authority: evident in the extension of administrative control, the assertion of military and strategic interests, the orientation of economic activity towards global markets, and the redefinition of the region as an environment vulnerable to industrial society. Since the 1970s, critiques of this intersection have been implicit in political developments: the devolution of authority to local governments, renewed interest in local resources to meet local needs, and views of the Arctic environment that incorporate Indigenous perspectives. Throughout, scientists have defined the Arctic in diverse ways: as distinctive (hence requiring special ideas and techniques) or as a place similar to more familiar places elsewhere (albeit colder or less productive) (Bocking and Martin 2017).

Throughout much of the history of the Arctic, Indigenous knowledge occupied a tacit position: essential, yet often unremarked upon. Explorers had commonly lived and traveled with Indigenous people, with some of the most celebrated, including Robert Peary and Vilhjalmur Stefansson, considering Indigenous knowledge a requirement for survival even as they also, for various reasons, including racial attitudes, a sense of cultural superiority, or the need to assert (white) credibility, obscured the roles of Indigenous people in exploration (Dick 2001; Martin 2020). In the early twentieth century, scientists' attitudes towards the Arctic might be grounded in science, yet also constitute, at times, a reaction against modernity and an openness to "indigenous wisdom" – evident in how Inuit ingenuity became a staple of Arctic travel accounts. But after the Second World War, and for the reasons detailed below, the history of Indigenous knowledge in the Canadian Arctic became a history of erasure. No government agency, and few scientists, accorded more than token acknowledgement to Indigenous knowledge. It was viewed as anecdotal and subjective, failing to meet scientific standards. Government policies commonly incorporated a dismissal of Indigenous knowledge. They did so, in part, by extending a tradition of intervention in hunting cultures, justified in terms of conservation. Intervention also encompassed relocation of communities, which both denied a community's knowledge and rendered it irrelevant. Hunting crises, incidents of extreme hunger, and transformation of settlement encouraged the view that whatever knowledge Indigenous peoples once possessed could not have survived the transformation of postwar Arctic society. Instead, the culture of knowledge of the land was seen as a vestigial remnant awaiting assimilation into the dominant culture.

Wildlife science played a central role in postwar perceptions of Arctic Indigenous knowledge. The Canadian Wildlife Service became essential to an interventionist program of wildlife management, asserting expert supervision of Indigenous-wildlife relations, while denying the existence of Indigenous knowledge and practices. Wildlife science and management contradicted Indigenous perspectives through a focus on populations, physiology, and range ecology, and by assuming that each hunter acts as an individual in competition with others and so must be regulated by an expert authority. Scientific practices reinforced this view, emphasizing aerial survey of populations – an approach consistent with practices within the North American research community, but that reinforced the boundary between scientific and Indigenous knowledge. Indigenous hunters were also thought to be prone to waste: corrupted by exposure to modern society, they were likely to hunt to excess once no longer limited by primitive technology (Kulchyski and Tester 2007; Sandlos 2007).

Changing social relations of Arctic science influenced perceptions of Indigenous knowledge. New scientific facilities, including aircraft and research laboratories, meant that scientists no longer needed to live among Indigenous people, learn their techniques for travel and survival, or, indeed, have any contact with them. The "view from above" provided by aircraft enabled scientists to extend their perspective over larger areas, breaking with a previous reliance on ground study. In doing so, scientists asserted a new "white" identity: seemingly objective, disengaged from local social and cultural circumstances – and one from which Indigenous people were excluded. Just as whiteness could be constructed as a category lacking racial identity, science could be asserted as pure and objective knowledge, as the standard against which other forms of knowledge could be evaluated, and as excluding aspects of other forms of knowledge that had an explicit racial identity (Kobayashi 2003; Garner 2007). And so, as scientists took to the air and to laboratories, Indigenous knowledge itself was commonly dismissed as an historical relic: subjective and partial.

These views of hunting, social conditions, and Indigenous knowledge were bound up with ideas about race. Interventions in community life reflected a desire, informed by attitudes regarding race and space, to transform and modernize Arctic society (Usher 2004). Arctic racial identities were defined not only in terms of ancestry but also, particularly for Inuit, in terms of relationships with the land. The perception that these relationships had been ruptured by colonization and economic and technological transformation could thus justify their assimilation into the dominant society: a process encouraged by the close association between white identity and state objectives, and, more generally, between whiteness and progress itself. In contrast, wildlife scientists argued that Indigenous hunters had little capacity for conservation. Wastefulness towards wildlife became, therefore, a racial characteristic. The argument that science was the only reliable source of knowledge was also framed, in part, in racial terms: it was a nonracial perspective, something that was exclusive to non-Indigenous (that is, white) people. This racialization of knowledge could then contribute to justifying authority over Indigenous livelihoods (Kobayashi 2003). Race was also implicated in contrasting views of whether Inuit adaptation to Arctic environmental conditions was inherent in their biology or the product of their ingenuity.

Arctic Indigenous peoples have long attracted the interest of anthropologists: unsullied by modernity, but able to adapt to a difficult environment – in Harvey Feit's words, research subjects at the "absolute zero of human culture" (quoted in Anderson 2004, 1). In the 1880s, Franz Boas saw the Arctic and the geographical isolation of Inuit as an opportunity to test theories of environment-culture relations (Powell 2015). During the interwar period, Knud Rasmussen studied Inuit intellectual culture, tracing their shared identity across much of the circumpolar region (Bown 2015). Other anthropological interests specific to the region have included Inuit travel and mapmaking, the impacts of European contact on Indigenous cultures, and their transition to a modern economy. Arctic anthropology has also reflected wider currents in the discipline: an interest in adaptation, in Indigenous human-environment relations, and in research methods that require immersion in community life through listening and participating in activities. The region became a site for exploring and applying ideas developed within cultural ecology, ethnoscience, and other fields of anthropology, and for developing new roles such as that of mediator between states and Indigenous populations (Balikci 1989; Wenzel 1999).

By the 1960s, some anthropologists in the Arctic had begun to examine Indigenous knowledge on its own terms. A significant early work was Richard Nelson's *Hunters of the Northern Ice*: a study of Inuit knowledge and adaptation (Nelson 1969). Eleanor Leacock also drew attention to Indigenous knowledge, noting the "prodigious knowledge" of elders. Milton Freeman studied Inuit perceptions and knowledge of the environment, and their use of this knowledge in decisions regarding settlement and hunting. In other research, he demonstrated the value of Indigenous knowledge as a source of information about species that was often superior to that provided by science. Studies of Arctic Indigenous knowledge proliferated during the 1970s. Henry Lewis described the importance of fire to Indigenous strategies of adaptation to the boreal forest. Harvey Feit worked with the Quebec Cree, describing their knowledge of wildlife populations and other ecological phenomena, and situating this knowledge within their worldview, in which animals and other aspects of nature are seen as "like persons," causality is personal, and hunting success is influenced by the hunter's previous behavior. Julie Cruikshank's studies of stories of Athapaskan women demonstrated the importance of landscape and species to their knowledge; she also found a remarkable persistence and conservatism in these accounts, and an incommensurability with science – as expressed by the question, "Do glaciers listen?" Fikret Berkes described how subarctic fishers rely on both knowledge and cooperation to adapt to unpredictable environmental changes. In the following decades, Berkes would study many aspects of Indigenous knowledge and resource practices, in Canada and elsewhere. Finally, Hugh Brody described how Indigenous cultures and social systems drew on environmental knowledge to construct maps of meaning (Berkes 1999; Bocking 2011; Cruikshank 2005). These anthropologists therefore pursued a variety of strategies: emphasizing the scientific credibility or incommensurability of Indigenous knowledge, its adaptive significance or its assertion of ethical relations between humans and other species.

The political history of the region, including land claim negotiations, influenced studies of Indigenous knowledge. Because an effective negotiating position required establishing that certain lands had long been occupied, there arose a demand to document Indigenous land use and knowledge, drawing on information provided by hunters, trappers, and other land users. The Inuit Land Use and Occupancy Study, initiated in 1973 and directed by Freeman, used map biographies and other techniques to document land use. These techniques served not so much to represent the cultural dimensions of knowledge as to translate it into a format suited to negotiations with government. They therefore demonstrated the complexities of recognizing Indigenous knowledge: it combined the technical process of mapping, the empowerment of people to enable them to assert their regional authority, and their incorporation within a still-dominant system of territorial authority (Bryan 2011). Land claim negotiations also provided opportunities for anthropologists to construct their role as mediator between the state and Arctic peoples. This role was encouraged by changes in the institutional position of anthropology, including its increased study in universities (permitting greater independence from federal policies). But while anthropologists pursued this role, distrust rooted in the history of northern colonialism, as well as a desire to speak directly to society generated resistance among Arctic peoples (Asch 2001). More generally, the shifting political contexts of research have meant that anthropologists no longer have the power to decide themselves where and how they will do fieldwork; instead, research strategies must be negotiated locally and be based on collaboration. Overall, anthropological perspectives on Indigenous knowledge have continually evolved, in response to both wider disciplinary concerns and institutional and political conditions specific to the Arctic.

These factors shaping the work of anthropologists have also been important for environmental scientists. They too saw themselves not merely as Arctic scientists but as members of larger scientific disciplines. This cosmopolitan identity has had diverse implications for relations with Indigenous knowledge. Scientists seeking to develop ideas particular to the Arctic have generally been more receptive to Indigenous knowledge. In contrast, researchers within disciplinary frameworks defined outside the region have tended to be less receptive. Thus, while the colonial regime in the Arctic (as elsewhere) drew on universalizing discourses of progress, modernity, and rationality, it was also shaped by local factors and by scientific communities that had distinctive views on the relations between Indigenous people and science. In short, this regime, in terms of the evolving status of Indigenous knowledge, was not monolithic, but fractured, heterogeneous, and open to diverse influences from both within and outside the scientific community. Collaborations between science and Indigenous knowledge have exhibited how historical inequalities between knowledge systems can be perpetuated. This has been evident in comanagement institutions: although intended to bring together these knowledge systems, they have also, in practice, embedded their unequal authority, with Indigenous knowledge obliged to frame its advice in scientific terms (Nadasdy 2003).

Empirical experience has been a significant local factor shaping Arctic scientists' perspectives on Indigenous knowledge. For example, in the 1970s, Indigenous observations of caribou began to be compared against scientific predictions. The result was often diminished confidence in science and more credibility for Indigenous knowledge. Once biologists began to learn from hunters that, for example, older males play an important role in protecting the herd against predators, they became more open to accepting the empirical validity of Indigenous knowledge. Overall, Indigenous knowledge provided for scientists an increasingly significant source of empirical information (Ferguson et al. 1998; Freeman 1992).

Other Arctic researchers also drew on aspects of Indigenous knowledge. One example was the ecologist William Pruitt. In his research, he used Inuit terms to designate different types of snow. Describing them as the "wisest instructors" on the nature of snow, he argued that their terms enabled far more precise description than was possible with English words (Pruitt 1978). Polar bear researchers used Indigenous knowledge to describe bear behavior, locate their dens, and generate insights into the behavior of seals (Smith and Stirling 1975). Thus, scientists often viewed Indigenous knowledge as a source of knowledge and observations not available through their own research. They did so by shifting the boundary of science so as to include Indigenous knowledge, even while excluding those aspects that would not contribute to their scientific goals, such as its social, cultural, and racial dimensions. Perspectives on race and whiteness were implicated in this shifting boundary between science and Indigenous knowledge. Thus, scientists' recognition of the value of Indigenous knowledge did not come merely through force of evidence. Acceptance of knowledge claims also required a supportive framework of institutions and disciplines, new research practices, including collaborative fieldwork involving scientists and Inuit, and a shift in perceptions regarding the relations between "white" science and knowledge that is tied to racial identity.

The history of Arctic resource development also illustrates some factors influencing perceptions of Indigenous knowledge. Much of the documentation of Indigenous knowledge in the region has been tied to development, including land claims and regulatory activities, as well as efforts to deal with local hazards. It has therefore exhibited the problematic ways in which Indigenous knowledge has been assembled and translated into forms compatible with modern perspectives. Some accounts have acknowledged the disruptions implicit in removing knowledge from its social context, coding and reifying it as TEK and incorporating it within bureaucratic approaches to decision-making. This has been especially evident in environmental impact assessments, with divergent views of these assessments – as a scientific or a civic process – implying different roles for Indigenous knowledge (O'Faircheallaigh 2007). Thomas Berger's Mackenzie Valley Pipeline Inquiry of 1974–1977 emphasized the deep experience and insights of Indigenous peoples regarding their homeland, combined with critical perspectives on other aspects of knowledge, including the roles of scientists in relation to public policy, and the relations between science and values (Berger 1977). It therefore provided an influential demonstration of how attitudes towards Indigenous knowledge can be interpreted in relation to other perspectives on science in its social contexts.

Political changes across the Arctic have encouraged communities to collect their own knowledge, and in doing so, to assert their rights to the land (McDonald et al. 1997).

These studies have shown that northern land and waters remain crucially important to Indigenous communities, as a basis for both material well-being and cultural and social integrity. In doing so, they have challenged a strictly biological perspective on renewable resources and impacts of industrialization (and its accompanying denial, consistent with attitudes of whiteness, of the significance of racial and cultural identities). This has also implied shifting views of the Arctic itself: from an "empty" resource frontier or a "pristine" wilderness, to that of homeland, long occupied by humans, and populated with species that have their own intentions and social relations. These changes demonstrate the political significance of local control over documentation of Indigenous knowledge (Nuttall 2010). The politics of identity have thus in the Arctic often been negotiated in terms of knowledge.

Communities also began to insist that their knowledge be incorporated into research, with scientists required to demonstrate an awareness of Indigenous knowledge before receiving permission to do research. Concern for research ethics and community relations is also producing new models of research that contribute to local self-determination by transmitting knowledge to younger generations and challenging assumptions about who has the right to define the Arctic environment. Community histories are based on local memories and oral histories – forming an alternative model of historiography, and illustrating how changing social and power relations have many implications for how Indigenous knowledge is understood (QTC 2014).

Finally, the Arctic presents distinctive issues of scale that have become evident in the relations between Indigenous knowledge and the global earth and environmental sciences. Inuit gained in the late 1950s a presence in early Cold War understanding of the global environment with concerns regarding the local health effects of radioactive fallout (see chapter "▶ Cold War and Earth Sciences" by Ron Doel). By the 1980s, concerns about contaminants that had traveled long distances focused on persistent organic pollutants. Research on these hazards initially framed Inuit as objects of research. However, by the 1990s, amidst changing political circumstances and research practices, they were asserting an increasing role in these studies: setting priorities and drawing on their own knowledge of the environment. Although global models, satellite-based remote sensing, and ice core studies have framed knowledge of the Arctic climate in terms of large spatial and temporal scales, Inuit have also gained a role in climate research, providing insights into weather, sea ice, and other environmental features on scales more relevant to human activities (Huntington 2011). Inuit participation in circumpolar environmental and political affairs, through the Inuit Circumpolar Council, the Arctic Environmental Protection Strategy, and the Arctic Council, has complemented their presence in studies of environmental change. They have stressed in these fora not merely the relevance of Indigenous knowledge, but that climate change is not just a physical transformation but a moral and ethical issue that can be understood in terms of Inuit concepts such as *Sila* (a worldview that encompasses weather, climate, the Arctic environment, and their relations to Inuit identity and ways of life), as well as concepts understood in universal terms, including human rights and environmental justice (Krupnik and Jolly 2002; Watt-Cloutier 2015).

However, scientific perspectives on Arctic and global environmental change can also marginalize Indigenous knowledge. This is seen, for example, in discussions of "opening up" the Arctic Ocean to resource development as it loses its permanent ice cover: a prospect that risks returning the region to its colonial image as an otherwise uninhabited place open to claims by outside interests. It is also seen in the continuing tendency to present Indigenous knowledge of the Arctic environment separately from the dominant scientific narrative of global environmental transformation. As in other arenas of global environmental science, this marginalizing of Indigenous knowledge through epistemic dominance can constitute its own form of slow violence (O'Lear 2016). The continuing challenge of including Indigenous knowledge in perspectives on the Arctic environment speaks to how different ideas about what counts as knowledge and history are at stake.

7 Conclusion

The historiography of Indigenous knowledge parallels its evolving roles in contemporary environmental affairs, where it asserts intellectual and moral authority but still exhibits the consequences of a history of erasure. This ambiguous status has been reinforced by the deeply rooted idea of science as the foundation of modernity, setting the terms by which rationality itself is defined. This historiography also aligns with a significant theme in the history of science: the relations between forms of knowledge with contrasting methodological, epistemological, and political implications, evident in efforts to extract and combine knowledges, as well as assert boundaries, amidst evolving political conditions and contested racial identities. Throughout, Indigenous knowledge has not been "discovered," but constructed, most often in forms compatible with scientific knowledge (including its reification as distinct from its social contexts), but also, in Indigenous contexts, in ways that challenge this knowledge.

The history of perceptions of Indigenous knowledge is relevant to study of other aspects of the history of knowledge of the earth and environment. Among these are the roles of disciplines and disciplinary boundaries in shaping scientific attitudes (evident in how there is no single "scientific" mind regarding Indigenous knowledge); of particular places and regions (as seen in the distinctive significance of Indigenous knowledge in the Arctic and in the diverse scales and geographies of nature, peoples and knowledge more generally); of racial identities and their consequences for knowledge; and of the relations between knowledge and power – evident in the contexts of empires, international development, and modernity. However, this history also challenges the agenda of the history of science. As a form of knowledge inseparable from issues of identity and of access to and control over nature and peoples, it calls into question the study of the history of knowledge as distinct from its contexts, as well as assumptions about history itself: its evidence, exclusions, and purpose.

References

Adams, William M. 2003. Nature and the colonial mind. In *Decolonizing nature: Strategies for conservation in a post-colonial era*, ed. William M. Adams and Martin Mulligan, 16–50. Abingdon: Earthscan.

Agrawal, Arun. 1995. Dismantling the divide between Indigenous and scientific knowledge. *Development and Change* 26: 413–439.

———. 2002. Indigenous knowledge and the politics of classification. *International Social Science Journal* 54: 287–297.

Anderson, David G. 2004. Reindeer, caribou and "fairy stories" of state power. In *Cultivating Arctic landscapes*, ed. David G. Anderson and Mark Nuttall, 1–16. New York/Oxford: Berghahn Books.

Anderson, Warwick. 2009. From conjugated knowledge to conjugated subjects: Science and globalization, or postcolonial studies of science? *Postcolonial Studies* 12 (4): 389–400.

———. 2018. Remembering the spread of western science. *Historical Records of Australian Science* 29: 73–81.

———. 2020. Finding decolonial metaphors in postcolonial histories. *History and Theory* 59 (3): 430–438.

Anderson, Warwick, and V. Adams. 2008. Pramoedya's chickens: Postcolonial studies of technoscience. In *The handbook of science and technology studies*, ed. E. Hackett, 181–204. Cambridge, MA: MIT Press.

Asch, Michael. 2001. Indigenous self-determination and applied anthropology in Canada: Finding a place to stand. *Anthropologica* 43: 201–220.

Balikci, A. 1989. Ethnography and theory in the Canadian Arctic. *Études/Inuit/Studies* 13 (2): 103–111.

Beattie, James, and Ruth A. Morgan. 2021. From history of science to history of knowledge? Themes and perspectives in colonial Australasia. *History Compass* 19: e12654. https://doi.org/10.1111/hic3.12654.

Benson, Etienne S. 2020. *Surroundings: A history of environments and environmentalisms*. Chicago: University of Chicago Press.

Berger, Thomas R. 1977. *Northern frontier, northern homeland: The report of the Mackenzie Valley Pipeline Inquiry*. Ottawa: Minister of Supply and Services.

Berkes, Fikret. 1999. *Sacred ecology: Traditional ecological knowledge and resource management*. Philadelphia: Taylor and Francis.

Bocking, Stephen. 2011. Indigenous knowledge and the history of science, race, and colonial authority in northern Canada. In *Rethinking the great white north: Race, nature, and the historical geographies of whiteness in Canada*, ed. A. Baldwin, L. Cameron, and A. Kobayashi, 39–61. Vancouver: UBC Press.

Bocking, Stephen, and Brad Martin, eds. 2017. *Ice blink: Navigating northern environmental history*. Calgary: University of Calgary Press.

Bown, Stephen R. 2015. *White eskimo: Knud Rasmussen's fearless journey into the heart of the Arctic*. Madeira Park: Douglas and McIntyre.

Brokensha, D., D. Warren, and O. Werner, eds. 1980. *Indigenous knowledge systems and development*. Washington, DC: University Press of America.

Brosius, J. Peter. 1997. Endangered forest, endangered people: Environmentalist representations of Indigenous knowledge. *Human Ecology* 25 (1): 47–69.

Bryan, Joe. 2009. Where would we be without them? Knowledge, space and power in Indigenous politics. *Futures* 41: 24–32.

———. 2011. Walking the line: Participatory mapping, Indigenous rights, and neoliberalism. *Geoforum* 42: 40–50.

Cameron, Emilie, Sarah de Leeuw, and Caroline Desbiens. 2014. Indigeneity and ontology. *Cultural Geographies* 21 (1): 19–26.

Chambers, David Wade, and Richard Gillespie. 2001. Locality in the history of science: Colonial science, technoscience, and Indigenous knowledge. *Osiris* 15: 221–240.

Cooper, Alix. 2007. *Inventing the Indigenous: Local knowledge and natural history in early modern Europe*. Cambridge: Cambridge University Press.

Cruikshank, Julia. 2005. *Do glaciers listen? Local knowledge, colonial encounters, and social imagination*. Vancouver: UBC Press.

Davis, Diana K. 2016. *The arid lands: History, power, knowledge*. Cambridge, MA: MIT Press.

Delbourgo, James. 2019. The knowing world: A new global history of science. *History of Science* 57 (3): 373–399.

Denevan, William M. 2011. The "pristine myth" revisited. *The Geographical Review* 101 (4): 576–591.

Dick, Lyle. 2001. *Muskox land: Ellesmere Island in the age of contact*. Calgary: University of Calgary Press.

Drayton, Richard. 2000. *Nature's government: Science, imperial Britain, and the improvement of the world*. New Haven: Yale University Press.

Dunaway, Finis. 2015. *Seeing green: The use and abuse of American environmental images*. Chicago: University of Chicago Press.

Escobar, Arturo. 1995. *Encountering development: The making and unmaking of the Third World*. Princeton: Princeton University Press.

Fairhead, James, and Melissa Leach. 1996. *Misreading the African landscape: Society and ecology in a forest-savanna mosaic*. Cambridge: Cambridge University Press.

Ferguson, M., R. Williamson, and F. Messier. 1998. Inuit knowledge of long-term changes in a population of Arctic tundra caribou. *Arctic* 51 (3): 201–219.

Freeman, M. 1992. Ethnoscience, prevailing science, and Arctic co-operation. *Canadian Papers in Peace Studies* 3: 79–93.

Garner, S. 2007. *Whiteness: An introduction*. London: Routledge.

Gieryn, Thomas. 1999. *Cultural boundaries of science*. Chicago: University of Chicago Press.

Goldman, Mara, Paul Nadasdy, and Matthew Turner, eds. 2011. *Knowing nature: Conversations at the intersection of political ecology and science studies*. Chicago: University of Chicago Press.

Goss, Andrew, ed. 2021. *The Routledge handbook of science and empire*. London: Routledge.

Hayden, Cori. 2004. *When nature goes public: The making and unmaking of bioprospecting in Mexico*. Princeton: Princeton University Press.

Hodge, Joseph Morgan. 2007. *Triumph of the expert: Agrarian doctrines of development and the legacies of British colonialism*. Athens: Ohio University Press.

Huntington, Henry. 2011. The local perspective. *Nature* 478: 182–183.

Kobayashi, A. 2003. The construction of geographical knowledge: Racialization, spatialization. In *Handbook of cultural geography*, ed. K. Anderson, M. Domosh, S. Pile, and N. Thrift, 544–556. London: Sage.

Kohler, Robert. 2011. History of field science: Trends and prospects. In *Knowing global environments: New historical perspectives on the field sciences*, ed. Jeremy Vetter, 212–240. New Brunswick: Rutgers University Press.

Kovach, Margaret. 2009. *Indigenous methodologies: Characteristics, conversations, and contexts*. Toronto: University of Toronto Press.

Krech, Shepard. 1999. *The ecological Indian*. New York: W. W. Norton.

Krupnik, Igor, and Dyanna Jolly, eds. 2002. *The earth is faster now: Indigenous observations of Arctic environmental change*. Fairbanks: Arctic Research Consortium of the United States.

Kulchyski, Peter, and Frank J. Tester. 2007. *Kiumajut (talking back): Game management and Inuit rights, 1900–70*. Vancouver: UBC Press.

Kull, Christian. 2000. Deforestation, erosion, and fire: Degradation myths in the environmental history of Madagascar. *Environment and History* 6: 423–450.

Latulippe, Nicole, and Nicole Klenk. 2020. Making room and moving over: Knowledge co-production, Indigenous knowledge sovereignty and the politics of global environmental change decision-making. *Current Opinion in Environmental Sustainability* 42: 7–14.

Liboiron, Max. 2021. *Pollution is colonialism*. Durham: Duke University Press.
Martin, Peter R. 2020. Indigenous tales of the Beaufort Sea: Arctic exploration and the circulation of geographical knowledge. *Journal of Historical Geography* 67: 24–35.
Mavhunga, Clapperton. 2018. *The mobile workshop: The tsetse fly and African knowledge production*. Cambridge, MA: MIT Press.
McCook, Stuart. 2011. Managing monocultures: Coffee, the coffee rust, and the science of working landscapes. In *Knowing global environments: New historical perspectives on the field sciences*, ed. Jeremy Vetter, 87–107. New Brunswick: Rutgers University Press.
———. 2013. Introduction. *Isis* 104: 773–776.
McDonald, M., L. Arragutainaq, and Z. Novalinga. 1997. *Voices from the bay*. Ottawa/Sanikiluaq: Canadian Arctic Resources Committee/Environmental Committee of Municipality of Sanikiluaq.
Menon, Minakshi. 2022. Indigenous knowledges and colonial sciences in South Asia. *South Asian History and Culture* 13 (1): 1–18.
Munro, Paul. 2020. *Colonial seeds in African Soil: A critical history of forest conservation in Sierra Leone*. New York: Berghahn.
Nadasdy, Paul. 2003. *Hunters and Bureaucrats*. Vancouver: UBC Press.
———. 2005. Transcending the debate over the ecologically noble Indian: Indigenous peoples and environmentalism. *Ethnohistory* 52 (2): 291–331.
Nelson, Richard. 1969. *Hunters of the northern ice*. Chicago: University of Chicago Press.
Nelson, Mark. 2005. Paradigm shifts in aboriginal cultures? Understanding TEK in historical and cultural context. *Canadian Journal of Native Studies* 25: 289–310.
Neumann, Roderick. 1998. *Imposing wilderness: Struggles over livelihood and nature preservation in Africa*. Berkeley: University of California Press.
Nixon, Rob. 2011. *Slow violence and the environmentalism of the poor*. Cambridge: Harvard University Press.
Nuttall, Mark. 2010. Epistemological conflicts and cooperation in the circumpolar north. In *Globalization and the circumpolar north*, ed. Lassi Heininen and Chris Southcott, 149–178. Fairbanks: University of Alaska Press.
Oba, Gufa. 2020. *African environmental crisis: A history of science for development*. New York: Routledge.
O'Brien, William, E. 2002. The nature of shifting cultivation: Stories of harmony, degradation, and redemption. *Human Ecology 30* (4): 483–502.
O'Faircheallaigh, Ciaran. 2007. Environmental agreements, EIA follow-up and aboriginal participation in environmental management: The Canadian experience. *Environmental Impact Assessment Review* 27: 319–342.
O'Lear, Shannon. 2016. Climate science and slow violence: A view from political geography and STS on mobilizing technoscientific ontologies of climate change. *Political Geography* 52: 4–13.
Orr, Yancey, J. Stephen Lansing, and Michael R. Dove. 2015. Environmental anthropology: Systemic perspectives. *Annual Review of Anthropology* 44: 153–168.
Osseo-Asare, Abena Dove. 2013. *Bitter roots: The search for healing plants in Africa*. Chicago: University of Chicago Press.
Parreñas, Juno Salazar. 2020. From colonial Indigenous knowledges to vernacular ideas in southeast Asia. *History and Theory* 59 (3): 413–420.
Powell, Richard. 2015. The study of geography? Franz Boas and his canonical returns. *Journal of Historical Geography* 49: 21–30.
Proctor, Robert N., and Londa Schiebinger, eds. 2008. *Agnotology: The making and unmaking of ignorance*. Stanford: Stanford University Press.
Pruitt, W. 1978. *Boreal ecology*. London: Edward Arnold.
Pualani Louis, Renee. 2007. Can you hear us now? Voices from the margin: Using Indigenous methodologies in geographic research. *Geographical Research* 45 (2): 130–139.
Qikiqtani Truth Commission. 2014. *Community histories 1950–1975: Iqaluit*. Iqaluit: Inhabit Media.

Radcliffe, Sarah A. 2017. Geography and indigeneity I: Indigeneity, coloniality and knowledge. *Progress in Human Geography* 41 (2): 220–229.
———. 2017b. Decolonising geographical knowledges. *Transactions of the Institute of British Geographers* 42: 329–333.
Raffles, Hugh, and Antoinette M.G.A. Winkler Prins. 2003. Further reflections on Amazonian environmental history. *Latin American Research Review* 38 (3): 165–187.
Rosier, Paul C. 2013. "Modern America desperately needs to listen": The emerging Indian in an age of environmental crisis. *Journal of American History* 100 (3): 711–735.
Rudwick, Martin J.S. 1985. *The great Devonian controversy: The shaping of scientific knowledge among gentlemanly specialists*. Chicago: University of Chicago Press.
Sandlos, John. 2007. *Hunters at the margin: Native people and wildlife conservation in the Northwest Territories*. Vancouver: UBC Press.
Schiebinger, Londa. 2008. West Indian abortifacients and the making of ignorance. In *Agnotology: The making and unmaking of ignorance*, ed. Robert N. Proctor and Londa Schiebinger, 149–162. Stanford: Stanford University Press.
———. 2011. Prospecting for drugs: European naturalists in the West Indies. In *The postcolonial science and technology studies reader*, ed. Sandra Harding, 110–126. Durham: Duke University Press.
Scott, James. 1998. *Seeing like a state*. New Haven: Yale University Press.
Sillitoe, P. 1998. The development of Indigenous knowledge: A new applied anthropology. *Current Anthropology* 39 (2): 223–252.
Sim, Timothy. 2021. What should decolonization mean for history of science? https://www.hps.cam.ac.uk/files/decolonise-sim.pdf. Accessed 8 June 2022.
Smith, Linda Tuhiwai. 2021. *Decolonizing methodologies: Research and Indigenous peoples*. 3rd ed. London: Zed Books.
Smith, Thomas G., and Ian Stirling. 1975. The breeding habitat of the ringed seal (*Phoca hispida*). The birth lair and associated structures. *Canadian Journal of Zoology* 53: 1297–1305.
Smithers, Gregory D. 2015. Beyond the "ecological Indian": Environmental politics and traditional ecological knowledge in modern North America. *Environmental History* 20: 83–111.
TallBear, Kim. 2013. *Native American DNA: Tribal belonging and the false promise of genetic science*. Minneapolis: University of Minnesota Press.
Tilley, Helen. 2010. Global histories, vernacular science, and African genealogies; Or, is the history of science ready for the world? *Isis* 101: 110–119.
———. 2011. *Africa as a living laboratory: Empire, development, and the problem of scientific knowledge, 1870–1950*. Chicago: University of Chicago Press.
Tuck, Eve, and K. Wayne Yang. 2012. Decolonization is not a metaphor. *Decolonization: Indigeneity, Education & Society* 1 (1): 1–40.
Usher, Peter J. 2004. Caribou crisis or administrative crisis? Wildlife and aboriginal policies on the barren grounds of Canada, 1947–60. In *Cultivating Arctic landscapes*, ed. D. Anderson and M. Nuttall, 172–199. New York: Berghahn Books.
Vetter, Jeremy. 2011. Introduction. In *Knowing global environments: New historical perspectives on the field sciences*, ed. Jeremy Vetter, 1–16. New Brunswick: Rutgers University Press.
Watt-Cloutier, Sheila. 2015. *The right to be cold*. Toronto: Penguin.
Wenzel, George. 1999. Traditional ecological knowledge and Inuit reflections of TEK research and ethics. *Arctic* 52 (2): 113–124.
White, Richard. 1991. *The middle ground: Indians, empires and republics in the Great Lakes region, 1650–1815*. New York: Cambridge University Press.
Whyte, Kyle. 2017. Indigenous climate change studies: Indigenizing futures, decolonizing the Anthropocene. *English Language Notes* 55 (1–2): 153–162.
Yusoff, Kathryn. 2018. *A billion black anthropocenes or none*. Minneapolis: University of Minnesota Pre.

Earth Sciences from the Perspectives of Science and Technology Studies

Caitlin D. Wylie

Contents

Abstract

The discipline of science and technology studies (STS) can provide historians with methodological approaches and theoretical frameworks to access key aspects of the Earth sciences, such as scientific practice, technology, and social values. This chapter outlines these tools as demonstrated by exemplary STS studies of the Earth sciences.

Keywords

Scientific practice · Technology · Social values

1 Introduction

Analyses of how people construct knowledge and the social implications of this scientific work are key elements of histories of the Earth sciences. The discipline of science and technology studies (STS) has a lot to offer historians investigating these

C. D. Wylie (✉)
Program in Science, Technology, and Society, University of Virginia, Charlottesville, VA, USA
e-mail: wylie@virginia.edu

E. Aronova et al. (eds.), *Handbook of the Historiography of the Earth and Environmental Sciences*, Historiographies of Science,
https://doi.org/10.1007/978-3-031-40799-4_23

topics in the form of theoretical frameworks, research methods, and published sources. In this chapter, I discuss examples of STS-informed studies of the Earth sciences to call attention to their usefulness for historians. In particular, studies of scientific practice, technology, and social values in the Earth sciences offer important methodological and theoretical paths for historians to follow. I hope that these studies and other STS research can inspire more historians of Earth sciences to ask questions about labor, tools, and ethics. Embracing these topics enriches the histories of the Earth sciences and helps those histories inform our understanding of today's global problems as well as the realities of scientific work and knowledge.

Many scholars demonstrate the insights gained by applying STS methods and mindsets to interpret the Earth sciences. Specifically, STS relies on reflexive methodologies that encourage critical engagement with scientists, such as participant observation, community-engaged research, and interdisciplinary collaborations to understand, analyze, and – ideally – improve science (e.g., Felt et al. 2016; Sismondo 2010). These methods involve close analysis of relationships among people, things, and ideas, and raise questions about how those relationships could be otherwise, such as more inclusive, more equitable, and more socially beneficial.

As an example of the benefits of this approach, let us consider my study of how people process fragmented, rock-embedded vertebrate fossils to become scientific specimens from which scientists can learn about past life. This question requires close attention to the workers who do this manual, dirty, seemingly simple work. Based on observation and interviews of fossil preparators, I found that they have no formal training. Instead, they rely on their embodied judgment, creativity, and values (e.g., aesthetics, their own control over their work, and the long-term conservation of fossils) to transform a fossil into an irreplaceable research specimen (Wylie 2015, 2018, 2019a, b, 2021). Yet this dynamic and crucial work goes undocumented in scientific papers and specimen records. These workers, like many others who prepare scientific data, are underrecognized for their contributions to knowledge. Institutions and scientists deem them "technicians" and "support staff," rather than researchers, and thereby justify giving them little credit in publications as well as lower pay and prestige than scientists. Furthermore, the invisibility of this work and these workers erases important insights about fossil specimens, limiting how scientists can interpret them as evidence.

This research began as a typical STS study of a scientific community's work of constructing knowledge. As I collected ethnographic data, though, I wanted to say more than what was happening and my interpretations for why. I wanted to make arguments for how research *should* happen based on what I was seeing, specifically, how workers should be treated, how scientific work should be documented, and, more generally, who should be doing scientific work. I relied on theories from STS and other fields to conceptualize power dynamics in scientific workplaces and ethical issues of credit and documentation in scientific work, such as theories of invisible labor (Shapin 1989, 1994; Law 1994), social capital and expertise (Bourdieu 1993; Collins and Evans 2007), communities of practice (Lave and Wenger 1991), and the classic idea of social construction of knowledge (e.g., Latour and Woolgar 1986; Sismondo 1996). This approach made it possible to apply my evidence about vertebrate paleontology laboratory communities to make broader arguments about the hands-on skills that science relies on, the creativity that informs every step of scientific work and not just theory-building, and the context-dependent forms of power and autonomy that workers negotiate in their everyday interactions. Based on these arguments, I raised questions about who *should* count as a researcher, and how to fairly recognize scientific work that is routinely dismissed as menial. These questions have implications for diversity, equity, and inclusion in science, for instance, as well as a richer understanding of how we know what we know about extinct animals. Thanks to STS, I could think about fossil preparators as a revealing case study of data-constructing workers and the challenges they both face and raise for scientific work and knowledge.

Similar to how observers of the Earth have long used specific specimens and landscapes to represent broader phenomena and global systems, I ask here what histories of the Earth sciences can teach us about social systems of work, knowledge, and responsible research in various places and times. STS provides powerful tools to frame these kinds of questions. In this chapter, I discuss the insights that scholars have already provided by applying STS to analyze the Earth sciences, and I point out opportunities for important future work.

2 STS of the Earth Sciences

Here is a tour of STS-informed studies of the Earth sciences to illustrate the most valuable methods and theoretical frameworks. This is not an exhaustive literature review; instead, I highlight studies that I consider exemplars and inspirations for future work. These authors do not necessarily consider themselves to be STS scholars, demonstrating the versatility of STS methods and theories. In particular, I discuss STS studies that investigate scientific practice, technology, and social benefit of the Earth sciences. These sources are primarily about Western contexts, highlighting a need for more studies of Earth sciences based in the rest of the world.

2.1 Scientific Practice

A hallmark focus of STS is scientific practice, i.e., the actual work that people do to investigate the world. The idea is that we must understand how people collect, define, analyze, and interpret evidence to understand the knowledge that they propose. I took this approach by studying how vertebrate fossils are chiseled out of rock and reconstructed to be studied and, crucially, who does this expert work. The observations and interviews I did are possible only for recent history, but there are other ways to access scientific work in the archives. Historians Paul Brinkman (2010) and Lukas Rieppel (2019) use field notebooks and museum archives to similarly study the people behind the dinosaurs, particularly fossil collectors and fieldwork assistants in the nineteenth-century United States. Their work illuminates the critical importance of fieldwork and specimens – and thus the people who make them possible – to the Earth sciences. Historian Samantha Muka (2023) likewise draws on archives as well as interviews to reveal the importance of technicians to knowledge about oceans, particularly aquarists who care for live specimen collections in aquariums and who create ingenious technologies of tanks and plumbing systems that sustain "oceans of glass" for research and display. Without these workers' expertise and creativity, scientists' ability to observe and experiment with oceanic ecosystems would be limited to fieldwork.

Amateurs, hobbyists, and volunteers have long contributed their labor and ideas to knowledge about the Earth. Residents guided European naturalists around their homelands, such as working-class men in northern England (Secord 1994) and Indigenous groups in the Americas (e.g., for renowned collectors and theorists Charles Darwin [(1839) 1913] and Hans Sloane [Delbourgo 2017]). These guides provided crucial access to local landscapes, experience-based knowledge, and protection and food. Likewise, volunteer earthquake observers (Coen 2013) and weather-spotters (Daipha 2015) have served for centuries as witnesses of natural phenomena for scientists, who continue today to build observers' report data into their decision-making models and research protocols. In other circumstances, witness accounts are ignored or forgotten. This has been especially true when the witnesses came from marginalized groups, such as the survivors of the 1811–1812 earthquakes in New Madrid, Missouri, who were primarily poor rural residents, tall-tale-telling hunters, Indigenous people, and enslaved people (Valencius 2013). By unearthing this "lost history" thanks to these ignored witnesses' preserved stories, historian Conevery Bolton Valencius (2013) reveals fascinating insights on how race and social class shaped seismic knowledge.

As these studies of ignored collaborators and "lost" histories demonstrate, researchers tend to get the credit for knowledge about nature because their names are on published papers. However, looking more closely at the work required to study nature reveals a much broader cast of characters who have contributed to what we know about the world. These contributors tend to have different education, backgrounds, and identities than scientists do, suggesting that jobs for non-researchers can be a powerful pipeline to include more diverse kinds of people in scientific practice today. A more diverse scientific community will produce scientific knowledge enriched by a wider variety of worldviews and values, which is likely to make that knowledge more comprehensive and socially relevant. Thus, histories of the Earth sciences that consider all contributors enrich our understanding of scientific work and knowledge as well as the social systems that underlie them, such as education, skill, prestige, recognition, and equitable access to participation in knowledge-making. It also awards contributors their due credit for important scientific work that may otherwise be missing from published or archival sources.

Along with discovering *who* is doing scientific work, studying *what* they are doing transports us to a particular time and place, as a powerful reminder that science is human work. Producing knowledge about nature thereby relies on all the creativity, wisdom, skill, failures, and foibles that make us human. For example, historian Martin Rudwick's foundational work on history of geology includes insightful assessments of how eighteenth- and nineteenth-century European researchers represented geological evidence (e.g., 1976a, 1985, 2005, and 2008). A geological feature such as a fault or an anticline cannot be carried home to be analyzed, except as small samples of rocks and minerals. To solve this problem, natural philosophers created a "visual language" for geology, which included conventions for mapping, use of color, and standard indicators of elevation and rock type (Rudwick 1976b). These meticulously drawn diagrams could then be carried home to be analyzed in more depth, compared with othe evidences, and shared with people who had never visited the feature itself. Famous fossilist Georges Cuvier extended this pictorial capture of geology to fossils, thereby building a "paper museum" of drawings of fossil specimens that were much easier to store and share than the heavy, fragile bones (Rudwick 2000). Scientific illustration and representation, like many aspects of scientific practice, are not documented in publications. Rudwick found it in Cuvier's archives, which capture aspects of the creation stories of the images that Cuvier had published without explanations for how they were made. It is easy to overlook the importance of skilled draftsmanship to this approach to research, as well as the complexity of building consensus around standards of representation, particularly when those representations serve as the primary evidence for a knowledge claim (e.g., Lynch and Woolgar 1990; Latour 1999; Coopmans et al. 2014). Rudwick calls attention to these everyday tasks of studying the Earth, alongside his analyses of grand theories and global collaborations among natural philosophers. After all, the skillful, tedious work of making evidence reliable and sharable is what makes theorizing possible, effective, and worthwhile.

Like Rudwick's oeuvre, the best histories of the Earth sciences link developments in scientific knowledge and practice to their social and political contexts. Rudwick's analyses of the French Revolution's epistemic repercussions in geology are masterful (e.g., Rudwick 2005). Historian Jim Secord (1986) uses a rather obscure controversy among nineteenth-century British natural philosophers to powerfully reveal the social values underlying the stratification of Victorian science, society, and imperialism. Historians have also long situated twentieth-century Earth sciences as a key geopolitical tool, from monitoring countries' geological resources to searching oceans for military craft to applying seismological tools to surveil nuclear tests (Turchetti and Roberts 2014). Historian David Sepkoski (2020) traces theories about extinction to their accompanying social milieu, such as by arguing that the previously rejected idea of mass extinction gained support in the 1980s alongside the "asteroid bomb" theory of the dinosaurs' extinction due to widespread fears of nuclear warfare. Studies of polar sciences in particular reveal complex social and political dynamics among scientists and among nations. Historian Michael Bravo (2018) traces a centuries-long history of how people have thought about the North Pole, as a reflection of their own cultures' beliefs and values about nature and society. Anthropologist Jessica O'Reilly (2017) examines Antarctic science as a site of performing global politics and environmental policies through the everyday work of ecological, geological, and climate research. These authors demonstrate that taking far-reaching perspectives on Earth sciences' methods and knowledge can yield insights into global politics, including the social, epistemic, and ethical values that underlie them.

Strikingly, Earth sciences are closely linked with beyond-Earth sciences. Studying how scientists study outer space offers insights into how they see themselves and their own planet (see ▶ "Planets: A History of Observing Worlds and Changing Worldviews"). STS scholar Valentina Marcheselli, anthropologist Lisa Messeri, and sociologist Janet Vertesi, respectively, conducted ethnographic studies of researchers as they searched for extraterrestrial microbes (Marcheselli 2020), explored exoplanets and Martian geology (Messeri 2016), and designed vehicles to explore Mars (Vertesi 2015, 2020), all *by studying Earth*. Earth thus serves as an analog or a testbed for investigating otherworldly phenomena. Likewise, knowledge these scientists produce about other worlds reveals insights about Earth. By asking what these researchers are really learning, Marcheselli, Messeri, and Vertesi flip the narrative from studying planets to studying those who study planets. Their insights illuminate the human side of seemingly inhuman topics. For example, Vertesi (2015) analyzes the "Rover dance" that NASA engineers unconsciously perform as they use their bodies to imitate the robot rover on Mars during their planning sessions for where to next move the machine. They are imagining themselves as the rover as a strategy to understand the machine and its Martian environment so that they can better control the rover and thus produce better research data. Classifying body movements as a component of research broadens the typical notion of scientific practice and reveals the many kinds of knowing that researchers rely on to make sense of this world through the study of other worlds.

2.2 Technology

Technology is a critical component of scientific practice and thus a primary focus of STS research and theories (e.g., Bijker et al. 2012; Wyatt 2008). Technologies in the Earth sciences range from stonemasons' hammers and Cuvier's paper fossils to precise measurement instruments and digital models that inform complex systems of simulation and prediction. For example, sociologist Phaedra Daipha (2015) highlights the embodied judgments of meteorologists as they work to advise governments and the public about future weather. Based on an ethnography of US National Weather Service staff, Daipha demonstrates how those workers' skillful judgments become incorporated into automated systems of weather prediction and public alerts – or not. Her study examines the linked constructions of knowledge about weather, technical systems to monitor and manage weather's effects on society, and social and professional roles of meteorologists, amateur weather trackers, and policymakers.

Simulation and prediction technologies are powerful and revealing sites of interaction between science and society, as scientists try to translate knowledge about natural processes into recommendations to help protect people and infrastructure. For example, sociologist Benjamin Sims (1999) followed a team of scientists, engineers, and construction workers trying to create machines to simulate earthquakes and thus provide testing scenarios for building designs for earthquake-vulnerable communities. Anthropologist Elizabeth Reddy (2023) followed Mexico's earthquake prediction system to illuminate its sociotechnical and socioecological foundations of citizen trust, scientific credibility, and political support for disaster planning. In these cases, studies of people studying earthquakes provide insights about the everyday teamwork, knowledge integration, individual skill, political and social values, and luck that underlie the epically important scientific and social knowledge of how to protect communities from geological disasters.

In the case of climate science, the need for long-term data calls into action a variety of technologies, including centuries of salt-soaked ships' logbooks containing daily weather notes, local and global measurements of temperature and greenhouse gases, and satellite images of polar ice cover. These diverse data inform many kinds of models to simulate and predict current and future climate scenarios and thereby inform policy and social responses (Edwards 2010; Oreskes and Conway 2010; Bokulich and Oreskes 2017). Historian Deborah Coen (2018) argues that today's climate science derives from a long history of "scaling," i.e., the skillful and fraught practice of fitting together various paradigms of understanding the Earth. This combination of simple and complex technologies that have originated in various time periods and places – as well as the ongoing negotiation of the "scale" of data, natural phenomena, and knowledge – may be signature components of the Earth sciences, which span local to global features, events, and patterns and which range from the deep past to the teeming present.

2.3 Earth Sciences for a Better World

STS prides itself on critical engagement with science and society (e.g., Downey and Zuiderent-Jerak 2021), a commitment that is present among historians and, I hope, growing. This engagement includes STS scholars facilitating public outreach and engagement in knowledge-making, and participating on research teams as a way of directly intervening in scientific practice and knowledge production while also studying it. Another method – one that may be more accessible to historians – involves interpreting evidence about science and technology to make normative arguments about the way science, technology, and/or society *should* be. This work includes, for example, calls for social justice, environmental sustainability, and more inclusive participation in science. We have already seen that the Earth sciences can be a very productive site for thinking about how nonscientists contribute to scientific work and knowledge, thanks to the field's reliance on a variety of expertise and skill. Furthermore, the Earth sciences' explanatory and predictive power about natural and anthropogenic phenomena makes them supremely relevant to many communities, as does their attention to a wide range of times and places. After all, where better to seek justice, equity, and social benefit than in a planet-level collective of sciences?

Social studies of mining are a powerful example of calls for social justice. Literature scholar Allison Bigelow (2020) examines sixteenth- and seventeenth-century Spanish texts about mining as evidence of how Spanish colonists assimilated Indigenous and African knowledges about minerals in the Americas, even as they oppressed the Indigenous peoples and enslaved Africans they were learning from. Bigelow's study of technical mining language offers fascinating insights into colonialism, racism, and power. Likewise, anthropologist Jessica Smith's (2021) study of the relationships between today's petrochemical companies, engineers, and mining communities raises crucial questions about the many meanings of "corporate social responsibility" and, in response, recommends new approaches in engineering education to better equip future mining engineers to serve society. To support the growing field of justice-centered studies of geological resources, Smith, sociologist Abby Kinchy, and environmental studies scholar Roopali Phadke created STS Underground, a network of scholars studying mining and the energy industry as a site of scientific practice, community participation, and ethical planning for the future (see the special issue introduced by Kinchy et al. 2018). Their approach is an admirable example of the importance of understanding scientific practice and knowledge in the Earth sciences in order to propose ways to make those systems more socially beneficial, such as for environmental conservation, climate change mitigation, economic well-being, and safety. It is also an excellent model for bringing together interdisciplinary scholars to think through a topic collaboratively. Historians could adopt the same approach to make histories of the Earth sciences more comprehensive, broader in geographic and temporal scope, more willing to recommend ways to do science otherwise, and, perhaps, more socially relevant and beneficial.

3 A Concluding Call for Future Histories

Here I have discussed three topics – scientific practice, technology, and social good – through which STS methods and studies can inform historians' investigations of the Earth sciences. There are many more. These three are merely helpful starting points because they apply well to the Earth sciences and there are already excellent examples of studies to demonstrate the value of these approaches for historians.

Other STS topics that deserve more development in histories of the Earth sciences include expertise, power (e.g., feminism, critical race theory, and decolonialism), policy and governance, relationships between science and the public, knowledge infrastructures, and ethics (for more ideas, see Felt et al. 2016 and chapter ▶ "History of the Earth Sciences from the South"). Furthermore, historians of the Earth sciences can serve as experts for the scientific community, the communities of STS scholars and historians of science, and for us as citizens of the Earth, to help achieve trustworthy knowledge about how our planet works, how we can enable equitable access to natural resources and disaster protection, and how to pursue environmental and climate justice. For example, we need spokespeople to help us all understand conflicted knowledges and policies surrounding natural disasters, fossil fuel production, pollution, remote sensing, and the global socioecological systems of weather, climate, and natural resources. As sciences that have been practiced for so long and about such a wide variety of places and time periods, the Earth sciences can serve both as an informative precedent for current social and scientific problems and as a model for how to integrate diverse kinds of knowledge and practitioners to address global social and scientific problems.

References

Bigelow, Allison M. 2020. *Mining language: Racial thinking, Indigenous knowledge, and colonial metallurgy in the early modern Iberian world*. Chapel Hill: University of North Carolina Press.

Bijker, Wiebe E., Thomas Parke Hughes, and Trevor Pinch, eds. 2012. *The social construction of technological systems*. Anniversary edition. Cambridge, MA: MIT Press.

Bokulich, Alisa, and Oreskes, Naomi. 2017. Models in geosciences. In Springer Handbook of Model-Based Science, ed. L. Magnani and T. Bertolotti. Springer: Cham. https://doi.org/10.1007/978-3-319-30526-4_41.

Bourdieu, Pierre. 1993. *Sociology in question*. London: Sage.

Bravo, Michael. 2018. *North pole: Nature and culture*. London: Reaktion Books.

Brinkman, Paul D. 2010. *The second Jurassic dinosaur rush: Museums and paleontology in America at the turn of the twentieth century*. Chicago: University of Chicago Press.

Coen, Deborah R. 2013. *The earthquake observers: Disaster science from Lisbon to Richter*. Chicago: University of Chicago Press.

———. 2018. *Climate in motion: Science, empire, and the problem of scale*. Chicago: University of Chicago Press.

Collins, Harry, and Robert Evans. 2007. *Rethinking expertise*. Chicago: University of Chicago Press.

Coopmans, Catelijne, Janet Vertesi, Michael Lynch, and Steve Woolgar, eds. 2014. *Representation in scientific practice revisited*. Cambridge, MA: MIT Press.

Daipha, Phaedra. 2015. *Masters of uncertainty: Weather forecasters and the quest for ground truth*. Chicago: University of Chicago Press.

Darwin, Charles. [1839] 1913. *A naturalist's voyage round the world: Journal of researches into the natural history and geology of the countries visited during the voyage round the world of H.M.S. Beagle under the command of captain Fitz Roy, R.N.* London: John Murray.

Delbourgo, James. 2017. *Collecting the World: Hans Sloane and the Origins of the British museum*. Cambridge, MA: Harvard University Press.

Downey, Gary, and Teun Zuiderent-Jerak, eds. 2021. *Making & doing: Activating STS through knowledge expression and travel*. Cambridge, MA: MIT Press.

Edwards, Paul. 2010. *A vast machine: Computer models, climate data, and the politics of global warming*. Cambridge, MA: MIT Press.

Felt, Ulrike, Rayvon Fouche, Clark A. Miller, and Laurel Smith-Doerr, eds. 2016. *The handbook of science and technology studies*. 4th ed. Cambridge, MA: MIT Press.

Kinchy, Abby J., Roopali Phadke, and Jessica M. Smith. 2018. Engaging the underground: An STS field in formation. *Engaging Science, Technology, and Society* 4: 22–42. https://doi.org/10.17351/ests2018.213.

Latour, Bruno. 1999. Circulating reference: Sampling the soil in the Amazon forest. In *Pandora's hope: Essays on the reality of science studies*, ed. Bruno Latour, 24–79. London: Harvard University Press.

Latour, Bruno, and Steve Woolgar. 1986. *Laboratory life: The construction of scientific facts*. 2nd ed. Princeton: Princeton University Press.

Lave, Jean, and Etienne Wenger. 1991. *Situated learning: Legitimate peripheral participation*. Cambridge: Cambridge University Press.

Law, John. 1994. *Organizing modernity: Social ordering and social theory*. Cambridge, MA: Blackwell.

Lynch, Michael, and Steve Woolgar, eds. 1990. *Representation in scientific practice*. Cambridge, MA: MIT Press.

Marcheselli, Valentina. 2020. The shadow biosphere hypothesis: Non-knowledge in emerging disciplines. *Science, Technology, & Human Values* 45 (4): 636–658. https://doi.org/10.1177/0162243919881207.

Messeri, Lisa. 2016. *Placing outer space: An earthly ethnography of other worlds*. Duke University Press.

Muka, Samantha. 2023. *Oceans under glass: Tank craft and the sciences of the sea*. Chicago: University of Chicago Press.
O'Reilly, Jessica. 2017. *The technocratic Antarctic: An ethnography of scientific expertise and environmental governance*. Ithaca: Cornell University Press.
Oreskes, Naomi, and Erik M. Conway. 2010. *Merchants of doubt: How a handful of scientists obscured the truth on issues from tobacco to climate change*. New York: Bloomsbury.
Reddy, Elizabeth. 2023. *Alerta! Engineering on Shaky Ground*. Cambridge, MA: MIT Press.
Rieppel, Lukas. 2019. *Assembling the dinosaur: Fossil hunters, tycoons, and the making of a spectacle*. Cambridge, MA: Harvard University Press.
Rudwick, Martin J.S. 1976a. *The meaning of fossils: Episodes in the history of paleontology*. 2nd ed. Chicago: University of Chicago Press.
———. 1976b. The emergence of a visual language for geological science 1760–1840. *History of Science* 14 (3): 149–195.
———. 1985. *The great Devonian controversy: The shaping of scientific knowledge among gentlemanly specialists*. Chicago: University of Chicago Press.
———. 2000. Georges Cuvier's paper museum of fossil bones. *Archives of Natural History* 27 (1): 51–68.
———. 2005. *Bursting the limits of time: The reconstruction of geohistory in the age of revolution*. Chicago: University of Chicago Press.
———. 2008. *Worlds before Adam: The reconstruction of geohistory in the age of reform*. Chicago: University of Chicago Press.
Secord, James A. 1986. *Controversy in Victorian geology: The Cambrian-Silurian dispute*. Princeton: Princeton University Press.
Secord, Anne. 1994. Science in the pub: Artisan botanists in early nineteenth-century Lancashire. *History of Science* 32: 269–315.
Sepkoski, David. 2020. *Catastrophic thinking: Extinction and the value of diversity from Darwin to the Anthropocene*. Chicago: University of Chicago Press.
Shapin, Steven. 1989. The invisible technician. *American Scientist* 77 (6): 554–563.
———. 1994. *A social history of truth: Civility and science in seventeenth-century England*. Chicago: University of Chicago Press.
Sims, Benjamin. 1999. Concrete practices: Testing in an earthquake-engineering laboratory. *Social Studies of Science* 29 (4): 483–518. https://doi.org/10.1177/030631299029004002.
Sismondo, Sergio. 1996. *Science without myth: On constructions, reality, and social knowledge*. Albany: SUNY Press.
———. 2010. *An introduction to science and technology studies*. 2nd ed. Chichester: Wiley-Blackwell.
Smith, Jessica M. 2021. *Extracting accountability: Engineers and corporate social responsibility*. Cambridge, MA: MIT Press.
Turchetti, Simone, and Peder Roberts, eds. 2014. *The surveillance imperative: Geosciences during the cold war and beyond*. New York: Palgrave Macmillan.
Valencius, Conevery Bolton. 2013. *The lost history of the New Madrid earthquakes*. Chicago: University of Chicago Press.
Vertesi, Janet. 2015. *Seeing like a rover: How robots, teams, and images craft knowledge of Mars*. Chicago: University of Chicago Press.
———. 2020. *Shaping science: Organizations, decisions, and culture on NASA's teams*. Chicago: University of Chicago Press.
Wyatt, Sally. 2008. Technological determinism is dead: Long live technological determinism. In *The handbook of science and technology studies*, ed. Edward J. Hackett, Olga Amsterdamska, Michael E. Lynch, and Judy Wajcman, 3rd ed., 165–180. Cambridge, MA: MIT Press.
Wylie, Caitlin Donahue. 2015. 'The Artist's piece is already in the stone': Constructing creativity in paleontology laboratories. *Social Studies of Science* 45 (1): 31–55. https://doi.org/10.1177/0306312714549794.

———. 2018. Trust in technicians in paleontology laboratories. *Science, Technology, & Human Values* 43 (2): 324–348. https://doi.org/10.1177/0162243917722844.
———. 2019a. Overcoming the underdetermination of specimens. *Biology and Philosophy* 34 (24). https://doi.org/10.1007/s10539-019-9674-2.
———. 2019b. The plurality of assumptions about fossils and time. *History and Philosophy of the Life Sciences* 41 (21). https://doi.org/10.1007/s40656-019-0260-3.
———. 2021. *Preparing dinosaurs: The work behind the scenes*. Cambridge, MA: MIT Press.

Cold War and Earth Sciences

Ronald E. Doel

Contents

Abstract

In this chapter, I review scholarly literature on the history of the earth sciences in the Cold War, focusing particularly on developments in North America and Europe. I explore the impact of military patronage and national security concerns on this research community, the influence of the International Geophysical Year on scientific programs and scientific internationalism, and the ways that funding for the physical environmental sciences shaped the production and dissemination of knowledge. I conclude with ideas about potentially fruitful new research questions.

Keywords

Earth sciences · Cold War · Military patronage · National defense · Environmental sciences · Control of nature · Anthropocene

R. E. Doel (✉)
Florida State University, Tallahassee, FL, USA
e-mail: rdoel@fsu.edu

E. Aronova et al. (eds.), *Handbook of the Historiography of the Earth and Environmental Sciences*, Historiographies of Science,
https://doi.org/10.1007/978-3-031-40799-4_25

1 Introduction

With few exceptions, earth scientists in North America do not talk about a "Cold War" period in their field in the way that physicists refer to the era of classical physics or evolutionary biologists talk about the time before the Modern Synthesis in the first half of the twentieth century. They see the decades after World War II as a prosperous and intellectually vigorous time, when abundant and stable government patronage allowed geophysicists and geochemists to collaboratively explore the planet—including the singularly important and influential International Geophysical Year of 1957–1958, which gave birth to the space age. New tools they developed in these decades unlocked deep insights into planetary processes—including global atmospheric circulation, ocean convection and circulation, plate tectonics, planetary impact events, and climate change.

Historians of recent science acknowledge that the Cold War decades were prosperous ones for earth sciences. But they also see patronage—particularly *military* patronage—as key for understanding the spectacular rise of the earth sciences in the second half of the twentieth century. Their work has addressed which research questions gained priority over others (and why); explored how secrecy and classification affected the production and circulation of knowledge; and analyzed how research programs and institutions devoted to the earth sciences and physical oceanography, including planetary sciences, took shape and expanded amid the Cold War. They have also examined various ways that earth scientists threaded new relations with the state involving scientific intelligence and foreign policy (including treaties involving Antarctica, Outer Space, and efforts to limit nuclear testing) and addressed how leading figures in the field treated conceptions about Earth's environment and its management, among them the control of nature and the precautionary principle.

For the term "Cold War Earth Sciences" to have meaning (rather than simply referring to all that took place between 1947 and the late 1980s), we need to recognize that certain advances in the earth sciences *happened to occur* during the Cold War period but were not particularly shaped by burgeoning military patronage or national security policies. For instance, many geophysicists remained active participants in identifying and exploiting petroleum reserves. Oil was certainly seen as a key strategic global resource after 1945 (Cantoni 2017), but the *field* of exploration geophysics, embracing seismic techniques as prospecting tools, continued to follow patterns already well-established in the 1920s and 1930s (Anduaga 2016). In addition, not all prominent earth scientists worked on topics relating to military interests. M. King Hubbert, the influential geophysicist who developed the concept of peak oil and helped direct research for the Shell Oil Company from 1943 to 1963, had no Pentagon ties, and embraced a holistic vision of the environmental sciences already familiar to biologists and ecologists before the onset of World War II (Doel 2007; Vance and Doel 2010). Pentagon and government priories profoundly affected many realms of the earth sciences in the Cold War era, but not all fields equally.

A robust outflow of scholarship on the earth sciences after 1945 began even before the Cold War ended. Walter Sullivan, the politically astute *New York Times* science journalist, wrote a book-length contemporary account of the scientific aims and global politics of the International Geophysical Year, illuminating the rapid ascendant of the earth sciences (Sullivan 1961). Earth scientists and historians joined forces to create the History Committee of the American Geophysical Union (AGU) in 1981; in 1988, the journal *Earth Sciences History* was launched. AGU History committee members produced a series of edited books that broadly addressed the earth sciences (Glen 1986) as well as individual works on post-World War II geophysics (e.g., Stern 1989); several important contributions, focused on geophysics generally and ionospheric research more specifically, came from C. Stewart Gillmor, an AGU History member and Princeton-trained historian of science based at Wesleyan University (Gillmor 1984, 1986). Gillmor's work aside, many of these earlier studies focused on the rational development of scientific ideas with limited attention to how social, political, and cultural factors shaped the questions that researchers pursued and the growth of institutions which supported them. In part, this reflected the divide between "internal" and "external" approaches to history of science (whether one emphasized "the trajectory of ideas" or "the social and cultural forces that produced and molded such ideas," as Janet Browne artfully put it) that came into sharp focus in the last decades of the twentieth century (Browne 2021, 15). But what most distinguishes this body of work from what might be termed scholarship on the Cold War physical environmental sciences was a deliberate focus on *patronage*. That is, how did military agencies and other governmental institutions concerned with national security (including advances in weapons systems) and foreign policy (including national prestige and treaty-making) incorporate earth scientists into their operations and planning amid the Cold War? How did these affinities affect the practice, aims, outlooks, achievements, and professional ambitions of geophysicists?

2 Links to National Security

The rise of "externalist" approaches to the history of American science in the 1980s—itself a new, bold focus of inquiry—made it possible for historians to perceive ways to map how Cold War pressures and opportunities influenced earth sciences' growth and institutionalization. After focusing on military influences on the discipline of physics—a natural and understandable fascination, given the centrality of nuclear weapons in global history after 1945—historians of science began to appreciate that the earth sciences were right behind them. As the Cold War began, military and governmental funding for the fields of earth sciences was surpassed only by that for physics (Oreskes and Doel 2002; Doel 2003a; Weart 2013). Military fascination with these fields, as Michael Aaron Dennis has observed, caused "a profound reconfiguration of the disciplines, institutions, and practices used to understand the Earth" (Dennis 2003, 809).

What made the earth sciences so crucial for military leaders (in the USA, but also in other advanced industrial nations around the globe, including the Soviet Union) were the weapons systems of greatest importance at the start of the Cold War. These were what is still called the "triad" in the nuclear age: advanced long-distance aircraft, submarines, and guided missiles (all capable of carrying nuclear weapons by the end of the 1950s). Military officials came to see the natural environment as a bounded, dynamic space through which the new weapons of the Cold War era would pass. Understanding physical phenomena became a necessity to accurately operate these weapon systems. Addressing unsolved problems in the earth sciences—in geodesy, atmospheric electricity, terrestrial magnetism, ionospheric physics, physical oceanography, and atmospheric sciences—would aid scientists as well as military planners by advancing fundamental research that simultaneously offered utilitarian advantages. While some branches of the biological environmental sciences also interested military planners in the early Cold War—including the atmospheric dispersion of radioactive fallout, the sound-detection abilities of marine mammals, and the migration patterns of whales (to distinguish them from hostile submarines)—these were narrow, occasional slats in the broad spectrum of fields constituting postwar fields of environmental biology (Kwa 1993; Doel 2009; Ritts and Shiga 2016; Kangas 2023).

A trickle of scholarship began addressing this issue in the late 1980s, becoming a faster and faster stream by the end of the century. Early contributions helped shape questions that later researchers pursued (e.g., Mukerji 1989). It is interesting, in hindsight, to see how rapidly the field blossomed. David H. DeVorkin's exploration of the postwar US V-2 program, revealing the importance of physical environmental factors of critical importance in operating missiles (DeVorkin 1987, 1992), certainly influenced my own preliminary assessment of post-WWII military influence on the physical environmental sciences (Doel 1997a). Those of us called to the recent history of the earth sciences and oceanography sensed fertile uncharted territory ahead; we also soon discovered that similarly interested scholars ranged from recent PhDs to senior figures, and that various patrons and organizations were saying yes to underwriting dedicated workshops, conferences, and gatherings. A workshop convened by the History Committee of the American Geophysical Union in 1998 led to a volume of *Historical Studies in the Physical and Biological Sciences* devoted to "Military Patronage and the Geophysical Sciences in the United States" (including Hacker 2000; Oreskes 2000a; Rainger 2000a; Warner 2000). The Second Maury Conference on the History of Oceanography, held in Woods Hole, Massachusetts, in 1999, resulted in a special edition of *Earth Sciences History* that also encompassed the Cold War earth sciences (including Oreskes 2000b; Rainger 2000b; van Keuren 2000; Levin and Doel 2000). A separate 2001 conference sponsored by the Peace Studies Program at Cornell University yielded another relevant contribution: a special issue on "Earth Sciences in the Cold War" in *Social Studies of Science* (including Barth 2003; Cloud 2003; Dennis 2003; Doel 2003a; Harper 2003). These papers collectively explored the intersection of the earth sciences and oceanography with national security policy, including how the experiences of members of this field differed from that of physicists, how abundant funding shaped the institutional growth of these fields and the unanswered questions its practitioners felt most important to pursue, how Cold War international tensions affected the practice of science, and how military concerns affected the production and dissemination of knowledge.

Two additional lines of inquiry—closely related to military funding for the Cold War earth sciences but centered on distinct spheres of inquiry—arose at this same time. One was cartography's intersection with geophysics. Earth scientists had long played important roles in determining *where* things were—the first technical bureau in the US government was the Coast & Geodetic Survey, founded in 1807—but geodetic and gravity data became particularly vital for emerging weapons systems after 1945 (Cloud 2000; Warner 2000, 2002). Several scholars have argued that military needs shaped the fields of geodesy, cartography, and geography even more profoundly: For John Cloud, military needs and demands for secrecy led to "a great *re-convergence* of these disciplines" during the Cold War "at spatial scales, extents and tolerances necessary to either wage or prevent nuclear war" (Cloud 2002, 262), while Trevor J. Barnes and Matthew Farish claimed these war-inspired shifts led to a new conception of science, resulting in a novel, theoretical understanding of *regions* (Barnes and Farish 2006, 807). Not all scholars agreed that military influences exclusively inspired these changes (Johnston et al. 2008), but many accept that the practices of map-making reflected familiar Cold War pressures. Secret maps of large portions of Earth's surface emerged from the clandestine US space satellite mission Corona, begun in 1959 (Cloud 2001, 2002); Cold War mapping gained new centrality in efforts to understand and project US geostrategic power (Cloud 2013; Barney 2015); and earth scientists, frustrated by military classification of seafloor depth data, grew inspired to create a physiographic map of ocean basins—what became the celebrated Heezen-Tharp maps of the world oceans, first published in 1957 (Doel et al. 2006; Frankel 2012a). In recent years, Marie Tharp—who coproduced these maps with Bruce Heezen—has gained attention from scholars and writers intrigued by her contributions as a women oceanographer and cartographer (Felt 2013; Trethewey 2023).

Another issue was the intersection of the earth sciences with agencies concerned with national security and scientific intelligence. Scholars intrigued by the military's postwar fascination with the earth sciences quickly realized that other realms of the state, including the National Security Council and the Central Intelligence Agency, were equally interested in what physical environmental scientists could share about the achievements of foreign researchers in Western Europe as well as in East Bloc nations. Earth scientists were well-suited to these roles for several reasons: Their research required access to diverse parts of the globe; they were largely politically conservative or moderates, not seen as likely security risks; and they were not physicists, a community famous for its embrace of internationality and liberal politics. The CIA quickly began to expand earth sciences intelligence-gathering in the early 1950s (Doel and Needell 1997; Doel 2015). Allan A. Needell revealed that the US architect of the International Geophysical Year, the Carnegie Institution of Washington geophysicist Lloyd V. Berkner, appended a "Secret Supplement" to a 1950 study of science in the US Department of State, extolling how science attaches in foreign embassies could aid crucial intelligence-gathering (Needell 2000;

Goossen 2020). Scientific intelligence is best seen as related to familiar practices of keeping tabs on the competition—something scientists routinely do to understand the state of play in their field (Doel 2020). But the Cold War intensified and institutionalized this practice, in the USA and other scientifically advanced nations. As I write, the CIA's CREST database of declassified materials now lists 946 documents matching "geophysics" and "Cold War," and 54 documents that include "earth sciences" and "Cold War." Not all potential stories merit historical attention, but undoubtedly a good number of them will.

3 Exploring the Planet: Expeditions, Global Initiatives, and Geostrategic Concerns

National security concerns after World War II supported earth science research programs on scales far beyond what had been possible in the interwar years. Expeditions reached all parts of the planet, deep into the oceans, and high into the atmosphere. As in prior decades and indeed centuries, researchers involved in these expeditions sought environmental knowledge that mattered both for their science and for their patrons. What emerged in the Cold War, historian Stephen J. Pyne has written, was a "third great age of discovery," which "claimed as its geographic domain the solar system, beginning with planet Earth" (Pyne 2016, 113; see also Pyne 2021).

The most far-reaching global initiative was the International Geophysical Year of 1957–1958, involving some 60,000 scientists in 67 nations and still the largest transnational scientific activity yet undertaken. Scholars have struggled to find adequate words to describe it: in their still authoritative survey of the IGY, Christy Collis and Klaus Dodds, borrowing from author Matthew Brzezinski, termed the enterprise a "scientific Olympics of sorts"; more recently, Benjamin Goossen found the IGY shared commonalities with high modernist initiatives such as the chartering of the United Nations and the Marshall Plan (Collis and Dodds 2008; Goossen 2020). Scholars have also struggled to judge the IGY as exemplifying Cold War tensions and practices or an exception to them (Doel 1997b). Michael Aaron Dennis termed it "arms control by other means" (Dennis 2006, 184), while Adrian Howkins found the IGY had a "Jekyll and Hyde personality," being both "an unprecedented example of international scientific co-operation and idealism" and "a worldwide stage for scientific competition and political rivalry" (Howkins 2008, 597). In recent years, scholars have generally stressed the IGY's global cooperative character (Korsmo 2004; McCray 2006; Korsmo 2007; Aronova et al. 2010) while noting various brush fires—including the challenge of incorporating both East and West German scientist-participants, and mainland China's decision to withdraw from the IGY after covert US political intervention—did reflect familiar Cold War tensions (Wang 1999; Doel et al. 2005; Wang and Zhang 2010; Olšáková 2017; Barrett 2022).

Key aspects of the IGY—at least within North American and European contexts—have received critical and sustained attention. Historians of science have written detailed overviews of the origins of this global project and the scientific ambitions of its leaders (Agar 1998; Bulkeley 1991, 2000, 2008; Collis and Dodds 2008; Dodds et al. 2010; Korsmo 2007; Kunkel 2021; Needell 2000; Powell 2008), as well as in-depth examinations of research efforts enhanced by IGY activities, particularly those focused on geophysical phenomena at the Earth's poles (Heidt and Bocking 2019; Herzberg et al. 2018; Josephson 2014; Korsmo and Graham 2002; Martin-Nielsen 2012). Efforts to create International Data Centers to safeguard and share IGY data—an early instance of the challenges of Big Data—have become a focus of attention (Korsmo 2007, 2008; Aronova et al. 2010; Aronova 2017a) as has the unexpected contribution of amateur scientists to the IGY (McCray 2006, 2008; Dora 2023).

Geographical dimensions of the IGY have also received considerable attention from historians. IGY leaders focused particular attention on Earth's polar regions. Scholars have argued that this emphasis was overdetermined. The predecessors of the IGY were both polar-focused. Political leaders, like military officials, recognized that both the Arctic and Antarctic by the early Cold War had become crucial geostrategic regions (the Arctic because it was the geographic center between North America and the Soviet Union, hence a possible locale for World War III; Antarctica because of unresolved territorial claims that threatened Cold War allegiances). As a result, understanding a wide range of geophysical phenomena at the poles, from atmospheric dynamics, glacial movements, and magnetic variations, were judged vital for science and national security (Doel et al. 2014). The ice-covered island of Greenland, halfway between the United States and the Soviet landmass and then a territory of Denmark, became a beehive of earth sciences research amid the IGY, as scientists understood environmental knowledge obtained there would also benefit military operations in the far north (Heymann et al. 2010; Ries 2012; Doel et al. 2016). Scholars have focused particularly on the construction of a nuclear-powered military base called Camp Century, built under the Greenland icecap, that served both as a cover for a proposed secret US Army plan to place medium-range nuclear-tipped missiles in tunnels bored through the ice, and as a drilling site for deep ice cores (intended to reveal the physical characteristics and stability of glacial ice) that also yielded profoundly important clues about Earth's past climatic history (Martin-Nielsen 2013; Nielsen and Nielsen 2021).

The Antarctic situation was distinct: Two South American nations, Argentina and Chile, claimed overlapping sectors of the Antarctica. To avoid conflicts between proposed IGY expeditions to the southern continent by various nations that maintained territorial claims, a political solution was proposed and adopted: the Antarctic Treaty of 1959 put existing territorial claims 'on ice,' and amid the IGY, Antarctica became a "continent for science" (Elzinga 2009; Dodds 2010a; Howkins 2011; Berkman 2011; Launius 2017). Because Antarctic IGY activities so deeply overlapped with Cold War foreign policy, historians have been particularly intrigued by national and international contexts of research in Antarctica, including transnational collaborations and how these affected the production of knowledge and the character of scientific research (Dodds 2005; Turchetti et al. 2008; Naylor et al. 2008; Roberts 2011; Roberts et al. 2016).

More recently, as additional archival sources have become available, scholars are discovering additional revealing topics, such as the creation of Antarctica's Atom-Free Zone (Musto 2018).

An even more important outcome of the IGY—and perhaps its best-remembered contribution—was its promotion of artificial satellites to better understand the near-Earth environment. The space age began when the Soviet Sputnik I—intended to help determine the density of the upper atmosphere—rocketed into orbit on October 4, 1957, becoming Earth's first artificial satellite. Because the instauration of the space age serves as a crucial moment in twentieth-century history, a robust historiography emerged decades ago, much of it neither focusing on Cold War factors nor on their influence on earth and planetary research. But certain studies do address the Cold War, including John Logsdon's most recent account of President John F. Kennedy's decision to go to the Moon (Logsdon 2010); Audra Wolfe's examination of US-Soviet Cold War rivalries (Wolfe 2013); Walter McDougall's now-classic account of the origins of the space program (McDougall 1985); Erik Conway's recounting of atmospheric sciences research (Conway 2008); Andrew J. Butrica's study of planetary radio astronomy (Butrica 1996); Stephen J. Pyne's narrative of the Voyager outer planetary mission (Pyne 2010); and several works by former NASA History Office chief Roger Launius (Launius 2010, 2014, 2016). Two historians of science have also offered insightful institutional histories of the Jet Propulsion Laboratory, a key NASA facility closely tied to earth and planetary exploration initiatives (Koppes 1982; Westwick 2008).

In the aftermath of the IGY, government policymakers continued to value and promote earth science initiatives to aid foreign policy aims. Because physical environmental scientists were ideally positioned as valuable and useful players in the Cold War—given their continuing contributions to national security; their role in gathering global data; their ability to navigate between the ideals of science and the demands of the American Cold War state; and their utility in operating field stations that demonstrated sovereignty claims (Dodds 2010b; Korsmo and Sfraga 2003; Sörlin 2014)—it is not surprising that members of this community became involved in various visible diplomatic initiatives and activities, both overt and covert. Scholars have begun to explore the ways that earth scientists contributed to highly visible efforts to promote desalination plants in various nations, including Israel and Saudi Arabia (Krige 2014; Krige and Sarkar 2018; Low 2020), and carried out a secret mission in India in the mid-1960s, initiated at the highest levels of the US government, to mitigate the Bihar drought through weather modification (Doel and Harper 2006). Both undertakings were conceived as ways to limit nuclear proliferation. But as Simone Turchetti and Peder Roberts have shown, government officials particularly valued earth scientists because gathering new information on potential enemies meant gathering information about the Earth: Surveillance networks owed their existence "to the dramatic expansion of funding to the geosciences after 1945" (Turchetti and Roberts 2014, 2). Leaders of transnational security organizations, including the North Atlantic Treaty Organization (NATO), similarly valued the earth sciences (Turchetti 2012, 2020; Camprubi and Robinson 2016; Robinson 2018, 2021).

4 Controlling Nature (and Planetary Pollution)

Earth scientists in the Cold War era faced a dilemma. Many felt that their research tools, and the epistemic standing of physical sciences, made them the most reliable interpreters of the planet's environmental conditions. They believed their instruments and measurements were yielding important insights into deep circulation in the oceans (setting limits on proposed nuclear waste disposal at sea) and the distribution of nuclear fallout over land. Some also believed that using science-based technology to control nature—chemical pesticides for agriculture, and weather control—would help solve major material problems. By the second half of the Cold War, however, biological scientists increasingly asserted the primacy of their own knowledge about earth's environment, emphasizing ecological concepts and conservation while critiquing the "technological fix" mentality underlying control of nature approaches. This clash did not solely involve earth scientists and physical oceanographers but was part of a profound shift in the dominance of science disciplines. While biology as a whole did not surpass the physical sciences until the 1990s, years after the Cold War ended, the biological environmental sciences began gaining increased funding, influence, and authority within the US scientific landscape much earlier in the Cold War (Kevles 1997; de Chadarevian 2002).

Scholars have fruitfully pursued several instances where earth scientists early in the Cold War felt confident that the control of nature would soon be within their purview. A particularly revealing case involved weather control, an undertaking with roots in the nineteenth century that gained new-found respect (and abundant military patronage) by 1947 (Fleming 2006, 2010; Harper 2008, 2017). Other physical scientists believed that nuclear weapons promised benefits for peaceful geological engineering, best illuminated by the Atomic Energy Commission's 1960s Project Plowshares (Kirsch 2005), which included a plan to employ hydrogen bombs to create a new harbor in northwest Alaska (O'Neill 2007). Earth scientists early in the Cold War also generally perceived the benefits of using radioactive fallout—effective as tracers to better understand atmospheric and oceanic circulation—outweighed concerns raised by geneticists and biological oceanographers about increased levels of radioactivity in the ecosystem (Hamblin 2008; Higuchi 2020). Debates over whether the biological or physical environmental sciences provided the most relevant, reliable, and robust knowledge of terrestrial processes continued to reverberate in academic institutions amid the wide-ranging social, political, and cultural unrest of the late 1960s (Kingsland 2005; Vance and Doel 2010; Lajus 2020).

Assessing planetary-level changes from Earth orbit further enhanced physical environmental scientists as interpreters of environmental change, including emerging threats from resource exploitation and pollution (McNeill 2000; Conway 2014). Several scholars have explored the impacts of specific satellite programs including Landsat (Mack 1990; Maher 2019) and the US-French environmental surveillance satellite Argos, launched in the late 1970s (Benson 2012). Pursuing similar lines of inquiry, other researchers have explored how NASA's earth-orbiting satellites enhanced the burgeoning realm of atmospheric sciences (Conway 2008; see also Fleming 2016) and gave shape to the emerging interdisciplinary field of Earth System Science (Barton 2020; Launius 2016). More recently, L. Ruth Rand studied a new form of pollution in near-earth space: orbiting space junk that litters the planetary environment, threatens functioning satellites and spacecraft, and occasionally crashes back to Earth's surface (Rand 2019).

Another aspect of pollution, initially lumped under the framework of "atmospheric pollution," involved climate change. It is at once both a Cold War concern and one that also happened to take place during the Cold War years, as greenhouse gases released into the atmosphere after the Industrial Revolution were recorded during the IGY and warming temperatures became evident in national and global measurements (Weart 1992; Fleming 1998; Edwards 2006). Already in 1947, polar melting (then widely interpreted as natural variation) was labeled a national security concern during a secret Pentagon-led briefing, as warming benefitted Soviet shipping as well as agricultural production (Doel 2009; Sörlin 2011). A decade later, in 1957, earth scientist Roger Revelle became lead author on an influential study about industrial-caused (anthropogenic) global warming and arranged for the geochemist Charles David Keeling to measure atmospheric carbon dioxide concentrations as part of the IGY (Weart 2008). So much rich scholarship now focuses on the discovery and politics of climate change (the most widely used term by the last decades of the Cold War) that only the briefest survey of emerging topics and themes is possible here. Researchers have explored climate change as a quintessential Cold War story (Oreskes and Conway 2008; Oreskes 2014b; Oldfield 2018; Sörlin and Lane 2018; Dörries 2020; Doose 2021); they have also examined the growth of global infrastructure essential for providing climate data (Miller 2001; Edwards 2013); the ozone hole threat (Grevsmühl 2018); the response of US presidential administrations to emerging knowledge about climate change (Howe 2014); how scientific consensus about climate change emerged (Oreskes 2018); and how media portrayals shaped public perceptions of this field (Christensen et al. 2013). A burgeoning area of investigation addresses how leaders of major fossil fuel industries and their ideological allies, including politicians, sought to create doubt about robust scientific findings, reinforcing climate skeptics and undermining possible remedies. In *Merchants of Doubt*, Naomi Oreskes and Erik Conway (Oreskes and Conway 2010) placed efforts to discredit climate science among similar campaigns to discredit health risks such as tobacco smoking, acid rain, DDT, and the ozone hole—where industry profits were threatened—but also specifically within Cold War geopolitical frameworks, including anti-Communist and antienvironmental commitments.

A separate but important issue for historians is: How did the disciplinary makeup of core earth sciences fields—particularly geology, geochemistry, and geophysics—shift over the decades of the Cold War? Borders between fields are not permanent and essential but rather dynamic and responsive to external pressures and factors (Good 2000); as Peter Bowler has noted in his now-classic *The Norton History of the Earth Sciences*, "The unity of the 'environmental sciences' is not created by the sciences themselves, it is imposed by the public's growing awareness of the threat posed to the environment by our activities" (Bowler 1992, 2; also Poole 2014). Since then, scholars have begun to explore how geophysicists and geochemists after 1945 gained greater influence and presence in leading university departments of geology and challenged the evidentiary and epistemic traditions long embraced by geologists (Oreskes and Doel 2002; Oldroyd 2002). The growing stature of geophysics (Dennis 2003; Doel 2003a; Launius 2016) and geochemistry (Kragh 2000; Shindell 2010, 2014) reflected both the perceived importance of new tools their practitioners could apply to major unsolved problems and the tsunami of military and state patronage (including NASA-related planetary science) that supported their applications. Similar dynamics affected oceanography during the Cold War (Hamblin 2005; Mills 2009), including ocean-focused institutions (Rainger 2003). Scholars have also explored how program managers in major foundations influenced strategies of knowledge production (Rozwadowski and van Keuran 2004; van Keuren 2000) and how the concept of "environment" came to be attached to these fields (Warde et al. 2021). Nevertheless, intellectual and professional clashes between traditional geologists and their geophysics and geochemistry colleagues in the post-1945 era—which transformed the practice of geology—remain remarkably underexplored.

5 Cold War Influences on the Production of Knowledge

Historians of science attracted to the question of how military patronage influenced knowledge production amid the Cold War are well-acquainted with a fruitful debate between two prominent historians of science at the end of the twentieth century, Paul Forman and Daniel J. Kevles. Forman argued that military funding for physics after World War II led to "a qualitative change in its purpose and character," resulting in "an enlistment and integration of the bulk of its practitioners and its practice in the nation's pursuit of security through ever more advanced military technologies" (Forman 1987, 150). Kevles countered that "physics is what physicists do": No normative path for research exists; all forms of patronage affect the production of knowledge. The debate, initially confined to the postwar physics community, soon motivated researchers interested in the Cold War earth sciences, who took up a related question Forman had posed: "What *direction* of the advance of science, and thus what *kind* of science, resulted from military sponsorship?" (Quoted in Doel 2003a, 637; see also Oreskes 2021).

To this day, interpreting military influence on the practice of geophysics and geochemistry remains a lively issue in Cold War historiography. In recent years, historians have noted limits in what both Forman and Kevles argued, while affirming crucial intersections between postwar American science and defense agencies (Oreskes 2014a). As John Krige observed, if Forman erred in claiming that state patronage was synonymous with state control and the exploitation of scientists, Kevles failed "to address the specificity of the state-science relationship in the Cold War (or any other historical period)," thus removing "critical tools" that historians needed to address so sharp a dichotomy (Krige 2014, 431). Just how military patronage—including data classification—affected the physical environmental sciences remains a central concern for the field.

Studies of the significance of Cold War military patronage for the earth sciences have gone in several fruitful directions. One initial focus involved the leading intellectual advance in postwar geophysics, the theory of plate tectonics. Almost all scholars who have traced the rapid emergence and widespread acceptance of plate tectonics in the 1960s agree that national security concerns involving undersea warfare released a flood of defense dollars for new instruments and unprecedented surveys of the ocean's floors, providing data about undersea earthquakes and magnetic patterns alongside midoceanic ridges (their planet-encircling extent itself a postwar discovery resulting from these investments in physical oceanography) (LeGrand 1988; Oreskes and LeGrand 2001; Oreskes and Doel 2002; Frankel 2012b). But Forman's first question still loomed: Had military patronage for seismology—in the immediate postwar years a small research community—distorted this field when the US military, in the late 1950s, invested heavily in seismology and global seismological networks? These networks mattered, for within a decade, they made it possible for earth scientists to distinguish between earthquakes and atomic blasts, a prerequisite for future nuclear arms control treaties to be verifiable and enforceable. Exploring the evolution of Cold War seismology, Kai-Henrik Barth argued that "seismologists did not lose control of their discipline, despite massive DOD interests" (Barth 2003). Fellow historian of science Michael Aaron Dennis disagreed: Seismology had until then always been an applied field (exploration geophysics or earthquake prediction), defined by an instrument (the seismograph), and thus "inseparable from practical technical problems–there was no field to distort" (Dennis 2003, 815). Recently Naomi Oreskes, bringing decades of research into focus, has offered a wide-ranging analysis of Navy patronage for oceanography in the Cold War that revisits many of these issues, including the emergence of plate tectonics and ways that patronage structures influenced the questions that earth scientists felt most worthy of pursuit (Oreskes 2021). Returning to the Forman-Kevles debate, Oreskes suggested a key issue had been overlooked: While "military support for oceanography and marine geophysics was both enabling and constricting," such funding "resulted in the creation of important domains of knowledge, but it also created significant, lasting, and consequential domains of ignorance" (Oreskes 2021, 8). These topics merit continuing inquiry (Antonello 2021).

Concerns aside, historians of science owe Forman and Kevles a great deal, for their claims encouraged scholars to explore a wide range of ways that patronage (state as well as military) influenced both the production and diffusion of knowledge in the Cold War era. This effort remains a work-in-progress: Diplomatic historians have highlighted a "hall of mirrors" dynamic underlying numerous foreign policy and intelligence-gathering undertakings throughout the Cold War, and scholars studying the earth sciences in this time, sensing subtle but profound influences, have faced similar challenges in articulating them. As Matthew Dörries has argued (Dörries 2011), the debate over "nuclear winter" in the 1980s reflected both the emerging science of climate change and a reaction to then US President Ronald Reagan's policies on nuclear warfare. Similarly, the intense mass extinction debate of the late twentieth century, fueled by discoveries in astrophysics and paleontology, also resonated with the then-pervasive awareness of atomic bomb disasters (Clube 1994), while understanding of the impacts of asteroids and comets on planetary bodies (part of the mass extinction debate, but also at the heart of interpreting the cause of lunar craters) was crucially aided by direct study of nuclear explosion craters at the Nevada Test Site (Doel 1996). That it took nearly two decades for the Stommel Diagram—a three-dimensional conceptual tool developed by the Woods Hole physical oceanographer Henry Stommel to affirm why scale mattered in studying the ocean's physical properties—to become an equally significant means for studying oceanic ecosystems reflects limitations on the diffusion of knowledge even within institutions employing both physical and biological oceanographers (Vance and Doel 2010). "Oceanography as a discipline still, in some ways, does not quite exist," Naomi Oreskes (2014c) recently argued. The extent to which this reflects Cold War practices—including classification policies, professional and disciplinary divides—remains a fruitful issue to explore (Hamblin 2005).

Scholars also have begun exploring a wide range of related topics, among them the growth of earth science and oceanographic institutions after 1945 (Wallace 2016), the recruitment of graduate students to these institutions, and limits imposed on the international practices of these fields amid Cold War global fractures. Navy interest in gaining new knowledge about the physical environment led to remarkable expansion of US oceanographic facilities, including the Scripps Institution of Oceanography, the Woods Hole Oceanographic Institution, and Columbia University's Lamont Geological Observatory (now Lamont-Doherty Earth Observatory) (Doel 2003a, b; Oreskes and Rainger 2000; Rainger 2000b, 2003; Weir 2000, 2001; Oreskes 2021). Valuable new insights have come from researchers addressing how the earth sciences expanded, not only in the Soviet Union (Aronova 2017b, c), but also within smaller advanced nations amid the Cold War, in particular Scandinavian countries (Wicken 2008; Sörlin 2009, 2014; Vollset et al. 2018). Other scholars have addressed a distinct realm of Cold War oceanography, human activities in underseas environments (Rozwadowski and van Keuran 2004; Benson and Rozwadowski 2007; Alder 2019). What has remained particularly elusive are detailed narratives of international associations of earth scientists in the second half of the twentieth century, in contrast to those for astronomers and physicists (for instance, Lalli 2019). We do not yet have in-depth historical accounts of either the American Geophysical Union or the International Union of Geodesy and Geophysics, although one recent overview of the IUGG makes occasional mention of global political challenges that hindered its activities (Joselyn and Ismail-Zadeh 2019).

6 Opportunities and Conclusions

In their 2000 article "Why Geophysics?" Naomi Oreskes and James R. Fleming asked, "Why devote energy to studying a discipline that many view as derivative, others as peripheral and unimportant? Why should historians and philosophers of physics pay attention?" (Oreskes and Fleming 2000, 253). Back then, Oreskes and Fleming were celebrating that emerging scholarship had begun to reveal that "collaboration with military organizations" was far from "a distinctive feature of physics" (255). Since then, much has changed. It is worth remembering that 30 years earlier, in the 1970s, what then was new under the sun was the history of science in America (Reingold 1976). Hence, it is satisfying, but not surprising, that historians by now have outlined key issues in Cold War geophysics and geochemistry (Masco 2010).

What we do not yet know about the Cold War earth sciences nevertheless remains vast. Some issues are perhaps particularly worth pursuing. Let me note four that promise to broaden our understanding of the evolution of the earth and environmental sciences in the second half of the twentieth century and beyond.

How did the earth sciences function globally during the Cold War? The International Geophysical Year remains an undertaking that is remarkably well-studied—and also just barely explored. Historians have struggled with the enormity of the IGY for decades: analyzing the full scales and dynamics of this enterprise, spanning some 66 nations, seems similar to the challenge the blind face when encountering an elephant (Collis and Dodds 2008; Elzinga 2009; Goossen 2020; Lehman 2020). In recent years, scholars have begun exploring how the IGY impacted nations beyond the USA, Britain, France, Belgium, and the Soviet Union, whose scientists set the IGY's agenda: New studies explore developments in Cold War Czechoslovakia, Poland, and the German Democratic Republic (Olšáková 2017), Australia (Moss 2020), Spain (Batlló 2008), and South Africa (van der Watt and Swart 2016). Such investigations are welcome and important additions to understanding research practices and the evolving status of science in advanced nation-states around the globe. But we still know little about what happened in the majority of countries that formally joined the IGY. As Barbara Silva has observed, we know almost nothing about the IGY's impact on Latin and South America apart from studies focusing on conflicts between Argentina and Chile over Antarctic territorial claims, before the Antarctic Treaty put such tensions to rest (Silva 2020). And apart from the already-advanced nation of South Africa, we know virtually nothing about what the IGY meant for scientists and political leaders in sub-Saharan Africa as the profound transition from colonial rule to independent nations began to unfold (Krige 2019, 439).

We also know too little about the large transnational transient institutions modeled on the IGY launched in the last decades of the Cold War: In what ways were these shaped by Cold War and national security pressures; in what ways were the environments in which they operated isolated from political, cultural, and intelligence-gathering influences? Thanks to recent scholarly work, we now are gaining insights into the International Geosphere and Biosphere Program, launched in the last years of the Cold War, that ambitiously sought to coordinate research on biological, chemical, and physical processes at global and regional scales (Kwa 2005; Uhrqvist and Linnér 2015; Rispoli and Olšáková 2020). But we still have few historical insights into such major undertakings as the Indian Ocean Expedition (1959–1965) and the International Year of the Quiet Sun (1964–1965) apart from how international data-sharing policies affected them (Aronova 2017a). Other major initiatives launched amid the Cold War—such as the International Symposia on Equatorial Aeronomy, first convened in 1962, born of a burgeoning Latin America and sub-Saharan African research infrastructure distinct from its European progenitors—await their historians.

What broader relationships exist between the earth sciences, foreign policy initiatives, and diplomacy amid the global Cold War? Much of what we know about how earth scientists became involved in foreign policy initiatives has come from studies of scientific and state activities at the highest levels. This approach has proven fruitful for decades for historians of science as well as diplomatic historians. As Allan A. Needell has shown, US leaders of the IGY, particularly Lloyd V. Berkner, negotiated the frequently fraught relationships between scientists, political leaders, and intelligence agencies in the 1950s, and such scholars as Zuoyue Wang, Roger Launius, and Paul Berkman, among others, have explored international treaty negotiations where earth scientists were central players, including the Limited Nuclear Test Ban Treaty, the Antarctic Treaty, and the Outer Space Treaty (Wang 2008; Berkman 2011; Launius 2017; see also Oreskes 2014a). More recently, Audra Wolfe has explored how leading scientists, working with policymakers and officials, helped shape American cultural diplomacy throughout the Cold War (Wolfe 2020).

But have we missed crucial insights by focusing almost exclusively on top-level discussions and negotiations, ignoring the efforts of working scientists who actually undertook these negotiations, or participated in transnational exchange programs that high-level officials sought to put into play? Emerging scholarship suggests this is the case. Lower-level scientific contacts allowed Swedish and Soviet climatologists to collaborate before and after World War II (Lajus and Sörlin 2014). Sustained interactions between US and Soviet earth scientists made possible by the 1972 US-U.S.S.R. Agreement to Protect the Environment created enduring midlevel professional and personal connections, networks that could not be discerned from top-level documentary evidence (Amramina 2015, 2021). Other scholars from Denmark and Russia recently explored how seismologists in the USA and Soviet Union found ways to directly exchange instruments in the 1960s. Here, seismographs served as diplomatic objects that furthered the aim of arms control by aiding nuclear detection efforts (Jacobsen et al. 2021). Examining the earth sciences through the broad lens of science diplomacy also allows renewed focus on the circulation of scientific knowledge on global, national, and regional scales (Roberts 2009; Turchetti et al. 2012; Martínez-Rius 2020; Jacobsen and Olšáková 2020; Adamson and Lalli 2021; Kunkel 2021; Robinson et al. 2023).

What fruitful intersections can be explored at the intersection of Cold War earth sciences and environmental history? We have, at present, two distinct historiographies of the environmental sciences in the Cold War United States: One that explores the well-funded earth sciences, constituted by military interest in their applications to national security; and a second, focused on the rise of the Environmental Movement, pollution, resource exploitation, and the creation of agencies dedicated to environmental protection, where biology (particularly ecology) is the central scientific field. These narratives exist side-by-side, rarely overlapping, even if they both address what we would consider sciences of the environment (Doel 2009; Aronova et al. 2010; Goossen 2020).

What can we gain by foregrounding questions that environmental historians have asked, for instance, about sustainable practices, the social and political commitments of scientists, and emerging conceptions of global systems? (Heyck and Kaiser 2010; Aronova 2017b, c).

Fortunately, much good work of this kind is already underway. Perrin Selcer has explored how the United Nations promoted soils research around the planet (Selcer 2018); Jacob Hamblin and Toshihiro Higuchi have both charted how potential and actual catastrophes, including radioactive fallout, stimulated public concern about the planet's environment (Hamblin 2008, 2013; Higuchi 2020), and Elena Aronova tackled the challenges of conceiving environmental monitoring, broadly conceived, in the Soviet Union at the height of the Cold War (Aronova 2015). Other scholars succeeded in addressing the full range of the environmental sciences by focusing on particular geographic regions, including the North American Arctic (Heidt and Bocking 2019), Eastern Europe (Olšáková and Oldfield 2012; Oldfield and Shaw 2015), the Soviet Union (Josephson 2014), and Earth's poles (Howkins 2015). But many questions remain. We need to better appreciate how these scientists, "as individuals, understood and negotiated their relationships with the national security state," as John Krige put it (Krige 2019, 439), and better understand who participated in the production of knowledge in the earth sciences through the Cold War, including minorities, members of Indigenous communities, and women (Oreskes 1996, 2000; Sörlin 2009; Doel et al. 2014; Fleming 2020). Jessica Wang's provocative study of Merle Tuve, Director of the Department of Terrestrial Magnetism at the Carnegie Institution of Washington, suggests one way to address Krige's question might be to probe the emotional lives of scientists (Wang 2012).

In what ways can the history of Cold War earth science inform the emerging concept of the Anthropocene? We now live in a new geological epoch, according to an ever-strengthening consensus of environmental scientists, where we can recognize distinct human impacts on planetary ecologies, that is, human activities have sufficient agency to alter the planetary environment at significant scales, evidenced by climate change and biodiversity loss (McNeill and Engelke 2016). As I write in summer 2023, a team of natural scientists has proposed that Crawford Lake, outside Toronto, Canada, could serve as a natural marker for the start of the Anthropocene: Plutonium accumulations at the bottom of this deep, pristine lake are likely to identify a year between 1950 and 1954 to tag the beginning of this epoch (Prillaman 2023). Can narratives of Cold War earth science—and the historians who produced them—contribute to the increasingly robust discussions among ecologists, social scientists, philosophers, and environmental scientists about our collective history, present and future? (Malhi 2017)

Scholars have suggested novel ways that historians of science can contribute to evolving concepts of the Anthropocene. Jessica Lehman, for instance, has argued that the IGY, as the first attempt to understand the Earth as a planet through widespread synoptic data collection, allows historians and historical geographers to critique what is known and yet unknown about the Earth. Creating such histories will require scholars to work collaboratively, in teams, to produce new narratives of big science at very large scales (Lehman 2020; see also Uhrqvist and Linnér 2015). Helmuth Trischler goes further: The Anthropocene, he believes, makes it possible "to rethink the nature-culture relation and to re-assess the narratives that historians of science, technology, and the environment have written until now" (Trischler 2016, 309). Applauding Dipesh Chakrabarty's efforts to de-center human-based narratives in order to create an "integral geohistory" that links human history and natural history, Trischler declares the Anthropocene concept should inspire new forms of interdisciplinarity between scientists and humanists (Trischler 2016, 328–329; Chakrabarty 2009). Reading a different lesson from the Anthropocene, Kathryn Yusoff has argued that thinking about extractive practices since the start of the industrial age, associated with dominant European activities, can inspire too-simplistic conceptions of the Anthropocene that erase understanding of deep-rooted racial inequities and imbalances. "No geology is neutral," wrote Yusoff, concluding her book with an insight already well familiar to many historians of science (Yusoff 2018, 108). No less familiar to historians are speculative geoengineering remedies central to certain Anthropocene imaginaries (LeCain 2015; McNeill and Engelke 2016).

In 2003, Kai-Henrik Barth, echoing the view of many colleagues, commented that "the history of the geophysical sciences remains largely unexplored" (Barth 2003, 745). No longer. The challenge of understanding our past and present planetary environment seems to promise this field continued relevance.

References

Adamson, M., and R. Lalli. 2021. Global perspectives on science diplomacy: Exploring the diplomacy-knowledge nexus in contemporary histories of science. *Centaurus* 63 (2): 1–16.

Agar, J. 1998. *Science & spectacle: The work of Jodrell Bank in post-war British culture.* New York: Harwood Academic Publishers.

Alder, A. 2019. *Neptune's laboratory: Fantasy, fear and science at sea.* Cambridge, MA: Harvard University Press.

Amramina, A. 2015. Political seismology or seismological politics: Natural Resources Defense Council–USSR experiments in underground nuclear test verification. *Seismological Research Letters* 86 (2A): 451–457.

———. 2021. *A common language of the earth: American-Soviet scientific collaborations during the Cold War.* Ph.D. Dissertation, University of Minnesota.

Anduaga, A. 2016. *Geophysics, realism, and industry: How commercial interests shaped geophysical conceptions, 1900–1960.* New York: Oxford University Press.

Antonello, A. 2021. The fortunes of modern oceanography. *Metascience* 30: 451–454.

Aronova, E. 2015. Environmental monitoring in the making: From surveying nature's resources to monitoring nature's change. *Historical Social Research* 40 (2): 222–245.

———. 2017a. Geophysical datascapes of the Cold War: Politics and practices of the World Data Centers in the 1950s and 1960s. *Osiris* 32 (1): 307–327.

———. 2017b. Recent trends in the historiography of science in the Cold War. *Historical Studies in the Natural Sciences* 47 (4): 568–577.

———. 2017c. Citizen seismology, Stalinist science, and Vladimir Mannar's Cold Wars. *Science, Technology, & Human Values* 42 (2): 226–256.

Aronova, E., K.S. Baker, and N. Oreskes. 2010. Big science and big data in biology: From the International Geophysical Year through the international biological program to the Long Term Ecological Research (LTER) Network, 1957—Present. *Historical Studies in the Natural Sciences* 40 (2): 183–224.

Aronova, E., S. Leonelli, G. Davies, and B. Rappert. 2017. Citizen seismology, stalinist science, and Vladimir Mannar's Cold Wars. *Science, Technology & Human Values* 42 (2): 226–256.
Barnes, T.J., and M. Farish. 2006. Between regions: Science, militarism, and American geography from World War to Cold War. *Annals of the Association of American Geographers* 96 (4): 807–826.
Barney, T. 2015. *Mapping the cold war: Cartography and the framing of America's international power*. Chapel Hill: The University of North Carolina Press.
Barrett, G. 2022. *China's cold war science diplomacy*. New York: Cambridge University Press.
Barth, K.H. 2003. The politics of seismology: Nuclear testing, arms control, and the transformation of a discipline. *Social Studies of Science* 33 (5): 743–781.
Barton, J.P. 2020. *Wiring the world: A history of the earth system concept in the US earth sciences, 1982–1989*. Ph.D. Dissertation, University of Toronto.
Batlló, J. 2008. Notes on the history of the International Polar Years in Catalonia and Spain. *Tethys* 5: 47–57.
Benson, E. 2012. One infrastructure, many global visions: The commercialization and diversification of Argos, a satellite-based environmental surveillance system. *Social Studies of Science* 42 (6): 843–868.
Benson, K.R., and H.M. Rozwadowski, eds. 2007. *Extremes: Oceanography's adventures at the poles*. Sagamore Beach: Science History Publications.
Berkman, P.M. 2011. President Eisenhower, the Antarctic treaty, and the origin of international spaces. In *Science diplomacy: Antarctica, science, and the governance of international spaces*, ed. P.A. Berkman, M.A. Lang, W.H. David, and O.R. Young, 17–27. Washington, DC: Smithsonian Contributions to Knowledge.
Bowler, P.J. 1992. *Norton history of the environmental sciences*. New York: Norton.
Browne, J. 2021. Charles Darwin and the Darwinian tradition. In *Handbook of the historiography of biology*, ed. M.R. Dietrich, M.E. Borrello, and O. Harman, 7–31. New York: Springer.
Bulkeley, R.I.P. 1991. *The sputniks crisis and early United States space policy: A critique of the historiography of space*. New York: Springer.
———. 2000. The sputniks and the IGY. In *Reconsidering sputnik: Forty years since the soviet satellite*, ed. R.D. Launius, J.M. Logsdon, and R.W. Smith, 117–124. London: Routledge.
———. 2008. Aspects of the Soviet IGY. *Russian Journal of Earth Sciences* 10: 1–17.
Butrica, A.J. 1996. *To see the unseen: A history of planetary radio astronomy*. Washington, DC: NASA.
Camprubi, L., and S. Robinson. 2016. A gateway to ocean circulation: Surveillance and sovereignty at Gilbraltor. *Historical Studies in the Natural Sciences* 46 (4): 429–459.
Cantoni, R. 2017. *Oil exploration, diplomacy, and security in the early Cold War: The enemy within*. New York: Routledge.
Chakrabarty, D. 2009. The climate of history: Four theses. *Critical Inquiry* 35 (2): 197–222.
Christensen, M., A.E. Nilsson, and N. Wormbs. 2013. *Media and the politics of Arctic climate change: When the ice breaks*. New York: Palgrave Macmillan.
Cloud, J. 2000. Crossing the Olentangy River: The figure of the earth and the military-industrial-academic complex, 1947–1972. *Studies in History and Philosophy of Modern Physics* 31 (3): 371–404.
———. 2001. Imaging the world in a barrel: CORONA and the clandestine convergence of the earth sciences. *Social Studies of Science* 31 (2): 231–251.
———. 2002. American cartographic transformations during the Cold War. *Cartography and Geographic Information Science* 29 (3): 261–282.
———. 2003. Introduction [Special Guest-edited issue on the earth sciences in the Cold War]. *Social Studies of Science* 33 (5): 629–633.
———. 2013. American cartographic transformations during the Cold War. *Cartography and Geographic Information Science* 29 (3): 261–282.
Clube, S.V.M. 1994. Hazards from space: Comets in history and science. In *The mass extinction debate: How science works in a crisis*, ed. W. Glen, 152–169. Stanford: Stanford University Press.
Collis, C., and K. Dodds. 2008. Assault on the unknown: The historical and political geographies of the International Geophysical Year (1957–8). *Journal of Historical Geography* 34 (4): 555–573.

Conway, E.M. 2008. *Atmospheric science at NASA: A history*. Baltimore: The Johns Hopkins University Press.
———. 2014. Bringing NASA back to earth: A search for relevance during the Cold War. In *Science and technology in the global Cold War*, ed. N. Oreskes and J. Krige, 251–272. Cambridge, MA: The MIT Press.
De Chadarevian, S. 2002. *Designs for life: Molecular biology after World War II*. Cambridge, UK: Cambridge University Press.
Dennis, M.A. 2003. Earthly matters: On the cold war and the earth sciences. *Social Studies of Science* 33 (5): 809–819.
———. 2006. Secrecy and science revisited: From politics to historical practice and back. In *The historiography of contemporary science, technology, and medicine: Writing recent science*, ed. R.E. Doel and T. Söderqvist, 189–203. London: Routledge.
DeVorkin, D.H. 1987. Organizing for space research: The V-2 rocket panel. *Historical Studies in the Physical and Biological Sciences* 18 (1): 1–24.
———. 1992. *Science with a vengeance: How the military created the U.S. space sciences after World War II*. New York: Springer.
Dodds, K. 2005. The great trek: New Zealand and the British/commonwealth 1955–58 trans-Antarctic expedition. *The Journal of Imperial and Commonwealth History* 33 (1): 93–114.
———. 2010a. Flag planting and finger pointing: The law of the sea, the Arctic and the political geographies of the outer continental shelf. *Political Geography* 29 (2): 63–73.
———. 2010b. Governing Antarctica: Contemporary challenges and the enduring legacy of the 1959 Antarctic Treaty. *Global Policy* 1 (1): 108–115.
Dodds, K., I. Gan, and A. Howkins. 2010. The IPY-3: The International Geophysical Year (1957–1958). In *The history of the International Polar Years (IPYs)*, ed. S. Barr and C. Lüdecke, 239–257. Berlin: Springer.
Doel, R.E. 1996. *Solar system astronomy in America: Communities, patronage, and interdisciplinary science, 1920–1960*. New York: Cambridge University Press.
———. 1997a. The earth sciences and geophysics. In *Science in the twentieth century*, ed. J. Krige and D. Pestre, 361–388. London: Harwood Academic Publishers.
———. 1997b. Scientists as policymakers, advisors, and intelligence agents: Linking diplomatic history with the history of contemporary science. In *The historiography of contemporary science and technology*, ed. T. Söderqvist, 215–244. New York: Harwood Academic Press.
———. 2003a. Constituting the Postwar earth sciences: The Military's influence on the environmental sciences in the USA after 1945. *Social Studies of Science* 33 (5): 635–666.
———. 2003b. Oral history of American science: A forty year review. *History of Science* 41 (4): 349–378.
———. 2007. Hubbert, Marion King. *New Dictionary of Scientific Biography* 21: 395–399.
———. 2009. Quelle place pour les sciences de l'environnement physique dan l'histoire environnementale? *Revue d'histoire moderne et contemporaine* 57 (4): 137–164.
———. 2015. Scientists, secrecy, and scientific intelligence: The challenges of international science in the Cold War era. In *Cold War science and the transatlantic circulation of knowledge*, ed. J. Dongen, 9–35. Leiden: Brill.
———. 2020. Internationalism in science after 1940. In *Cambridge history of science*, ed. H.R. Slotten, R.L. Numbers, and D.L. Livingstone, vol. 5, 60–74. New York: Cambridge University Press.
Doel, R.E., and K.C. Harper. 2006. Prometheus unleashed: Science as a diplomatic weapon in the Lyndon B. Johnson Administration. *Osiris* 21 (1): 66–85.
Doel, R.E. and A.A. Needell. 1997. Science, scientists, and the CIA: Balancing international ideals, national needs, and professional opportunities. *Intelligence and National Security* 12 (1): 59–81.
Doel, R.E., D. Hoffmann, and N. Krementsov. 2005. National states and international science: A comparative history of international science congresses in Hitler's Germany, Stalin's Russia, and Cold War United States. *Osiris* 20: 49–76.
Doel, R.E., T.J. Levin, and M. Marker. 2006. Extending modern cartography to the ocean depths: Military patronage, Cold War priorities, and the Heezen–Tharp mapping project, 1952–1959. *Journal of Historical Geography* 32 (3): 605–626.

Doel, R.E., R.M. Friedman, J. Lajus, S. Sörlin, and U. Wråkberg. 2014. Strategic Arctic science: National interests in building natural knowledge–interwar era through the Cold War. *Journal of Historical Geography* 44: 60–80.
Doel, R.E., K.C. Harper, and M. Heymann, eds. 2016. *Exploring Greenland: Cold War science on ice*. New York: Palgrave Macmillan.
Doose, K.A. 2021. Global problem in a divided world: Climate change research during the late Cold War. *Cold War History* 21 (4): 469–489.
Dora, V. 2023. From the radio shack to the cosmos: Listening to sputnik during the International Geophysical Year (1957–1958). *Isis* 114 (1): 123–149.
Dörries, M. 2011. The politics of atmospheric sciences: 'Nuclear winter' and global climate change. *Osiris* 26: 198–223.
———. 2020. Hot climate Cold War. *Historical Studies in the Natural Sciences* 50 (1–2): 67–73.
Edwards, P.N. 2006. Meteorology as infrastructure globalism. *Osiris* 21: 229–250.
———. 2013. *A vast machine: Computer models, climate data, and the politics of global warming*. Cambridge, MA: MIT Press.
Elzinga, A. 2009. Geopolitics, science and internationalism during and after the IGY. In *2nd SCAR workshop on the history of Antarctic research. Multidimensional exploration of Antarctica around the 1950s*, ed. C. Lüdecke, 71–81. Puenta Arenas: Boletin Antártico Chileno, Instituto Antártico Chileno.
Felt, H. 2013. *Soundings: The story of the remarkable woman who mapped the ocean floor*. New York: Holt.
Fleming, J.R. 1998. *Historical perspectives on climate change*. New York: Oxford University Press.
———. 2006. The pathological history of weather and climate control: Three cycles of promise and hype. *Historical Studies in the Physical and Biological Sciences* 37 (1): 3–25.
———. 2010. *Fixing the sky: The checkered history of weather and climate control*. New York: Columbia University Press.
———. 2016. *Inventing atmospheric science: Bjerknes, Rossby, Wexler, and the foundations of modern meteorology*. Cambridge, MA: MIT Press.
———. 2020. *First woman: Joanne Simpson and the tropical atmosphere*. New York: Oxford University Press.
Forman, P. 1987. Behind quantum electronics: National Security as basis for physical research in the United States, 1940–1960. *Historical Studies in the Physical and Biological Sciences* 18 (1): 149–229.
Frankel, H.R. 2012a. *The continental drift controversy, Vol. 3: Introduction of seafloor spreading*. New York: Cambridge University Press.
———. 2012b. *The continental drift controversy, Vol. 4: Evolution and plate tectonics*. New York: Cambridge University Press.
Gillmor, C.S. 1984. Aging of geophysicists. *Eos* 65 (20): 353–354.
———. 1986. Federal Funding and knowledge growth in ionospheric physics, 1945–81. *Social Studies of Science* 16 (1): 105–133.
Glen, W. 1986. Reviewed work(s): History of geophysics. vol 1. *Isis* 77 (2): 349–351.
Good, G.A. 2000. The assembly of geophysics: Scientific disciplines as frameworks of consensus. *History and Philosophy of Science Part B: Studies in History and Philosophy of Modern Physics* 31 (3): 259–292.
Goossen, B.W. 2020. A benchmark for the environment: Big science and 'artificial' geophysics in the global 1950s. *Journal of Global History* 15 (1): 149–168.
Grevsmühl, S.V. 2018. Revisiting the 'ozone hole' metaphor: From observational window to global environmental threat. *Environmental Communication* 12 (1): 71–83.
Hacker, B.C. 2000. Military patronage and the geophysical sciences in the United States: An introduction. *Historical Studies in the Physical and Biological Sciences* 30 (2): 309–313.
Hamblin, J.D. 2005. *Oceanographers and the Cold War: Disciples of marine science*. Seattle: University of Washington Press.

———. 2008. *Poison in the well: Radioactive waste in the oceans at the Dawn of the nuclear age*. New Brunswick: Rutgers University Press.
———. 2013. *Arming mother nature: The birth of catastrophic environmentalism*. New York: Oxford University Press.
Harper, K.C. 2003. Research from the boundary layer: Civilian leadership, military funding and the development of numerical weather prediction (1946–55). *Social Studies of Science* 33 (5): 667–696.
———. 2008. Climate control: United States weather modification in the Cold War and beyond. *Endeavour* 32 (1): 20–26.
———. 2017. *Make it rain: State control of the atmosphere in twentieth-century America*. Chicago: University of Chicago Press.
Heidt, D., and S. Bocking, eds. 2019. *Cold science: Environmental knowledge in the North American Arctic during the Cold War*. Abingdon-on-Thames: Taylor & Francis.
Herzberg, J., C. Kehrt, and F. Toma. 2018. *Ice and snow in the Cold War: Histories of extreme climatic environments*. New York: Berghahn Press.
Heyck, H., and D. Kaiser. 2010. Focus: New perspectives on science and the Cold War – Introduction. *Isis* 101 (2): 362–366.
Heymann, M., H. Knudsen, M.L. Lolck, H. Nielsen, K.H. Nielsen, and C.J. Ries. 2010. Exploring Greenland: Science and technology in Cold War settings. *Scientia Canadensis* 33 (2): 11–42.
Higuchi, T. 2020. *Political fallout: Nuclear weapons testing and the making of a global environmental crisis*. Stanford: Stanford University Press.
Howe, J.P. 2014. *Behind the curve: Science and the politics of global warming*. Seattle: University of Washington Press.
Howkins, A. 2008. Reluctant collaborators: Argentina and Chile in Antarctica during the International Geophysical Year, 1957–58. *Journal of Historical Geography* 34 (4): 596–617.
———. 2011. Melting empires? Climate change and politics in Antarctica since the International Geophysical Year. *Osiris* 26: 180–197.
———. 2015. *The polar regions: An environmental history*. Cambridge, UK: Polity.
Jacobsen, L.L., and D. Olšáková. 2020. Diplomats in science diplomacy: Promoting scientific and technological collaboration in international relations. *Berichte zur Wissenschaftsgeschichte* 43 (4): 465–472.
Jacobsen, L.L., I. Federova, and J. Lajus. 2021. The seismograph as a diplomatic instrument: The soviet-American exchange of instruments, 1958–1964. *Centaurus* 63 (2): 277–295.
Johnston, R., M. Fairbrother, D. Hayes, T. Hoare, and K. Jones. 2008. The cold war and geography's quantitative revolution: Some messy reflections on Barnes' geographical underworld. *Geoforum* 39: 1802–1806.
Joselyn, J.A., and A. Ismail-Zadeh. 2019. IUGG evolves (1940–2000). *History of Geo- and Space Science* 10: 45–72.
Josephson, P. 2014. *The conquest of the Soviet Arctic*. Cambridge, MA: Harvard University Press.
Kangas, P.C. 2023. *A history of radioecology*. New York: Routledge.
Kevles, D.J. 1997. Big science and big politics in the United States: Reflections on the death of the SSC and the life of the human genome project. *Historical Studies in the Physical and Biological Sciences* 27: 269–297.
Kingsland, S. 2005. *The evolution of American ecology, 1920–2000*. Baltimore: The Johns Hopkins University Press.
Kirsch, S. 2005. *Proving grounds: Project plowshare and the unrealized dream of nuclear earthmoving*. New Brunswick: Rutgers University Press.
Koppes, C.R. 1982. *JPL and the American space program: A history of the jet propulsion laboratory*. New Haven: Yale University Press.
Korsmo, F.L. 2004. Shaping up planet earth: The international geophysical year (1957–1958) and communicating science through print and film media. *Science Communication* 26 (2): 162–187.
———. 2007. The genesis of the International Geophysical Year. *Physics Today* 60 (7): 38–43.

———. 2008. The policy process and the International Geophysical Year, 1957–1958. In *Smithsonian at the poles: Contributions to International Polar Year science*, ed. I. Krupnik, M.A. Lang, and S.E. Miller, 23–34. Washington, DC: Smithsonian Institution Press.

Korsmo, F.L., and A. Graham. 2002. Research in the American north: Action and reaction. *Arctic* 55 (4): 319–328.

Korsmo, F.L., and M.P. Sfraga. 2003. From interwar to Cold War: Selling field science in the United States, 1920s through 1950s. *Earth Sciences History* 22 (1): 55–78.

Kragh, H. 2000. The chemistry of the universe: Historical roots of modern cosmochemistry. *Annals of Science* 57 (4): 353–368.

Krige, J. 2014. Diplomacy (Post-1945), science and technology. In *The Oxford encyclopedia of the history of American science, medicine, and technology*, ed. H.R. Slotten, 252–267. New York: Oxford University Press.

———., ed. 2019. *How knowledge moves: Writing the transnational history of science and technology*. Chicago: University of Chicago Press.

Krige, J., and J. Sarkar. 2018. US technological collaboration for nonproliferation: Key evidence from the Cold War. *The Nonproliferation Review* 25 (3–4): 249–262.

Kunkel, S. 2021. Science diplomacy in the twentieth century: Introduction. *Journal of Contemporary History* 56 (3): 473–484.

Kwa, C. 1993. Radiation ecology, systems ecology and the management of the environment. In *Science and nature: Essays in the history of the environmental sciences*, ed. M. Shortland, 213–250. Oxford: Alden Press.

———. 2005. Local ecologies and global science: Discourses and strategies of the International Geosphere-Biosphere Programme. *Social Studies of Science* 35 (6): 923–950.

Lajus, J. 2020. Soviet official critiques of the resource scarcity prediction by limits to growth report: The case of Evgenii Fedorov's ecological crisis rhetoric. *European Review of History* 27 (3): 321–341.

Lajus, J., and S. Sörlin. 2014. Melting the glacial curtain: The politics of Scandinavian–Soviet networks in the geophysical field sciences between two polar years, 1932/33–1957/58. *Journal of Historical Geography* 44: 44–59.

Lalli, R. 2019. Patterns to scientific internationalism: What can a comparative history of IAU and IUPAP teach us? In *Under one sky: The IAU centenary symposium proceedings IAU symposium*, ed. C. Sterken, J. Hearnshaw, and D. Valls-Gabaud, vol. 349, 189–196.

Launius, R.D. 2010. An unintended consequence of the IGY: Eisenhower, Sputnik, the founding of NASA. *Acta Astronautica* 67 (1–2): 254–263.

———. 2014. *Frontiers of space exploration*. Westport: Greenwood Press.

———. 2016. 'We will learn more about the earth by leaving it than by remaining on it:' NASA and the forming of an earth science discipline in the 1960s. In *Innovation in science and organizational renewal*, ed. T. Heinze and R. Münch, 211–242. New York: Palgrave Macmillan.

———. 2017. Establishing open rights in the Antarctic and outer space: Cold War rivalries and geopolitics in the 1950s and 1960s. In *Handbook on the politics of Antarctica*, ed. K. Dodds, A.D. Hemmings, and P. Roberts, 217–231. Northampton: Edward Elgar Publishing.

LeCain, T.J. 2015. Against the Anthropocene: A neo-materialist perspective. *International Journal of History, Culture, and Modernity* 3: 1–28.

LeGrand, H.E. 1988. *Drifting continents and shifting theories: The modern revolution in geology and scientific change*. New York: Cambridge University Press.

Lehman, J. 2020. Making an Anthropocene Ocean: Synoptic geographies of the International Geophysical Year (1957–1958). *Annals of the American Association of Geographers* 110 (3): 606–622.

Levin, T.J., and R.E. Doel. 2000. The Lamont-Doherty earth observatory oral history project: A preliminary report. *Earth Sciences History* 19 (1): 26–32.

Logsdon, J.M. 2010. *John F. Kennedy and the race to the moon*. New York: Palgrave Macmillan.

Low, M.C. 2020. Desert dreams of drinking the sea, consumed by the Cold War: Transnational flows of desalination and energy from the Pacific to the Persian Gulf. *Environmental History* 26 (2): 145–174.

Mack, P.E. 1990. *Viewing the earth: The social construction of the Landsat satellite system*. Cambridge, MA: MIT Press.
Maher, N.M. 2019. Brining the environment back in: A transnational history of Landsat. In *How knowledge moves: Writing the transnational history of science and technology*, ed. J. Krige, 201–224. Chicago: University of Chicago Press.
Malhi, Y. 2017. The concept of the Anthropocene. *Annual Review of Environment and Resources* 42: 77–104.
Martínez-Rius, B. 2020. For the benefit of all men: Oceanography and Franco-American scientific diplomacy in the Cold War, 1958–1970. *Berichte zur Wissenschaftsgeschichte* 43: 581–605.
Martin-Nielsen, J. 2012. The other Cold War: The United States and Greenland's ice sheet environment, 1948–1966. *Journal of Historical Geography* 38 (1): 69–80.
———. 2013. 'The deepest and most rewarding hole ever drilled': Ice cores and The Cold War in Greenland. *Annals of Science* 70 (1): 47–70.
Masco, J. 2010. Bad weather: On planetary crisis. *Social Studies of Science* 40 (1): 7–40.
McCray, W.P. 2006. Amateur scientists, the International Geophysical Year, and the ambitions of Fred Whipple. *Isis* 97 (4): 634–658.
———. 2008. *Keep watching the skies! The story of operation moonwatch & the dawn of the space age*. Princeton: Princeton University Press.
McDougall, W.A. 1985. *...the heavens and the earth: A political history of the space age*. New York: Basic Books.
McNeill, J.R. 2000. *Something new under the sun: An environmental history of the twentieth century world*. New York: Norton.
McNeill, J.R., and P. Engelke. 2016. *The great acceleration: An environmental history of the Anthropocene since 1945*. Cambridge, MA: Harvard University Press.
Miller, C.A. 2001. Scientific internationalism in American foreign policy: The case of meteorology (1947–1958). In *Changing the atmosphere: Expert knowledge and environmental governance*, ed. C.A. Miller and P.N. Edwards, 167–218. Cambridge, MA: MIT Press.
Mills, E.L. 2009. *The fluid envelope of our planet: How the study of ocean currents became a science*. Toronto: University of Toronto Press.
Moss, T. 2020. 'There are many other things more important to us than space research:' The Australian Government and the dawn of the space age, 1956–1962. *Australian Historical Studies* 51 (4): 442–458.
Mukerji, C. 1989. *A fragile power: Scientists and the state*. Princeton: Princeton University Press.
Musto, R.A. 2018. Cold calculations: The United States and the creation of Antarctica's atom-free zone. *Diplomatic History* 42 (4): 640–668.
Naylor, S., K. Dean, and M. Siegert. 2008. The IGY and the ice sheet: Surveying Antarctica. *Journal of Historical Geography* 34 (4): 574–595.
Needell, A.A. 2000. *Science, Cold War, and the American state: Lloyd V. Berkner and the balance of professional ideals*. New York: Harwood Academic Publishers.
Nielsen, K.H., and H. Nielsen. 2021. *Camp century: The untold story of America's secret Arctic military base under the Greenland ice*. New York: Columbia University Press.
O'Neill, D. 2007. *The firecracker boys: H-bombs, Inupiat Eskimos and the roots of the environmental movement*. New York: Basic Books.
Oldfield, J.D. 2018. Imagining climates past, present and future: Soviet contributions to the science of anthropogenic climate change, 1953–1991. *Journal of Historical Geography* 60: 41–51.
Oldfield, J.D., and D.J.B. Shaw. 2015. *The development of Russian environmental thought: Scientific and geographical perspectives on the natural environment*. New York: Routledge.
Oldroyd, D.R. 2002. *The earth inside and out: Some major contributions to geology in the twentieth century*. London: The Geological Society.
Olšáková, D. 2017. Between stalinism and infrastructural globalism: The international geophysical year (1957–8) in Czechoslovakia, Poland and the German Democratic Republic. *Acta Poloniae Historica* 115: 97–122.

Olšáková, D., and J. Oldfield. 2012. Environmental histories of the Visegrad countries: Cold War and the environmental sciences. *Bohemia* 52: 359–361.
Oreskes, N. 1996. Objectivity or heroism? On the invisibility of women scientists. *Osiris [second series]* 11: 87–113.
———. 2000a. *Laissez-tomber*: Military patronage and women's work in mid-20th-century oceanography. *HSPBS* 30 (2): 373–392.
———. 2000b. Getting oceanography done. *Earth Sciences History* 19 (1): 36–43.
Oreskes, N. ed with LeGrand, H.E. 2001. *Plate tectonics: An insider's history of the modern theory of the earth*. Boulder: Westview Press.
Oreskes, N. 2014a. Science in the origins of the Cold War. In *Science and technology in the global Cold War*, ed. N. Oreskes and J. Krige, 1–10. Cambridge, MA: The MIT Press.
———. 2014b. Changing the mission: From the Cold War to climate change. In *Science and technology in the global Cold War*, ed. N. Oreskes and J. Krige, 141–188. Cambridge, MA: The MIT Press.
———. 2014c. Scaling up our vision. *Isis* 105 (2): 379–291.
———. 2018. The scientific consensus on climate change: How do we know we're not wrong? In *Climate change: What it means for us, our children, and our grandchildren*, ed. J.F.C. DiMento and P. Doughman, 65–99. Cambridge, MA: MIT Press.
———. 2021. *Science on a mission: How military funding shaped what we do*. Cambridge, MA: Harvard University Press.
Oreskes, N., and E.M. Conway. 2008. Challenging knowledge: How climate science became a victim of the Cold War. In *Agnotology: The making and unmaking of ignorance*, ed. R.N. Proctor and L. Schiebinger, 55–89. Stanford: Stanford University Press.
———. 2010. *Merchants of doubt: How a handful of scientists obscured the truth on issues from tobacco smoke to global warming*. New York: Bloomsbury Press.
Oreskes, N., and R.E. Doel. 2002. The physics and chemistry of the earth. In *The Cambridge history of science*, ed. M.J. Nye, vol. 5, 538–582. Cambridge: Cambridge University Press.
Oreskes, N., and J.R. Fleming. 2000. Why geophysics? *Studies in History and Philosophy of Modern Physics* 31B: 353–357.
Oreskes, N., and R. Rainger. 2000. Science and security before the atomic bomb: The loyalty case of Harald U. Sverdrup. *Studies in History and Philosophy of Science Part B: Studies in History and Philosophy of Modern Physics* 31 (3): 309–369.
Poole, R. 2014. What was whole about the whole earth? Cold War and scientific revolution. In *The surveillance imperative: The rise of the geosciences during the Cold War*, ed. S. Turchetti and P. Roberts, 213–236. New York: Palgrave Macmillan.
Powell, R.C. 2008. Science, sovereignty and nation: Canada and the legacy of the International Geophysical Year, 1957–1958. *Journal of Historical Geography* 34 (4): 618–638.
Prillaman, M. 2023. Canada's Crawford Lake could mark the beginning of the Anthropocene. *Science News* 204 (2): TBC.
Pyne, S.J. 2010. *Voyager: Seeking newer worlds in the third great age of discovery*. New York: Viking.
———. 2016. Space: A third great age of discovery. *Space Policy* 37 (3): 113–119.
———. 2021. *The great ages of discovery: How Western civilization learned about a wider world*. Tucson: University of Arizona Press.
Rainger, R. 2000a. Science at the crossroads: The Navy, Bikini Atoll, and American oceanography in the 1940s. *Historical Studies in the Physical and Biological Sciences* 30 (2): 349–371.
———. 2000b. Patronage and science: Roger Revelle, the U.S. Navy, and Oceanography at the Scripps Institution. *Earth Sciences History* 19 (1): 58–89.
———. 2003. Adaptation and the importance of local culture: Creating a research school at the Scripps Institution of oceanography. *Journal of the History of Biology* 36: 461–500.
Rand, L.R. 2019. Falling cosmos: Nuclear reentry and the environmental history of earth orbit. *Environmental History* 24 (1): 78–103.
Reingold, N. 1976. Reflections on 200 years of American science. *Nature* 262: 9–13.

Ries, C.J. 2012. On frozen ground: William E. Davies and the military geology of northern Greenland 1952–1960. *Polar Journal* 2 (2): 334–357.
Rispoli, G., and D. Olšáková. 2020. Science and diplomacy around the earth: From the man and biosphere programme to the international geosphere-biosphere program. *Historical Studies in the Natural Sciences* 50 (4): 456–481.
Ritts, M., and J. Shiga. 2016. Military cetology. *Environmental Humanities* 8 (2): 196–214.
Roberts, L. 2009. Situating science in global history: Local exchanges and network of circulation. *Itinerario* 33 (1): 9–30.
Roberts, P. 2011. *The European Antarctic: Science and strategy in Scandinavia and the British Empire*. New York: Springer.
Roberts, P., L.M. der Watt, and A. Howkins, eds. 2016. *Antarctica and the humanities*. London: Palgrave Macmillan.
Robinson, S.A. 2018. *Ocean Science and the British Cold War State*. Cham, Switzerland: Palgrave Macmillan.
———. 2021. Scientific imaginaries and science diplomacy: The case of ocean exploitation. *Centaurus* 63 (1): 150–170.
Robinson, S., et al. 2023. The globalization of science diplomacy in the early 1970s: A historical exploration. *Science and Public Policy* 50 (4): 749–758.
Rozwadowski, R.M., and D.K. van Keuran, eds. 2004. *The machine in Neptune's garden: Historical perspectives on technology and the marine environment*. Sagamore Beach: Science History Publications.
Selcer, P. 2018. *The postwar origins of the global environment: How the United Nations built spaceship earth*. New York: Columbia University Press.
Shindell, M. 2010. Instruments and practices in the development of planetary geology. *Spontaneous Generations* 4 (1): 191–230.
———. 2014. From the end of the world to the age of the earth: The Cold War development of isotope geochemistry at the University of Chicago and Caltech. In *Science and technology in the global Cold War*, ed. N. Oreskes and J. Krige, 107–140. Cambridge, MA: The MIT Press.
Silva, B.K. 2020. *Chile: A Center of Global Astronomy, 1850–2019. Oxford research encyclopedia of Latin American history*. https://doi.org/10.1093/acrefore/9780199366439.013.878
Sörlin, S. 2009. Narratives and counter-narratives of climate change: North Atlantic Glaciology and Meteorology, c. 1930–1955. *Journal of Historical Geography* 35 (2): 237–255.
———. 2011. The anxieties of a science diplomat: Field coproduction of climate knowledge and the rise and fall of Hans Ahlmann's 'polar warming'. *Osiris* 26: 66–88.
———. 2014. Circumpolar science: Scandinavian approaches to the Arctic and North Atlantic, ca. 1920–1960. *Science in Context* 27 (2): 275–305.
Sörlin, S., and M. Lane. 2018. Historicizing climate change – Engaging new approaches to climate and history. *Climatic Change* 151 (1): 1–13.
Stern, D.P. 1989. A brief history of magnetospheric physics before the spaceflight era. *Reviews of Geophysics* 27 (1): 103–114.
Sullivan, W. 1961. *Assault on the unknown: The International Geophysical Year*. New York: McGraw-Hill.
Trethewey, L. 2023. *The deepest map: The high-stakes race to chart the world's oceans*. New York: Harper Wave.
Trischler, H. 2016. The Anthropocene: A challenge for the history of science, technology, and the environment. *NTM Zeitschrift für Geschichte der Wissenschaften, Technik und Medizin* 24 (3): 309–335.
Turchetti, T. 2012. Sword, shield, and buoys: A history of the NATO Subcommittee on Oceanographic Research, 1959–1973. *Centaurus* 54 (3): 205–231.
Turchetti, S. 2020. The (science diplomacy) origins of the cold war. *Historical Studies in the Natural Sciences* 50 (4): 411–432.

Turchetti, S., and P. Roberts. 2014. Introduction: Knowing the enemy, knowing the earth. In *The surveillance imperative: Geosciences during the cold war and beyond*, ed. S. Turchetti and P. Roberts, 1–19. New York: Palgrave Macmillan.
Turchetti, S., S. Naylor, K. Dean, and M. Siegert. 2008. On thick ice: Scientific internationalism and Antarctic affairs, 1957–1980. *History and Technology* 24 (4): 351–376.
Turchetti, S., N. Herran, and S. Boudia. 2012. Introduction: Have we ever been 'transnational'? Towards a history of science across and beyond borders. *Journal for the History of Science* 45 (3): 319–336.
Uhrqvist, O., and B.O. Linnér. 2015. Narratives of the past for future earth: The historiography of global environmental change research. *Anthropocene Review* 2 (2): 159–173.
van der Watt, L.M., and S. Swart. 2016. The whiteness of Antarctica: Race and South Africa's Antarctic history. In *Antarctica and the humanities*, ed. P. Roberts, M.L. van der Watt, and A. Howkins, 125–156. London: Palgrave Macmillan.
Van Keuren, D. 2000. Building a new foundation for the ocean sciences: The National Science Foundation and oceanography, 1951–1965. *Earth Sciences History* 19 (1): 90–109.
Vance, T.C., and R.E. Doel. 2010. Graphical methods and cold war scientific practice: The Stommel Diagram's intriguing journey from the physical to the biological environmental sciences. *Historical Studies in the Natural Sciences* 40 (1): 1–47.
Vollset, M., R. Hornnes, and G. Ellingsen. 2018. *Calculating the world: The history of geophysics as seen from Bergen*. Bergen: Fakbokforlaget.
Wallace, M.L. 2016. Reimagining the Arctic atmosphere: McGill University and Cold War politics, 1945–1970. *Polar Journal* 6 (2): 358–378.
Wang, Z. 1999. US-China scientific exchange: A case study of state-sponsored scientific internationalism during the Cold War and beyond. *Historical Studies in the Physical and Biological Sciences* 30 (1): 558–570.
———. 2008. *In Sputnik's shadow: The President's science advisory committee and Cold War America*. New Brunswick: Rutgers University Press.
Wang, J. 2012. Physics, emotion, and the scientific self: Merle Tuve's Cold War. *Historical Studies in the Natural Sciences* 42 (5): 341–388.
Wang, Z., and J. Zhang. 2010. China and the International Geophysical Year. In *Globalizing polar science: Reconsidering the polar and geophysical years*, ed. R. Launius, J.R. Fleming, and D.H. DeVorkin, 143–155. New York: Palgrave Macmillan.
Warde, P., L. Robin, and S. Sörlin. 2021. *The environment: History of an idea*. Baltimore: The Johns Hopkins University Press.
Warner, D.J. 2000. From Tallahassee to Timbuktu: Cold War efforts to measure intercontinental distances. *Historical Studies in the Physical and Biological Sciences* 30 (2): 393–415.
———. 2002. Political geodesy: The Army, the air force, and the world geodetic system of 1960. *Annals of Science* 59 (4): 363–389.
Weart, S.R. 1992. From the nuclear frying pan into the global fire. *Bulletin of the Atomic Scientists* 48 (5): 18–27.
———. 2008. *The discovery of global warming*. rev. updated ed. Cambridge, MA: Harvard University Press.
———. 2013. Rise of interdisciplinary research on climate. *Proceedings of the National Academy of Science* 110 (1): 3657–3664.
Weir, G. 2000. Fashioning a professional dialogue in oceanography: The U.S. Navy and the Ocean Science Community, 1924–1960. *Earth Sciences History* 19 (1): 110–129.
———. 2001. *An ocean in common: American Naval Officers, Scientists, and the Ocean Environment*. College Station: Texas A&M University Press.
Westwick, P.J. 2008. *Into the Black: JPL and the American Space Program, 1976–2004*. New Haven: Yale University Press.
Wicken, O. 2008. Space science and technology in the Cold War: The ionosphere, the military, and politics in Norway. *History and Technology: An International Journal* 13 (3): 207–229.

Wolfe, A.J. 2013. *Competing with the soviets: Science, technology, and the state in Cold War America*. Baltimore: The Johns Hopkins University Press.
———. 2020. *Freedom's laboratory: The Cold War struggle for the soul of science*. Baltimore: The Johns Hopkins University Press.
Yusoff, K. 2018. *A billion black Anthropocenes or none.* Minneapolis: University of Minnesota Press.

Historiography of the International Geophysical Year (1957–1958)

Daniella McCahey

Contents

Abstract

This chapter explores the International Geophysical Year (IGY) as a landmark moment in the history of geophysical sciences, examining its origins, execution, and lasting impact through the lens of existing scholarship. Initiated in the early 1950s, the IGY harnessed new technologies and geopolitical conditions to coordinate an unprecedented international scientific effort. Spanning from July 1957 to December 1958 and 67 nations, the IGY served as both a product of and response to the Cold War, while simultaneously intersecting with accelerating global decolonization movements. The chapter traces how participating nations used the IGY to assert scientific credibility, national legitimacy, and geopolitical relevance. By synthesizing themes from the existing scholarship, particularly the histories of geophysics, Cold War science, scientific internationalism, and decolonization, this chapter positions the IGY as a foundational event in modern global science.

D. McCahey (✉)
Texas Tech University, Lubbock, TX, USA
e-mail: dmccahey@ttu.edu

E. Aronova et al. (eds.), *Handbook of the Historiography of the Earth and Environmental Sciences*, Historiographies of Science,
https://doi.org/10.1007/978-3-031-40799-4_30

Keywords

International Geophysical Year · Geophysics · Colonialism · Cold War · Internationalism

1 Introduction: Historical Perspectives on the International Geophysical Year

On April 5, 1950, American physicist James Van Allen hosted a dinner to honor British geophysicist Sydney Chapman, gathering with fellow scientists to discuss the state of geophysical sciences. Their conversation reflected on how World War II had introduced transformative technologies like rockets, sonar, radar, weather balloons, and more that could now be used for peaceful scientific advancement. In the years following the war, growing geopolitical stability and a technological boom made international scientific collaboration both feasible and urgent. The assembled men agreed that a new, global scientific initiative was necessary to take full advantage of recent technological innovations and to foster international cooperation in a rapidly changing world. Though this dinner party story soon undertook somewhat mythologized origins, afterward, Chapman and American physicist Lloyd Berkner took advantage of this political and scientific situation to propose an International Polar Year, which they soon expanded beyond the Poles, naming their effort the International Geophysical Year (IGY). The International Council of Scientific Unions (ICSU) endorsed the proposal, encouraged member nations to form National Committees, and reached out to the USSR to join the West, despite ongoing Cold War tensions (Good 2010). In June 1953, at a meeting in Belgium, Chapman was elected president of the IGY's organizing committee, which set the official dates for the IGY July 1, 1957, to December 31, 1958.

The International Geophysical Year stands as a defining moment in the history of global scientific collaboration and exploration. This initiative, what one journalist famously called an "assault on the unknown" (Sullivan 1961), organized by the ICSU's Comité Spécial de l'Année Géophysique Internationale (CSAGI), brought together over 60,000 scientists from 67 nations to conduct a series of coordinated scientific studies aimed at understanding the Earth's geophysical properties in a year that coincided with peak sunspot activity. Those who lived through the IGY appeared aware of its significance, and several books were published in the immediate aftermath by journalists and scientists alike, expounding on its significance, or reflecting on their own participation in such an achievement (Fraser 1957; Marshack 1958; Chapman 1959; Laursen 1959; Sullivan 1961; Wilson 1961; Ross 1961). Often characterized as an unprecedented project in global science, considering the scope and significance of the IGY, it seems odd that as of this writing, there has been no significant scholarly monograph focused on the IGY, although it has been the subject of some edited volumes and journal special issues. It is almost as if the subject is too unwieldy as the single focused study of any scholar.

Yet scholars have produced a significant body of published work on the IGY, its historical context, and its wider implications. Indeed, in his chapter in this volume, Ron Doel addresses the manner in which this event has been historically situated in the context of Cold War science. But the IGY was relevant beyond Cold War science and geopolitics (Collis and Dodds 2008). This chapter serves as an introduction to the scholarship on the International Geophysical Year. While of course the IGY is of particular significance for the developments that characterized "Cold War science" and its accompanying geopolitical rivalries, its timing was also during a period of intense escalating decolonization movements around the world. Indeed, the Suez crisis, commonly used to mark the beginning of the rapid decline of British and French imperial power, occurred in late 1956, as intensive planning for the IGY was underway. The IGY thus was deeply implicated in decolonial geopolitics, as many countries used their participation in its research programs to indicate their vitality, modernity, and deservingness of independence, just as former imperial powers used their commitment to science to demonstrate the benevolence of their rule in a period when justifications for colonialization were increasingly untenable.

While the Cold War and decolonization demonstrate the significance of the IGY for geopolitical rivalries and tensions, it is also significant for the history of global scientific collaboration and international cooperation. Inspired by two earlier International Polar Years, the IGY may have spurred international rivalries, but it was also imagined as a project of scientifical globalization. It built upon similarly international collaborative projects of the past but also contributed to a continuing legacy of the development of similar projects. Finally, the scientific achievements of the IGY were profound. Significant discoveries were made in a variety of fields. It is no wonder that many studies of geophysical sciences during this period address the significance of the IGY. Furthermore, the IGY laid the groundwork for development of permanent scientific institutions and practices around the world. Through a focus on the historical themes covered in scholarship of the International Geophysical Year, specifically the histories of geophysics, scientific internationalism, the Cold War, and decolonization, this chapter demonstrates the far-researching historical impacts of this project.

2 Scientific Achievements and Expanding Knowledge

The scientific accomplishments of the IGY were groundbreaking and far-reaching. The event allowed for coordinated, worldwide studies of Earth's magnetic field, the ionosphere, the oceans, and atmospheric phenomena. The IGY led to new discoveries in fields such as geomagnetism, seismic activity, and atmospheric science. Notably, the IGY's emphasis on the study of the ionosphere and the upper atmosphere laid the foundation for understanding space weather, which had significant implications for communication systems and satellite technology.

Many scientific disciplines consider the IGY to be a significant moment in their own histories. After all, the IGY marked major investments and therefore major achievements in a broad variety of scientific fields, including seismology, geomagnetism, oceanography, glaciology, meteorology, and more. Notably, the IGY did not include a mandate for the study of geology, which some believed would endorse broader national interests in resource extraction (Belanger 2006). Additionally, many national governments invested in geophysical projects (including, but not limited to geological research) during the IGY, but that, for various reasons (including, but not limited to, military secrecy) were considered separate from IGY programs, such as the Commonwealth Trans-Antarctic Expedition (Fuchs and Hillary 1959), the New Zealand Geological Survey Antarctic Expedition (Harrington 2017), and Project Orion (Dyson 2003).

A key driving impetus of the IGY was large-scale data collection (Aronova et al. 2010; Lehman 2020). This led to the creation of World Data Centers, which established practices for data management and exchange that persist today (Aronova 2017). While not all such data was freely exchanged, given the geopolitical climate of the era (Oreskes 2021), and its control continued to be deeply embedded in politics (Dean et al. 2008), this data-focused approach became the center of many fields of geophysical sciences, shifting away from smaller-scale qualitative field observations toward big science. While in some cases it infused certain types of computational and instrument-focused research projects with new vigor at both local and international levels, (Hough 2008), it also had the effect of rendering some small-scale projects largely irrelevant in the changing fields.

This focus on wide-scale and (eventually, in many cases) long-term data collection had major implications for many later scientific discoveries. This concerted data collection was crucial for major scientific projects of the late twentieth and early twenty-first centuries, serving as benchmark measurements for a variety of research topics in earth and environmental sciences, including climate change (Edwards 2013). For example, Charles David Keeling's landmark series of measurements of atmospheric carbon dioxide at Mauna Loa, beginning during the IGY, was the first significant evidence of rapidly increasing carbon dioxide levels in the atmosphere (Weart 2008). Likewise, the IGY's inclusion of a widespread program of ice-core drilling demonstrated ability of ice sheets to store historic climate records (Heefner 2025). This climate-related research was supplemented by the international sharing of other climate data, such as the meteorological research of Harry Wexler (Fleming 2010). As such, the IGY was a key factor in the development of modern climate science (Morgan 2024). Additionally, the tracking of ozone in the Antarctic, beginning during the IGY, led to the discovery of the ozone hole in 1985 (Farman et al. 1985). Wide-scale mapping of the earth's geomagnetic field supported the hotly contested continental drift theory, eventually providing a basis for plate tectonics (Oreskes 2013). Though some data-gathering programs were not especially successful or useful in the short term, like the United States' moon camera program, this data became quietly useful in the years that followed (Dick 2010).

Notably, the IGY represents the start of large-scale concerted research of the Antarctic continent, its ice sheets, and its weather systems. While Argentina had maintained a meteorological station on Laurie Island since 1904 and the British Colonial Office's Discovery Investigations had conducted long-term monitoring of the Southern Ocean during the interwar period (McCahey 2023), scientific work in

Antarctica had generally been conducted in fits and starts. This changed during the IGY. Nearly every area of IGY research was conducted in Antarctica (Pyne 1998). Traverses over the Antarctic ice sheet revealed the depth of the ice sheet and the shape of the land beneath (Naylor et al. 2008b). The concerted and notably interrelated scientific work in this period led to the emergence of a holistic view of Antarctic science, whereby scientists understood various fields of geophysical sciences were not only linked to each other but also to biological systems (Fogg 1992).

The IGY's international but also scientifically interdisciplinary research agenda, conducting geophysical research in a variety of environments, likewise led to an explosion of research integrating studies of physical research with biological research programs. Legacies of the IGY can therefore be seen in the emergence of Earth systems science in academic curricula as well as projects like the 1987–2015 International Geosphere-Biosphere Program (Malone 1997; Rispoli and Olšáková 2020). Furthermore, the IGY led directly to the development of both a "planetary" consciousness and the concept of the Anthropocene (Lövbrand et al. 2009; Conway 2010; Goossen 2020).

In addition to its contributions to earth sciences, the IGY was pivotal in laying the groundwork for the space age. The IGY marked the launch of the first artificial Earth-orbiting satellites (Launius et al. 2002). These IGY satellites provided essential data about the upper atmosphere and Earth's magnetic field, which were critical for understanding the environment of space (Bulkeley 1991). The United States and the Soviet Union used the IGY as a springboard for their space exploration efforts, resulting in the rapid advancement of satellite technology and human spaceflight. Other countries beyond the USA and the USSR developed research programs to track satellites and build their own (Muir-Harmony 2010; Godwin 2010). These developments had profound implications for both scientific research and global communications, helping to create a more interconnected world (Slotten 2022). While satellites are perhaps the more famous legacy of space research during the IGY, James Van Allen, in whose living room the IGY was allegedly first conceptualized, used satellite data to discover belts of radiation encircling the Earth (Van Allen 1959). Known as the Van Allen belts, their discovery had immediate implications for the planning of future space travel. Indeed, Van Allen immediately worked with the US military to investigate how to disrupt them (Fleming 2011). The Van Allen belts have been characterized as the first discovery of the space age and have been a subject of constant study since (Li and Hudson 2019). Others used the now accessible regions of the ionosphere and the space between the earth and the sun to collect information on electromagnetic activity, evolving into what became known as the study of "space weather" (Good 2002, 2011; Cade and Chan-Park 2015).

Finally, the IGY boosted public interest in science and led to greater support for science education and investment into national science programs. The IGY's organizers recognized the need for selling the public on the value of large-scale scientific programs. IGY activity was widely covered in the popular press. Organizers made efforts to communicate the value of their work into classrooms, but also, in the West, into the increasingly popular medium of the television (Korsmo 2004). For instance, the *BBC* ran *The Restless Sphere* at the start of the IGY in 1957, a documentary featuring Prince Philip presenting the ideas driving the IGY, its major questions, and its global reach (Destefani et al. 1957). Furthermore, driven in part by public excitement for the IGY, in some areas of research, there was substantial contribution from scientific amateurs, expanding the roles of the public in "big science" endeavors and creating a model for the "citizen" scientist" (McCray 2006; Schröder 2007; della Dora 2023). For example, in the United States, National Science Foundation funding more than doubled in 1956, with much of it directed toward the IGY. NASA and the US Antarctic Program both were formally created to continue work started during the IGY. *Sputnik*'s launch, as well as other geopolitical considerations of scientific research in this period, spurred the creation of a presidential science adviser, strengthened federal science programs, and prompted education reforms, including the widespread development or expansion of university systems (Lowen 1997; Simpson 1998; Malone 1997; Labaree 2016).

The International Geophysical Year (IGY) was a transformative event in the history of modern science, catalyzing a shift toward large-scale, interdisciplinary, and internationally coordinated research. Its legacy extends across numerous scientific fields, from atmospheric and space sciences to oceanography and climate research, while also reshaping public engagement with science and influencing education policy and scientific infrastructure worldwide. The IGY's emphasis on systematic data collection, global collaboration, and the integration of physical and biological sciences laid the groundwork for Earth system science and shaped the trajectory of geophysical research for decades. More than just a scientific milestone, the IGY was a pivotal moment that helped redefine the relationship between science, society, and the planet.

3 Global Scientific Collaboration and International Cooperation

A central theme of the IGY was the remarkable international cooperation it fostered. This collaborative effort is often characterized as unprecedented, especially during a period characterized by Cold War tensions, colonial unrest, and the political restructuring of many nations. Despite these divides, the IGY established scientific collaboration as a powerful tool for global engagement. Yet, the IGY was not the first major international collaborative scientific program. Indeed, it was not the only attempt at international scientific collaborations in Cold War-era earth sciences (Hamblin 2000). In fact, through the twentieth century, proponents of scientific internationalism advocated that greater scientific cooperation and collaboration could contribute to a more peaceful world (Manzione 2000; Krige 2006). Viewed in this light, the IGY was less of a moment of scientific and diplomatic exceptionalism but instead fits within a broader history of global scientific collaborations.

While its organizers may have acclaimed the unprecedented scale of the IGY, they acknowledged that they were building on an older tradition of international scientific projects. Multiple scholars have demonstrated a longer tradition of international, global, and transnational collaboration within the tradition of Western science

(Rothenberg 2010; Turchetti et al. 2012; Rupke 2020; Gillin 2024). The IGY had more specific parentage. There had been two International Polar Years within the previous century, organized in Europe and including several countries in the Northern Hemisphere. Unsurprising perhaps, considering the general lack of interest in Antarctic exploration for much of the nineteenth century as well as its relative inaccessibility, Antarctica was almost completely neglected by the participating nations of the First International Polar Year in 1882. Only one research program was set up in the Antarctic, consisting of a German party to the sub-Antarctic island of South Georgia. South Georgia had a particular importance during the IPY period since the transit of Venus was to occur on 6 December 1882 and the island was well-placed to observe this phenomenon (Barr and Luedecke 2010; Cronenwett 2010).

During the Second International Polar Year (1932–1933), like its predecessor, the organizers planned to study phenomena on the largest possible scale. While allegedly bipolar in scope, the second IPY, just like the first, focused almost entirely on the Arctic. Additionally, due to the worldwide financial crisis, the Second International Polar Year was smaller than its organizers originally imagined it would be. But the Second International Polar Year was still lauded for the "international collaboration achieved in order to advance our knowledge of Nature" (Laursen 1959), even if the northern hemisphere was soon torn apart by the Second World War. Indeed, participating countries enthusiastically drew on the radio communications knowledge and weather maps compiled by the IPY to gain wartime strategic advantages, providing precedent for the keen military interest that many countries held for IGY activities (Sullivan 1961). Yet despite their purported "internationalist" bent, the IPYs were both dominated by Europe, North America, and Japan (Stevenson 2010).

In the Antarctic specifically, there was further precedent for international collaboration. The Norwegian-British-Swedish Antarctic Expedition (1949–1952) involved scientists from these three countries, with Norway mainly responsible for meteorology and topographical surveys, Britain for geology, and Sweden for glaciology. This expedition demonstrated that this type of international cooperation in the Antarctic could lead to fruitful research efforts (Roberts 2011). The spirit of collaboration dominated Antarctic science and exploration during the IGY (Sullivan 1957). For example, the United States and New Zealand closely collaborated, even establishing a joint research station at Cape Hallett (Templeton 2000). Chile, Argentina, and the United Kingdom, countries that had a long-standing sovereignty dispute over the Antarctic Peninsula, agreed to suspend their hostilities for the duration of the IGY (Howkins 2008). Indeed, several of the (invariably) men involved in polar research during the IGY penned memoirs that stressed the international collaboration driving their work (Dufek 1957; Siple 1959; Ronne 1961; Victor 1963; Behrendt 1998; MacDowall 1999).

The successful scientific collaboration in the Antarctic continued after its conclusion. The Antarctic Treaty, one of the most successful treaties of the twentieth century and the first nonnuclear proliferation treaty of the Cold War, was born out of the IGY (Pyne 1998; Musto 2018). Even if it is true that the IGY was not the sole factor contributing to the creation of the Treaty, it certainly played a major part in the Treaty's rhetoric (Bulkeley 2010; Dodds 2023). Indeed, Article II reads that

"Freedom of scientific investigation in Antarctica and cooperation toward that end, as applied during the International Geophysical Year, shall continue" (Antarctic Treaty 1959). Some have even argued that for all intents and purposes, the IGY never really ended in Antarctica (Belanger 2010). And this collaboration persisted, with exchanges of scientists between bases (including between the Soviet bloc and the United States/United Kingdom during some of the tensest moments of the Cold War) and international collaborative projects continuing in the region to the present day (Dewart 1989; Swithinbank 2002; Turchetti et al. 2008a; de Pomereu and McCahey 2022). In the wake of the IGY, Antarctic affairs have been characterized as models of scientific internationalism (Elzinga 1993; Turchetti et al. 2008b). The Scientific Committee on Antarctic Research (SCAR), the international body that coordinates scientific research in the Antarctic, was also born from the IGY, when the ICSU organized a permanent committee to oversee Antarctic research in 1957 (Summerhayes 2008). Ever since, SCAR has been at the center of international scientific efforts to protect Antarctic environments and promote biological as well as geophysical research in the region (Antonello 2019).

The global nature of the IGY was highlighted by the establishment of research stations in remote and contested areas. These areas, long considered uninhabited or difficult to access, were suddenly the focus of intense scientific inquiry. Antarctica, in particular, became a site of cooperation between Cold War rivals, with both the United States and the Soviet Union contributing to the research efforts. This cooperation laid the foundation for the 1959 Antarctic Treaty, which established Antarctica as a zone for peaceful scientific exploration, transcending geopolitical tensions. It served as an exemplar for international science not only in the Antarctic but also in the rest of the world as well (Spiller 2020). The IGY, therefore, was not only a scientific endeavor but also a moment for the potential for international diplomacy through science.

While, on the one hand, the launch of *Sputnik* during the IGY simultaneously launched the US-Soviet space competition through the 1960s, on the other hand, it also laid the groundwork for the joint Apollo-Soyuz in 1975 and its accompanied celebrated "handshake in space" (Volf 2025). Likewise, the Antarctic Treaty, born out of the IGY, and the 1958 United Nations Conference on the Law of the Sea served as models for the Outer Space Treaty (1968), which banned nuclear weapons in space and limited celestial bodies to peaceful science and exploration. As a successful instance of Cold War science diplomacy, the IGY therefore demonstrated that scientific collaboration could contribute to thawing of otherwise tense geopolitical situations. Furthermore, the IGY led to the creation of new international networks, such as satellite tracking, some of which remained intact at the IGY's conclusion, making it a useful diplomatic tool for many of the participating countries (Muir-Harmony 2010).

Though the IGY has perhaps taken on an outsized significance in the literature on international science, it was also certainly not the last project of this sort. Only a few years later, scientists around the world dedicated resources to other international collaborative geophysics projects including the Upper Mantle Project, the International Indian Ocean Expedition, and the International Years of the Quiet Sun, all meant to complement the data collection during the IGY. In 2007–2008, marking the 50th anniversary of the IGY, the World Meteorological Organization and the International

Council for Science launched another International Polar Year, coordinating a year of intensive and targeted research in polar regions, particularly related to questions about the global climate (Barr and Luedecke 2010). Concurrently, the United Nations sponsored the International Heliophysical Year, to foster international collaboration in solar-terrestrial-planetary-heliospheric physics, in addition to the Climate and Weather in the Sun-Earth System and Electronic Geophysical Year (Davila et al. 2004; Simpson 2004). Further, other scientific disciplines took inspiration from the IGY. The International Biological Program (1964–1974), for example, used the IGY as a model for promoting international collaboration for environmental and ecological research, including human adaptability to environmental change (Kwa 1987; Aronova et al. 2010; Heggie 2019; Duffin 2019).

The legacy of the IGY is not only found in its scientific achievements but also in its role in promoting peaceful global cooperation in science. Despite the Cold War and the ongoing processes of decolonization, the IGY demonstrated that scientific collaboration could transcend political and ideological divides. In 1982, the United States House of Representatives Science and Technology Committee voted unanimously to formally mark the 25th anniversary of the IGY as well as reaffirm its commitment to international scientific cooperation, despite the heightening of Cold War tensions in this period (Richman 1982). The IGY's success established a model for future international scientific efforts, such as the World Climate Research Programme and the International Space Station, which continue to promote global cooperation today. In the decades following the IGY, organizations such as the United Nations Committee on the Peaceful Uses of Outer Space (COPUOS) and the World Meteorological Organization (WMO) were created to ensure that scientific research benefited humanity as a whole. These organizations helped to ensure that scientific advancements were used for peaceful purposes and that the benefits of these advancements were shared globally.

The International Geophysical Year stands as a landmark in the history of science diplomacy and internationalism, exemplifying how scientific collaboration can serve as a unifying force even amid geopolitical divides. While it drew from a longer tradition of international cooperation in science, the IGY distinguished itself through its unprecedented scope, global participation, and enduring institutional legacies. Its influence extended beyond scientific discovery to shape frameworks for peaceful collaboration in contested spaces like Antarctica and outer space. The IGY not only demonstrated the feasibility of large-scale cooperative research but also laid the groundwork for future international scientific initiatives and treaties. As such, it represents both a culmination of earlier internationalist ideals and a springboard for the collaborative global science infrastructure that persists today.

4 The Cold War Context and Geopolitical Rivalries

The IGY occurred at the height of the Cold War, a period of intense ideological and political competition between the United States and the Soviet Union. Many scholars have demonstrated the relationship with earth and environmental sciences and the

Cold War more broadly (Doel 2003; Farish 2010; Josephson 2014; Bruno 2016; Bocking and Heidt 2019; Herzberg et al. 2018). While the two superpowers engaged in scientific cooperation during the IGY, their rivalry also played a significant role in shaping the outcomes of the year. Indeed, it was generally known that many of the subjects scheduled for study during the IGY had considerable military applications in addition to being topics of scientific importance (Bulkeley 1991). For example, gravimetric data gathered in this period had direct applications to missile guidance programs (Warner 2005). While most contemporaries cheered a general lack of geopolitical tensions in the organization of the IGY, one fly in the ointment was the People's Republic of China's withdrawal from the programming to protest the Republic of China's participation, an outcome largely orchestrated by the United States (Wang and Zhang 2010; Barrett 2022).

The most iconic event of the IGY, and one that resonated deeply within the context of the Cold War, was the Soviet Union's successful launch of *Sputnik 1*, the first artificial satellite, on October 4, 1957 (Siddiqi 2003). The launch of Sputnik not only marked the dawn of the space age but also intensified the Cold War rivalry. The United States, caught off guard by the Soviet achievement, accelerated its own space program, eventually leading to the Apollo missions and the first human landing on the Moon. The space race, which was directly influenced by the IGY, symbolized the technological and ideological competition between the two superpowers. The scientific advancements of the IGY, and the space exploration efforts that followed, were seen as crucial elements of national prestige, power, and influence during this period of global rivalry (Muir-Harmony 2020).

Space research was not the only avenue of research undertaken during the IGY that had consequences within the context of the Cold War. For example, during the IGY, oceanographers were interested in studying the physical movement of water (deep currents, surface currents, and tides), the chemical and physical interchanges between the seas and the atmosphere, and the causes and effects of sea-level rise, among other subjects. Many of these studies were conducted in collaboration with naval logistics and funding, as the strategic benefits of understanding the physical properties and geographies of the ocean were widely apparent, for the purposes of better underwater navigation and surveillance (Weir 2001; Hamblin 2005; Roberts 2014; Oreskes 2021). This was also true for countries apart from the two superpowers (Camprubí and Robinson 2016; Robinson 2018).

Polar regions as well were subject to Cold War tensions. A variation on the space race played out in the Antarctic, as the United States built a research station at the geographic South Pole, for the Soviet Union to counter with their own stations at the geomagnetic South Pole and the so-called "Pole of Inaccessibility" (Fogg 1992; Belanger 2006; Naylor et al. 2008a; Dodds 2010). It was during the IGY that the 1820 Russian imperial expedition to Antarctica took on particular geopolitical significance as the Soviet Union pushed forward an agenda that Russians had been the first to discover the seventh continent (Tammiksaar 2021). Likewise, it was at least partially due to concerns that the Antarctic might become a Cold War battlefield that pushed relatively nearby Australia and New Zealand to support the Antarctic Treaty (Haward 2017; Antonello 2013).

Cold War rivalries also were at the center of Arctic research programs developed or expanded during the International Geophysical Year. Both possessing massive Arctic frontiers, the United States and the Soviet Union were canny to the strategic and geopolitical significances of their research programs (Lyubovtseva et al. 2020). Further, considering its unique geographical position in terms of possessing a massive Arctic frontier of its own, but also sandwiched between the two superpowers, the International Geophysical Year was significant for Canadian Cold War geopolitics in the Arctic, as Canada used the IGY to assert its own national identity and scientific agenda in the region, just as it collaborated heavily with the United States and, to a lesser extent, with other NATO allies (Powell 2008; Jones-Imhotep 2017; Heidt and Lackenbauer 2022; Wiseman 2024). The intensification of US military-science in the Arctic during the Cold War during the IGY contributed to the development of the American Camp Century base in Greenland beginning in 1959, a highly publicized scientific research program that provided cover for a secret nuclear missile base (Doel et al. 2016). The IGY had provided a model for large-scale scientific research providing a cover for military activities. Further, through the IGY, outer space, the ocean, and polar regions were converted into frontiers for Cold War geopolitical expansion and control (Althoff 2007; Spiller 2015; Adler 2019; Squire 2021).

While the International Geophysical Year is often celebrated as a triumph of scientific cooperation, it was deeply entangled with the geopolitical dynamics of the Cold War. The IGY provided a platform for both collaboration and competition between superpowers, serving scientific advancement and strategic interests in equal measure. From the launch of Sputnik to the militarized study of oceans and polar regions, the IGY demonstrated how scientific endeavors could be mobilized for ideological and nationalistic purposes. Its research projects in space, the Arctic, and Antarctica blurred the lines between peaceful exploration and military ambition, transforming these remote frontiers into stages for geopolitical assertion. Ultimately, the IGY stands as a powerful example of how Cold War rivalries shaped and were shaped by global scientific enterprise.

5 Imperialism, Decolonization, and the Role of Developing Nations

While the Cold War is often considered the key geopolitical context of the IGY, it is important to note that it also unfolded during another critical moment in global history: the era of decolonization. In the 1950s, many former colonies in Africa, Asia, and the Caribbean were gaining independence from Western powers, and the geopolitical landscape was shifting rapidly. For many newly independent nations, the IGY represented both an opportunity and a challenge. While the event brought together countries from across the world, the participation of colonized nations in the IGY was limited, reflecting the global inequalities of the time.

In general, the Global North controlled the majority of the resources and technological capabilities used for the IGY's ambitious research initiatives. As a result, many of the scientific benefits of the IGY were not equally shared, and many of the global South's concerns, particularly about the distribution of scientific knowledge and resources, were sidelined. Yet, the IGY also marked the beginning of a shift in the dynamics of imperialism. Countries that had been under colonial rule began to gain more agency and representation in international scientific forums, although it would take decades before truly equitable participation became widespread. Likewise, declining and former imperial powers like the United Kingdom, France, and Japan used the IGY to platform national rather than imperial science on a global stage, as they sought a role for themselves in the bipolar world of the Cold War.

During the IGY itself, participants from newly independent nations highlighted the contribution that their country was making to the largest scientific project in human history. The CSAGI held regional meetings in the Global South, in cities including Rio de Janeiro and Bukavu (in the still-Belgian-ruled Democratic Republic of the Congo). Both the participants and other observers extoled the government investments into scientific research and development, the role of their scientists in major international research, and the public support and participation in these endeavors (Hennessey 1956). Indeed, the IGY saw the creation of geophysical research centers in many new countries. In this way, investment in science was seen as a symbol for the modernity and vitality of these new nations (Kochhar 2008). This was also true in the countries that had a more liminal status in 1957–1958, such as those members of the British Commonwealth. Indeed, while Canada (see above), South Africa, Australia, and New Zealand collaborated with the United Kingdom during the IGY, they also launched their own national projects (Bullen 1957; de Beer 2016).

Indeed, nothing perhaps demonstrated the IGY's contribution to the development of these countries' independent national identities, distinctive from those of the British Empire, more than the Commonwealth Trans-Antarctic Expedition (1955–1958). The first successful overland crossing of the Antarctic continent, this joint United Kingdom-New Zealand project, which also involved some contributions of funds and personnel from South Africa and Australia, occurred concurrently with the IGY (Fuchs and Hillary 1959). While many in all four participating countries insisted that the TAE, which had the appearance of focusing less on science and more on geopolitics, was entirely separate from the IGY, in reality, the two projects were inextricably intertwined, particularly in the United Kingdom and New Zealand contexts, sharing personnel, supplies, transport, and even, in the case of New Zealand, facilities. These countries considered data gathered from the TAE to be part of their national contributions to the IGY. When Edmund Hillary, leader of the New Zealand support party, beat the British crossing party to the South Pole, this expedition became a showcase for New Zealand patriotism and its own burgeoning imperialistic relationship with the Antarctic (Dodds 2005; Pickles 2016).

In Antarctica, the IGY marked the beginning of a new period of imperialism and colonization; twelve countries contributed to scientific research programs, which evolved into the permanent occupation of research sites all over the Antarctic continent. Many of these countries had been some of the biggest imperial powers of the first half of the century, using scientific research programs to justify their incursion into this new frontier (Roberts 2011; Martin-Nielsen 2023). This new "wave of imperialism" saw wide-reaching investment in the Antarctic, spearheaded by the United States (Scott 2019). Further, the IGY saw countries like New Zealand, South Africa, and Australia develop independent Antarctic research programs separate from the United Kingdom's work in the region.

Yet, despite the IGY occurring in a moment of a shifting world order, in some ways, the IGY reflects the stratification of divides between the Global North and the Global South. The much-exulted World Data Centers established during the IGY were located in the United States, the Soviet Union, and then split between Western Europe and Japan (Korsmo 2007). Additionally, the IGY highlighted the real inequalities between geophysical science between the countries in the Global North and the Global South, as many countries in Africa and Asia lacked equipment, facilities, and personnel to compete on the global stage, leading some to worry that these countries would simply become sites of data collection rather than centers of science (Osborne 1967). Furthermore, not all countries could equally participate in IGY research. For example, only three African nations participated in oceanographic research, of thirty-six, with such programs; apartheid South Africa was the only of these three to include work beyond the shore (Lehman 2020). Only twelve countries carried out Antarctic research programs during the IGY. These countries were the original signatories of the Antarctic Treaty, which other countries have accused of reaffirming colonial power (Tepper and Haward 2005). Indeed, during the IGY, imperial powers counted research in their colonies toward their national contributions (British National Committee for the International Geophysical Year 1957). Finally, many of the international collaborative programs between the Global North and Global South, launched during the IGY and in its wake, involved joint research projects between Western powers and either former colonies or countries in their direct spheres of influence (Selcer 2018), thus reifying imperial hierarchies in the postcolonial world.

The significance of the IGY in the context of decolonization lies in the shifting role of science in the postwar world order. As newly independent nations sought to assert themselves on the global stage, participation in global scientific projects, such as the IGY, offered an opportunity for engagement with the world beyond the colonial powers. Over time, the IGY became an important stepping stone for these countries to engage with global science and technology, and many later contributed to subsequent international scientific collaborations. Yet, as the history of the IGY remains generally understudied in regions outside the Global North, even the historiography of the IGY threatens replicate historic inequalities rooted in the politics of decolonization. A future challenge for historians may be to "decolonize the IGY," and therefore more broadly address global disparities and colonial legacies in the geophysical sciences (Turchetti 2023).

The International Geophysical Year unfolded not only within the geopolitical context of the Cold War but also amid the profound global transformations of decolonization. While it offered a symbolic platform for scientific unity, the IGY also exposed and, in many ways, reinforced the deep structural inequalities between the Global North and South. Scientific leadership, funding, infrastructure, and prestige remained concentrated in former imperial powers and Cold War superpowers, often marginalizing the contributions and capacities of newly independent nations. Yet, participation in the IGY allowed these emerging states to stake claims to scientific modernity, national identity, and geopolitical relevance, even if on unequal terms. The IGY thus embodied both the promises and limitations of international science during a moment of global realignment, serving as a gateway for broader participation by postcolonial nations, while also reflecting the persistent legacies of empire in global scientific systems and narratives.

6 Conclusion

In conclusion, the International Geophysical Year was more than just a landmark in the history of science: It was a reflection of a rapidly changing world grappling with the aftermath of war, the onset of the Cold War, and the global unraveling of colonial empires. These contexts all underscore the complexity of the IGY as both a geopolitical and scientific phenomenon. While it was inspired by earlier international collaborations, the IGY expanded the scale and scope of global scientific enterprise, integrating new nations into the scientific community and fostering institutions that would outlast the moment itself. The IGY exemplified the potential of scientific diplomacy to bridge political and ideological divides, while also highlighting the tensions between imperial powers and newly independent states, between the Soviet bloc and NATO-aligned countries, and between the Global North and the Global South. Through its scientific achievements and its role in promoting global cooperation, the IGY helped to shape the modern scientific landscape and set the stage for future global collaborations in space exploration, environmental science, and beyond. As this chapter has shown, scholarship on the IGY reveals it as an intersection of geophysics, diplomacy, and decolonization, an "assault on the unknown" not only in terms of Earth sciences but also in reimagining global cooperation. The IGY stands as a pivotal case study for understanding the entangled histories of science and international relations in the twentieth century, and its legacy continues to shape scientific collaboration in our own time.

Competing Interest Declaration The author(s) has no competing interests to declare that are relevant to the content of this manuscript.

References

Adler, Antony. *Neptune's Laboratory: Fantasy, Fear, and Science at Sea*. Harvard University Press, 2019.

Althoff, William F. *Drift Station: Arctic Outposts of Superpower Science*. Washington, DC: Potomac Books, 2007.

Antarctic Treaty. 1959.

Antonello, Alessandro. "Australia, the International Geophysical Year and the 1959 Antarctic Treaty." *Australian Journal of Politics & History* 59, no. 4 (2013): 532–46.

Antonello, Alessandro. *The Greening of Antarctica: Assembling an International Environment.* Oxford/New York: Oxford University Press, 2019.

Aronova, Elena. "Geophysical Datascapes of the Cold War: Politics and Practices of the World Data Centers in the 1950s and 1960s." *Osiris* 32, no. 1 (2017): 307–27.

Aronova, Elena, Karen S. Baker, and Naomi Oreskes. "Big Science and Big Data in Biology: From the International Geophysical Year through the International Biological Program to the Long Term Ecological Research (LTER) Network, 1957–Present." *Historical Studies in the Natural Sciences* 40, no. 2 (2010): 183–224.

Barr, Susan, and Cornelia Lüdecke, eds. *The History of the International Polar Years*. New York: Springer, 2010.

Barrett, Gordon. *China's Cold War Science Diplomacy.* Cambridge: Cambridge University Press, 2022.

Behrendt, John C. *Innocents on the Ice: A Memoir of Antarctic Exploration, 1957.* Niwot: University Press of Colorado, 1998.

Belanger, Dian Olson. *Deep Freeze: The United States, the International Geophysical Year, and the Origins of Antarctica's Age of Science.* Boulder: University of Colorado Press, 2006.

Belanger, Dian Olson. "The International Geophysical Year in Antarctica: A Triumph of 'Apolitical' Science, Politics, and Peace." In *Globalizing Polar Science: Reconsidering the International Polar and Geophysical Years*, edited by Roger D. Launius, James Rodger Fleming, and David H. DeVorkin, 265–78. New York: Palgrave Macmillan US, 2010.

Bocking, Stephen, and Daniel Heidt, eds. *Cold Science: Environmental Knowledge in the North American Arctic during the Cold War.* Abingdon: Routledge, 2019.

British National Committee for the International Geophysical Year. *The United Kingdom Contribution to the International Geophysical Year, 1957–58.* London: Royal Society, 1957.

Bruno, Andy. *The Nature of Soviet Power: An Arctic Environmental History.* Cambridge: Cambridge University Press, 2016.

Bulkeley, Rip. *The Sputniks Crisis and Early United States Space Policy.* London: Palgrave Macmillan UK, 1991.

Bulkeley, Rip. "The Political Origins of the Antarctic Treaty." *Polar Record* 46, no. 1 (2010): 9–11.

Bullen, K. E. "The International Geophysical Year." *The Australian Quarterly* 29, no. 1 (1957): 16–25.

Cade, William B., and Christina Chan-Park. "The Origin of 'Space Weather.'" *Space Weather* 13, no. 2 (2015): 99–103.

Camprubí, Lino, and Sam Robinson. "A Gateway to Ocean Circulation: Surveillance and Sovereignty at Gibraltar." *Historical Studies in the Natural Sciences* 46, no. 4 (2016): 429–59.

Chapman, Sydney. *IGY: Year of Discovery: The Story of the International Geophysical Year.* Ann Arbor: University of Michigan Press, 1959.

Collis, Christy, and Klaus Dodds. "Assault on the Unknown: The Historical and Political Geographies of the International Geophysical Year (1957–8)." *Journal of Historical Geography* 34, no. 4 (2008): 555–73.

Conway, Erik M. "The International Geophysical Year and Planetary Science." In *Globalizing Polar Science: Reconsidering the International Polar and Geophysical Years*, edited by Roger D. Launius, James Rodger Fleming, and David H. DeVorkin, 331–342. New York: Palgrave Macmillan US, 2010.

Cronenwett, Philip N. "Publishing Arctic Science in the Nineteenth Century: The Case of the First International Polar Year." In *Globalizing Polar Science: Reconsidering the International Polar and Geophysical Years*, edited by Roger D. Launius, James Rodger Fleming, and David H. DeVorkin, 37–46. New York: Palgrave Macmillan US, 2010.

Davila, Joseph M., Arthur I. Poland, and Richard A. Harrison. "International Heliophysical Year: A Program of Global Research Continuing the Tradition of Previous International Years." *Advances in Space Research* 34, no. 11 (2004): 2453–58.

de Beer, Johan. "South African and the International Geophysical Year." In *The History of Geophysics in Southern Africa*, edited by Johan de Beer, 158–164. Stellenbosch: Sun Press, 2016.

de Pomereu, Jean and Daniella McCahey. *Antarctica: A History in 100 Objects*. London: Conway, 2022.

Dean, Katrina, Simon Naylor, Simone Turchetti, and Martin Siegert. "Data in Antarctic Science and Politics." *Social Studies of Science* 38, no. 4 (2008): 571–604.

della Dora, Veronica. "From the Radio Shack to the Cosmos: Listening to Sputnik during the International Geophysical Year (1957–1958)." *Isis* 114, no. 1 (2023): 123–49.

Dewart, Gilbert. *Antarctic Comrades: An American with the Russians in Antarctica*. Columbus: Ohio State University Press, 1989.

Dick, Steven J. "Geodesy, Time, and the Markowitz Moon Camera Program: An Interwoven International Geophysical Year Story." In *Globalizing Polar Science: Reconsidering the International Polar and Geophysical Years*, edited by Roger D. Launius, James Rodger Fleming, and David H. DeVorkin, 307–328. New York: Palgrave Macmillan US, 2010.

Dodds, K. 2023. The International Geophysical Year and the Antarctic Treaty System. In A. Howkins & P. Roberts (eds.), *The Cambridge History of the Polar Regions* (pp. 536–562). Cambridge University Press

Dodds, Klaus. "The Great Trek: New Zealand and the British/Commonwealth 1955–58 Trans-Antarctic Expedition." *The Journal of Imperial and Commonwealth History* 33, no. 1 (2005): 93–114.

Dodds, Klaus. "Assault on the Unknown: Geopolitics, Antarctic Science and the International Geophysical Year." In *New Spaces of Exploration: Geographies of Discovery in the Twentieth Century*, edited by Simon Naylor and James R. Ryan. London/New York: I.B. Tauris, 2010.

Doel, Ronald E. "Constituting the Postwar Earth Sciences: The Military's Influence on the Environmental Sciences in the USA After 1945." *Social Studies of Science* 33, no. 5 (2003): 635–66.

Doel, Ronald E., Kristine C. Harper, and Matthias Heymann, eds. *Exploring Greenland: Cold War Science and Technology on ice*. New York: Palgrave Macmillan, 2016.

Dufek, George John. *Operation Deepfreeze*. Harcourt, Brace and Company, New York, 1957.

Duffin, Jacalyn. *Stanley's Dream: The Medical Expedition to Easter Island*. Montreal: McGill-Queen's University Press, 2019.

Dyson, George. *Project Orion: The True Story of the Atomic Spaceship*. Macmillan, New York, 2003.

Edwards, Paul N. *A Vast Machine: Computer Models, Climate Data, and the Politics of Global Warming*. Cambridge, MA: The MIT Press, 2013.

Elzinga, Aant. "Antarctica: The Construction of a Continent by and for Science." In *Denationalizing Science: The Contexts of International Scientific Practice*, edited by Elisabeth Crawford, Terry Shinn, and Sverker Sörlin, 73–106. Dordrecht: Springer Netherlands, 1993.

Farish, Matthew. *The Contours of America's Cold War*. University of Minnesota Press, Minneapolis 2010.

Farman, J. C., B. G. Gardiner, and J. D. Shanklin. "Large Losses of Total Ozone in Antarctica Reveal Seasonal ClOx/NOx Interaction." *Nature* 315, no. 6016 (1985): 207–10.

Fleming, James Rodger. "Polar and Global Meteorology in the Career of Harry Wexler, 1933–62." In *Globalizing Polar Science: Reconsidering the International Polar and Geophysical Years*, edited by Roger D. Launius, James Rodger Fleming, and David H. DeVorkin, 225–241. New York: Palgrave Macmillan US, 2010.

Fleming, James Rodger. "Iowa Enters the Space Age: James Van Allen, Earth's Radiation Belts, and Experiments to Disrupt Them." *The Annals of Iowa* 70, no. 4 (2011), 301.

Fogg, G.E. *A History of Antarctic Science*. Cambridge: Cambridge University Press, 1992.

Fraser, Ronald. *Once Round the Sun: The Story of the International Geophysical Year*. New York: Macmillan, 1957.

Fuchs, Vivian, and Edmund Hillary. *The Crossing of Antarctica: The Commonwealth Trans-Antarctic Expedition, 1955–1958*. Boston: Little, Brown, 1959.

Gillin, Edward J. *An Empire of Magnetism: Global Science and the British Magnetic Survey in the Age of Imperialism*. Oxford: Oxford University Press, 2024.

Godwin, Matthew. "'Britnik': How American Made and Destroyed Britain's First Satellite." In *New Spaces of Exploration: Geographies of Discovery in the Twentieth Century*, edited by Simon Naylor and James R. Ryan, 173–195. London/New York: I.B. Tauris, 2010.
Good, Gregory A. "From Terrestrial Magnetism to Geomagnetism: Disciplinary Transformation in the Twentieth Century." *Geological Society, London, Special Publications* 192, no. 1 (2002): 229–39.
Good, Gregory A. "Sydney Chapman: Dynamo behind the International Geophysical Year." In *Globalizing Polar Science: Reconsidering the International Polar and Geophysical Years*, edited by Roger D. Launius, James Rodger Fleming, and David H. DeVorkin, 177–203. New York: Palgrave Macmillan US, 2010.
Good, Gregory A. "Measuring the Inaccessible Earth: Geomagnetism, In Situ Measurements, Remote Sensing, and Proxy Data." *Centaurus* 53, no. 2 (2011): 176–89.
Goossen, Benjamin W. "A Benchmark for the Environment: Big Science and 'Artificial' Geophysics in the Global 1950s." *Journal of Global History* 15, no. 1 (2020): 149–68.
Hamblin, Jacob Darwin. "Visions of International Scientific Cooperation: The Case of Oceanic Science, 1920–1955." *Minerva* 38, no. 4 (2000): 393–423.
Hamblin, Jacob Darwin. *Oceanographers and the Cold War: Disciples of Marine Science*. Seattle: University of Washington Press, 2005.
Harrington, H. J. *Return to the Antarctic: The New Zealand Geological and Survey Antarctic Expedition 1958–1959*. Nelson: Eugene Brian Fitzgerald, 2017.
Haward, Marcus. "The Originals: The Role and Influence of the Original Signatories to the Antarctic Treaty." In *Handbook on the politics of Antarctica*, edited by Klaus Dodds, Alan Hemmings, and Peder Roberts, 232–240. Edward Elgar Publishing, Northampton 2017.
Heefner, Gretchen. *Sand, Snow, and Stardust: How US Military Engineers Conquered Extreme Environments*. Chicago: University of Chicago Press, 2025.
Heggie, Vanessa. *Higher and Colder: A History of Extreme Physiology and Exploration*. Chicago: University of Chicago Press, 2019.
Heidt, Daniel, and P. Whitney Lackenbauer. *The Joint Arctic Weather Stations: Science and Sovereignty in the High Arctic, 1946–1972*. Calgary: University of Calgary Press, 2022.
Hennessey, James J. "The Planet Earth: General Program and Philippine Participation in the International Geophysical Year." *Philippine Studies* 4, no. 4 (1956): 535–51.
Herzberg, Julia, Christian Kehrt, and Franziska Torma, eds. *Ice and Snow in the Cold War: Histories of Extreme Climatic Environments*. Berghahn Books, New York 2018.
Hough, Susan E. "Seismology and the International Geophysical Year." *Seismological Research Letters* 79, no. 2 (2008): 224–31.
Howkins, Adrian. "Reluctant Collaborators: Argentina and Chile in Antarctica during the International Geophysical Year, 1957–58." *Journal of Historical Geography* 34, no. 4 (2008): 596–617.
Jones-Imhotep, Edward. *The Unreliable Nation: Hostile Nature and Technological Failure in the Cold War*. Cambridge, MA: The MIT Press, 2017.
Josephson, Paul R. *The Conquest of the Russian Arctic*. Cambridge, MA: Harvard University Press, 2014.
Kochhar, Rajesh. "Science as a Symbol of New Nationhood: India and the International Geophysical Year 1957–58." *Current Science* 94, no. 6 (2008): 813–16.
Korsmo, Fae L. "Shaping Up Planet Earth: The International Geophysical Year (1957–1958) and Communicating Science Through Print and Film Media." *Science Communication* 26, no. 2 (2004): 162–87.
Korsmo, Fae L. "The Genesis of the International Geophysical Year." *Physics Today* 60, no. 7 (2007): 38–43.
Krige, John. "Atoms for Peace, Scientific Internationalism, and Scientific Intelligence." *Osiris* 21, no. 1 (2006): 161–81.
Kwa, Chunglin. "Representations of Nature Mediating Between Ecology and Science Policy: The Case of the International Biological Programme." *Social Studies of Science* 17, no. 3 (1987): 413–42.

Labaree, David F. "Learning to Love the Bomb: The Cold War Brings the Best of Times to American Higher Education." In *Educational Research: Discourses of Change and Changes of Discourse*, edited by Paul Smeyers and Marc Depaepe, 101–117. Cham: Springer International Publishing, 2016.
Launius, Roger D., John M. Logsdon, and Robert W. Smith, eds. *Reconsidering Sputnik: Forty Years Since the Soviet Satellite*. London: Routledge, 2002.
Laursen, V. *The Histories of the International Polar Years and the Inception and Development of the International Geophysical Year.* Annals of the International Geophysical Year. London: Pergamon Press, 1959.
Lehman, Jessica. "Making an Anthropocene Ocean: Synoptic Geographies of the International Geophysical Year (1957–1958)." *Annals of the American Association of Geographers* 110, no. 3 (2020): 606–22.
Li, W., and M.K. Hudson. "Earth's Van Allen Radiation Belts: From Discovery to the Van Allen Probes Era." *Journal of Geophysical Research: Space Physics* 124, no. 11 (2019): 8319–51.
Lövbrand, Eva, Johannes Stripple, and Bo Wiman. "Earth System Governmentality: Reflections on Science in the Anthropocene." *Global Environmental Change* 19, no. 1 (2009): 7–13.
Lowen, Rebecca S. *Creating the Cold War University: The Transformation of Stanford*. University of California Press, 1997.
Lyubovtseva, Y. S., Gvishiani, A. D., Soloviev, A. A., Samokhina, O. O., & Krasnoperov, R. I. "Sixtieth anniversary of the International Geophysical Year (1957–2017) – Contribution of the Soviet Union." *History of Geo- and Space Sciences* 11, no. (2), (2020): 157–151.
MacDowall, Joseph. *On Floating Ice: Two Years on an Antarctic Ice-Shelf South of 75's*. Edinburgh: Carnegie Publishing, 1999.
Malone, Thomas. "Building on the Legacies of the International Geophysical Year." *Eos, Transactions American Geophysical Union* 78, no. 18 (1997): 185–91.
Manzione, Joseph. "'Amusing and Amazing and Practical and Military': The Legacy of Scientific Internationalism in American Foreign Policy, 1945–1963." *Diplomatic History*, 24, no. 1 (2000): 21–55.
Marshack, Alexander. *The World in Space: The Story of the International Geophysical Year*. New York: T. Nelson, 1958.
Martin-Nielsen, Janet. *A Few Acres of Ice: Environment, Sovereignty, and "Grandeur" in the French Antarctic*. Ithaca: Cornell University Press, 2023.
McCahey, Daniella. "Britain's Polar Empire, 1769–1982." In *The Cambridge History of the Polar Regions*, edited by Adrian Howkins and Peder Roberts, 302–324. Cambridge: Cambridge University Press, 2023.
McCray, W. Patrick. "Amateur Scientists, the International Geophysical Year, and the Ambitions of Fred Whipple." *Isis* 97, no. 4 (2006): 634–58.
Morgan, Ruth A. *Climate change and international history: Negotiating Science, Global Change, and Environmental Justice*. London: Bloomsbury Academic, 2024.
Muir-Harmony, Teasel. "Tracking Diplomacy: The International Geophysical Year and American Scientific and Technical Exchange with East Asia." In *Globalizing Polar Science: Reconsidering the International Polar and Geophysical Years*, edited by Roger D. Launius, James Rodger Fleming, and David H. DeVorkin, 279–305. New York: Palgrave Macmillan US, 2010.
Muir-Harmony, Teasel. *Operation Moonglow: A Political History of Project Apollo*. New York: Basic Books, 2020.
Musto, Ryan A. "Cold Calculations: The United States and the Creation of Antarctica's Atom-Free Zone." *Diplomatic History* 42, no. 4 (2018): 640–68.
Naylor, Simon, Martin Siegert, Katrina Dean, and Simone Turchetti. "Science, Geopolitics and the Governance of Antarctica." *Nature Geoscience* 1, no. 3 (2008a): 143–45.
Naylor, Simon, Katrina Dean, and Martin Siegert. "The IGY and the Ice Sheet: Surveying Antarctica." *Journal of Historical Geography* 34, no. 4 (2008b): 574–95.
Oreskes, Naomi. "Earth Science: How Plate Tectonics Clicked." *Nature* 501, no. 7465 (2013): 27–29.

Oreskes, Naomi. *Science on a Mission: How Military Funding Shaped What We Do and Don't Know About the Ocean*. Chicago: University of Chicago Press, 2021.

Osborne, Denis B. "Geophysics in the Tropics." *Eos, Transactions American Geophysical Union* 48, no. 1 (1967): 267–72.

Pickles, Katie. "Southern Outreach: New Zealand Claims Antarctica from the 'Heroic Era' to the Twenty-First Century." In *New Zealand's Empire*, edited by Katie Pickles and Catharine Coleborne, 229–244. Manchester University Press, Manchester 2016.

Powell, Richard C. "Science, Sovereignty and Nation: Canada and the Legacy of the International Geophysical Year, 1957–1958." *Journal of Historical Geography* 34, no. 4 (2008): 618–38.

Pyne, Stephen J. *The Ice: A Journey to Antarctica*. Seattle: University of Washington Press, 1998.

Richman, B. T. IGY celebrations. *Eos, Transactions American Geophysical Union, 63*(34) (1982): 617–617.

Rispoli, Giulia, and Doubravka Olšáková. "Science and Diplomacy Around the Earth: From the Man and Biosphere Programme to the International Geosphere-Biosphere Programme." *Historical Studies in the Natural Sciences* 50, no. 4 (2020): 456–481.

Roberts, Peder. *The European Antarctic: Science and Strategy in Scandinavia and the British Empire*. New York: Palgrave Macmillan, 2011.

Roberts, Peder. "Scientists and Sea Ice under Surveillance in the Early Cold War." In *The Surveillance Imperative: Geosciences during the Cold War and Beyond*, edited by Simone Turchetti and Peder Roberts, 125–144. New York: Palgrave Macmillan US, 2014.

Robinson, Samuel A. *Ocean Science and the British Cold War State*. New York: Palgrave Macmillan, 2018.

Ronne, Finn. *Antarctic Command*. Bobbs-Merrill, 1961.

Ross, Frank. *Partners in science: The Story of the International Geophysical Year*. New York: Lothrop, Lee & Shepard, 1961.

Rothenberg, Marc. "Making Science Global? Coordinated Enterprises in Nineteenth-Century Science." In *Globalizing Polar Science: Reconsidering the International Polar and Geophysical Years*, edited by Roger D. Launius, James Rodger Fleming, and David H. DeVorkin, 23–35. New York: Palgrave Macmillan US, 2010.

Rupke, Nicolaas A. 2020. The Geomagnetic Project: Internationalism in Science between the French Revolution and the Franco-Prussian War. In *The Cambridge History of Science: Volume 8: Modern Science in National, Transnational, and Global Context*, edited by David N. Livingstone, Hugh Richard Slotten, Ronald L. Numbers, 32–42. Cambridge: Cambridge University Press.

Schröder, Wilfried. "Amateur Observations of Atmospheric Phenomena during the IGY." *Eos, Transactions American Geophysical Union* 88, no. 12 (2007): 141–43.

Scott, Shirley V. "Imperialism in Antarctica." In *The Palgrave Encyclopedia of Imperialism and Anti-Imperialism*, edited by Immanuel Ness and Zak Cope, 1–8. Cham: Springer International Publishing, 2019.

Selcer, Perrin. *The Postwar Origins of the Global Environment: How the United Nations Built Spaceship Earth*. New York: Columbia University Press, 2018.

Siddiqi, Asif A. *Sputnik and the Soviet Space Challenge*. Gainesville: University Press of Florida, 2003.

Simpson, Christopher, ed. *Universities and Empire: Money and Politics in the Social Sciences during the Cold War*. New York: New Press, 1998.

Simpson, Sarah. Space Weather Research to Benefit from IGY Anniversary. Space Weather 2, no. 7 (2004). https://doi.org/10.1029/2004SW000098.

Siple, Paul. *90 Degree South: The Story of the American South Pole Conquest*. New York: Putnam, 1959.

Slotten, Hugh R. *Beyond Sputnik and the Space Race: The Origins of Global Satellite Communications*. Johns Hopkins University Press, Baltimore 2022.

Spiller, James. *Frontiers for the American Century: Outer Space, Antarctica, and Cold War Nationalism*. New York: Palgrave Macmillan, 2015.

Spiller, James. "International Science in Antarctica." In *The Cambridge History of Science: Volume 8: Modern Science in National, Transnational, and Global Context*, edited by David N. Livingstone, Hugh Richard Slotten, Ronald L. Numbers: 75–89. Cambridge: Cambridge University Press, 2020.

Squire, Rachael. *Undersea Geopolitics: Sealab, Science, and the Cold War*. Blue Ridge Summit: Rowman & Littlefield Publishers, 2021.

Stevenson, W. R. The Polar Years and Japan. In R. D. Launius, J. R. Fleming, & D. H. DeVorkin (Eds.), *Globalizing Polar Science: Reconsidering the International Polar and Geophysical Years* (pp. 123–141). Palgrave Macmillan US, 2010.

Sullivan, Walter. *Quest for a Continent*. New York: McGraw-Hill, 1957.

Sullivan, Walter. *Assault on the Unknown: The International Geophysical Year*. New York: McGraw-Hill, 1961.

Summerhayes, Colin P. "International Collaboration in Antarctica: The International Polar Years, the International Geophysical Year, and the Scientific Committee on Antarctic Research." *Polar Record* 44, no. 4 (2008): 321–34.

Swithinbank, Charles. *Vodka on ice: A Year with the Russians in Antarctica*. Sussex: Book Guild Ltd, 2002.

Tammiksaar, Erki. "The Russian South Pole Expedition in the Context of Political Interests of the Soviet Union during the Cold War Era." In *The Russian Cold*, edited by Julia Herzberg, Andreas Renner, and Ingrid Schierle, 74–94. New York: Berghahn Books, 2021.

Templeton, Malcolm. *A Wise Adventure: New Zealand and the Antarctic 1920–1960*. Wellington: Victoria University Press, 2000.

Tepper, Rohan, and Marcus Haward. "The Development of Malaysia's Position on Antarctica: 1982 to 2004." *Polar Record* 41, no. 2 (2005): 113–24.

The Restless Sphere: The Story of the International Geophysical Year. 1957. BBC Television.

Turchetti, Simone. "Decolonize the IGY! Tales about Past Episodes of Scientific Collaboration and the Importance of Challenging Them in the Future." *Presented at the Lyne Starling Trimble Public Event Series*, American Institute of Physics, College Park, 2023.

Turchetti, Simone, Katrina Dean, Simon Naylor, and Martin Siegert. "Accidents and Opportunities: A History of the Radio echo-Sounding of Antarctica, 1958–79." *The British Journal for the History of Science* 41, no. 3 (2008a): 417–44.

Turchetti, Simone, Simon Naylor, Katrina Dean, and Martin Siegert. "On Thick Ice: Scientific Internationalism and Antarctic Affairs, 1957–1980." *History and Technology* 24, no. 4 (2008b): 351–76.

Turchetti, Simone, Néstor Herran, and Soraya Boudia. "Introduction: Have We Ever Been 'Transnational'? Towards a History of Science Across and Beyond Borders." *The British Journal for the History of Science* 45, no. 3 (2012): 319–36.

Van Allen, James A. "Radiation Belts Around the Earth." *Scientific American* 200, no. 3 (1959): 39–47.

Victor, Paul-Emile. *Man and the Conquest of the Poles*. New York: Simon and Schuster, 1963.

Volf, Darina. "The Thorny Road to a Handshake: The Apollo-Soyuz Test Project as a Challenge to the US and Soviet Space Programs." *Kritika* 26, no. 1 (2025): 63–90.

Wang, Zuoyue, and Jiuchen Zhang. "China and the International Geophysical Year." In *Globalizing Polar Science: Reconsidering the International Polar and Geophysical Years*, edited by Roger D. Launius, James Rodger Fleming, and David H. DeVorkin, 143–155. New York: Palgrave Macmillan US, 2010.

Warner, Deborah J. "A Matter of Gravity: Military Support for gravimetry during the Cold War." In *Instrumental in War: Science, Research, and Instruments Between Knowledge and the World*, edited by Steven Walton, 339–362. Boston: Brill, 2005.

Weart, Spencer R. 2008. *The Discovery of Global Warming: Revised and Expanded Edition*. Cambridge: Harvard University Press.

Weir, Gary E. *An Ocean in Common: American Naval Officers, Scientists, and the Ocean Environment.* College Station: Texas A&M University Press, 2001.
Wilson, J. Tuzo. *IGY: The Year of the New Moons.* New York: Alfred A. Knopf, 1961.
Wiseman, Matthew. *Frontier Science: Northern Canada, Military Research, and the Cold War, 1945–1970.* Toronto: University of Toronto Press, 2024.

Philosophy of the Earth Sciences

Derek D. Turner

Contents

Abstract

Over the last 20 years or so, much work on the philosophy of the geosciences has privileged epistemological questions about confirmation, explanation, and underdetermination. Taking a cue from nineteenth-century geologist Edward Hitchcock's reflections on his own practice, this chapter argues that philosophers should broaden their understanding of geoscientific practice to include its esthetic as well as its epistemic dimensions. Recent work in environmental esthetics – particularly Arnold Berleant's esthetics of engagement and Allen Carlson's scientific cognitivism – can help elucidate the esthetic aspects of the earth sciences. This broadening of our understanding of the values embedded in scientific practice could also help reinvigorate the relationship between philosophy and history of the geosciences.

Keywords

Esthetics · Earth science · Geology · Historical science

D. D. Turner (✉)
Connecticut College, New London, CT, USA
e-mail: dtur@conncoll.edu

E. Aronova et al. (eds.), *Handbook of the Historiography of the Earth and Environmental Sciences*, Historiographies of Science,
https://doi.org/10.1007/978-3-031-40799-4_3

1 Introduction

Edward Hitchcock is best known today for his systematic study of fossil footprints from the Connecticut River valley in western Massachusetts in the 1830s and 1840s. Late in his career, in 1863, he published a set of *Reminiscences of Amherst College*, where he had served both as professor of natural theology and geology and as president of the college. In these reminiscences, he offered the following reflections on his work as a geologist:

> I reckon, and who does not reckon, among the purest pleasures of life, the opportunity to gaze upon the beautiful, the bizarre, and the sublime in natural scenery. Such occasions form delightful oases among life's barren sands. We never forget them. They have few or no drawbacks, and we enjoy them by retrospection over and over again, and with increasing relish. But though such scenes lie not exclusively within the province of the geologist, he is prepared better than others to enjoy them. His home is among them. (1863, pp. 402–403)

This passage points to a distinctively esthetic way of thinking about geological research. Hitchcock emphasizes the esthetic pleasure of geological fieldwork. And he reaches for esthetic concepts – "the beautiful, the bizarre, and the sublime" – in order to say something about the source of that esthetic pleasure. It is tempting to speculate that when he mentions "the bizarre," he might be referring to the fossil footprints that he studied so intensively. Especially important is Hitchcock's claim that the geologist "is prepared better than others" to enjoy "such scenes."

What Hitchcock says about his own geological research suggests a new direction for philosophical reflection on the geosciences. Philosophers of science have, until very recently, mostly neglected the esthetic dimension of the geoscientific work. For complex reasons having to do with the contingencies of disciplinary history, philosophers tend to privilege epistemic questions about scientific knowledge, explanation, and empirical success. Recent work in environmental esthetics can help develop Hitchcock's suggestions more fully. Recognizing that geoscientific research is a form of esthetic engagement could also contribute to a reinvigoration of the relationship between history of science and philosophy of science.

Section 2 fills in some additional background about Edward Hitchcock and his spouse, Orra White Hitchcock, who was (not coincidentally) a pathbreaking scientific artist and illustrator. Section 3 provides an overview of philosophical work on the earth sciences over the last 20 years, highlighting the ways in which philosophers have privileged epistemological questions, to the neglect of the esthetic. Section 4 introduces two helpful approaches to environmental esthetics: Allen Carlson's scientific cognitivism, and Arnold Berleant's esthetics of engagement. Each of these approaches can help elucidate different aspects of Hitchcock's geological esthetics. Finally, Sect. 5 circles back to the geosciences and shows how these approaches to environmental esthetics can help enrich our understanding of geoscientific practice in ways that might be useful for the historiography of the earth sciences.

2 Learning from Hitchcock's Geological Esthetics

One way to understand Hitchcock's reference to "the beautiful, the bizarre, and the sublime in natural scenery" is to think about traditions of landscape painting. Hitchcock's esthetic sensibilities place him in a tradition that emphasizes the picturesque in nature. According to Allen Carlson:

> [T]he idea of the picturesque gave rise to a mode of aesthetic appreciation in which nature is experienced as if divided into a set of scenes – into blocks of scenery. Such scenes aim in subject matter and composition at ideals dictated by the arts, especially landscape painting. (Carlson 2010, p. 291)

One representative of this type of landscape painting is Thomas Cole's 1836 "View from Mt. Holyoke, Northampton, Massachusetts, After a Thunderstorm" (Fig. 1). This painting captures the scenery of the Connecticut River valley, where Hitchcock was, around that very time, collecting fossil footprints, research that he later compiled in (Hitchcock 1858). Cole is explicitly contrasting the wilder ridgeline of the Holyoke range with the cultivated river valley below. The receding thunderstorm creates a classic impression of the sublime, or of nature's ferocity and indifference as seen from a safe remove.

When Hitchcock exhorts his readers to go out and do geological fieldwork, he writes about natural scenery in a way that evokes this tradition of landscape painting:

Fig. 1 Thomas Cole's "View from Mt. Holyoke" (1836). (Image in the public domain)

> Gladly at such times would I entice you into the wild scenes of Nature. The mountains, the valleys, the gorges, the beetling cliffs, the caverns, the mines, the wild cataracts, the deep solitudes, stand ready to welcome you, to inspire you with fresh vigor, and to feast you with their beauties and sublimities, as much as if none before you had reveled upon them. (Hitchcock 1863, p. 406)

This could almost serve as a caption for Thomas Cole's painting, which offers us the perspective gained after ascending from the more domesticated valley to the wilder ridge of the Mt. Holyoke Range. But Hitchcock is not urging his readers merely to go out and experience nature. He sees something distinctively valuable in studying the landscape:

> [L]et me exhort you to go forth, not with fishing tackle and fowling-piece, (the meagre resort of many), but with minds well stored with scientific principles, a hammer in hand, and an aneroid barometer by your side, and laying your course for the mountains, learn the character of the rocks, their origin and fossil contents. (1863, p. 406)

This introduces a theme that I will develop further, later on: Putting it vaguely for now, we might say that geological research is a way of deepening one's connection with places, a way of cultivating sense of place (Tuan 1975). Studying nature, "learning the character of the rocks, their origin and fossil contents," cannot be distinguished sharply from appreciating nature's "beauties and sublimities."

Like many nineteenth-century naturalists, Hitchcock's scientific practice included drawing, and his published work on the fossilized trackways includes hand-drawn diagrams (Fig. 2).

Drawing was an important aspect of Hitchcock's scientific practice, in part because representations such as that in Fig. 2 served as data. But sketch work of this kind also helps to develop perceptual skills. Practice at drawing can help the naturalist get better at seeing fine-grained features of the fossils.

Hitchcock's wife, Orra White Hitchcock, was herself an accomplished artist and naturalist who was well-known at the time for her botanical illustrations (D'Arienzo 2010). She also produced many drawings of landscapes for Edward Hitchcock's published works, and large-scale geological diagrams for his lectures at Amherst College – the nineteenth-century version of Powerpoint slides. Figure 3 is an example of one of Orra Hitchcock's geological diagrams, showing a cross-section of the Connecticut River Valley (see also Oldroyd 2013 on the rise of geological diagrams in the late 1700s to early 1800s).

In Orra Hitchcock's work, we can see the idealized representation of geological history. And Orra Hitchcock is here representing exactly the same landscape that Thomas Cole painted. The oxbow in Cole's painting is just north of Mt. Tom, depicted in Fig. 2. Both focused on the place where the Connecticut River cuts a gorge between the Mt. Holyoke Range, to the east, and Mt. Tom, to the southwest.

For Edward and Orra Hitchcock, geological research was thoroughly woven together with artistic practice and with esthetic appreciation. Edward Hitchcock also saw nature itself as God's work – literally, as a work of art. He encouraged a

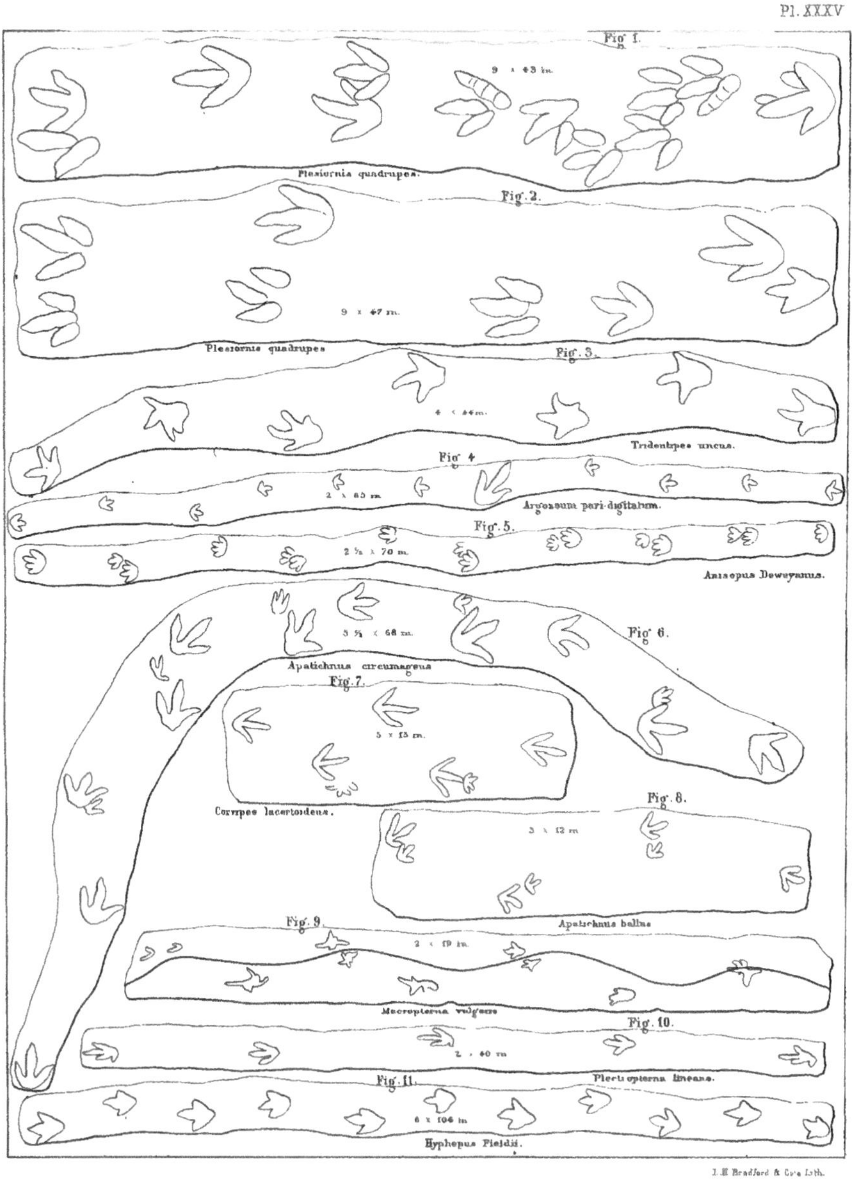

Fig. 2 An example of Edward Hitchcock's drawing of fossilized trackways. (From Hitchcock 1858, Plate XXXV. Image in the public domain)

"sanctified, Christian love of Nature" (1863, p. 407) and consistently sought to integrate his geological work with his Christian views, especially in his *Religion of Geology* (1851). Together, the Hitchcocks' work is an excellent example of what Wragge-Morley (2020) has called "aesthetic science." Wragge-Morley points out

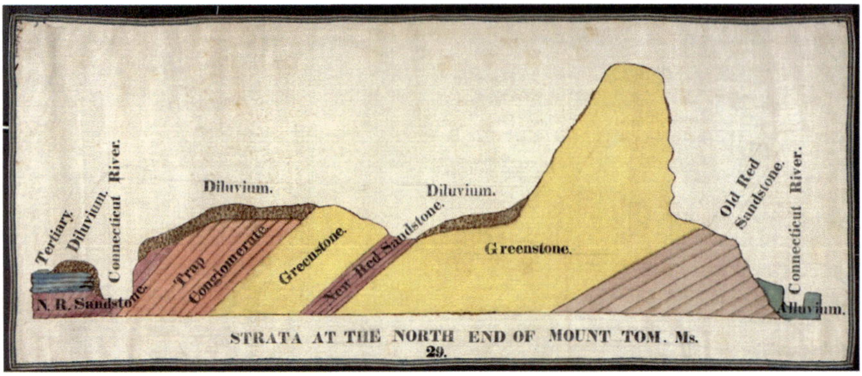

Fig. 3 One of Orra White Hitchcock's geological diagrams. (Image in the public domain)

that the tradition of natural theology invites one to think of natural science as a way of engaging with and appreciating God's creation. That tradition treats nature as a divine artwork.

One other aspect of Hitchcock's geological esthetics that deserves attention is what we might call the geological imagination, or perhaps "stereotemporal experience" (Turner 2019a). In the passage quoted at the start, Hitchcock says that the geologist is "better prepared than others" to enjoy the beauty, the bizarreness, and the sublimity in nature. But why? One answer is that geological knowledge makes it possible to experience the present landscape while also, at the same time, imaginatively placing oneself in a landscape that existed long ago. Here is one place where Hitchcock tries to articulate this stereotemporal experience:

> In this country we speak of an Ante-Columbian history as giving a hoary aspect to our annals, and making them therefore of intense interest. But the geologist can point to beds of gravel and sand near the tops of most of the New England mountains, and to the erosion of their summits by icebergs and glaciers, showing that once the ocean stood for ages above them, and that an arctic climate prevailed; nay, he can show where the Connecticut and the Hudson, by the slow action of their streams, have cut gorges into the rocks a thousand feet deep. (1863, p. 404)

Where Thomas Cole captured a landscape contrasting the wilder ridgeline with the cultivated and domesticated river valley, Hitchcock's geological imagination takes him to a time when the Mt. Holyoke range was covered by ice. He is imagining the time it took the Connecticut River to cut the gorge between Mt. Holyoke and Mt. Tom. Orra White Hitchcock's stratigraphic diagram (Fig. 3 above) also represents the same landscape in a way that informs this stereotemporal experience.

Herbert (2010) argues that Edward Hitchcock was a "romantic naturalist." According to Herbert, "Hitchcock's devotion to the aesthetics of landscape distinguishes him from most fellow geologists of his era …" (2010, p. 72). In particular, Herbert thinks that Hitchcock's geological interpretations of the volcanic ridges and hills around Amherst, Massachusetts, are informed by the romantic notion of the sublime. Wry (2019, p. 99) also argues that Hitchcock was probably very familiar with Edmund Burke's treatment of the beautiful and the sublime, and that this familiarity shows up in some of Hitchcock's writings, for example, when he claims that "nature everywhere is fitted up in a lavish manner with all the elements of the sublime and the beautiful" (Hitchcock 1851, p. 157). To Herbert and Wry's observations, we might add, though, that some of Hitchcock's writings suggest that the sublime might involve temporal as well as spatial distance from the observer. From a geological perspective, the deep past has a kind of sublimity – it is awe-inspiring, harsh, and even terrifying, though appreciated only from a safe temporal distance. The volcanic ridges around Amherst point to dangerous upheavals in the past, which we can contemplate and appreciate from a distance of millions of years, like a distant mountain range or the receding thunderstorm in Cole's painting. It is just that nature's ferocity is receding into the past, rather than into the distance.

Emily Dickinson was a younger contemporary of Edward Hitchcock in Amherst, Massachusetts. Dickinson may well have attended Hitchcock's lectures at Amherst College, and she likely studied his geology textbook in school at the Amherst Academy (Uno 1998; Wry 2019). From Hitchcock's writings, she would have learned that the landscape around Amherst was shaped by volcanism. The Mount Holyoke range, just to the south of Amherst, was formed by basalt flows that date to the beginning of the Jurassic, the very moment in time when so many of Hitchcock's dinosaur tracks were formed. Some of Dickinson's poetry reflects this geological knowledge:

> Volcanoes be in Sicily
> And South America
> I judge from my Geography
> Volcanoes nearer here
> A Lava step at any time
> Am I inclined to climb
> A Crater I may contemplate
> Vesuvius at Home

Dickinson could literally have climbed "a Lava step at any time," near her home, if she hiked up the traprock ridges of the Mt. Holyoke Range, to the very place where Thomas Cole situates the viewer. Her poetry affords a good illustration of how geological knowledge of the deep past can contribute to sense of place. The poem also expresses a sense of the temporal sublime, of a terrifying, cataclysmic event – a volcanic eruption – that the geologist can contemplate from a safe temporal remove, just as we might contemplate geographically distant volcanoes in Sicily or South America.

To summarize, Hitchcock clearly saw geological research as an esthetic activity. His geological esthetics has a number of different themes: (1) the use of esthetic concepts, such as the beautiful, the sublime, and the bizarre; (2) the idea that geological knowledge better positions us to appreciate nature; (3) the romantic interest in the picturesque, and in natural scenery; (4) the interest in embodied engagement with the landscape, i.e., in grabbing your rock hammer and heading into the field; (4) the sense that nature is literally God's work, and hence a work of art; (5) an interest in stereotemporal experience, as facilitated by geological imagination; (6) temporalizing the esthetic concept of the sublime; and (7) conceiving of geological research as activity that helps cultivate a sense of place in the sense of Tuan (1975), i.e., the "at home" in Dickinson's "Vesuvius at home."

It is difficult to understand what Hitchcock thought he was up to, or how he conceived of his own scientific work, or his relationship to those whom he influenced, such as Dickinson, without help from esthetics. This, in turn, points to a serious shortcoming in recent philosophy of science. Although things are beginning to change (see Wylie 2015, 2021; Turner 2019; Currie 2017, 2023), for decades, philosophers of science have privileged epistemic questions about the earth sciences, to the neglect of esthetic considerations.

3 Philosophy of the Geosciences: 2002–2024

Carol Cleland's (2002) paper, "Methodological and Epistemic Differences between Historical and Experimental Science" is a good place to commence an overview of recent work in the philosophy of the geosciences. One reason for this is that Cleland has also published her ideas in geoscientific venues (e.g., Cleland 2001, 2013), where her thinking has had some influence. Furthermore, her way of framing the issues has set the tone for much of the subsequent discussion. Some aspects of her view – such as her reliance on the "asymmetry of overdetermination" – have not caught on. However, Cleland's critics have tended to adopt some of her opening distinctions and questions (see Turner 2005, 2007, 2009; Kleinhans et al. 2005; Forber and Griffith 2011; Tucker 2004, 2011; Currie 2018).

Cleland (2002, 2011, 2013) distinguishes "prototypical historical science" from "classical experimental science," treating these two as ideal types. What makes historical science distinctive, she proposed, is its methodology. Experimental researchers test hypotheses about regularities among event-types by trying to rule out false positives and false negatives. Historical scientists, by contrast, are more focused on testing hypothesis about token events in the past. They do this by generating multiple hypotheses that would, if true, explain the presently observable traces. Then they search for the proverbial "smoking gun," a trace or set of traces that discriminates between rival historical hypotheses. Tellingly, Cleland used a geoscientific example – the confirmation of the Alvarez impact hypothesis – to illustrate how historical science works. In her account, for example, the Chicxulub crater is a smoking gun that confirms the impact hypothesis. Cleland also deployed a highly original argument to help establish the epistemic credentials of historical natural science. She drew upon David Lewis' (1979) idea that causal overdetermination is temporally asymmetrical: Downstream effects typically overdetermine their upstream causes, but the opposite is rarely true. Aside from some unusual examples (such as firing squads), upstream causes do not usually overdetermine downstream effects. Cleland argued that historical and experimental scientists both exploit this asymmetry of overdetermination, though from different directions. The overdetermination of upstream causes by their downstream effects means that usually, there will be some set of downstream traces that can serve as a smoking gun, even if we have not found it yet.

Cleland's account of the distinctive methodology of historical science as involving the search for a "smoking gun" differs somewhat from Frodeman's (1995) earlier suggestion that geology is fundamentally an interpretive or hermeneutic science (see also Frodeman 2003). Both Frodeman and Cleland agree that geology does have a distinctive methodology, and both are interested in the epistemological upshots of their observations about method (see Raab and Frodeman 2002).

Other philosophers subsequently pushed back against various pieces of Cleland's view. For example, Adrian Currie (2015, 2018) argues that we should resist the idea that historical science has a distinctive methodology. He argues, contra Cleland, that historical investigators are often successful precisely because they are "methodological omnivores," who take advantage of a wide variety of modeling techniques and inferential tools. In some of my own earlier work (e.g., Turner 2005, 2007), I argued that Cleland's epistemic optimism did not actually follow from the asymmetry of overdetermination, and that there might be good reasons for a more pessimistic outlook. And there are important kinds of historical scientific research that do not fit Cleland's account of the search for the smoking gun (Turner 2009). Meanwhile, Forber and Griffith (2011) reexamined Cleland's case study and argued that in that case, what really matters is consilience among distinct, independent lines of evidence. Currie, for his part, never accepted Cleland's idea that historical science has a distinctive methodology. Neither does he accept her argument from the asymmetry of overdetermination. But he has developed a different line of argument for an optimistic view of the epistemology of historical science. Indeed, much of the discussion of historical science over the last 20 years has focused on this question about the prospects for overcoming epistemic underdetermination with respect to hypotheses about the deep past.

At the center of Cleland's work, and much of the work of other philosophers who have responded to her, is an interest in traditional epistemological questions about confirmation, inference, and underdetermination. Underdetermination was also a central theme in discussions of scientific realism. The deep past is unobservable, and so historical science raises questions about access to the unobservable past, and about the status of what geoscientists tell us about the deep past (Turner 2007, 2018). This has led some philosophers to explore questions about realism in connection with the geosciences. For example, Rossetter (2018) uses James Hutton's work as a test case for thinking about novel predictive success. For another example, Miyake's (2017) work on seismology puts some pressure on the traditional classification of geology as historical science and so pushes back against one aspect of Cleland's framing. Nevertheless, Miyake is also interested in questions about access to the unobservable parts of our own planet. Scientific realists have always been interested in the project of explaining the empirical success of science. Even Currie (2018), who is generally skeptical about the value of the realism debate, frames his project as an effort to explain how historical scientists are able to achieve some epistemic successes even in unlucky circumstances.

Cleland's distinction between "prototypical historical science" and "classical experimental science" harkens back to the older distinction that Wilhelm Windelband (1980/1894) drew between *idiographic science* and *nomothetic science*. Idiographic science is all about reconstructing historical particulars, whereas nomothetic science aims to discover general patterns, regularities, and/or laws. Cleland's distinction between hypotheses about token events and hypotheses about regularities among event types closely tracks the idiographic/nomothetic distinction. Many philosophers have also taken this distinction to point to different modes or styles of explanation. Some explanations, such as classic Hempelian deductive-nomological (D-N) ones, invoke laws, or perhaps causal generalizations. Hempel, notoriously, saw historical explanations as derivative, incomplete sketches of iterated D-N explanations. In response to Hempel (1942, 1965), other philosophers (e.g., Gallie 1955) argued that narrative or historical explanation – or what they sometimes called "genetic explanation" – stands on its own as a legitimate mode of explanation that is not reducible to covering-law explanations. More recently, philosophers of science have shown a great interest in questions about narrative explanation (see Hull 1975; Beatty 2016, 2017; Currie 2014; Currie and Sterelny 2017; Roth 2019; Ereshefsky and Turner 2020; Velleman 2003). This discussion is closely related to questions about historical contingency, with some philosophers, especially John Beatty, arguing that narrative is especially good at expressing relationships of historical contingency. This discussion of explanation also has an epistemic dimension, since one of the main things at issue is whether narrative explanation is (epistemically) as good as other forms of scientific explanation. Ever since Gould and Lewontin's (1979) critique of "just so stories" in evolutionary biology, many philosophers have worried that narrative explanations are too easy to generate and too difficult to subject to empirical test. For that reason, one goal of much of this philosophical work has been to help establish the epistemic credentials of narrative explanation.

Currie (2018) has also revived discussion about the importance of speculation in historical science, where speculative claims are generally thought of as claims that outrun the current empirical evidence. The role of speculation has long been of interest to geoscientists (Davis 1926). And Currie's defense of speculation is also epistemic. He argues that speculation often has indirect investigative payoff. The main worries about speculation are also epistemic (Turner 2019b).

Finally, other theorists have focused on issues in the geosciences that Cleland's account of the search for the smoking gun does not capture very well. For example, Bokulich (2020) and Dresow (2021) explore questions about geochronology. Tamborini (2020) focuses on the implications of technological change for the epistemology of the deep past. And some philosophers have sought to shift the emphasis from the confirmation (or disconfirmation) of hypotheses and theories – the central issue in Cleland's account – to a different set of questions about modeling practices in the geosciences (Bokulich and Oreskes 2017; Currie 2018; Bokulich 2021; cf. Oreskes et al. 1994). Again, however, this recent work tends to emphasize epistemological questions.

This section offers something of an aerial survey of recent work in philosophy of science that bears on the geosciences. This survey surely leaves out a great deal, but the goal is just to identify some of the main contours of the discussion among philosophers. These include:

1. An interest in the distinction between historical and experimental science, with the geosciences seen as falling mostly under the "historical" heading.
2. An interest in the relative epistemic status of historical science, including the geosciences.
3. An interest in historical modes of scientific explanation.
4. An interest in how scientists succeed in "overcoming underdetermination" and "gaining access" to the unobservable deep past.

The larger theme here is a privileging or foregrounding epistemic question. If we go along with this privileging of epistemic questions and concerns, that leaves us in the awkward situation of having to downplay much of what Edward Hitchcock says about his own geological research. His reflections on the beautiful, the bizarre, and the sublime in nature turn out to be merely ancillary. According to the prevailing picture that we get from recent philosophy of science, what is really going on is that Hitchcock is collecting data in the form of historical traces, such as fossilized footprints. Then he is doing his best to synthesize different lines of evidence and to "overcome underdetermination" – that is, to draw some solid inferences about the deep past of the Connecticut River Valley. This is all fine as far as it goes, but it also ignores a great deal of what Hitchcock himself is telling us about his work. If we are going to take seriously what he says about his geologizing, we might do better to turn to environmental esthetics.

4 Help from the Direction of Environmental Esthetics

The work of two philosophers, in particular, can help us to better understand Edward Hitchcock's esthetic reflections on his geological research. This section focuses on Arnold Berleant's *esthetics of engagement*, as well as Allen Carlson's *scientific cognitivism*. Berleant and Carlson are sometimes seen as developing rival views in environmental esthetics. However, while differing in emphasis, one can understand both views as highlighting different aspects of our esthetic engagement with landscapes and places. There is some danger of anachronism here, as Hitchcock's own esthetic sensibility was probably a bit different – a bit of romanticism, an interest in the picturesque, and an admixture of Christian commitment. Nevertheless, here and there in Hitchcock's writing, we can find hints that point toward more contemporary views in esthetics. Following up on these hints can give us new ways of looking at the history of the geosciences.

To start with, it might help to begin by thinking about the notion of the picturesque. According to that model, esthetic experience involves some distance between the perceiver and, let us say, the landscape or scenery that one is "taking in." Perhaps one familiar sort of experience is stopping at a pullout or scenic viewpoint at a national park and appreciating some rock formation from a distance. According to this model, we experience a natural scene almost as if it were a landscape painting, or perhaps a postcard. The viewer, in this model, remains largely passive – basically a spectator who is viewing nature from a distance rather than interacting with it. When Hitchcock relishes "the opportunity to gaze upon the beautiful, the bizarre, and the sublime in natural scenery," he is clearly working within this model of the picturesque. Another feature of this way of understanding the esthetic experience of nature is that it is unimodal. It emphasizes visual experience, to the neglect of other sensory modalities. One way to think about both Berleant and Carlson's work is that they are both seeking to move beyond the limitations of this "natural scenery" model. Berleant argues that this model of the picturesque leaves out much of what is important about our perceptual experience of nature. Carlson, for his part, argues that it leaves out the cognitive dimension of esthetic experience.

Berleant (1992, 2010) wants to resist the suggestion that nature or natural landscapes are something "out there" to be appreciated from a distance, as we might view a landscape painting in a museum. Instead, the experiencer is *in* nature, immersed in it, constantly engaging with their surroundings. Berleant remains consistent with the model of the picturesque in his emphasis that esthetic engagement is fundamentally perceptual, but he departs from that model in other important ways. According to him, esthetic engagement is multimodal, embodied, and interactive. It requires all of one's senses, and it involves a give and take with the landscape. Rather than taking in a view from a distance, the experiencer is moving around, taking in sights and sounds from different angles. The visual experience of a scene in the desert cannot be separated from the feeling of the sandy trail under one's feet, or the sun on one's face. Berleant writes:

> The concept of aesthetic engagement signifies human embeddedness and active participation in the experience of appreciation. Rather than adopting a sense of distance in contemplating a landscape or an art object, engaged appreciation encourages a close involvement characterised by experiential reciprocity. (2010, p. 344)

That reciprocity is central to what Berleant means by "engagement." Think about something as simple as a hike along a trail. The walker actively traverses the landscape, but the topography and features of the place, including the contours of the trail, also influence the walker. There is a sense in which the land dictates, for example, that an uphill trail will be more strenuous. Berleant writes that "just as people impose themselves on things, so, too, to things exercise an influence on people" (2010, p. 341). The walk itself can almost be seen as an esthetic collaboration between the walker and the landscape. Thus, Berleant's emphasis on esthetic engagement shifts the focus from the esthetic qualities of the landscape as an object of appreciation, to the interactive experiential process.

When Edward Hitchcock exhorted his reader to "go forth ... a hammer in hand, and an aneroid barometer by your side ..." we can read that as an injunction to go out and engage esthetically with nature, in Berleant's sense of "engage." Traditional philosophy of science, as we saw in Sect. 2, tends to see geological fieldwork as mere data collection, where the rationale for doing fieldwork is narrowly epistemic. The goal is to assemble traces, to document historical evidence. Berleant's work, however, invites us to see scientific fieldwork as a form of esthetic engagement with places, where that esthetic engagement might turn out to be intrinsically rewarding, independently of its epistemic payoff.

Whereas Berleant, as we have seen, challenges the model of the picturesque by immersing the experiencer in nature, Carlson (2000, 1977, 1981) challenges that same model from a different direction. Carlson argues that esthetic appreciation of artwork is always mediated by background knowledge. What is more, knowing something about historical context better positions us to appreciate a work of art. Consider, by way of example, Thomas Cole's painting of the view from Mt. Holyoke (Fig. 1 above). Of course, someone who knows nothing at all about the painting – who painted it, or where, or when – can still enjoy it. However, learning a bit more about the context can deepen our appreciation of what we are seeing. For example, suppose you learn that Cole, like Emily Dickinson, was interested in volcanoes. A few years later, he painted Mt. Etna, in Sicily – the volcano that Dickinson also references obliquely in her poem. This makes us wonder if both painter and poet were interested in comparing older, American volcanoes – the traprock ridge of the Holyoke range – with younger, European ones? Or suppose you learn that Thomas Cole painted his view from Mt. Holyoke around the same time he was working on his famous series, *The Course of Empire*. Did he see the cultivated Connecticut Valley as one moment in the longer cycle depicted in that series of paintings? The point is that learning a bit about the historical and artistic context affects how we experience the artwork. According to Carlson (2010), one weakness of the model of the picturesque is that it leaves out this cognitive aspect of esthetic experience.

The guiding idea of Carlson's scientific cognitivism in esthetics is that what goes for art also goes for nature: Our experience of plants, animals, and landscapes is also affected by what we know.

> If to aesthetically appreciate art we must have knowledge of artistic traditions and styles within those traditions, to aesthetically appreciate nature we must have knowledge of the different environments of nature and of the systems and elements within those environments. In the way in which the art critic and the art historian are well equipped to aesthetically appreciate art, the naturalist and the ecologist are well equipped to aesthetically appreciate nature. (Carlson 1977, p. 273)

To the naturalist and the ecologist, we might add the geoscientist. Indeed, we can read Carlson as providing the theoretical underpinnings for Hitchcock's remark that the geologist is "better prepared than others" to enjoy nature's scenes. Carlson's point about the importance of background knowledge is well illustrated by

Hitchcock's fossilized three-toed trackways. One can marvel at and appreciate the trackways without knowing anything about them. But even just knowing something about their age cannot help but affect our esthetic appreciation of them. Suppose we learn that the trackways were made around the same time as the volcanic activity that created the Mt. Holyoke range (Raymo and Raymo 2001). Even this modicum of historical knowledge – i.e., that they belong to the deep past, and that they are related to other landscape features from the same era – will make a difference to how we appreciate them. One way to see this is by invoking Korsmeyer's (2016) argument that the value of "real old things" has to do with how they place us into causal contact with earlier things and events. The idea that esthetic judgments are sensitive to knowledge of causal history is also a running theme of work in environmental philosophy (Krieger 1973; Elliot 1982; Katz 1992).

Carlson's scientific cognitivism has some fascinating consequences for how we think about the relationship between esthetic and epistemic values in science (Turner 2019a). For example, philosophers of science have tended to think of goals such as knowledge and understanding as anchoring values that organize and structure scientific activity. Carlson's view turns things upside down by suggesting that scientific knowledge might be instrumental to richer or deeper esthetic appreciation of nature.

Neither Berleant nor Carlson ever set out with the express aim to shed light on scientific practice. They and others working in environmental esthetics are often more focused on whether esthetic values can undergird arguments for environmental protection or conservation (for an overview, see Hettinger 2008). And while some of Berleant and Carlson's respective insights are compatible, this brief survey has downplayed their disagreements. For example, Carlson sees his scientific cognitivism as a way of vouchsafing some objectivity when it comes to esthetic judgments. Berleant, on the other hand, sees his esthetics of engagement as challenging pretensions to objectivity. Their work does, however, suggest ways of further developing and refining some of the esthetic hints in Edward Hitchcock's reflections on his own scientific practice. For present purposes, I mainly want to make the fairly modest point that recent work in environmental esthetics is relevant to understanding the esthetic dimensions of the practice of the geosciences. More work is surely needed to further develop these suggestions in ways that counterbalance the privileging of the epistemic that I discussed in Sect. 3. There remains plenty of room for disagreement about details. For example, Turner (2019a) and Currie (2023) have different views about the relationship between the epistemic and the esthetic dimensions of scientific practice. Turner (2019a), building on Carlson's scientific cognitivism, argues that the relationship is conceptual, whereas Currie (2023) argues that the relationship is a merely contingent one, and that scientific training attunes people's esthetic capacities to the traditional epistemic goals of science. Both agree, however, that more work is needed to do justice to the esthetic dimensions of science, and particularly of the geohistorical sciences.

Some researchers whose work focuses on the role of field experience in earth science teaching are already in touch with the esthetic dimensions of geoscientific practice. For example, Mogk and Goodwin (2012) argue that "immersion in nature" – an idea that echoes Berleant on the esthetics of engagement – is an important aspect of field experience for students of the earth sciences. They also stress the "affective aspects of field instruction." Approaches to place-based geoscience education explicitly link scientific training to cultivation of sense of place (Semken et al. 2017). That geological practice has an esthetic dimension is nothing new to scientists. Philosophical work in environmental esthetics can help us to better understand that esthetic dimension.

5 Historiography with Esthetics in Mind

To summarize the argument up to this point: Section 2 explored the esthetic dimensions of Edward and Orra White Hitchcock's geological work. Section 3 surveyed recent work in the philosophy of historical science, with its privileging of the epistemic. Section 4 introduced some ideas from environmental esthetics. The suggestion is that environmental esthetics can help serve as a corrective to the oversights of recent philosophy of science. The challenge is to bring the esthetic and epistemic considerations into contact, and to clarify how they are related. This project also has some interesting ramifications for historiography.

To help set the stage, Currie and Walsh (2018) observe that historical narratives typically involve decisions about which details to foreground, and which to background. For example, there are many ways of telling the story of Edward Hitchcock. One could focus on his contributions to dinosaur science; on his interest in science and religion; on his institutional relationship to Amherst College; on his relationship to European contemporaries; or on any number of other historical strands. One possible way in which philosophy of science can contribute to historiography is to suggest priorities for foregrounding. Of course, given the prevailing interest in epistemological issues, philosophers tend to bring certain sorts of questions to the conversation. The argument developed here suggests that philosophers could contribute more to historiography by broadening their interests to include esthetic considerations as well as the relationship between the esthetic and the epistemic dimensions of scientific practice. One way to bring this into sharper focus is to consider another moment from the history of North American geology.

Figure 4 is a field sketch that the nineteenth-century American geologist, G.K. Gilbert, included in his 1877 work, *Geology of the Henry Mountains*. Gilbert had signed on with John Wesley Powell's survey of the Rocky Mountain region in 1874. His published report on the Henry Mountains, in Utah, includes some wonderful examples of field sketches. Figure 4 depicts a geological unconformity. The younger horizontal strata above are rocks from the Paleogene, while the inclined strata below are from the Cretaceous. Thus, Gilbert is offering us a look at the K-Pg boundary.

To begin with, consider the many things involved in the production of field sketches like this one:

Fig. 4 G.K. Gilbert's sketch of an unconformity near Salina, Utah (Gilbert 1877, p. 15). Image in the public domain

1. The sketchwork is an *embodied artistic practice*, done with certain tools (pencils, paper, etc.)
2. The sketchwork is a form of *embodied interaction with the landscape*. Importantly, Gilbert had to position himself in the landscape in a way that gave him a useful perspective on the outcrop. The sketch thus involves esthetic engagement, in Berleant's sense.
3. The artistic practice *contributes to the development of perceptual skills*. The practice of drawing can make one a better perceiver of contrasts and relationships, and a better judge of what one is seeing in the landscape.
4. The sketchwork involves *abstraction*. Gilbert is leaving out lots of detail, while foregrounding details that he judges to be geologically important. In this respect, the field sketch is somewhat like a map. Maps always leave out some details in order to foreground others. For example, a subway map typically does not contain information about surface roads.
5. The sketchwork is *mediated by background theory*. Gilbert's decisions about how to represent the landscape reflect certain background geological assumptions and theoretical interests. For example, he relies on the familiar principle of superposition (i.e., the principle that younger rock layers sit on top of older ones), and his artwork reflects the long-standing interest in unconformities as challenges to geological theory. Thus, the sketchwork is cognitively loaded, in Carlson's sense.
6. The sketchwork evinces a *receptivity to the landscape*. There is a sense in which the landscape itself is doing something to the artist. Without wanting to overstate

this, there is nevertheless a sense in which the landscape "has a say" in how it gets represented. For example, Gilbert's sketch requires open sight lines and accessible outcrops, and other sorts of landscapes (e.g., heavily forested ones, or flat prairie) might not be so amenable.

7. Gilbert's sketch is a good example of what Alisa Bokulich (2018) calls a *data model*. It is not raw data but is rather a model or representation of the data, and one that involves some interpretation, correction for biases, and noise reduction. For example, where the boundary between strata might be difficult to discern in practice when the sun is at the wrong angle, the geologist can correct that by drawing the boundary sharply.
8. Gilbert's field sketch is an example of a *hybrid scientific object*. On the one hand, it is an esthetic object, since it is quite literally an artwork. But it is just as much an evidential object, since it may serve as data for the construction of a narrative history of the Henry Mountains.
9. The sketchwork *involves esthetic judgment*. Although the main purpose of the field sketch is not to capture the scenic beauty of a landscape, esthetic judgment (in the narrow, traditional sense) may still come into play. For example, the artist might decide that the outcrop "looks better" from a particular angle. Or the artist might judge that a particular unconformity is especially dramatic. Indeed, its strikingness may contribute to the sense that it poses an explanatory challenge.
10. Given certain background assumptions about stratigraphy, the artwork also *provides structure for a geological narrative* of the history of the site.
11. The artwork also *activates the imagination*, and in particular it involves imaginative picturing of the geological history of the site.

The practice of sketching an important geological site is thus a highly complex one, with both epistemic and esthetic dimensions. The epistemic and esthetic aspects of the practice blur together both in the process of sketching the landscape, and in the product itself, which is both an artwork and an evidential object.

From this all too brief discussion, we can draw a couple of conclusions. When we look at a historical moment such as Gilbert's report on the geology of the Henry Mountains, there are good reasons to be careful about privileging epistemic questions. In Gilbert's practice – i.e., going out into the landscape and producing representations of important features – epistemic and esthetic considerations weave together. Just as Wylie (2015) emphasizes that prepared fossil specimens are both data and, in a sense, artworks created by preparators, Gilbert's field sketch is at one and the same time a geological data model and a bit of landscape art. Trying to isolate the epistemic dimension of this work from the esthetic and artistic will lead to a distorted understanding of the practice. Moreover, Berleant and Carlson's work in environmental esthetics can help to clarify the esthetic dimensions of Gilbert's practice. On the one hand, his field sketch is the result of immersive esthetic engagement in a particular place, in Berleant's sense. It is also theoretically mediated in ways suggested by Carlson's scientific cognitivism.

6 Conclusion

One way to think about some of the philosophical work summarized in Sect. 3 is that it was pushing back against ideas about science that were too narrow. Philosophers of science had inherited a picture of science from the logical empiricists that never fit the theory or practice of the geosciences too well. In various ways, Cleland and other philosophers argued that questions about methodology, explanation, confirmation, underdetermination, etc. look different when we shift our focus to scientific efforts to reconstruct the deep past. In spite of disagreements about details, philosophers working on historical science often shared a sense that traditional questions about the epistemology of science had been too narrowly framed. This contribution has argued that much of the work of these philosophers over the last 20 years has also been too narrowly framed. Edward Hitchcock's reflections on "the beautiful, the bizarre, and the sublime" in nature point the way to a rethinking of geoscientific practice as suffused with both esthetic and epistemic values. In addition to showing how recent work in environmental esthetics – work that has mostly remained siloed from philosophy of science, in spite of its obvious relevance – can contribute to broadening our understanding of geoscientific practice, this expansion of philosophers' field of vision could also generate more productive exchange between philosophers and historians of science.

References

Beatty, J. 2016. What are narratives good for? *Studies in History and Philosophy of Biology and Biomedical Sciences* 58: 33–40.

———. 2017. Narrative possibility and narrative explanation. *Studies in History and Philosophy of Science A* 62: 31–34.

Berleant, A. 1992. *The aesthetics of environment*. Philadelphia: Temple University Press.

———. 2010. Reconsidering scenic beauty. *Environmental Values* 19 (3): 335–350.

Bokulich, A. 2018. Using models to correct data: Paleodiversity and the fossil record. *Synthese* 198 (supp 24): 5919–5940.

———. 2020. Calibration, coherence, and consilience in radiometric measures of geologic time. *Philosophy of Science* 87: 425–456.

———. 2021. Taming the Tyranny of scales: Models and scales in the geosciences. *Synthese* 199 (5–6): 14167–14199.

Bokulich, A., and N. Oreskes. 2017. Models in the geosciences. In *Springer handbook of model-based science*, ed. L. Magnani and T.W. Bertolotti, 891–911. Springer.

Carlson, A. 1977. Appreciation and the natural environment. *Journal of Aesthetics and Art Criticism* 37 (3): 267–275.

———. 1981. Nature, aesthetic judgment, and objectivity. *Journal of Aesthetics and Art Criticism* 40 (1): 15–27.

———. 2000. *Aesthetics and the environment: The appreciation of nature, art, and architecture*. London: Routledge.

———. 2010. Contemporary environmental aesthetics and the requirements of environmentalism. *Environmental Values* 19 (3): 289–314.

Cleland, C. 2001. Historical science, experimental science, and the scientific method. *Geology* 29 (11): 987–990.

———. 2002. Methodological and epistemic differences between historical and experimental science. *Philosophy of Science* 69 (3): 474–496.
———. 2011. Prediction and explanation in historical natural science. *British Journal for the Philosophy of Science* 62 (3): 551–582.
———. 2013. Common cause explanation and the search for a smoking gun. In *125th Anniversary volume of the Geological Society of America: Rethinking the fabric of geology*, ed. V. Baker, Special Paper 502, 1–9.
Currie, A.C. 2014. Narratives, mechanisms, and Progress in historical science. *Synthese* 191: 1163–1183.
———. 2015. Marsupial lions and methodological Omnivory. *Biology and Philosophy* 30: 187–209.
———. 2017. Paleoart as science. *Extinct: The philosophy of Palaeontology Blog*. February 27, 2017. www.extinctblog.org/extinct/2017/2/27/paleoart-asscience.
———. 2018. *Rock, bone, and ruin: An optimist's guide to the historical sciences*. Cambridge, MA: MIT Press.
———. 2023. Epistemic engagement, aesthetic value, and scientific practice. *British Journal for the Philosophy of Science* 74 (2): 313–334.
Currie, A., and K. Sterelny. 2017. In Defence of storytelling. *Studies in History and Philosophy of Science A* 62: 14–21.
Currie, A., and K. Walsh. 2018. Frameworks for historians and philosophers. *Hopos: The Journal of the International Society for the History of the Philosophy of Science* 9 (1). https://doi.org/10.1086/699797.
D'Arienzo, D. 2010. The Union of the beautiful with the useful: Through the eyes of Orra White Hitchcock. *The Massachusetts Review* 51 (2): 294–336.
Davis, W.M. 1926. The value of outrageous geological hypotheses. *Science* 63 (1636): 436–468.
Dresow, M. 2021. Measuring time with fossils: A start-up problem in scientific practice. *Philosophy of Science* 88 (5): 940–950.
Elliot, R. 1982. Faking Nature. *Inquiry* 26 (1): 81–93.
Ereshefsky, M., and D. Turner. 2020. Historicity and explanation. *Studies in History and Philosophy of Science A* 80: 47–55.
Forber, P., and E. Griffith. 2011. Historical reconstruction: Gaining access to the deep past. *Philosophy and Theory in Biology* 3: 1–19.
Frodeman, R. 1995. Geological reasoning: Geology as an interpretive and historical science. *Geological Society of America Bulletin* 107: 960–968.
———. 2003. *Geo-logic: Breaking ground between philosophy and the earth sciences*. Albany: SUNY Press.
Gallie, W.B. 1955. Explanations in history and the genetic sciences. *Mind* LXIV (254): 160–180.
Gilbert, G.K. 1877. *Report on the geology of the Henry Mountains. US geological survey unnumbered series*. Washington, DC: US Government Printing Office.
Gould, S.J., and R.C. Lewontin. 1979. The Spandrels of San Marco and the Panglossian paradigm: A critique of the Adaptationist Programme. *Proceedings of the Royal Society of London* 205: 581–598.
Hempel, C.G. 1942. The function of general Laws in history. *Journal of Philosophy* 39 (2): 35–48.
———. 1965. *Aspects of scientific explanation and other essays in the philosophy of science*. New York: The Free Press.
Herbert, R.L. 2010. The sublime landscapes of Western Massachusetts: Edward Hitchcock's romantic naturalism. *Massachusetts Historical Review* 12: 70–99.
Hettinger, N. 2008. Objectivity in environmental aesthetics and environmental protection. In *Nature, aesthetics, and environmentalism: From beauty to duty*, ed. A. Carlson and S. Lintott. New York: Columbia University Press.
Hitchcock, E. 1851. *The religion of geology and its connected sciences*. Boston: Phillips, Sampson, and Company.
———. 1858. *Ichnology of New England: A report on the sandstone of the Connecticut Valley, especially its fossil footmarks*. Boston: William White, Printer to the State.

———. 1863. *Reminiscences of Amherst college: Historical, scientific, biographical, and autobiographical*. Northampton: Bridgman & Childs.
Hull, D.L. 1975. Central subjects and historical narratives. *History and Theory* 14 (3): 253–274.
Katz, E. 1992. The big lie: Human restoration of nature. *Research in Philosophy and Technology* 12: 231–241.
Kleinhans, M.G., C.J.J. Buskes, and H. De Regt. 2005. Terra incognita: Explanation and reduction in earth science. *International Studies in the Philosophy of Science* 19 (3): 289–317.
Korsmeyer, C. 2016. Real old things. *British Journal of Aesthetics* 56 (3): 219–231.
Krieger, M. 1973. What's wrong with plastic trees? *Science* 179 (4072): 446–455.
Lewis, D. 1979. Counterfactual dependence and time's arrow. *Nous* 13 (4): 455–476.
Miyake, T. 2017. Scientific realism and the earth sciences. In *The Routledge handbook of scientific realism*, ed. J. Saatsi, 333–344. London: Routledge.
Mogk, D.W., and C. Goodwin. 2012. Learning in the field: Synthesis of research on thinking and learning in the geosciences. *Geological Society of America Special Papers* 486: 131–163. https://doi.org/10.1130/2012.2486(24).
Oldroyd, D.A. 2013. Maps as pictures or diagrams: The early development of geological maps. In *Rethinking the fabric of geology. Geological Society of America*, ed. V.A. Baker, vol. 502. https://doi.org/10.1130/2013.2502(04).
Oreskes, N., K. Shrader-Frechette, and K. Belitz. 1994. Verification, validation, and confirmation of numerical models in the earth sciences. *Science* 263 (5147): 641–646.
Raab, R., and R. Frodeman. 2002. What is it like to be a geologist? A phenomenology of geology and its epistemological implications. *Philosophy and Geography* 5: 69–81.
Raymo, C., and M. Raymo. 2001. *Written in stone: A geological history of the northeastern United States*. Hensonville: Black Dome Press.
Rossetter, T. 2018. Realism on the rocks: Novel success and James Hutton's theory of the earth. *Studies in History and Philosophy of Science Part A* 67: 1–13.
Roth, P. 2019. *The philosophical structure of historical explanation*. Oxford: Oxford University Press.
Semken, S., E.G. Ward, S. Moosavi, and P.W.U. Chinn. 2017. Place-based education in geoscience: Theory, research, practice, and assessment. *Journal of Geoscience Education* 65: 542–562.
Tamborini, M. 2020. Technoscientific approaches to deep time. *Studies in History and Philosophy of Science* 79: 57–67.
Tuan, Y.-F. 1975. Place: An experiential perspective. *Geographical Review* 65 (2): 151–165.
Tucker, A. 2004. *Our knowledge of the past*. Cambridge: Cambridge University Press.
———. 2011. Historical science, over- and underdetermined: A study of Darwin's inference of origins. *British Journal for the Philosophy of Science* 62 (4): 805–829.
Turner, D. 2005. Local underdetermination in historical science. *Philosophy of Science* 72: 209–230.
———. 2007. *Making prehistory: Historical science and the scientific realism debate*. Cambridge, UK: Cambridge University Press.
———. 2009. Beyond detective work: Empirical testing in Paleobiology. In *The Paleobiological revolution: Essays on the growth of modern paleontology*, ed. M. Ruse and D. Sepkoski, 201–214. University of Chicago Press.
———. 2018. Three kinds of realism about historical science. In *The Routledge handbook of scientific realism*, ed. J. Saatsi, 321–332. London: Routledge.
———. 2019a. *Paleoaesthetics and the practice of paleontology*. Cambridge, UK: Cambridge University Press.
———. 2019b. Speculation in the historical sciences. In *Philosophy, theory, and practice in biology* 11. https://doi.org/10.3998/ptpbio.16039257.0011.011.
Uno, H. 1998. Chemical conviction; Dickinson, Hitchcock, and the poetry of science. *The Emily Dickinson Journal* 7 (2): 95–111.
Velleman, J.D. 2003. Narrative explanation. *Philosophical Studies* 112 (1): 1–25.

Windelband, W. (1980/1894). Rectorial address, Strasbourg, 1894. *History and Theory* 19: 169–185.

Wragge-Morley, A. 2020. *Aesthetic science: Representing nature in the Royal Society of London, 1650–1720.* Chicago, IL: University of Chicago Press.

Wry, J.R. 2019. Deep mapping in Edward Hitchcock's geology and Emily Dickinson's poetry. *Textual Cultures* 12 (1): 95–119.

Wylie, C.D. 2015. The artist' s piece is already in the stone: Constructing creativity in paleontology laboratories. *Social Studies of Science* 45 (1): 31–55.

———. 2021. *Preparing dinosaurs: The work behind the scenes.* Cambridge, MA: MIT Press.

Women Geologists 1780–1840: Re-reading Charlotte Murchison

Adelene Buckland

Contents

Abstract

It has long been established that many women were interested participants in the earth sciences in the first half of the nineteenth century. Until now, much of the scholarship has focused on the figure of the "geological wife" (or daughter or sister), whose skills in languages, drawing, and fossil collecting often provided invaluable assistance to their husbands. But since they are viewed as helpmates, or assistants, these women's stories have mostly been told through the lens of the careers of their male family members, and they only rarely have been considered to have played any creative role in the research. This chapter reconsiders that scholarship by looking at the work of Charlotte Murchison (née Hugonin). Re-evaluating Charlotte's work, the chapter demonstrates the key role her fossil collection and illustrations played in the development of key theoretical ideas (including Charles Lyell's uniformitarianism) in the 1820s, and argues that the significance of her work was a key factor in pushing for women's access to academic geological lectures in the 1830s. Charlotte's story – told on its own terms – exposes the limitations of viewing women's participation as familial assistance. It enables us to see how this paradigm has obfuscated our

A. Buckland (✉)
King's College London, London, UK
e-mail: adelene.buckland@kcl.ac.uk

E. Aronova et al. (eds.), *Handbook of the Historiography of the Earth and Environmental Sciences*, Historiographies of Science,
https://doi.org/10.1007/978-3-031-40799-4_7

understanding of women's intellectual autonomy, and of their significant, independent research achievements.

Keywords

Women · Earth sciences · Geology · Fossil collecting · Stratigraphy · Scientific illustration · Helpmates

1 Introduction

As several historians have demonstrated, women played an important role in the development of British geology between 1780 and 1850 (see Burek and Higgs (eds.) 2007; Creese 1998). M.R.S. Creese (2007), for instance, points out that British women contributed five times as many papers to scientific journals in this period as women in Russia/Poland, the next-ranked nation for female contributors. To take a different measure, as scientific publishing for broad audiences took off in the 1820s and 1830s, women like Jane Marcet, Jane Kilby Welsh, Rosina Zornlin, and Maria Hack expounded new geological knowledge to broad audiences, including children, ladies, or those with a predisposition to Paleyan natural theology (see Larsen 2018). In fossil collecting, women were especially well-represented: Lady Gordon Cumming of Altyre collected fossil fishes on the shores of Moray Firth in the 1840s, while Lady Eliza Maria Gordon Cumming's fossil collection supplied gentlemen geologists like William Buckland, Louis Agassiz, and Roderick Murchison, aided by pen and watercolor sketches of the fossils made by her daughter, Lady Seymour (Creese 2007: 39–40). Lady Mary Cole and "the Misses Talbot" mapped for William Buckland in Glamorganshire, South Wales (Buckland in Gordon 1894: 16). Barbara Marchioness of Hastings had "the finest collection of Speeton clay shells I have ever met with," according to one male geologist, who noted that "in vertebrates she beats all the private collections" (he took away an oolitic starfish to write up for a paper) (Edward Forbes in Kölbl-Ebert 2004: 76, 77). Hastings also regularly corresponded with Richard Owen and a host of other leading naturalists, produced a detailed and original stratigraphical section of the Eocene strata near Milford in Hampshire, published three geological papers, and sold her collection to the British Museum in 1855 (Kölbl-Ebert 2004; Kölbl-Ebert 2002: 17–18). As these examples show, women also participated in geological mapmaking and the drawing of sections: Etheldred Benett's stratigraphical description of Upper Chisgrove was one of the earliest detailed quarry sections produced in England, and she made many contributions of fossil specimens to James Sowerby's *Mineral Conchology of Great Britain* (which contained the work of 29 separate female contributors, in fact). Benett's specimens were also added to the collections of the Geological Society of London, and Benett contributed key information to George Bellas Greenough's *Geological Map of England and Wales* (see Kölbl-Ebert 2002: 16–17; Torrens 1985; Torrens et al. 2000; Spamer, Bogan and Torrens, 1989). In yet another example of women's work in the earth sciences, Maria Graham publicly challenged

Greenough on his interpretation of the Chilean earthquake of 1822, dramatizing the importance of firsthand witnessing over distant masculine authority (Thompson 2002). And, though this list is by no means exhaustive, of course the work of Mary Anning (and by extension Elizabeth, Mary, and Margaret Philpot) in Dorset is well-known (see esp. Torrens, 1995).

For Martina Kölbl-Ebert (2007a), this relative abundance of women in early British geology is explained by the nation's unique combination of early industrialization and comparatively late professionalization. Compared with other European nations, British geology relied on "an informal workforce" (Kölbl-Ebert 2007a: 158) – comprising the amateur "gentlemen geologists" long since identified by Roy Porter (1978), but which also enabled women to play an active role in the science. At the same time, "a comparatively liberal gender model" in Britain, "which did not rigidly discriminate between a private, family sphere of life for women and a public sphere of men," as well as the relatively practical nature of women's fashions before the advent of the crinoline in the 1850s, was conducive to British women of science in a way that was unparaleled on the continent (Kölbl-Ebert 2007a: 158). If this was the "heroic age" of geology, in which the major stratigraphic divisions and the reality and character of an immensely ancient earth were worked out, we might also argue that it was also a golden age of women's involvement, too.

This chapter will reconsider the historiography by focusing on one particular type of woman geologist – the geological wife (or, in an extended permutation, the geological sister or daughter). Key women in this category include Elizabeth C. Cary, wife of Louis Agassiz (Gerassi-Navarro 2017), and Mary Hughes, married to the Woodwardian Professor of Geology, Thomas McKenny Hughes (Hart 2007). They include Anne Phillips, who discovered what is known as the "Anne Phillips conglomerate" while accompanying her brother on geological fieldwork in 1842, which disproved Roderick Murchison's theory of the intrusive origin of the Malvern Hills (Morgan 2006, 2007). Phillips also sorted and catalogued boxes of fossils, colored geological maps, conducted geological correspondence, and took care of scientific instruments. Perhaps more famously, Mary Morland was a draughtswoman for Georges Cuvier (whose daughter Clementine, and stepdaughter Sophie Duvaucel, were employed as his research assistants at the Jardin des Plantes (Orr 2007)). Afterward, Morland married the Oxford geologist William Buckland and became his amanuensis, draughtswoman, curator, editor, cataloguer, model-maker, and later expert in marine bryozoans, even as she gave birth to, and raised, ten children (Kölbl-Ebert 1997a, 2012). Mary Elizabeth Lyell, wife of Charles, was a skilled expert in the taxonomy of fossil shells and undertook scientific translation, correspondence, and writing (Kölbl-Ebert 2004). The list goes on. As Jim Secord has put it, participation by female family members in scientific work was so common as to feel almost routine: "after the last Geological Society meeting of the spring season," he writes, "the researchers gathered up their hammers and their wives and set off on extensive stratigraphical tours" (1990: 25).

This focus on female assistantship – particularly in the form of wives, sisters, and daughters – has tended to emphasize two key, relentlessly familiar points. Firstly, that women were *there*. They may not have been admitted to the Geological Society until 1919, but they were in the field, in the cabinets, and working in diverse roles from geology's earliest inception. Secondly, however, it is also often argued that women are either almost invisible from the record and/or were almost entirely inhibited from contributing to geology's *intellectual* history (and, by extension, to the general history of science) (see Burek et al., 2010). As Kölbl-Ebert puts it, "British women at that time were allowed, or even desired, to assist and to contribute observations and data. They certainly also talked to their husbands or non-related mentors about their work, but a public display of female creativity was beyond the invisible boundary of social conduct" (2007a: 162). For Patricia Fara, more generally, "exhibiting varying degrees of meekness, they [women contributors to science] accepted being also co-opted as menial assistants who received no acknowledgement for their work" (2004). According to David Oldroyd, until the late twentieth century there were only two female "idea producers" in the history of geology: Inge Lehmann (who in 1936 suggested there might be a solid core at the center of the earth) and Marie Tharp (who proposed the idea of rift valleys in mid-ocean ridges on the Pacific Ocean Floor in 1952) (see Creese, 2007: 47). This means that there were *none* in the nineteenth century (and even then, Tharp's research was dismissed as "girl talk" by one of the leaders of her research project at Columbia). As such, the overriding story we have told about the history of geology has been this: in a subordinate function to their male friends and relatives, women were permitted to color maps, pack boxes of fossils, or write letters. They contributed to the *body* of geological science – to its "observations and data." But they were not allowed to be responsible for anything that emanated from its brain.

Perhaps, however, it is time to throw out this distinction between "idea" and "data" as little more than a restatement of the old, outmoded Cartesian mind/body distinction on which intellectual history once thrived. This is a history in which male nineteenth- and twentieth-century geologists were understandably invested, given that it put them on the right side of an intellectual tradition that automatically distinguished men's work from the supposedly embodied and inferior knowledge of women and other disenfranchised groups. But, of course, the idea/data distinction has, in any case, been roundly demolished by a whole series of subfields of the history of science for several decades now. For a start – and just to take an example pertinent to earth sciences – early nineteenth-century geologists began by propounding a method of "enumerative induction" – of a radically empirical method that admitted only of limited observations stripped bare of damagingly speculative (and sometimes heretical) theoretical implication (see Laudan 1987: 168–69). This had class as well as gender implications, since navvies, quarrymen, and engineers would be permitted to send their observations to the gentlemen of the Geological Society, who alone had the education and sense to begin to make measured theoretical speculations on their significance (see Buckland, 2013: 9–10). But as is well known, this distinction between "observation" and "theory" was almost impossible to sustain in practice: to take just one key example, which will be of consequence later, the use of fossils to define stratigraphic boundaries – which looks like a resolutely taxonomic and descriptive enterprise – clearly had theoretical and ideological significance. As Secord famously described it, Roderick Murchison's work defining the Silurian system "the method of correlation by fossils was tied to the notion of geographical expansibility, for the Silurian system could be spread internationally through the use of palaeontology" (1982: 417). Murchison advocated for the use of fossils in stratigraphic taxonomy in part because it helped him extend the territorial range of his system – which in turn, garnered the system prestige enough to make it standard. To put it another way, fossil collecting and taxonomic description might seem to be purely inductive on the surface, but in this period they were often tied into important debates about the age of the earth, the mechanisms by which events in earth history had largely taken place, the value of fossils in unveiling that story, and theoretical disputes about boundaries between geographical and temporal regions. There was no "data" without theory. There is no body of evidence without ideas. For that reason, "data-producers" and "idea-producers" cannot be so rigidly separated.

Perhaps even more importantly, since Martin Rudwick's seminal 1976 essay on the topic, it has been clear that in geology, such activities as sketching, illustrating, fossil collecting, and the coloring and drawing of maps were key activities of geology's theoretical work. Illustrations offered visual confirmation of unseen landscapes, argument for high-level ideas, and gave formalized expression to complex ideas about such ideas as erosion, the intrusion or deposition of molten veins in rock, or the age of the earth. As Rudwick puts it:

> as the science emerged, … the causal and temporal interpretation of the observed structural configurations required – and was perhaps made possible by – the development of ever more abstract, formalized and theory-laden modes of representation. By about 1840, these forms of visual communication in geology no longer functioned as supplements to verbal description and verbal concepts; still less were they merely decorative in function. They had become an essential part of an integrated visual-and-verbal mode of communication. (1976, 152)

Since Rudwick's article and his subsequent work in *Scenes from Deep Time* (1992), a wide variety of scholars have explored an even broader range of the theoretical and intellectual contributions of illustration to nineteenth-century science. This has included, for instance, exploring the tensions between a phenomenalist (or perhaps something like "empiricist") commitment to drawing particular landscapes in situ and developing more formalized, theoretical models of illustration, to the use of illustrations to form natural-historical communities with shared sets of ideological and scientific agendas (see Klonk, 1996, esp. Chap. 3; Belknap 2018; Smith 2006; Adams 2017). We now also have a large body of evidence that *writing* was a tool and an instrument of scientific practice – from note-taking and description to the construction of lectures and teaching aids, to the elaboration of complex and sophisticated narratives (which themselves often constituted theoretical or hypothetical arrangements of the evidence itself) (see Shapin and Schaffer, 1985; Buckland 2013; Secord 2000, 2018; Heringman 2004; O'Connor 2007; on lectures see James 2004; Lightman 2007; Secord 2007; Carroll 2008; on the scientific article,

Bazerman 2000; Gross et al. 2002; and on fieldnotes, for example, see Sanjek 1990; for the scientific book more generally see, for examples, Frasca-Spada and Jardine, 2007; Topham 2009). And of course, there is also a distinguished and vibrant body of work on museum collections, and private collections, as well as the significance of fossils as evidence in nineteenth-century geology, as critical and constitutive modes of its research agendas (e.g., Alberti 2002; Berkowitz and Lightman 2017; Rudwick 1972).

It should hardly need pointing out then, that if writing, illustrating, mapmaking, and collecting were fundamental, and theoretically sophisticated, activities, on which the science of geology was premised in this period, then there can be no certain ground on which we continue to uphold a data/ideas distinction in our analysis of gendered contributions to the science. Such a distinction is rarely – if ever – upheld in any other context except in the continuing assumption that women (and sometimes, by extension, working-class practitioners) were excluded from participation in geology's theoretical dimensions. But it is precisely on this distinction that the narrative of the female helpmate – wife, sister, or daughter, subservient to the intellectual impulses of her husband, father, brother – continues to rest. At the very least, this demands further scrutiny.

Once we begin to give it this scrutiny, immediate questions emerge. Do we always need to emphasize, for instance, that Anne Phillips was working "under her brother's instructions" (Morgan 2007: 265)? Might we not argue, instead, that she – like any self-respecting geologist – was working on a particular geological problem, the terms of which were *obviously* and *necessarily* dictated by the terms of a geological debate which included the voices of many others, as would be the case for anybody of any gender contributing to a pre-established field of knowledge? She had a brother in the same way that John Phillips had a sister, or friends, or an uncle (William Smith) who were also his interlocutors and collaborators. Why should the fact of a woman's relation to a man immediately cause us to emphasize her lowly status in the intellectual hierarchy to which she was contributing? At least since Peter Bowler wrote of the "non-Darwinian revolution" (1988), we have dismantled the idea of the heroic, singular, original man of science working in a vacuum rather than in an entangled set of networks of support, cooperation, collaboration, and culture – so why should we perpetuate the demand for singular, solitary genius when it comes to women? It is of course significant that many women gained access to scientific participation through familial relations, and that this may have shaped the particular modes of scientific inquiry and endeavor in which they were engaged. But the assumptions embedded in the idea that they were mere "helpmates" or "assistants" because of those familial relations might mean we persist in upholding those old standards by proxy, accrediting women only insofar as they were part of the network of a masculine hero, or wishing for them to be heroes themselves.

This chapter seeks to reframe our understanding of gender in nineteenth-century geology by using the early work of Charlotte Murchison as a case study in applying these studies of discipline – of the relations between the theory, method, communication, and practice of science – to the reconsideration of questions of gender. In particular, I argue that the idea of the "geological wife," "familial assistant," or "female helpmate" is deeply embedded in the historiography but has taken us as far as it can as a set of (rather blunt) category of analysis. It has absorbed nineteenth-century gender stereotypes into the understanding of women's scientific collaborations and contributions, perpetuating the long (slow)-dying heroic "man of science" motif.

Charlotte Murchison is a particularly good case study in this regard because her story is so emblematic of the accepted narrative: multiple sources tell us that it was *her* scientific interest that lured her husband into the field, but that illness and age saw her contributions fade as her husband took the inevitable center stage. The fact that her husband, Roderick, was not shy of the limelight, and that she seemed to be happy in a supportive role, has only underscored this narrative. As Roderick's first biographer, the geologist Archibald Geikie, put it of Roderick's marriage to Charlotte:

> hitherto he had lived at his own free will. From this time he came under the influence of a thoughtful, cultivated, and affectionate woman. Quietly and imperceptibly that influence grew, leading him with true womanly tact into a sphere of exertion where his uncommon powers might find full scope. To his wife he owed his fame, as he never failed gratefully to record, but years had to pass before her guidance had accomplished what she had set before her as his aim (Geikie 1875, I: 70).

There could be no more stereotypical image of the Victorian angel in the house, with her quiet, imperceptible "womanly tact" designed to direct the energies and unleash the "uncommon powers" of the all-conquering male. But freeing Charlotte from the shackles of the category of "wife" and taking her contributions on their own terms – as collaborations and contributions to a shared network of ideas – it becomes possible to see a new interpretation of the accepted events of her scientific life, and to begin, at least, to articulate a different vision of the ways in which we might frame and interpret women's scientific work in this period. In particular, I focus on three key moments in her early career: her first geological tour of 1825; her tour with Charles Lyell and her husband to France and Italy in 1828; and the admission of women (including Charlotte) to academic (rather than public-facing) geological lectures at King's College London and the Oxford meeting of the British Association for the Advancement of Science in 1832. Doing so, it becomes possible to see Charlotte in a new light. Looking at these moments as part of *her* narrative, she emerges as a key contributor to one of the most significant and influential theoretical debates of the century and as a key interlocutor of many of those at the forefront of earth sciences in Britain the 1820s and 1830s. Re-examining the historiographic materials on a single figure – Charlotte Murchison – throws new light on the historiography of women in the earth sciences in general.

2 1825

The story of Charlotte "converting" her husband to science, as outlined above, relies largely on the *Lives and Letters of Roderick I. Murchison* by Archibald Geikie. And, while archival research does not substantially transform that story, it is important to tell it in detail to begin to appreciate the full measure of the intellectual and practical work involved in Charlotte and Roderick Murchison's geological lives.

When Charlotte Hugonin met Roderick Murchison on the Isle of Wight in the summer of 1815, she was already a serious-minded (or, in Roderick's words, "attractive, piquante, clever, highly educated and about three years my senior" (as quoted in Geikie 1875, I: 68)) 27-year-old with a father (General Hugonin) recently returned from Waterloo and who shared astronomical enthusiasms with his daughter; her mother was a gifted botanist whose gardens at Nursted House in Hampshire were an experimental playground for rare and dazzling cuttings, plants, and seeds. Roderick had spent the Battle of Waterloo in a depot at Ipswich, having spent (by his own account) most of his 6 years in the army taking singing lessons, flirting with women, reading trashy novels, playing whist, and drinking whisky. He could speak only a smattering of any language, while Charlotte, brought up during the Napoleonic Wars when travel to Europe was impossible for almost all, was fluent in several languages, including French and Italian. Her notebooks from their honeymoon a year later from this time are filled with joy and wonder at the "magnificent Alpine scenery," the lake at Geneva ("nothing can be more enchanting than the first view of the Lake descending the mountains"), "the glaciers of Grindelwald and the Giesbach falls," and the "volcanic and scattered rocky scenery" of Italy (C. Murchison 1816–18: 8, 7, 9, 25). On this tour Charlotte climbed Mont Blanc, was dragged through the Swiss snow in a carriage gone out of control, almost drowned in a felucca, and was surrounded by "a hundred vagabonds" she thought would smash her carriage into pieces (C. Murchison, 1816–18: 16).

In Italy, however, Charlotte contracted malaria and almost died. The couple returned to England to settle down at the much less glamorous location of Barnard Castle, County Durham, which was owned by Charlotte's uncle. Charlotte made a herbarium and started a mineral collection and began to entertain the local gentry. Or, as Roderick put it, "my wife was always striving to interest me in something more intellectual than the chase, and began to teach herself mineralogy and conchology" (in Geikie 1875, I: 333). Roderick was not convinced about the "something more intellectual," and instead sold his unprofitable estates in Scotland and began spending money on foxhounds, horses, and parties, speculating on risky foreign investments and drawing on his dwindling capital to fund it. Eventually, Roderick moved the household – against Charlotte's wishes – to the expensive hunting ground of Melton Mowbray in Leicestershire and took on the upkeep of ten horses. The financial strain began to take its toll.

In 1825, under all this financial pressure, the couple moved to London. Charlotte acquainted Roderick with Humphry Davy and persuaded him to join the Geological Society, whose all-male parliamentary-style debates after the delivery of its papers at Somerset House and raucous dinners at the nearby Crown and Sceptre suited her husband perfectly. She also took control of the finances, began to keep records of Roderick's spending, and set up a summer of geologizing for them both on the south coast of England, where they took lessons from William Fitton, President of the Geological Society. And, as the 2021 film *Ammonite* – starring Kate Winslet and Saoirse Ronan – suggests (before departing from the facts in service of a very different story), Charlotte spent part of this tour – several weeks, in fact – honing her fossil-hunting skills in Lyme Regis with Mary Anning.

However, Roderick didn't stay in Lyme with the women. Instead, the couple set the pattern of a lifetime in terms of geological collaboration:

> Driving, boating, walking, or scrambling, the enthusiastic pair signalised their first geological tour by a formidable amount of bodily toil. Mrs. Murchison specially devoted herself to the collection of fossils, and to sketching the more striking geological features of the coast-line, while her husband would push on to make some long and laborious detour. In this way, while she remained quietly working at Lyme Regis, he struck westward for a fortnight into Devon and Cornwall, to make his first acquaintance with the rocks to which in after years Sedgwick and he were to give the name by which they are now recognised all over the world. (Geikie 1875, I: 127).

This account is, of course, shot through with gendered assumptions: Roderick's work is goal-directed, driven – he "push[es] on," the work is "long and laborious." Charlotte stays behind – an image of the domestic heroine waiting (though in this case "working") "quietly" while her hero heads off into battle. Beneath that, nonetheless, is a clear account of Charlotte's work fossil-collecting, making sketches, and "working at Lyme Regis" with a world-renowned expert in Anning. Roderick's "long and laborious detour" into Devon and Cornwall has little to do with the work that would immediately emerge from this field trip – a paper on the geology of Hampshire and Surrey that would mark Roderick's inaugural contribution to the Geological Society. This doesn't matter to Geikie, because he is teeing up his readers for one of the key triumphs of Roderick's career, the establishment of the Silurian System, which would be based partly on tours in the West Country. In the telling of *that* story, it isn't of much concern what may or may not have been happening at Lyme Regis, or in Hampshire and Sussex, in these months.

But in the telling of *Charlotte*'s story, the south coast of England is crucial. Charlotte was not only learning how to collect fossils, but also putting together her own collection, and she had a specific agenda in mind. The cliffs of Lyme Regis are split roughly into two. Anning had made her name with her "old passion for bones," as she described it, working in the oolitic blue lias rocks on the foreshores of Lyme Regis. Here she made her discoveries of the first ichthyosaurus specimen in 1812 (and further ichthyosaur specimens throughout the next two decades), as well as the first complete plesiosaurus in 1823. (She would find the first British example of a "flying dragon," or pterodactyl, in 1828.) Charlotte's interest, though, lay not in the gigantic lizards of the lias rocks, but in conchology – the study of fossil shells – and in particular in the yellowish greensands that overlay the lias at the top of the cliffs. These shells matched those of the collection she had started of similar rocks around her childhood home in Hampshire. Letters record that Charlotte and Anning went "beating bits of greensand to peices [sic] for shells" (Anning, N.D.). But they helped each other in other ways, too. In 1829, for instance, Anning wrote to Charlotte, asking her to act as a broker for "a very good head of an Ichthyo-vulgaris about two feete in length, which I value at five £" and also an ammonite "which I value at one pound" and which she hoped Charlotte could get Lyell to take (Anning 1829). As a woman of privileged social standing with regular contact with leading men of science, Charlotte acted as a go-between for Anning and the gentlemen of the Geological Society.

Correspondence between these two women demonstrates that their relationship operated on a quid pro quo basis. Anning also writes, "I have taken the liberty to send you a few specimens which I hope you will do me the honour to accept of, the shells are all from the …greensand Mr Sowerby must name these for you and also the Ammonites" (Anning 1829). It is not clear that Sowerby *did* name any of these shells in her name, though he did name the *Ammonite Murchisonae* after her in 1827 "as a just tribute for the ardour with which she has pursued the study of Fossil Conchology, the pleasing effects of which those who are so happy as to be acquainted with her know how to appreciate" (Sowerby 1829, 96). (Charlotte had found the type specimen of that species on a trip to the Isle of Skye in 1826.) Charlotte brokered Anning's oolitic specimens among the wealthy elite in London; in return, Anning found specimens for Charlotte's own collection in the younger greensands that were of interest to her. Indeed, Anning told Charlotte that "since I wrote to you I have not found much. I soon lost the greensand mania, and returned to my old passion for bones the other day, having found a part of a plesiosaurus & so intent in getting it out that I had like to have been drowned …" (Anning 1829).

In a later letter, Anning again expressed her thanks for Charlotte's "kind invitation to town" and said, "I think should anything occur to prevent my accepting it will be the death of me," and thanked Charlotte for her "kindness in mentioning my skeletons to differant [sic] collectors" (Anning 1833). On the envelope of the same letter she wrote "… forgot to say should I get anything good and small that I will not forget that your Cabinet is not full, and also the specimens you name." The bulk of the letter elaborates on what might be going into Charlotte's cabinet: "you and Mr De la B.[eche] between you have given me the green Sand Mania, both Miss Philpot & myself have been beating bits of greensand to peices [sic] to find shells, and after our return we sit down an [sic] turn over the leaves of Sowerby untill our poor heads is compleate jumble of Bivalves univalves & Mr D having been so kind as to lend me his Sowerby – I think to be able to get you many specimens that are not named in this Book." Indeed, she had "not discovered anything, since I am left Lyme, as I have been to [sic] much taken up with greensand" (Anning 1833).

All this matters because it is usually glossed over in a sentence or two in a history of Roderick's story. The wife is credited with "converting" an errant husband to a new faith (science), and with saving his soul from his reckless, gambling ways in the process. But here we see her engaged in her own projects: building a cabinet of shells from the Cretaceous chalks and greensands of Hampshire and Dorset, advocating for her work with Sowerby, and acting as a broker for Anning in return. Indeed, the Sussex geologist Gideon Mantell wrote to Roderick in 1830, mentioning that, "Mrs Murchison was so kind as to interest herself about some dozen [?] Lias fossils for me from Mary Anning," and asks her to try to procure some more (Mantell 1830). The geographer Joseph Barclay Pentland corresponded with Charlotte during this period, writing on 14 May that Cuvier, "requests me to present his kind thanks for your trouble respecting" a specimen "and to express his desire, that you would purchase for him the specimens mentioned in your letter, the price of which I shall record to

Mr Murchison." His daughters (mentioned above) "also to recall them to your recollection, & to say how happy they would be to see you again in Paris" (Pentland 1829). Six years later Pentland promised – again from Paris – to "get the shells for Lord Munster" and asked Charlotte to encourage Roderick to support Frederic Cuvier's election to the Royal Society (Pentland 1835). Here we have ample evidence that Charlotte was a respected naturalist in her own right, cultivating her own networks and projects, and creating a fossil collection of international standing and reputation.

These activities mean that Charlotte was also, in Oldroyd's terms, emerging as an "idea producer" in geology in this period – and in more ways than we might expect. Roderick gave "his" first paper to the Geological Society on 16 December 1826, following this tour along the South coast of England. The paper attempted "an exact delineation" of the rocks of Hampshire, which Roderick said he had first "examined a portion of… with my friend Dr. Fitton" (Murchison 1826: 97). Roderick presented a newly colored Ordnance map "together with such illustrative specimens of the strata and their fossils as I have been able to collect in the course of last summer" (97). And the paper described crushed and fragile shells in the crumbling sands of the upper greensand and valleys of gault beneath them that swelled into the hills of the lower greensand: three bands of rock that had once been considered just one. Below these formations was the "Weald Clay" and the "Hastings or Iron Sand." Overall, this was an "extensive inland tract, in which all the members of the series, from the chalk to the beds below the Weald clay, appear to be more largely and uninterruptedly displayed than in any other part of England" (98).

Much of the paper is concerned with the order of superposition of the rocks in the region. But fossil evidence is offered as crucial support for the delineation of the strata and its division into smaller, more clearly conceptualized categories. Here Charlotte's research is significant. "The most abundant and characteristic" fossils in the "firestone" or "upper greensand" – the *Ammonites rostratus* and the *Pecten orbicularis* – were used to suggest a correlation between the rocks in Hampshire and those on the Isle of Wight. As such, they contributed to the detailed clarification of an emerging history of the Cretaceous rocks of the entire south of England. The fossils "I have collected are chiefly from Nursted and Buriton, Hants," Roderick notes. Nursted House was Charlotte's family home, with Buriton just over 2 miles away, and it was where they had begun their tour. Even the starting point was dictated by Charlotte's connections. But more importantly, we know that it was in fact Charlotte who did the bulk of the fossil collecting. Though Roderick doesn't mention it in the paper, as Geikie puts it, on this tour, Roderick.

> had for companion his own wife. Many a deep lane and rocky dingle did they explore together for fossils. Dr. Fitton came down to visit them and joined in the pursuit, tracing out by degrees the well-marked succession of cretaceous strata shown in that district (Geikie 1875, I: 127).

Evidence from volume 6 of Sowerby's *Mineral Conchology* clearly corroborates Geikie's claim and locates Charlotte's centrality in this work. Sowerby figured in that volume "a Lyas fossil" ammonite "from Lyme … about 3 ½ inches in diameter," which had been "liberally presented by Mrs. Murchison"; an "Ammonite Catullus" "found in the Malm-rock of Sussex (Upper Greensand) in a quarry at Nursted near Petersfield," for which "we are indebted to Mrs. Murchison for the discovery"; and an "extraordinary shell," the "Inoceramus pictus," was one of which "we have only seen but one specimen, for the use of which we are indebted to our very scientific and liberal friend Mrs. Murchison." That shell "was found in the Chalk Marl at Guildford," on the same tour of Sussex and Hants (Sowerby 1829, 106, 123, 215). Sowerby himself identified many of the fossils listed in the appendix to Roderick's paper, but there was also an Avicula – a "new species not yet figured" – and presumably discovered by Charlotte (R. Murchison, 1826: 99).

Roderick's "I," therefore, subsumes the work of his wife within it. This should not come as a surprise, but as a simple acknowledgement of the system of intellectual coverture at play here. Only men were allowed to speak at the Geological Society. As such, a wife's work was likely to be, without question, subsumed into the first-person "I" of the husband's account. That is not to say that she did not perform that labor – nor to say that all did not tacitly understand that she had contributed substantial material or ideas to her husband's work. It is simply a reminder of an intellectual convention that a woman's work belonged to her husband, and that his proclamation of the work "he" had done straightforwardly included the work she had done too.

Furthermore, though the 1825 paper was only a first foray into geological publication for Roderick, it was part of a bigger research project in which he and Charlotte were engaged. Immediately following the Murchisons' paper was one by De la Beche, also on the greensands, this time in Lyme Regis and Devon. As such, when De la Beche and Charlotte drove Anning to the "greensand mania" she described in her letters, they did so likely because both were working on a shared problem at the behest of Fitton himself. Fitton's work in this field would be presented to the Geological Society in a series of papers between 1824 and 1836, culminating in his classic *The Strata Below the Chalk* (1835): a 300-page, 17-plate work that painstakingly worked out the complex and confusing succession of Cretaceous rocks in the south of England. Fitton's work – proven partly by the detailed research of people like Charlotte and De la Beche – demonstrated (as Roderick's paper described) that there was an *upper* greensand, a layer of gault, and then *lower* greensand rocks (all containing marine fossils). Below that was Weald Clay, not only in the part of Kent after which it was named, but also in the Isle of Wight and Dorset. These layers contained fossils of freshwater origin. Then came the ferruginous Hastings Sands (also freshwater). The Murchisons' paper can be seen as a direct contribution to this research project – partly because Fitton himself acknowledged Charlotte's role in this emerging set of ideas. At the back of *Strata Below the Chalk* is a list of the key fossils in Hampshire and Western Sussex that elaborated this sequence of strata and another list of fossils from the greensands near Nursted House (see Fig. 1): "the following List of Fossils from the Upper green-sand," Fitton writes, "in the vicinity of Petersfield, in addition to those mentioned in Mr. Murchison's paper on that district … I owe to the kindness of Mrs. Murchison, in whose collection the specimens are placed" (Fitton 1836: 156).

Furthermore, the detailed elaboration of these strata was important because it established the credence of what had, in the early 1820s, appeared to be a startling theoretical supposition. In 1822, the geologist Gideon Mantell had argued, in his *Illustrations of the Geology of Sussex*, that there were no ammonites, belemnites, echinites, or corals (all marine fossils) in the Weald, though they were abundant in the Cretaceous rocks above them and the Oolitic rocks below (Mantell 1822: 67–68). As Charles Lyell would later put it, "the evidence of so unexpected a fact as the infra-position of a dense mass of purely freshwater origin to a deep-sea deposit (a phenomenon with which we have since become familiar, in other chapters of the earth's autobiography) was received, at first, with no small doubt and incredulity" (Lyell 1852: 226). With regular reference to Fitton's work, Lyell notes that "the relative position of the beds is unequivocal; the Weald Clay being distinctly seen to pass beneath the Lower Greensand in various parts of Surrey, Kent, and Sussex, and to re-appear in the Isle of Wight at the base of the Cretaceous Series, being, no doubt, continuous far beneath the surface" (Lyell 1852: 226). This mattered because it suggested something perhaps even more significant and which would be crucially important to Lyell in his groundbreaking work in *Principles of Geology* (1830–1833). For here was clear evidence that the earth, in a single place, had transformed itself from sea, to land (with rivers and lakes), back to sea again. Furthermore, the fossil evidence revealed that many of the species alive at the time of these great transformations had *survived* the changes. If that was true, the changes had to have happened relatively gradually (no comets, visitations from God, or Biblical-scale deluges complete with entirely new species epochs) – though also quickly enough that the species were roughly similar throughout all those events.

(75.) ***List of Fossils from the Beds below the Chalk, in part of Hampshire and Western Sussex.***

[*Upper Green-sand.*]

The following List of Fossils from the Upper green-sand in the vicinity of Petersfield, in addition to those mentioned in Mr. Murchison's paper on that district, (Geol. Trans. 2nd Series, vol. ii. p. 99.) I owe to the kindness of Mrs. Murchison, in whose collection the specimens are placed.

Ammonites catillus.	*Fishes.* **Scales and vertebræ.**
A——— splendens.	*Gryphæa columba.*
Arca carinata.	*G——— vesiculosa.*
Avicula **gryphæoides. (See Pl. XI. f. 3.)**	*G——— sinuata.*
Cucullæa. **Much compressed.**	*Nautilus elegans* ?? **Much compressed.**
Echino-spatangus. **(Three new species.)**	*Pecten asper.*

Fig. 1 'List of fossils from the beds below the chalk, in part of Hampshire and Western Sussex' as it appears in Fitton's *Strata Below the Chalk* (1836), pp. 156–159. All the fossils from the "Upper green-sand" are from "Mrs. Murchison's … collection," while many individual "Lower green-sand" specimens are also accredited to her

Dr. Fitton *on the Strata below the Chalk.* 157

Pecten Beaveri.
P—— *nitidus.*
P—— *orbicularis.*
P—— *quinquecostatus.* Besides one or two other species, indistinct.
Plicatula inflata.
Solarium granulatum. Mantell.
S—— or *Trochus.* Another species.
Thetis major.
Vegetable impressions. Like those of the upper chalk near Lewes. Mantell, t. 9. p. 157.

[*Lower Green-sand.*]

Ammonites dentatus. Stopham brick-yard, West of Pulborough, Sussex; in a mass of phosphate of lime. This, with the other fossils from the same place, was found by Mr. Martin covered by green-sand, in an excavation at the top of the Weald clay. Martin MSS.
A———. (Near to *Nutfieldiensis*, but a distinct species.) Lowest member of the sands near Pulborough: Martin.
A———. (Impressions of one or two other species.) Stopham brick-yard: Martin MSS. One of these is in a roundish mass of phosphate of lime, precisely like specimens found near Copt Point, Kent: the outer surface is whitish, and is eroded in small tortuous grooves, as if eaten by worms: the interior brown.
Auricula (new species). Martin, *ibid.*
Avicula? Martin, *ibid.*
Cardium? (New species; delicately striated.) Stopham brick-yard: Martin MSS.
Coprolite. Stopham brick-yard: Martin MSS.
Corbula gigantea. Bowyer's Common: Mrs. Murchison.
C—— *striatula.* Bowyer's Common, near Petersfield: Mrs. Murchison. Parham: Mantell, and Martin.
C—— (new). Near Pulborough: Martin, p. 32.
Crustacea. "At Bognor Common, in Fittleworth parish, there is a vein of friable "green-stone, rich in Pectens, Trigoniæ, and the remains of a crustaceous fossil "like a shrimp." Martin, p. 34, note.
Cucullæa decussata. Parham: Mantell, and Martin.
C——— *glabra?* Bowyer's Common: Mrs. Murchison.
C———. (Another species—or two?) Stopham brick-yard: Martin MSS.
Cyprina angulata. (*Venus*, Min. Con.) Stopham: Martin MSS.
Cytherea parva. (*Venus*, Min. Con.) Parham: Mantell. Bowyer's Common, and

Fig. 1 (continued)

The fossils of the greensands that Charlotte collected were part of an emerging vision of the earth as capable of undergoing gradual, perpetual transformation.

As such, it should be clear that Charlotte's specimens – as much as Roderick's headlong horsedash to Cornwall – were a crucial part of the evidence that gave Roderick his first paper at the Geological Society and inaugurated him as a member of the geological elite. Neither was this just a piece of administrative assistance Charlotte gave to Roderick, nor the work of a helpmate. The specimens she was collecting for that paper – and for her broader collection – were indisputably a key source for the growing evocation of a Cretaceous world on the south coast of England. And, though none of this would be said in the form of a Geological Society paper, Charlotte's work played a small but significant role in the highest-level

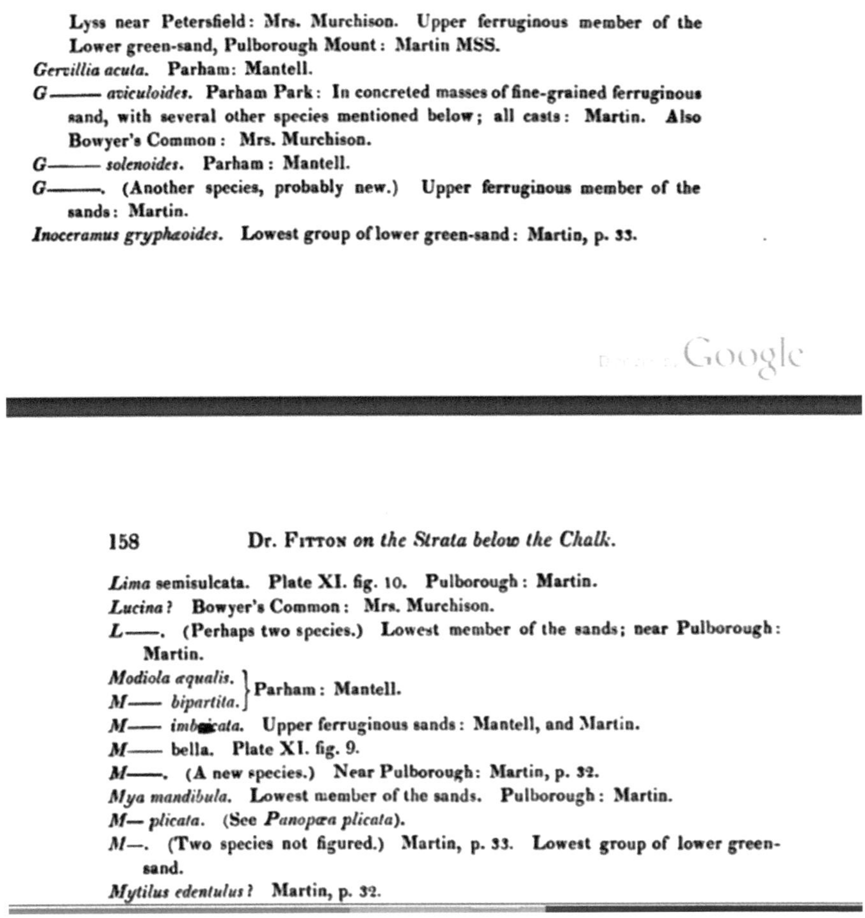

Lyss near Petersfield: Mrs. Murchison. Upper ferruginous member of the Lower green-sand, Pulborough Mount: Martin MSS.
Gervillia acuta. Parham: Mantell.
G—— aviculoides. Parham Park: In concreted masses of fine-grained ferruginous sand, with several other species mentioned below; all casts: Martin. Also Bowyer's Common: Mrs. Murchison.
G—— solenoides. Parham: Mantell.
G——. (Another species, probably new.) Upper ferruginous member of the sands: Martin.
Inoceramus gryphaoides. Lowest group of lower green-sand: Martin, p. 33.

158 Dr. Fitton *on the Strata below the Chalk.*

Lima semisulcata. Plate XI. fig. 10. Pulborough: Martin.
Lucina? Bowyer's Common: Mrs. Murchison.
L——. (Perhaps two species.) Lowest member of the sands; near Pulborough: Martin.
Modiola æqualis. } Parham: Mantell.
M—— bipartita. }
M—— imbricata. Upper ferruginous sands: Mantell, and Martin.
M—— bella. Plate XI. fig. 9.
M——. (A new species.) Near Pulborough: Martin, p. 32.
Mya mandibula. Lowest member of the sands. Pulborough: Martin.
M— plicata. (See *Panopæa plicata*).
M—. (Two species not figured.) Martin, p. 33. Lowest group of lower green-sand.
Mytilus edentulus? Martin, p. 32.

Fig. 1 (continued)

theoretical work going on at the Geological Society in the 1820s. Fossil collecting was a significant part of the highest-level "idea producing" or theoretical debate at the Geological Society, and at least one woman was a direct collaborator on that research.

In this way, fossil collections ought to be read as forms of publication, and as of especial significance for women. As a point of comparison, Cynthia Burek has demonstrated that Benet's fossil collection was highly regarded: she contributed the second highest number of specimens of all contributors to *Mineral Conchology*. "However," Burek writes, "she was not a prolific writer and during her lifetime she only published two articles, both fossil lists from her extensive collection of thousands of Cretaceous and Jurassic specimens, many from the Upper Greensand" – the exact area in which Charlotte was also working

M—— *lanceolatus*. Parham: Mantell.
Natica. Parham: Parkinson; Organic Remains, vol. 3. Pl. 6. fig. 2. Lyss near Petersfield: Mrs. Murchison.
Nautilus. Near Pulborough: Martin, p. 31.
Nucula impressa. Parham: Mantell. Lowest member of the sands; Pulborough: Martin.
N—— *antiquata*. Upper ferruginous member of the Lower green-sand; Pulborough Mount: Martin.
Orbicula. (Not figured.) In sand rock: lowest member of the sands; Pulborough and Stopham: Martin.
Panopæa plicata. Lower member of the sands; Parham Park, and Stopham brick-yard: Martin MSS.
Patella. Parham: Mantell.
Pecten obliquus. Parham: Mantell, and Martin.
P—— *orbicularis*. Pulborough: Martin. Parham: Mantell.
P—— *quadricostatus*. Parham: Mantell.
P——. (A new species.) Martin.
Pholadomya. Lowest member of the sand near Pulborough: Martin, p. 32. and MSS.
Pinna tetragona. In sand-rock; lowest member of the sands: Martin.
P——. (A cast.) Stopham brick-yard: Martin.
Rostellaria calcarata. Parham: Mantell.
R———— *Parkinsonii*. Parham: Mantell. Upper member of the sands; Pulborough Mount: Martin.
R————. (Very like *Pes Pelicani*.) Parham: Mantell.
Tellina æqualis. } Parham: Mantell, and Martin.
T—— *inæqualis*. }
T——? (A species not figured.) Lowest member of the sands, near Pulborough: Martin.
Terebratula lata. } Parham: Mantell.
T———— *ovata*. }

Fig. 1 (continued)

(Burek 2001, 193). The point is important. Given that Charlotte could not read a paper at the Geological Society, and thus formally publish in the *Transactions*, a fossil collection would have been a key means by which she engaged and participated in scientific discussion. And for several purposes, it also had distinct advantages over publication in print. If the rapidly progressing theories debated at the Geological Society were ever-shifting sands, which could be washed out with each new tide of information, a fossil collection could be instantly responsive to new directions in the field. A well-made collection was not the output of an assistant with no "ideas": it was both the constitution and the expression of the intellectual work of geology.

The most cursory glance at geological books and papers in the 1830s reveals traces of Charlotte's collection. It's not just her regular appearance in the appendix of Fitton's *Below the Chalk*, it's also a fossil ammonite and a sketch of a dragonfly collected by Charlotte in Gideon Mantell's *Wonders of Geology* (1838, see I: 460 and

Dr. Fitton *on the Strata below the Chalk.* 159

Terebratula nuciformis. Upper member of the sands; Pulborough Mount: Martin.
T——— depressa. Upper ferruginous member of the Lower green-sand; Pulborough Mount: Martin MSS.
T———. (Another species—uncertain.) Same place: Martin.
Thetis major. Lowest member of the sands; Pulborough: Martin.
T—— minor. Bowyer's Common, and Lyss, near Petersfield: Mrs. Murchison. Parham: Mantell. Upper member of the sands; Pulborough Mount: Martin.
T—— (new). Stopham brick-yard. Martin MSS.
Trigonia alæformis. } Bowyer's Common: Mrs. Murchison. Parham: Mantell. Stopham: Martin.
T——— clavellata? }
T——— dedalæa. Parham: Mantell.
T——— spinosa. Pulborough: Martin, p. 33.
Turbo rotundatus. Parham: Martin MSS.
T——. (Another species.) Near Pulborough: Martin, p. 31.
Turritella granulata. Parham, and Pulborough: Martin MSS.
T———. (Another species.) Near Pulborough: Martin MSS.
Venus angulata. See *Cyprina.*
V—— Faba. } Parham: Mantell.
V—— ovalis. }
V—— parva. See *Cytherea parva.*
V——. (Casts of two other species.) Stopham: Martin MSS.
Vermicularia concava. Parham-park, Parkmount-lane near Pulborough, Chiltington-common. Martin, pp. 33 and 34.

Fig. 1 (continued)

399 for descriptions of the plates); it's a nautilus striatus Charlotte gathered in Whitby and an ink bag, both illustrated and published in William Buckland's *Geology and Mineralogy with Reference to Natural Theology* (1836, plates 33; 45, fig. 4). Or, it's an Ophiura Egertoni from her collection from Whitby, cited in an 1835 paper given at the Geological Society by William John Broderip (Broderip 1835, plate XI, fig. 5). Trilobites Charlotte found in black schistose rock at Mount Pleasant, near Carmarthen in Wales, figured in Roderick's *Silurian System*, alongside her many illustrations for that book (1839, described on 664). And if we don't have a clear picture of Charlotte's collection and its significance now, it is literally because Charlotte was a woman: her collection was posthumously broken up and absorbed into Roderick's collection by the Geological Society, assumed to be of no significance beyond the name of the male with whom she was most closely associated. That does not mean, of course, that that assumption was correct. And in 1828, Charlotte's collection would be dramatically expanded and acquire new theoretical significance that clearly demonstrates her status as an "idea producer" in the earth sciences at this time.

3 1828

In his magisterial *Worlds Before Adam,* Martin Rudwick describes the tour Charlotte and Roderick undertook with Charles Lyell to France and Italy in the summer of 1828. As Rudwick puts it, Roderick "was ambitious for some more challenging fieldwork, which he tackled with military zeal. Charlotte had learned from Mary Anning how to be 'a good practical fossilist' or collector and like Buckland's and Mantell's wives she was a fine artist; both skills made her a valuable companion in the field" (Rudwick 2008: 259).

Though there is acknowledgment of Charlotte's labor here, there is also a subtle sidelining of Charlotte's agency and desire, which rings with echoes of Geikie. All the action is driven by Roderick's desires – his "ambition" for "challenge," the use of Roderick's (frankly lackadaisical) "military" career as a metaphor for his industry and activity. As in Geikie, Charlotte's link with Anning again appears to suggest a quieter, more contemplative mode of relation to science – she has "practical" skills, and is no different from other artistic "wives" and "companion[s]." Note the shift from past simple (Roderick "was ambitious" and "he tackled" the fieldwork with "zeal") to past perfect tense to describe Charlotte's actions (she "had learned" to collect fossils). Charlotte's skills are a settled fact, while Roderick's pursuit of new knowledge is continuous and ongoing. In the next sentence, Rudwick writes that not only was Charlotte a good companion, but "so was Lyell, as a more experienced and knowledgeable geologist and a much better linguist" than Roderick. It should be noted here that Charlotte was *also* a "much better linguist" than her husband, and that her field notebooks from this tour are written in English, French, and Italian, but since these are all presented as secondary skills this is perhaps, at this stage, a quibble (Rudwick 2008: 259).

More importantly, Rudwick notes that this "great tour on the Continent, lasting almost a year … was, as it were, Lyell's *Beagle* voyage," the journey on which he really learned to see the world in a new way, and which was critical in the production of *Principles of Geology* (2008: 260). This was the tour on which Lyell would witness for himself the power of simple, ongoing geological processes like erosion to carve out valleys and other large-scale structures: for Lyell, this would prove that geological processes operating at *present*-day intensities had been powerful enough to shape earth history on a grand scale. On this tour, he would also begin to use *percentages* of extant and extinct fossils to describe the *gradual* shifts between geological epochs, moving from the most recent fossil beds incrementally further into the past. In other words, on this tour he would come to his uniformitarian vision of the earth.

Kölbl-Ebert's account of what happened in 1828 is more attentive to Charlotte's position in the team that elaborated this new theory. She notes, for instance, that having ascended the Puy de Dome on 18 May 1828, the trio established the pattern of work that would be most efficient. They undertook multiple field expeditions together and, as on the south coast of England, Roderick (and here Lyell, too) kept "track of correlations and bigger structures," while Charlotte did "much of the time-consuming fossil-hunting, sketching landscape and geological structures, and –

speaking French fluently," visiting "local experts" (2007b: 111). As Lyell put it, Charlotte was engaged "in making panoramic sketches, receiving several of the gentry & Professors to whom we had letters in the neighbourhood & collecting plants & shells etc." (as cited in Kölbl-Ebert, 2007b: 111). Crucially, as Kölbl-Ebert also points out, Charlotte's work was then integrated with that of Roderick and Lyell: "On the 20th May," Roderick wrote, "we were making sections of the tertiary lacustrine deposits we had examined at Verduaison above Cusset & Vernet (see Blue book & Lyell's sections) & Lady M's notes" (as cited in Kölbl-Ebert, 2007b: 111). This was explicitly team-like behavior.

Nonetheless, even Kölbl-Ebert continues to use language like "assistant" and "helpmate" to describe Charlotte's participation, claiming that the men decided "route and research topics" (2007b: 109). In doing so, she supports the idea that women could do scientific *work* but were debarred from scientific *thinking* – or, at least, from being able to use their thoughts to direct research agendas. Partly this is because Lyell himself described Charlotte as "an invaluable assistant" and says that, "she is so much interested in the affair as to be always desirous of keeping out of the way when she would interfere with the work" (as cited in Kölbl-Ebert, 2007b: 111). And there are methodological issues: Charlotte's diaries are actually field notebooks and are sparse in color and detail, meaning that we have a much more vibrant picture of the tour as given by the men in their more voluminous and verbose correspondence. But we should not always take these men's words at face value. Was Lyell likely to have understood the delicate balancing act it required for Charlotte to secure her participation in the team – always deferring to masculine authority *in order* to gain access to the work in the first place? Furthermore, both men – like Geikie and others after them – were clearly invested in the narrative of masculine, military action in the field with female assistance in the domestic sphere. This narrative structured their understanding of the contributions made by each member of the team in gendered terms. But this does not mean that Charlotte was not, in fact, participating in the intellectual processes that contributed to the development of new knowledge in France in 1828 or anywhere else. It might rather mean that, even when she was doing serious amounts of the practical and intellectual toil, Roderick and Lyell nonetheless *understood* her role as mere assistance and therefore underplayed it even to themselves. This is not an accusation of a conspiracy: rather, these men had simply not been trained to see women's intellectual labor on equal terms with their own.

This is perhaps easiest to see in Lyell's account of an illness Roderick contracted on the 7 August while the party were at Fréjus, and which forced them to remain in Nice from 9 to 27 August:

> Murchison has not yet regained his strength from a severe attack at Fréjus, so as to be able to take the field again. But we have been hard at work in writing, from our materials, a paper on the excavation of valleys, which is at last finished, and after two evenings' infliction, is intended to reform the Geological Society, and afterwards the world, on this hitherto-not-in-the-least-degre e-understood subject. Besides this mighty operation, we have performed two jaunts with Mrs. Murchison, each at half-past four o'clock in the morning, to see certain deposits of fossil shells, and collecting these, with which she has been much pleased; and this

> and the cessation of eternal bustle, while the campaign was at its height, while were as yet only crossing the Balkan, and before our descent upon these hot latitudes has restored her health and spirits, which had failed sadly (as cited in Kölbl-Ebert, 2007b: 114).

The language of the "campaign" and "crossing the Balkan" draws on the well-rehearsed metaphor of fieldwork-as-battle, of course, but there are two other things of interest here. First, the subject of the paper – "intended to reform the Geological Society, and afterwards the world" – and second, the pronoun. This time it is not an "I" that subsumes female intellectual labor, but a "we" that obfuscates our understanding of it. Who is the "we" who undertook the "hard ... work" of "writing" this "paper on the excavation of valleys?" This is a matter of longstanding discomfort. Both Lyell and Roderick's names appear on the paper, so the "we" would appear to belong to them. But at the same time, throughout his career – both before and after this paper – Roderick was vehemently opposed to the explanation of geological processes the paper presents. When the three geologists were in Le Puy, Rudwick reminds us, "Murchison proved to be a staunch 'diluvialist,'" arguing that (as he put it himself) "gradual erosion could never produce such broken irregular effects," and instead arguing that geological processes (like wholesale floods) must have operated at much greater intensities in the deep geological past (Rudwick, 2008: 267). Kölbl-Ebert points out that, "years later, when he compiled his journal from his field notes, he thought it necessary to take refuge in more violent causes of change" to explain the valleys of Auvergne (as cited in Kölbl-Ebert, 2007b: 115). At this time, Murchison wrote:

> I must if I ever have time review what Lyell (for he was the chief penman) wrote in a joint paper on the excavation of the valleys of Auvergne. For, whilst I fully admit the re-excavation of certain valleys & the clearing out of the lava & basalt which filled them, I do not believe that the original water-courses were formed by the streams, but they originated in cracks & depressions formed by great & ancient Earthquakes, long anterior to the eruptions of the youngest volcanoes of Auvergne. (as cited in Kölbl-Ebert, 2007b: 115)

As Murchison's hint that Lyell "was the chief penman" suggests (alongside his desire to "review" a paper he was supposed to have written just to check what it actually said), these "great and ancient Earthquakes" were the exact opposite of the processes described in the paper. The paper itself wholly advances the position that simple erosion had created the valleys and that, by extension, "modern causes" (geological processes operating at present-day intensities) could explain earth history. This paper showed that the Sioule river had cut a gorge through basalt lava and the underlying gneiss. The lava was traceable to a recent volcano, and the cones of volcanic ash it had left behind would have been swept away had a large-scale flood (like Noah's Flood, as men at the Geological Society like William Buckland or Charles Daubeny believed) actually occurred. Furthermore, a bed of gravel underlying the rock showed that the lava had flowed down a valley that existed *before* the volcanic eruption. As the engraving that constituted the central proof of the paper clearly demonstrated (Fig. 2), these processes were longstanding, gradual, and continuous.

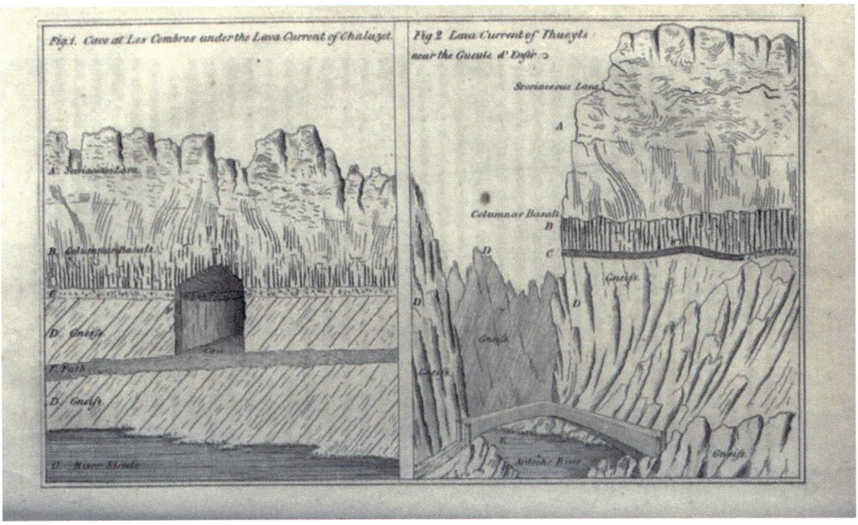

Fig. 2 Charlotte Murchison's important illustration to Lyell and Murchison's 1829 paper in the *New Edinburgh Philosophical Journal* (plate between pages 45 and 46)

As Rudwick describes the engraving, "this sketch was published by Lyell and Murchison in their article (1829) on the power of rivers to erode valleys in hard rocks, even in a geologically brief timespan, without invoking any exceptional 'diluvial' agency," and it clearly demonstrates the argument. "The case," Rudwick goes on, "was similar to several that [George Poulett] Scrope had already described and illustrated …; but this engraving, published with the article in Jameson's Edinburgh periodical, would have been seen by many more geologists than Scrope's expensive album" (Rudwick 2008: 268). Indeed, the image was crucial to the articulation and demonstration of the argument, as well as to its persuasiveness among the members of the Geological Society, because it *proved*, visually, a phenomenon that had already been noted by other geologists like Scrope and Croizet – but which words had not yet been persuasive enough to make scientific fact.

This particular engraving – with its crucial role in "reforming the Geological Society, and afterwards the world," as Lyell had put it – was one of three published with the paper in the *Edinburgh New Philosophical Journal* (Figs. 3 and 4). The second (Fig. 3) demonstrated that the "tertiary lacustrine formations of Auvergne" had "been excavated to a considerable depth prior to the operations which gave rise to the trachytic breccias and conglomerates overlying alluvial beds, containing bones of extinct quadrupeds" (Lyell and Murchison 1829, VI: 47). In other words, the valley had been eroded before the volcanic explosion. The third was "a transverse section of Mont Perrier, exhibiting in detail the character of all the strata of trachytic conglomerates and alternate with alluvions containing fossil bones of extinct quadrupeds" (VI: 47). Together, these images explained, demonstrated – and proved, to

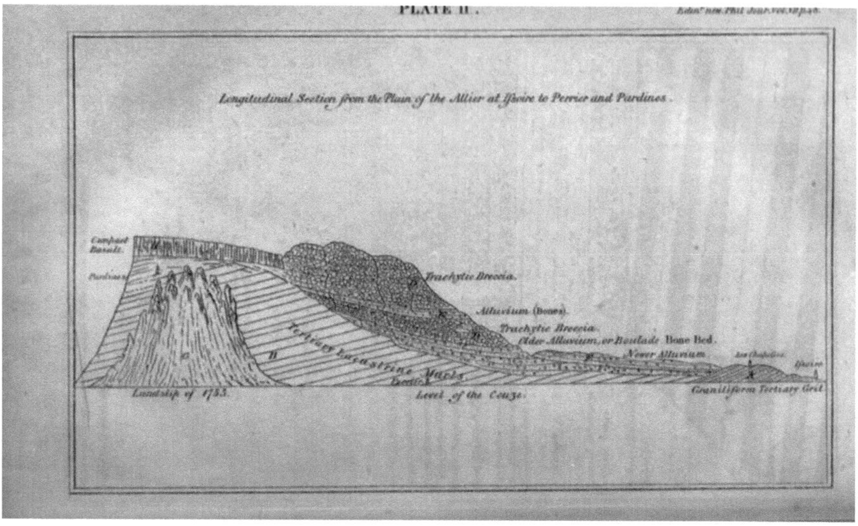

Fig. 3 Charlotte's second illustration to the 1829 paper, published as a plate between pages 48 and 49

those who had not been able to travel to Auvergne themselves – that fluviatile erosion had created these valleys, and that modern causes were at least hypothetically sufficient for explicating geological structures that seemed to have been created by more violent events.

And so we return to the "two evenings infliction" Lyell described – the "writing" of the paper undertaken by a "we" that has been, for two centuries, assumed to include Roderick Murchison. Rudwick chooses to focus on the two men, whom he credits with having "published" the images. Nobody credits the person who actually produced them: Charlotte. Still, members of the Geological Society widely understood her to have been their author: Scrope wrote in a letter to Roderick on 4 December 1828 that he very much enjoyed the "battery against the Diluvian Heresy" the paper represented. "At all events," he continued, "I much wish to see you and gossip of Auvergne and my old cronies the Puys … I am anxious to renew my recollections of Central France from your conversation and Mrs. Murchison's drawings" (Scrope 1828). Scrope explicitly uses the drawings as tools to rethink the Auvergne when he is not actually there and to re-engage in a theoretical debate in which he had much at stake. And in the 1914 edition of Lyell's *Antiquity of Man,* edited by the geologist R.H. Rastall, Charlotte is credited not only as the illustrator of these three figures, but as a co-author of the paper itself: Lyell's published oeuvre includes "three papers with Sir Roderick and Mrs. Murchison; *Edinburgh Philosophical Journal*, 1829; abstract in *Proceedings of the Geological Society* 1; *Annales des Sciences Naturelles* 1829" (1914, xi).

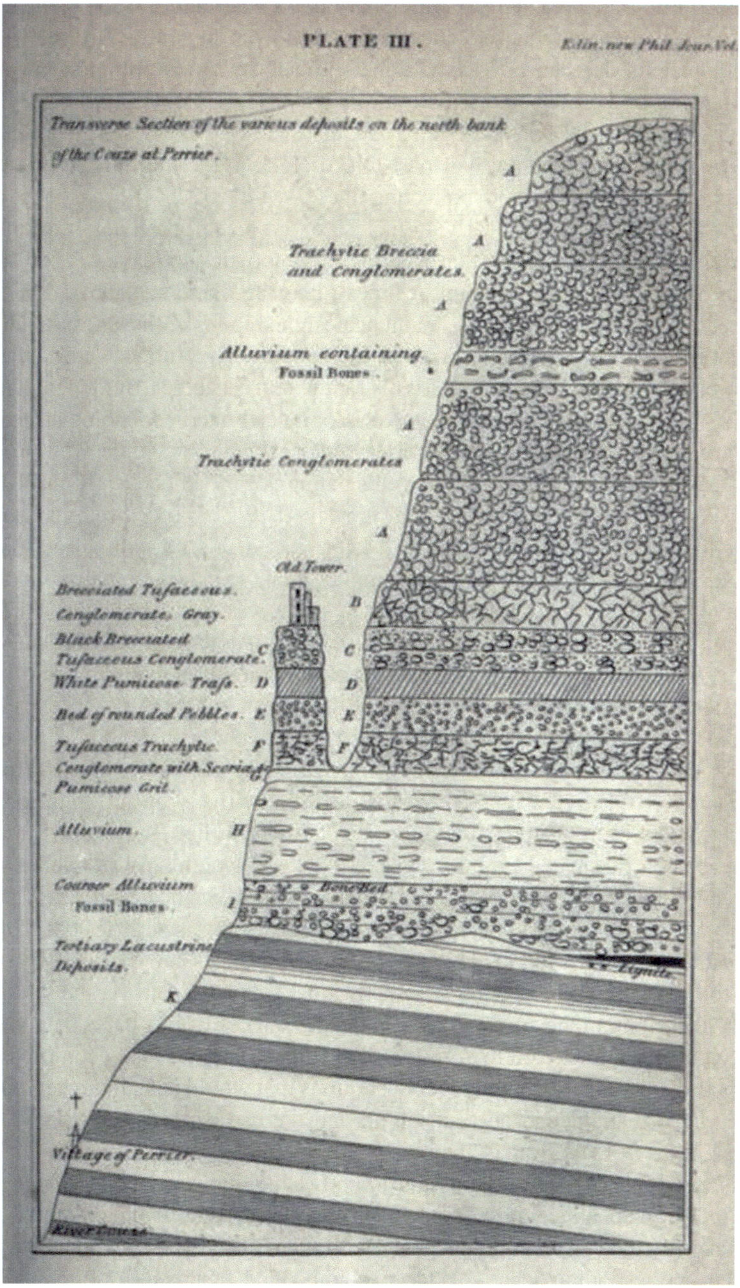

Fig. 4 Charlotte's third illustration, on a plate between pages 48 and 49

Rastall's bibliography casually names Charlotte as a co-author some 80 years after the fact, but it does so because it accepts illustration as constitutive of the work the paper represents. He can only have done so because it was common knowledge that Charlotte had produced the illustrations.

To this evidence, we can add Lyell's account that "we have performed two jaunts with Mrs. Murchison, each at half-past four o'clock in the morning, to see certain deposits of fossil shells, and collecting these, with which she has been much pleased." The "we" here is clearly an inclusive first-person pronoun, referring only to himself – we already know that Roderick was too sick to leave his bed. It goes hand in hand with Lyell's attempt at breezy language here – "jaunts," "pleased," "restored her health and spirits" – language designed to make these fossil-collecting expeditions as purely touristic, or even therapeutic. They almost appear as little holiday excursions to collect pretty fossils for the lady's cabinet. But in fact we know that they were serious: George Antonio Risso, a celebrated conchologist, led Lyell and Charlotte over the deposits at Nice – an area unusually rich in the well-preserved fossils that would later be critical in Lyell's naming of the Pliocene epoch (Rudwick 2008: 270). We also have evidence that in May 1829 Charlotte wrote to Croizet to ask for fossil eggs from Auvergne, and later chased this up with Joseph Barclay Pentland (though the letter we have implies that he never actually sent them) (Pentland 1829). So we can assume that Charlotte was fossil-collecting with Lyell in a way that was strategic to the aims of the 1828 paper, and building a rich and complex set of fossil evidence for her collection which was in significant dialogue with emerging geological theories.

In the face of all this evidence, the idea usually advanced for Roderick putting his name to the 1829 paper on valley erosion has been decidedly weak. "Roderick Murchison was uneasy," with the paper, Kölbl-Ebert writes, "even if in the presence of Lyell he had been persuaded to agree" (2007b: 115). For Rudwick, "Murchison may have been persuaded by his companion – not for nothing was Lyell a trained barrister – or perhaps by the sheer perceptual impact of 'the beautiful pet volcanoes' of Vivarais" – still, this was only a "nominally joint paper" (2008: 268). So Murchison was unwell in bed, later disavowed the paper's contents entirely, admitted he didn't really write it, or recall exactly what was in it. And yet the scholarship persists in attributing the article to him on the grounds of Lyell's persuasive rhetoric. Surely, it is less of a stretch to argue that, since Charlotte was collecting fossils for the paper with Lyell and drawing the illustrations that would constitute the bulk of its proofs, *she* was the actual co-contributor, and the use of Roderick's name on the paper signifies *her* labor rather than his.

This raises a second, interesting question. If Charlotte was a collaborator on this paper, did she also agree with its contents in a way that her husband did not? Perhaps Roderick Murchison was a diluvialist. That does not mean that his wife was, too, and only the unquestioning idea – so deeply embedded in the historiography – that women plodded unthinkingly in their husbands' intellectual footsteps, as "assistants" rather than genuine collaborators, can blind us from seeing that Charlotte may have had her own mind on the matter.

4 1832

With that point, we come to the final section of this chapter, and another point of difference between Charlotte and the two men with whom she went on tour in 1828. As is well-known, in 1832, women were included in academic geological lectures for the first time: first at Lyell's lectures at King's College London and then at the British Association for the Advancement of Science. Charlotte's presence at both these sets of lectures has often been noted, and Kölbl-Ebert goes so far as to say that "it was her wish to attend, that opened Lyell's geological lectures at King's College to women" (1997b, 41). But this has not been clearly linked to her intellectual contributions to the development of geology in this period. I want to finish this essay by arguing that it was Charlotte's fieldwork – in tandem with Mary Somerville's publication of *Mechanism of the Heavens* in 1831 – that made the issue of women's attendance at academic lectures so difficult to avoid.

In the autumn of 1831, Lyell was preparing his lectures for King's and he attended the first meeting of the BAAS in York, which women had attended by default, since the fledgling institution seemed to be of public rather than scholarly bent. In early November, he "went by invitation to drink tea with the Murchisons, and was thus kept to geology till half-past eleven o'clock at night … Mrs. Murchison was in high spirits, and has evidently enjoyed their tour, and above all the York meeting" of the British Association for the Advancement of Science (in K. Lyell, 1881, I: 350). "Her real delight in conchology is of great use to her in taking away the tedium of long sojourns in out-of-the-way places," he wrote to Mary Horner, the daughter of a former President of the Geological Society and the woman he would marry the following year, "and she had a fair opportunity of collecting" (in K. Lyell 1881, I: 350–51). With excitement, he added that the Murchisons had "found *recent* Irish Channel shells in abundance, 300 feet high on the Lancashire coast! I am delighted, because the identity of the living species of plants and animals in England and Ireland made me argue that some great changes in the distribution of land and sea had taken place in that part within the modern era" (in K. Lyell 1881, I: 351). Much like the interchange of marine and freshwater fossils on the south coast of England, here was another example of the shifting lands and seas created by ordinary geological processes, with further evidence supplanted by Charlotte.

At this time, Lyell was also planning to "use our joint and unpublished notes on Auvergne" in the second volume of *Principles* (in K. Lyell, 1881, I: 355). In honor of that fact, "and because we had together part of the best and longest geological tour which I ever made, my tour which made me what I am in theoretical geology," Lyell was planning on dedicating one of its volumes to Roderick (in K. Lyell 1881, I: 355). Once again, we should not take this to mean that he was only making a dedication to Murchison. Lyell's letters are full of admiration for women like Cordelia Mallet, married to a seismologist, and whose "considerable turn for natural history" he much admired (in K. Lyell, 1881, I: 370). Lyell also recalls fossil and fern-hunting expeditions with his sisters Marianne, Eleanor, and Caroline near their ancestral home of Kinnordy in Scotland (in K. Lyell 1881, I: 338, 341, 359). He repeated (albeit to his would-be wife, Mary Horner) the geologist Thomas Webster's supposition that women possess a greater "liberal curiosity" for science than men whose heads had been stuffed with the dead languages of Latin and Greek, and he greatly admired his friend Mary Somerville, whose technically audacious, mathematically dazzling translation of Pierre-Simon Laplace's *Mechanism of the Heavens* had just been published (1881, I: 323). "There are not five persons in this country that could have written that book, if that number," marveled Lyell, noting seriousness that most mathematicians at Cambridge weren't clever enough to understand it (1881, I: 374). Lyell's world was rich in intellectual women with whom he was willing to interchange ideas and labor. But, of course, he also wrote of Somerville that had she "been married to La Place, or some mathematician, we should never have heard of her work. She would have merged it in her husband's, and passed it off as his." For, while "a man may desire fame, reputation, and even glory, for the sake of sharing it with one he loves," not so "a woman," who "cannot share it with her husband," because "it will be the utmost she can do not to make him of less importance by it" (in K. Lyell, 1881, I: 325). The man who wrote this letter was more than capable of thanking *both* Murchisons in print by mentioning only the husband. Perhaps, as he began to contemplate matrimony himself, he had the model of the Murchison's marriage – as joint intellectual collaboration – in mind.

Nonetheless, the lectures at King's were supposed to be academic lectures and for men only. Lyell wrote that on a Wednesday in May 1832, there were "grand disputes at the Geological Society" – still a male-only institution –.

> about the propriety of admitting ladies to my lectures. Babbage most anxious to bring his mother and daughter and Lady Guildford; Harris to bring Lady Mary Kerr; and so on. I begged them all not to do so, and they promised; but at last Murchison said, "My wife, however, must come. I promised to bring her, and she would be much disappointed. I will not bring her till the doors are closed." Then they all declared they would too, and so bring the affair to a crisis one way or other. (1885, I: 381)

As Rudwick puts it, "Lyell himself was at first opposed to the request. His reason is clear: the admission of women to the lecture room would be 'unacademical'" (1975: 241–42). Rudwick, in fact, concurs: "This view is consistent with his great concern for the academic prestige of the College, and hence of the Chair that he held and of the lecture that he gave" (1975: 242). We know from Rebekah Higgitt and Charles Withers's study of female audiences at the British Association for the Advancement of Science that it was generally presumed that female audiences were there for social purposes, celebrity-spotting, or perhaps to show off a new bonnet (2007: 9–13). But this view is also inconsistent with what we know of the deep networks of scientifically inclined women with whom Lyell was acquainted. And it overlooks a key fact of the debate at the Geological Society itself – that the decisive intervention was made by Roderick Murchison. It appears from Lyell's accounts that other men at the Geological Society had agreed not to bring their wives and daughters *until* Roderick intervened. Partly this may be attributable to Roderick's famous sense of entitlement – if he and his wife wanted to come, nothing would stop them. And yet this intervention caused the other men to re-join the cause.

As such, we need to ask ourselves: what precipitated this crisis in the first place? Why were all these men so willing to argue the case for women at this point, and seemingly at no other? Why did Roderick say that he would bring Charlotte, ban or no ban, and sneak her in after the doors were closed, and why was it *this* intervention that made the crucial difference?

Two days after the battle at the Geological Society, "the ladies, Dr. Fitton, and others" went "assailing … the King's College Council to admit ladies," until the Council relented (in K. Lyell, 1881, I: 381–82). Fitton's name is perhaps not insignificant here, given that he was Charlotte's first teacher in geology and that he had credited her collection in *Strata Below the Chalk.* Furthermore, the "ladies" here included Betsy and Georgiana Babbage, Susan Countess of Guildford (daughter and joint heiress of the banker Thomas Coutts), Somerville, and Charlotte. Rejecting women like this would mean turning away some of the most illustrious and scientifically minded women in London, and we know from Lyell's correspondence that – along with "7 titled ladies," in attendance at the lectures were "Mrs Somerville regular and Mrs Murchison" (Lyell 1832).

The instant he heard the decision that women could be admitted after all, Lyell wrote to Charlotte Murchison. "The important affair of whether lady geologists are to be trusted within [the] walls" of the College had been settled, he told her. And, in a classically Lyellian move of diplomacy, he added that, though he "had always protested abt. such innovations as un-academical, of course" he did "not object" to their decision now it had been made. Still, there were conditions: the Council had agreed to "assign a separate entry & a particular part of the theatre to the ladies, & their chaperones," who were "not to be allowed too great propinquity to the caps and gowns." Since King's was situated in the theatre district, the College would also "make arrangements to know any lady who comes, in order to prevent any improper person from coming" (as cited in Wilson, 1972). It is hard to tell if this is meant as a reassurance to Murchison, who would not have to worry about the reputation of "academical" women being lowered by the presence of "improper person[s]" – or, instead, if it is intended to put her in her place and remind her of the privilege and the responsibility she bore in attending the lecture as one of her sex.

Nonetheless, Lyell was writing to Murchison, I suggest, because it was *her* wish to attend that was critical in this debate. A key part of Lyell's argument in his lectures was that "molluscs were the most suitable fossils for geological dating" (Rudwick 1975: 249–50) and in his ninth lecture he "presented and defined – for the first time in public – the periods of Tertiary time which this research had distinguished: the Recent, Newer Pliocene, Older Pliocene, Miocene and Eocene, in order of increasing disparity from present faunas and inferred increasing age" (Rudwick 1975: 250). This came complete with a Synoptical Table of these fossils and an example drawn from Auvergne, which provided a "perfect restoration of [the] Eocene period, [its] lakes, seas, volcanos, land quadrupeds, reptiles, testacea, plants, insects, Geography" (Lyell, as cited in Rudwick, 1975: 250). Lyell's argument partly addressed the theory of fossil evidence in geological dating: he believed that each period of the geological past should no longer be defined not by a single "characteristic" fossil which identified the age of the rock in which it was found – for what if that characteristic life form had left no fossils intact? – but by a statistical analysis of *all* the fossils which had miraculously survived in each formation. As such, in the most recent bands of Tertiary strata, 90–95% of the fossilized species were still extant; 30–35% of the fossils in the next band; 18% in the next oldest, and just 3% of the species found in the oldest (but still geologically recent) bands. The crucial point here is that, in the back of Murchison's field notebook from the Auvergne trip, key evidence that made this argument possible is on display: there is a list of Tertiary fossils in Auvergne, and also near Bordeaux, complete with information about the rocks in which the fossils had been found. Charlotte notes 120 terrestrial species of fossils across France, 36 species of lake-dwelling univalves (water snails and slugs), and 11 species of bivalves (molluscs with hinged shells, like oysters and mussels) (C. Murchison, 1828: 30–36). As such, Lyell was presenting work that Murchison had been intimately involved in collecting from the outset. How could he disbar her from hearing about the evidence she herself had contributed to? How could women be prohibited from hearing lectures whose content they had co-created?

To add further corroboration to this theory, in the following year, Lyell married Mary Horner: Rudwick simply says he "took that long-suffering lady on a geological honeymoon" (1975: 252). In fact, Mary was not "long-suffering" at all, but became a distinguished conchologist in her own right. It seems that Lyell had this intention in mind from the beginning: both he and his new wife, alongside his sisters, began training in conchology with George Sowerby (son of James) that same spring. "I get on in conchology somewhat," he wrote in May 1833, while the women "make rapid progress" (Lyell 1881, I: 397). Having spent a summer with Murchison and having been able to make a key set of theoretical and methodological arguments based on the analysis of fossil molluscs she had collected alongside him, Lyell knew the importance of a geologically trained partner and sought to improve his own conchological skills into the bargain. Perhaps an added bonus of marrying Mary was that he could replicate the pattern of the Murchisons' marriage and avoid the sticky problem of crediting other people's wives with work that did not really represent their husbands' views.

Murchison also pressed home the necessity of women's attendance at the second meeting of the British Association for the Advancement of Science, which took place at Oxford in the same year as the King's lectures. At the inaugural meeting at York in 1831, few scientific men had wanted to lend their names to the fledgling project and women had been let in as they usually were at public (rather than academic) occasions. But when the Association moved to Oxford in 1832, this seemed an uncomfortable precedent. The Oxford Professor of Geology and Mineralogy, William Buckland, wrote to Roderick, who had presumably raised the matter with him, to say that "everybody … agreed that if the meeting is to be of scientific utility Ladies ought not to attend the Reading of the papers – especially in a place like Oxford." The Association wasn't for hobbyists. It was "a serious Philosophical Union of Working Men." Still, women "at parties is quite another thing & in this I think the more Ladies the better," he added (as cited in Higgitt and Withers, 2008: 6).

Nonetheless, Murchison attended the BAAS enthusiastically and slept in Buckland's attic (she became a godparent to the Bucklands' most recent child during the meeting). At this and subsequent BAAS meetings she not only attended parties, like the one in 1833 in which she searched with William and Mary Buckland for

shake off the stereotypes and stories of the past and to remember that the "geological wife" or daughter was a figure of significant agency, purpose, and intellectual significance and is worthy of renewed attention and understanding.

References

Adams, M. 2017. Geological Illustration and the Geo-Humane Science, 1811–1840. *Nineteenth-Century Contexts*. https://doi.org/10.1080/08905495.2017.1311654.

Alberti, S. 2002. Placing nature: Natural history collections and their owners in nineteenth-century provincial England. *BJHS* 35 (3): 291–311.

Bazerman, C. 2000. *Shaping written knowledge: The genre and activity of the experimental article in science*. Madison: University of Wisconsin Press.

Belknap, G. 2018. Illustrating natural history: Images, periodicals, and the making of nineteenth-century scientific communities. *BJHS* 51 (3): 395–422.

Berkowitz, C., and B. Lightman, eds. 2017. *Science museums in transition: Cultures of display in nineteenth-century Britain and America*. Pittsburgh: University of Pittsburgh.

Bowler, P. 1988. *The non-Darwinian revolution. Reinterpreting a historical myth*. Baltimore: John Hopkins University Press.

Broderip, W. 1835. Description of some Fossil Crustacea and Radiata, found at Lyme Regis, in Dorsetshire. *Transactions of the Geological Society of London* 5: 171–174.

Buckland, W. 1836. *Geology and mineralogy with reference to natural theology*. London: Pickering.

Buckland, A. 2013. *Novel science: Fiction and the invention of nineteenth-century geology*. Chicago: Chicago University Press.

Burek, C. 2001. The first lady geologist? Or collector par excellence? *Geology Today* 17 (5): 192–194.

Burek, C., and B. Higgs, eds. 2007. *The role of women in the history of geology*. London: Geological Society.

Burek, C., S. Turner, and R. Moody. 2010. *Forgotten women in an extinct saurian (man's) world*. Vol. 343, No.1, 111–153. London: Geological Society Special Publications.

Carroll, V. 2008. *Science and eccentricity: Collecting, writing and performing science for early nineteenth-century audiences*. London: Pickering.

Creese, M.R.S. 1998. *Ladies in the Laboratory? American and British women in science, 1800–1900: A survey of their contributions to research*. Lanham: Scarecrow Press.

———. 2007. Fossil hunters, a cave explorer and a rock analyst: Notes on some early contributors to geology. In *The role of women in the history of geology*, ed. C. Burek and B. Higgs, 39–49. London: Geological Society.

Fara, P. 2004. *Pandora's breeches: Women, science and power in the enlightenment*. London: Pimlico.

Fitton, W. 1836. *Observations on some of the Strata between the Chalk and Oxford Oolite, in the South-East of England*. London: Taylor.

Frasca-Spada, M., and N. Jardine, eds. 2007. *Books and the sciences in history*. Cambridge: Cambridge University Press.

Geikie, A. 1875. *Life of Sir Roderick I. Murchison, based on his life and letters*. London: Murray. 2 vols.

Gerassi-Navarro, N. 2017. *Women, travel and science in nineteenth-century Americas: The politics of observation*. Cham: Palgrave Macmillan.

Gordon, E.O. 1894. *The life and correspondence of William Buckland*. London: Murray.

Gross, A., J. Harmon, and M. Reidy, eds. 2002. *The scientific article from the seventeenth century to the present*. Oxford: Oxford University Press.

Hart, J. 2007. *The role of women in British quaternary science*, ed. Burek and Higgs, 83–95.

Heringman, N. 2004. *Romantic rocks, aesthetic geology*. Ithaca: Cornell University Press.
Higgitt, R. and C. Withers, 2008. Science and sociability: Women as audience at the British Association for the Advancement of Science, 1831–1901. *Isis* 99 (1): 1–27.
James, F. 2004. Reporting Royal Institution lectures, 1826 to 1867. In *Science serialized: Representations of the sciences in nineteenth-century periodicals*, ed. S. Shuttleworth and G. Cantor, 67–79. Cambridge: Cambridge University Press.
Klonk, C. 1996. *Science and the perception of nature: British landscape art in the eighteenth and early nineteenth centuries*. New Haven: Yale University Press.
Kölbl-Ebert, M. 1997a. Mary Buckland (nee Morland), 1797–1857. *Earth Sciences History* 16 (1): 33–38.
———. 1997b. Charlotte Murchison (née Hugonin) 1788–1869. *Earth Sciences History* 16 (1): 39–43.
———. 2002. British geology in the early nineteenth century: A conglomerate with a female matrix. *Earth Sciences History* 21 (1): 3–25.
———. 2004. Barbara, Marchioness of Hastings (1810–1858). *Earth Sciences History* 23 (1): 75–87.
Kölbl-Ebert, M. 2007a. *The role of British and German women in early nineteenth-century geology: A comparative assessment*, ed. Burek and Higgs, 157–163.
Kölbl-Ebert, M. 2007b. The geological travels of Charles Lyell, Charlotte Murchison, and Roderick Impey Murchison in France and Northern Italy (1828). In *Four centuries of geological travel: The search for knowledge on foot, bicycle, sledge and camel*, ed. P.N.W. Jackson, 109–117. London: Geological Society Special Publications.
———. 2012. Sketching rocks and landscape: Drawing as a female accomplishment in the service of geology. *Earth Sciences History* 31 (2): 270–286.
Larsen, K. 2018. *The women who popularized geology in the nineteenth century*. Cham: Springer.
Laudan, R. 1987. *From mineralogy to geology: The foundations of a science, 1650–1830*. Chicago: University of Chicago Press.
Lightman, B. 2007. Lecturing in the spatial economy of science. In *Science in the marketplace: Nineteenth-century sites and encounters*, 97–133. Chicago: University of Chicago Press.
Lyell, C. 1852. *A manual of elementary geology; Or, the ancient changes of the Earth and its inhabitants*. 3rd ed. London: Murray.
Lyell, K.M., ed. 1881. *Life, letters and journals of Sir Charles Lyell*. London: Murray.
Lyell, C., and R. Murchison. 1829. On the excavation of valleys, as illustrated by the volcanic rocks of Central France. *The Edinburgh New Philosophical Journal* VII: 15–49. Edinburgh: Black; London: Longman, Rees, Orme, Brown & Green.
Mantell, G. 1822. *The fossils of the south downs: Or, illustrations of the geology of Sussex*. London: Relfe.
———. 1838. *The wonders of geology*. London: Relfe and Fletcher. 2 vols.
Morgan, N. 2006. Anne Phillips and the mystery of the Malverns. *Geoscientist* 16 (7): 6–15.
Morgan, N. 2007. *Anne Phillips: John Phillips' geological companion*, ed. Burek and Higgs, 265–275.
Murchison, R. 1826. Geological sketch of the North-Western extremity of Sussex, and the adjoining parts of Hants and Surrey. *Transactions of the Geological Society of London* 2: 97–108.
———. 1839. *The Silurian system: Founded on geological researches in the Counties of Salop, Hereford, Radnor, Montgomery, Caemarthen, Brecon, Pembroke, Monmouth, Gloucester, Worcester, and Stafford; With descriptions of the coal-fields and overlying formations*. London: Murray. 2 vols.
O'Connor, R. 2007. *Earth on show: Fossils and the poetics of popular science 1802–1856*. Chicago: University of Chicago Press.
Orr, M. 2007. *Keeping it in the family: The extraordinary case of Cuvier's daughters*, ed. Burek and Higgs, 277–286.
Porter, R. 1978. Gentlemen and geology: The emergence of a scientific career. *The Historical Journal* 21 (4): 809–836.
Rastall, R.H. 1914. In *Geological Evidences of the Antiquity of Man*, ed. C. Lyell. London/New York: Dent/Dutton.

Rudwick, M. 1972. *The meaning of fossils: Episodes in the history of palaeontology*. London: Macdonald.

———. 1975. Charles Lyell, F.R.S. (1797–1875) and his London lectures on geology, 1832–33. *Notes and Records* 29 (2): 231–263.

———. 1976. The emergence of a visual language for geological science 1760–1840. *History of Science* 14: 149–195.

———. 1992. *Scenes from deep time: Early pictorial representations of the prehistoric world*. Chicago: Chicago University Press.

———. 2008. *Worlds before Adam: The reconstruction of geohistory in the age of reform*. Chicago: University of Chicago Press.

Sanjek, R. 1990. The secret life of fieldnotes. In *Fieldnotes: The makings of anthropology*, ed. R. Sanjek, 187–210. Ithaca: Cornell University Press.

Secord, J. 1982. "King of Siluria": Roderick Murchison and British geology. *Victorian Studies* 25 (4): 413–442.

———. 1990. *Controversy in Victorian geology: The Cambrian-Silurian dispute*. Princeton: Princeton University Press.

———. 2000. *Victorian sensation*. Chicago: University of Chicago Press.

———. 2007. How scientific conversation became shop talk. In *Science in the marketplace: Nineteenth-century sites and encounters*, 23–59. Chicago: University of Chicago Press.

———. 2018. *Visions of science: Books and readers at the dawn of the Victorian Age*. Chicago: University of Chicago Press.

Shapin, S., and S. Schaffer. 1985. *Leviathan and the air-pump: Hobbes, Boyle, and the experimental life*. Princeton: Princeton University Press.

Smith, J. 2006. *Charles Darwin and Victorian visual culture*. Cambridge: Cambridge University Press.

Sowerby, J.d.C. and Charles Edward Sowerby, 1829. *Mineral conchology of Great Britain*. London: Meredith.

Spamer, E.E., A.E. Bogan, and H.S. Torrens. 1989. Recovery of the Etheldred Benett collection of fossils mostly from the Jurassic-Cretaceous strata of Wiltshire, England, analysis of the taxonomic nomenclature of Benett (1831), and notes and figures of type specimens contained in the collection. *Proceedings of the Academy of Natural Sciences of Philadelphia* 141: 115–180.

Thompson, C. 2002. Earthquakes and petticoats: Maria Graham, geology, and early nineteenth-century "polite" science. *Journal of Victorian Culture* 17 (3): 329–346.

Topham, J. 2009. Scientific Books, 1800–1830. In *The Cambridge History of the Book in Britain*, ed. M. Turner and M. Suarez, vol. 5, 827–833. Cambridge: Cambridge University Press.

Torrens, H. 1985. Women in geology 2, Etheldred Benett. *Open Earth* 21: 12–13.

———. 1995. Mary Anning (1799–1847) of Lyme; "The greatest fossilist the world ever knew". *British Journal for the History of Science* 28 (3): 257–284.

Torrens, H., E. Benamy, E.B. Daeschler, E.E. Spamer, and A.E. Bogan. 2000. Etheldred Benett of Wiltshire, England, the first lady geologist – Her fossil collection in the Academy of Natural Sciences of Philadelphia, and the rediscovery of "lost" specimens of Trigoniidae (Mollusca: Bivalvia), with their soft anatomy preserved. *Proceedings of the Academy of Natural Sciences of Philadelphia* 150: 59–123.

Wilson, L. 1972. *Charles Lyell: The years to 1841: The revolution in geology*. New Haven/London: Yale University Press.

Woodward, H. 1879. A memoir of Samuel Woodward. *Transactions of the Norfolk and Norwich Naturalists' Society* 2: 563–593.

Archival Sources

Anning, M. to C. Murchison. 5 Feb 1829. LDGSL 838/M/A7/1.

———. 11 Oct 1833. LDGSL/838/M/A7/3.

Lyell, C. to G. Mantell. 26 May 1832. https://ses.library.usyd.edu.au/bitstream/handle/2123/380/adt-NU1999.0011corresp4.pdf;jsessionid=9801B6E71934AAE9F3C2382EBDF4736C?sequence=10

Mantell, G. to R. Murchison. 21 Jan 1830. LDGSL/838/M/8/8.
Murchison, C. 1816–18. Field notebook in Italy, with some France and Switzerland. LDGSL/840/1 [LDGSL840/N158].
———. 1828. Field notebook in France, Italy, Austria, Switzerland, LDGSL/840/2 [LDGSL 840/N160].
———. 1833–35. Field notebook Central England and Wales. LDGSL/840/N162.
Pentland, J. to C. Murchison. 12 May 1829. LDGSL/838/)/10/2.
Pentland, J. to Charlotte from Paris. 19 Jan 1835. LDGSL/838/P/10/11.
Scrope, G.P. to R. Murchison. 4 Dec 1828. LDGSL/838/S/10/3.

Part V

Geographies

History of the Earth Sciences from the South

Jarrod Hore

Contents

Abstract

The history of the earth sciences has predominantly revolved around the accumulation of knowledge within north Atlantic institutions and networks. This chapter examines a range of recent attempts to resituate this narrative within wider frames and argues for the particular importance of southern perspectives. By exploring southern archives, materials, knowledges, and networks, it becomes evident that the development of modern thinking about the planet extends well beyond the European revelation of deep time. Incorporating these perspectives reveals the interplay between Indigenous geomythologies, imperial and colonial struggles, and concurrent investments in the earth and its pasts within a globalized world. While histories of geology and geoscience have traditionally focused on the construction of new temporalities, this chapter proposes that historians should consider spatiality alongside temporality. By embracing the spatial axis of earth science, historians can gain a more comprehensive understanding of the subfield, accounting for material, political, anthropological, and potentially postcolonial dimensions. I suggest that this exploration of a place-line – how horizons of possible thought differ from place to place – within the history of the earth sciences offers promising avenues for future research.

J. Hore (✉)
UNSW, Sydney, NSW, Australia
e-mail: j.hore@unsw.edu.au

E. Aronova et al. (eds.), *Handbook of the Historiography of the Earth and Environmental Sciences*, Historiographies of Science,
https://doi.org/10.1007/978-3-031-40799-4_19

Keywords

Geology · Earth science · Colonialism · Deep history · Southern perspectives · Place lines · Natural knowledge

1 Introduction

The modern earth sciences were born in an era of accelerating imperialism. Many of the unequal relations that characterized the expansion of north Atlantic empires were crystallized in the emerging tradition of geology, which eventually formed the foundation of the wider earth sciences. Along with biology, geology organized and then generated global modes of enquiry that accelerated both the exploitation of nature and its enlistment in imperial projects of racial categorization and stadial assessment. The outlines of this story have been well-known since at least the 1980s. Historians of science undergoing a social turn noted then that geology in particular had immediate applications throughout the middle of the nineteenth century as an instrument of empire and as a vector of capitalism (Stafford 1984). The agents who pursued these connected geologic and political economic ends – the European and North American geologists of the nineteenth century – were extraordinarily prone to heroic auto-mythmaking. Many embraced and enacted the territorial, military, and imperial metaphors of the late nineteenth and early twentieth centuries, which were worked into the very fibers of academic and practical earth science (Alam 1978; Rudwick 1976; Secord 1982). Early perspectives connecting extraction, science, and politics have remained relevant touchstones throughout the wider history of science, which now must be understood, as Suman Seth aptly notes, as always ideological and always colonial (Seth 2009).

So much is clear within the history of science, but the picture is less clear within the history of the earth sciences. As Kapil Raj has argued, the history of science has become an important intellectual battleground in the "post-colonial world," where Indigenous scientists, historians, and social scientists have sought to press their own claims of "scientificity" in the face of centuries of dismissal (Raj 2013; chapter ▶ "Indigenous Knowledge and Perspectives"). Historians of science and empire writing from this position have also long pointed out that imperialism, capitalism, and the earth sciences are indivisible. Drawing initially from Mao Zedong, scholars such as Anis Alam and Deepak Kumar understood that imperial and colonial political economies existed in a symbiotic relationship with the scientific study of the earth. European imperialism was a system that created "captive markets" and captured "raw materials" for industrial production (Alam 1978, 244; Kumar 1980). This "production science" was defined not just by its applied character, as Alam and others argued (Alam 1978, 241), but by its intimate involvement in the mobilization of natural materials as imperial resources. These relationships stand in high relief in a series of colonial and postcolonial contexts. From the point of their first encounters with continents outside of Europe naturalists and earth scientists were highly interested in minerals and their extraction because of their prospective value to industrializing societies. By the late nineteenth century, the intellectual and practical dimensions of this interest had congealed into an organized system of definition, training, surveying, and mining that spanned nations, empires, and the globe (Anthony 2018; MacLeod 2000). Yet, historians of the earth sciences have only belatedly sought to scrutinize the colonial conditions that inflected so-called universal knowledge of the planet and its subterranes. Lukas Rieppel, writing in this collection, is led to these conditions through an analysis of capital, imperialism, and extraction, all in the context of geological practice (chapter ▶ "Earth Science and Extractive Capitalism in the Age of Empire"). Far more than just identifying political economies, defining structuring characteristics, and charting intellectual developments, scrutinizing the colonial conditions of universal knowledge involves resituating the history of earth science in recognition of its long involvement with non-European geographies. It is tempting to fit this attempt to resituate modern thinking about the earth within a postcolonial tradition, but I suggest in this chapter that it is more productive to conceive of this body of scholarship as pursuing a history of the earth sciences from the south.

This chapter focuses on histories of earth science from and about non-European geographies, especially those in the southern hemisphere. I follow the historian of science Pratik Chakrabarti, who argues that deep history itself acquired a "geographical motif of 'southern-ness'" as European naturalism absorbed ideas about primitivity, myth, and place in the colonial contexts of South America, Africa, South Asia, and Australasia (Chakrabarti 2020, 62). I argue that histories of the earth sciences that are conducted within this mode amount to a kind of "southern theory" for the physical, rather than the social sciences (Connell 2007). Usefully employed by historians, anthropologists, and literary scholars, the category of the south has become a starting point for critiques of the Eurocentricity of habits of inquiry like "theory," of ideas like modernity and development, and of nineteenth-century literary cultures (Chakrabarty 2000; Comaroff and Comaroff 2012; Comyn and Fermanis 2021). And, much as "postcolonial ecocriticism" has revealed the complex tensions between environmentalism and various colonial projects, approaching earth science history from the south politicizes and historicizes deep histories of the planet that are often construed as "objective knowledge" (Mount and O'Brien 2019; Daggett 2019). Locally grounded histories of the earth sciences challenge a tradition fully inhabited by European agents and ideas and beg urgent historiographical questions. If not as a core concern, how have colonial sites and materials figured in the history of the earth sciences? When and where does this historiography turn to a critique of power relations? And how is this reflected (or not) in the major issues and debates that have defined the subfield? This chapter will explore these questions across two sections addressing the traditional preoccupations of earth science history and the challenges to these models that have recently been mounted from the south. The question of where and when postcolonial perspectives have influenced views from the south is an important one that surfaces periodically throughout the chapter. On balance, though, historians of the earth sciences are still yet to embrace postcolonialism widely. I propose that along with other approaches from outside the subfield such as deep history, political geology, and geological anthropology, there is significant scope to make postcolonialism a more central concern in future earth science histories.

The history of the earth sciences is peculiar scholarly terrain because it organizes thinkers who work not only across different fields but across two entirely different disciplines. It might seem that the history of the earth sciences is a kind of scholarly relic, one that unites the "two cultures" of the natural sciences and the arts (Snow 1959). In a way that has had important consequences for the manner in which the history of the earth sciences has been written, scientists and historians have been involved in a shared project. In the years after its foundation in 1982, the journal of record for the subfield, *Earth Sciences History* even admitted two notation styles: those of *Northeastern Geology* for the scientists and *Isis* for the historians. Both groups were most often concerned with geology, the most historical of all the earth sciences, although this focus regularly opened out into related topics such as the evolution of geophysics or the development of oceanography. Reflecting this dominant concern, this chapter will consider histories of the earth sciences that cluster around geology, which as the historian Martin Rudwick has argued, most effectively "transposed" the notion of "historicity" into the natural world. Geology was the discipline that "burst the limits of time" at the turn of the nineteenth century, and it has retained an abiding interest to all kinds of historians since (Rudwick 2014, 3–5; Bashford et al. 2023a).

Due to its constituent roles in imperial expansion and exploitation, geology is especially important in making sense of the ways that historians and scientists have challenged north Atlantic perspectives on the history of the earth sciences. These perspectives have typically concentrated on the emergence of geohistory within a set of influential metropolitan institutions and networks. They hold that earth history was put together in places like England, western Europe, and the northeastern United States by a new group of intellectuals and professionals. Institutions like the Geological Society, the Geological Survey of Great Britain, and the Royal School of Mines seeded a global network of naturalists, geologists, and engineers from the middle of the nineteenth century (Rudwick 1985; Secord 1986b). Even at its most sensitive, the history of the earth sciences has often been a kind of expansive intellectual history of the natural knowledge that these figures collected on behalf of metropolitan institutions or their colonial equivalents. It is partly this method that has led historians of the earth sciences into a view of earth history from the north, where records and specimens have accumulated since the nineteenth century. As Chakrabarti has argued though, the naturalism these institutions valued so highly was an essential part of the colonial project (Chakrabarti 2020). This chapter explores the alternatives to this northern view of the history of the earth sciences. Sometimes inadvertently, alternatives have been broached by those working squarely within northern traditions, but they have also been challenged by perspectives from marginal positions. I argue that modern earth science history looks different from these sites, where colonial pasts and presents have challenged historians to examine the situated politics of natural knowledge. The first part of the chapter addresses how the history of the earth sciences developed a northern center of gravity that absorbed materials and information from across the globe. The second part of the chapter explores how scholars have recently begun to critique and resituate this history, and considers how these new perspectives might be combined with others from outside the field of the history of the earth sciences to combat the inertia of prevailing global (north) frames.

2 Traditional Histories of the Earth Sciences and Their Northern Centers of Gravity

In general terms, historians of the earth sciences have typically focused on the development of Western knowledge about the planet at the expense of reckoning with the politics of that knowledge. From about the 1960s, the history of science more widely took a social turn and began exploring how forms of expert knowledge were produced, circulated, and validated. Scholars interested in these problems began investigating the practice of science and considering its wider social impacts. Throughout this period, the history of the earth sciences remained relatively attached to traditionally positivist understandings of scientific knowledge, even as a post-positivist challenge raised new and fundamental questions about the development of science away from the metropolitan contexts of the north Atlantic (Chakrabarti and Worboys 2020). In contrast to these latter developments, historians of the earth sciences remained preoccupied with what we might still think of as universal problems. Among the most abiding of these was the long tension between understandings of the earth sciences as either chronological or physical disciplines. Classifying these competing "emphases" as the natural historical tradition on the one hand and the natural philosophical tradition on the other, the historian Rachel Laudan explained that the former dealt with the "reconstruction of the history of the earth" and that the latter was "concerned with understanding the processes of change on earth" (Laudan 1982, 7). Another problem, particularly trenchant during the emergence of modern plate tectonic theory during the middle of the twentieth century, was how to integrate the philosopher Thomas Kuhn's assessment of the nature of "paradigm shifts" in scientific revolutions (Kuhn 1962; Marvin 1973; Laudan 1978). Often failing to divide themselves along disciplinary lines, both the historians and the scientists within the subfield seem to have accepted that these intellectual problems were highly important.

Although the core intellectual problems were reasonably agreed upon, there have been differences in the ways that scholars within the history of the earth sciences have gone about their work. While historians and social scientists have steadily sought to achieve a holistic understanding of modern geohistory, the physical scientists working within the subfield have tended to reflect more narrowly on the development of certain methodological traditions or ideas. In the 1980s geologist-historians bristled at the Kuhnian assumption that they worked in a state of "scientific stagnation" before the breakthrough of plate tectonics (Hallam 1973). One such figure, the metageologist David Kitts, argued that physical scientists already had the tools to understand the development their fields, without the approaches, concepts, and theories that historians and social scientists were gradually bringing to bear on the history of the earth sciences (Kitts 1986, 3–4). The narrative adopted by traditionalists in lieu of the critiques of social science explained that geological thought emerged from eighteenth century European naturalism. From here, geology was rationalized throughout the nineteenth century, as its practitioners absorbed causal thinking, experiment, and analogous comparison, and split natural historical endeavor into an array of "earth sciences." During the beginning of the twentieth century, these earth sciences turned toward "actualistic comparison" and then, increasingly toward mid-century, sophisticated mathematical quantification (Visher 1986). Various elements of this traditional narrative have been challenged by historians over the past 50 years, and especially since the late 1980s, when the historian of science David Oldroyd observed that the subfield was "scarcely touched by current controversies in sociology of science and even studies in philosophy of science." Since then, Oldroyd's desired "closer union between scientists and historians of science" has largely addressed this disconnect (Friedman 1990). In this tradition, the history of the earth sciences is a kind of intellectual history focused on the relationships between modern humans – as geo-logical thinkers – and the planet.

As historians have sought to explain or refute shifts in intellectual paradigms or understand the frequent controversies that erupted between specialists, they have been led to examine the agents and conditions of change. Classic works in the history of the earth sciences have situated key agents in England, Europe, or North America. As a result, the history of the earth sciences has been and is still predominantly a history of north Atlantic knowledge accumulation. Martin Rudwick's "great Devonian controversy" was played out by successive directors of the British Geological Survey and Royal School of Mines, Henry De la Beche, and Roderick Murchison. As we might expect of such figures, their dispute over a sequence of fossils in southern England took place in the gentlemanly context of Victorian London (Rudwick 1985). James Secord added the University of Cambridge, base of Adam Sedgwick, to this social background of "controversy in Victorian geology" (Secord 1986a). Rachel Laudan's study of the intellectual foundations of the tradition that De la Beche, Murchison, and Sedgwick reshaped, *From Mineralogy to Geology*, widened the frame slightly, integrating British geology with the broader European context defined by figures such as Abraham Werner and Georges Cuvier (Laudan 1987). It was well understood that North American earth science maintained linkages to this tradition, but, as Naomi Oreskes has explained, it advanced along a kind of parallel path guided by a commitment to fieldwork over theory and a local network of influential research institutions (Oreskes 1999). From Rudwick's Devonian controversy to the breakthrough of plate tectonics in the 1960s, the history of the earth sciences has presented the conditions of shifting geological knowledge as the product of a similar set of north Atlantic institutions governed by a relatively homogenous cast of white agents. Focusing on these conditions and agents has bound the history of the earth sciences to a view from the north. Information from places like Africa and South America has flowed in and out of these places – as it did during the 1920s when the South African geologist Alex Du Toit matched strata on both sides of the Atlantic for the Carnegie Institution of Washington, substantiating the theory of continental drift (Chetty 2021) – yet it has only seldom been used to redirect, reorient, or resituate the history of the earth sciences.

There are numerous tensions that run throughout this body of scholarship. For example, work within the history of the earth sciences typically seeks to balance an old tension between geology and biology. Among others, biogeographers, zoologists, and paleontologists, were often heavily involved in critical debates within earth science (Frankel 1985). The early twentieth century German climatologist and geophysicist Alfred Wegener initially trusted that Australian fauna would make an important contribution to "the overall problem of continental drift," a hunch that the English botanist Albert Seward and the Punjabi palaeobotanist Birbal Sahni took up in the 1920s and 1930s (Wegener 1915/1966, 110; Seward 1929; Sahni 1936). Du Toit, again, synthesized much of this work as a geologist in *Our Wandering Continents* in 1937. This "holistic outlook" also brought other competing tensions into a kind of balance (Du Toit 1937, 4). There are both theoretical and practical strands within the scholarship that focuses on this period, historiographical iterations of the traditional split between earth science as natural philosophy or as natural history. Perhaps most importantly for the purposes of this chapter, there is also a remarkable tension between localism and globalism that brings colonial contexts, knowledges, and objects into imperial earth science. Not only did figures like Sahni and Du Toit ground themselves in colonial worlds, but the objects of their study were necessarily local: specific fossil strata in India, South Africa, Brazil, and Australia. The reconfiguration of these local places as sites of global interest and the circulation of the knowledge generated in them quite literally transformed earth science, just as it did for many kinds of scientific knowledge throughout modern history (Raj 2013). Understanding this allows us to recognize how histories of the earth sciences must be resituated in both geopolitical and geographical ways to account for the place of southern agents, actors, perspectives, and materials.

From this perspective, even conventional histories of the emergence of the modern earth sciences are necessarily transnational and transcultural, even if they wear these approaches lightly. Even the basic contours of how geological surveying developed in some European colonies in the latter half of the nineteenth century reveal the trans-imperial dimensions of earth science. From the 1860s, Western science adopted a practical geological focus and European practices of surveying and mapping were extended over places like the Indian subcontinent, generating a return of intellectual information about the "mineral potential" of the colonies. James Secord wrote that these geologists were practitioners of "the ultimate cosmopolitan science," providing for the ordering, exploitation, and extraction of "the materials of the terrestrial globe" (Secord 2018, 401–402). Geologists and paleontologists positioned themselves as "sub-imperial" actors, mediating between the information accumulated in places like India or the American West and the knowledge generated at institutions like the Royal School of Mines or the American Museum of Natural History (Stafford 1989, 223). These earth scientists also became involved with an increasing number of organizations, institutions, and benefactors seeking to capitalize on the study of the deep past (Rieppel 2019). Even during these early stages, then, earth scientists working in colonial worlds were playing an "integral" role in the "development of Western science" (Chakrabarti and Worboys 2020, 9–15). The inclusion of these agents effectively extends and flattens the networks through which the modern earth sciences developed by confronting the false hierarchies implicit in histories focused on the intellectual contexts of Europe or North America. A challenge to this tradition might go further, though, by asking what kinds of history emerge when scholars focus, instead, on the colonial or southern contexts of earth science.

3 Challenging Earth Science History from the Colonial World

Historians writing about the colonial contexts of modern earth science have intermittently challenged northern perspectives by insisting on the locality of subjects, materials, and knowledges. By resisting the inertia that has kept European and North American agents and institutions at the heart of earth science history, these scholars have come to insist on the importance of concepts like marginality, materiality, and circulation (Finney et al. 2022). Since the early 2000s, scholars such as Silvia Figueirôa, Howard Plotkin, and Sumathi Ramaswamy have shown that one important future direction for the subfield lies in localized intellectual histories of how particular forms of landscape thinking and storytelling have come to relate to a universalized earth science. Accounts of how these local knowledges were adopted or marginalized within earth science proper have increasingly featured in the central forum of *Earth Sciences History*. As postcolonial perspectives reshaped histories of botany, geography, and medicine, this journal has published occasional articles that address the development of the earth sciences in colonial, borderlands, or non-Western contexts. Covering the modern history of the earth sciences across South Asia, East Asia, South America, Africa, North America, and Oceania, these articles have often focused on the ambivalent role of Western agents, the diffusion of expertise, the conditions of contact, or the way that non-Western scientists fed information back into global science. Good examples have examined the (geo) politics of meteorites in Central and North America, the histories of agricultural development, geological training, and mineral surveying in Brazil, Indigenous meteorological traditions in China, and the contested mineral economies of Southwest Africa (Plotkin 2005, 2014; Figueirôa 2007, 2016; Chen et al. 2014; Hearth 2021). Very few of these articles explicitly deploy the language of postcolonial critique, yet they all invert or upend the conventional northern story of the development of the earth sciences.

However, while this move to focus on locality and marginality and invert the northern history of the earth sciences is a relatively recent development within the subfield, it articulates with more established critiques within area studies and the history of imperialism. These critiques, much like the Marxist strand of the colonial history of science discussed earlier, have been especially useful for scholars interested in South Asia. For example, the historian of South Asia and empire, Sumathi Ramaswamy, has explained how the "global revelations of the paleo-sciences" were textured by interests in lost worlds and places that were at once highly localized and immediately global. Ramaswamy's investigation of the incarnation and disappearance of the proposed Indian Ocean continent of Lemuria was therefore a restitching of Tamil place-making and geomythology into the central story of northern earth science: the modern realization of continental change, (dis)connection, and movement over deep time (Ramaswamy 2004). Lemuria may have been "lost" but the narratives that congealed around it were inherited by plate tectonics, which, as geographer Adam Bobbette has shown was made real in the Indonesian archipelago. Scientists working within a social matrix including "mystics, colonial Christians and Muslims, Sanskritists, ethnologists, antiquarians, occultists, postcolonial revolutionaries, and ethnonationalist animists" (Bobbette 2023, ix–xviii). The key insight of these lines of inquiry is to reveal that the modern geological theories broached in north-Atlantic scientific and intellectual communities were always conditioned by southern knowledges. Southern or Indian Ocean continents, for instance, were much discussed within early paleogeography and geophysics. Zoologist Phillip Sclater, geophysicist Alfred Wegener, and geologist Eduard Suess all proffered hypotheses on southern evidence during a key period of the development of the earth sciences. Likewise, tropical Javanese volcanoes like Merapi made potent microcosms of the earth system because of the way humans could observe how they shaped terrain through eruption and erosion. The scientific theories that emerged during these engagements were a product of an exchange between the institutions of European geology and figures simultaneously conducting studies along a range of ethnological, zoological, botanical, spiritual, and even mystical lines in distant sites (Chakrabarti 2020, 20; Bobbette 2022; Fleetwood 2022). When historians have considered the emergence of modern earth science from within this tapestry of colonial archives and texts, they have discerned that it involved so much more than just the European revelation of deep time.

As we can see from southern studies of the paleo-sciences, the actual practice of assembling a history of the earth involved handling a range of multiple perspectives. For late nineteenth century British colonists in Central India, the geological dramas of a newly ancient earth played out amidst mythical stories of Indian Romance. As Pratik Chakrabarti explains, "Man and Nature" were joined in this moment and in this place as they were within older Orientalist visions of the subcontinent. Geological surveyors like Joseph Medlicott and William Blanford and missionaries like Stephen Hislop connected the traditional vision of a living earth with a newer set of colonial priorities, which included the assembly of an account of human prehistory, the definition of "primitivity," and above all the study of industrial resources like coal and other minerals (Chakrabarti 2019, 151–153). This latter part of the story enters conventional earth science history as the practical setting of super-continental theorizing in the late nineteenth century. Alan Leviton and Michele Aldrich argue that the concept of Gondwana-Land – later named by Suess in *The Face of the Earth* (1904) – grew out of this practical geological work on the Indian coalfields between the 1850s and 1870s (Leviton and Aldrich 2012, 247–269). Roughly a decade before Suess coined Gondwana-Land surveyors on the Geological Survey of India (GSI) and their colleagues in African and Australian colonies were proposing ancient lithic, fossil, and faunal linkages between the continents, extrapolating on the idea of lost Indian-Ocean continent of Lemuria. Importantly, not only did this practical work in the so-called "Gondwana System" – named after the *Adivasi* Gond people who lived in Central India – feed into Suess' theorizing about the earth but it also generated similar engagements with rocks, materials, and peoples in different southern sites.

These engagements with southern soils and climates sit on the margins of the history of the earth sciences but they nevertheless form part of the origins of later, more specialized geophysical research. Moving within imperial infrastructures and pathways to traverse India and Africa, various figures began to compare and connect the two sites from the 1840s. Building on the insights of the orientalist and naturalist Thomas Newbold, Blanford took part in a military expedition to Abyssinia in 1867 and observed similarities between the volcanic *regur* of central and western India and the soils of West Africa (Leviton and Aldrich 2004; Chakrabarti 2020, 180–181). The soils were one thing, but the political geologies of local peoples were also comparable. Robyn d'Avignon has described how Maninka peoples in nearby Bambuk conceived of mineral deposits as "spirited geobodies," claiming and politicizing subterranean space through an investment of spirituality in the underground (d'Avignon 2020, 22, 2022). Here, again, we can observe how southern mythologies were part of the context of northern earth science. The similarities were not simply geographical or anthropological either, and they extended to the fundamental characteristics of the fieldwork that natural historians and geological surveyors conducted in situ. In Southern Africa, the study of earth history, natural history, and prehistory converged in the intellectual work of several late nineteenth century geologists and paleontologists. In this context, geological knowledge, the South African historian Saul Dubow suggests, became the stable "intellectual substrate upon which cognate work in anthropology, archaeology, and ethnology was founded." In this context, insights were produced into continental movement and connection, the evolution of species, and of course the antiquity of man (Dubow 2004, 109). The key point here is not that earth science developed out of the master discipline of geology, but that in the late nineteenth century Cape Colony, as in many other southern contexts, this undisciplined complex of paleo-inquiry *was* earth science.

One of the most significant figures within mid-twentieth century earth science, the South African geologist Alex du Toit, was trained in this context, where fieldwork involved expert knowledge across multiple disciplines. Du Toit was appointed to the Geological Commission of the Cape of Good Hope in 1903, and over two decades, he developed an unrivalled knowledge of the entire sedimentary Karoo system, the most extensive and commercially valuable geological unit in Southern Africa. Modern earth science, here, is a conventional pursuit of knowledge about minerals and geography conducted within state institutions. Yet, as Suryakanthie Chetty has explained, du Toit and South African earth science was indivisible from anthropology and colonial history (Chetty 2021, 182). The Geological Commissions fieldwork in the Karoo put earth scientists in regular contact with the Indigenous Sān peoples of Southern Africa, but du Toit himself also dedicated the final chapter of his 1926 *The Geology of South Africa* to a discussion of "primitive man" and he served as President of the South African Archaeological Society in 1946 (Du Toit 1926/1939, 426–434). Throughout this whole period, earth science in Africa was also closely allied to the political project of South African settler nationalism, on the one hand, and the emerging academic discipline of paleoanthropology, on the other. By the 1920s, not only was the African continent being refigured by nationalist statesmen like Jan Smuts as the "mother-continent" of the southern hemisphere landmasses, but it was proposed as one of the "cradles of mankind" (Beinart and Dubow 2021, 193–200). Considering earth science history from this southern site leads us to follow an intriguing twisting trajectory. In Southern Africa, for instance, the itinerary of thinking about the earth and its history proceeds from the enchanted geos of peoples like the Maninka, through the imperial search for industrial resources in the nineteenth century, into new provincial, then national, then global investments in the earth that involved a wide array of political and scientific justifications.

The multiple valences and turns of the history of the earth sciences, viewed from the south, are especially significant in the context of the revelations that overthrew the orthodoxy of continental fixity during the middle of the twentieth century. Although some form of paleontological connection between the southern continents was apparent well before Suess' suggestion of Gondwana Land in 1885 and Alfred Wegener's promotion of continental drift from 1912, earth scientists were unable to devise an acceptable geophysical model of continental connection (LeGrand 1988, 55–99; Oreskes 1999). In this case, the vast quantity of facts about and materials from the earth that accumulated over the course of the nineteenth century ironically provided a store of evidence with which to reject the notion of continental drift (Oldroyd 1996). On the other hand, the information and networks forged during the global search for industrial resources and fossil energy provided a new starting point. From the early twentieth century, the key piece of evidence for continental connection in deep time shifted from the *Glossopteris* fossil leaf and other coal flora to the sedimentary systems underneath these deposits. Du Toit, who spent this period assembling evidence, powered this shift in his pilot study, *A Geological Comparison of South America with South Africa* (1927), and a new geohistory of the world, *Our Wandering Continents* (1937). These publications established southern continental connection in deep time as a "paleontological fact" that coexisted with widespread rejection of continental drift for other reasons (Oreskes 1999). In the 1950s, a new generation of scientists sought to use paleomagnetism to evaluate different models of earth history, culminating eventually in the ascendence of plate tectonic theory in the early 1960s, validating du Toit et al. (Hallam 1973). The outlines of this history of the earth sciences in the story of the rejection of continental drift and the development of plate tectonics have largely remained secure since the emergence of earth science history as a subfield in the 1970s. It is notable that du Toit is a central figure within this literature, even if the southern circumstances of the information and insights he gathered have remained peripheral topics until quite recently (Bashford et al. 2021).

Du Toit's vision of earth history was significantly broader than many of his colleagues in Europe or North America, and framing him as simply an insightful forerunner working within the limited idioms of survey geology and stratigraphy risks minimizing his contribution. This is an example of how the history of the earth sciences has either misunderstood or marginalized a range of contexts for the planetary knowledge-making that it deals with as a central concern. Technical histories of what we might label the "high" earth sciences essentially follow the British geologist come historian of science Anthony Hallam. They understandably begin with the paradigm shift of plate tectonics following the application of sophisticated instruments and modeling to earth science after World War Two. Such histories ask how prior controversies and their contexts connect to this moment of breakthrough and how a disciplinary view like geophysics became prioritized over one like geology (Hallam 1973, 1983; Oreskes 1999; Frankel 2012). These are important details; however, they lead historians of science and technology toward deterministic arguments (Anderson 2009) and leave little room for a consideration of how different physical, geopolitical, colonial, and even intellectual contexts became part of modern earth science. Chetty and Chakrabarti argue that these contexts were foremost for du Toit and his southern colleagues in India and Australia, who readily accepted the evidence for continental connection in deep time because of their shared (British) colonial imaginations (Chetty 2021; Chakrabarti 2019). Such arguments suggest that the temporal axis, which obviously has been critical within traditional histories of the emergence of modern earth science (Gould 1987), is no longer the most promising avenue for further research. Historians of geology have made time and temporality their specialty, but future historians of the earth sciences might consider making the landscape or spatial axis their primary concern.

Scholars focusing on the physical and material contexts of knowledge about the earth have found that where thinking takes place matters. This is largely because of the importance of fieldwork within the earth and environmental sciences. Even if the empirical data of "high" earth science came to rely on sophisticated instruments and models, its foundations were in the practices of fieldwork that essentially defined the nineteenth century study of environments and undergrounds (Griffiths and Robin 1997; Hore 2022). Well before du Toit began his studies in the Karoo and across a wide range of similar southern contexts, field science provided information about discreet sites and generated more holistic knowledge about the earth (Vetter 2016; MacGregor 2018). What the historian of science Neil Safier would call the "itineraries" that link "individuals, objects, and impulses between sites" and transform observations into empirical data provide a map of places and their impacts (Safier 2010, 137–139). So it was that early nineteenth century natural theologies were melded with Awabakal geomyths, the materiality of carbon, and the intellectual pursuit of the Permian period in colonial New South Wales. Colonial geologist William Branwhite Clarke pursued all these projects with vigor, but his correspondence with more influential geologists in Cambridge and London tended to omit the various First Nations cosmologies that often enough drew him and other colonists to resources like coal. Between Clarke and Murchison, who in 1841 named the part of the stratigraphic column in which these coals were eventually situated, these details were superfluous (Hore 2023; Benton and Sennikov 2022). Any history of the intellectual idea of the Permian period, which was largely substantiated across the southern hemisphere by figures like Clarke and then du Toit, should be flexible enough to admit the local histories of encounter and brokerage that were part of fieldwork in the colonial world. Again, it is worth noting that this *was* earth science. As Alison Bashford, Emily Kern, and Adam Bobbette explain, "human engagement with the origins, and meaning of the earth and cosmos" took place "among distinct but overlapping temporal regimes." Clarke, like many other geologists working in colonial sites, was necessarily a mediator of these orders. Earth science became, in such contexts, a "cosmopolitan" process of imagining and charting the deep history of the planet, including changing environmental, atmospheric, and biological orders, as well as (eventually) continental arrangements (Bashford et al. 2023b).

Earth scientists interested in these debates took their prompts from specific materials. Clarke spent most of his career examining and thinking about the coal and sedimentary formations in the eastern part of the Australian continent and du Toit became an expert on the sedimentary formations of the Karoo and the diamond-bearing kimberlite pipes of the Northern Cape. Geologists and naturalists took cues from specific objects too. For example, the distinctive fossil seed-fern *Glossopteris* became one of the key pieces of evidence for continental drift over the course of the late nineteenth and early twentieth centuries. First described by the Frenchman Adolphe Théodore Brongniart in the 1820s, by the 1860s, Thomas Oldham and the GSI overturned the northern consensus that the *Glossopteris* fern originated in the Jurassic, showing that in India and Australia, at least, the tongue-shaped fossil thrived during the earlier Permian period. Beyond a certain point, the material conditions of the Gondwana system confounded the GSI's attempts to match Indian strata with the European stratigraphic column. To determine the lower horizon of this valuable coal-bearing system, Oldham needed to triangulate, and to do this, he looked southeast to the British colony of New South Wales (Oldham 1860). By the turn of the twentieth century, a comparable problem had developed in South America, where a network of scientists had encountered similar fossils. This time, the specimens were housed in a network of increasingly parochial natural history museums (Lopes and Podgorny 2001; see also chapter ► "Collections and Museums in the Historiography of the Earth Sciences"), where earth scientists began to use the *Glossopteris* fossil to weigh into multiple debates about coal deposits, continental connection, and continental drift. During the 1890s, specimens were uncovered in Brazil and Argentina that, as Mariana Waligora has shown, spurred scientists such as Francisco Moreno and Friedrich Kurtz to connect stratigraphies in Bajo de Véliz to known deposits in the Cape of Good Hope, India, New South Wales, and Tasmania. As these researchers "followed the *Glossopteris* trail," they created new circuits through which information about the earth and its past changes could flow (Waligora 2020). It is interesting enough to follow these circuits, via du Toit's *A Geological Comparison of South America with South Africa* (1927), back into the northern debate over the movement of continents; however, a more novel approach might stay with the *Glossopteris* fossil itself and consider how the accumulation of these objects generated local intellectual histories.

Indeed, a wave of recent work has embraced the locality and spatiality of the geosciences, effectively grounding the intellectual history of the earth. It adopts a similar methodological approach as the so-called "new material history," which presents human existence and thinking as "the product of a dynamic and creative material environment" (LeCain 2017, 11). Geohistory itself, argues Alison Bashford, "folds into the present in ways material and living," and its "poetics" or "fictions" are traceable outside strictly scientific domains in travel literature, science fiction, fantasy, environmentalist ethics, Indigenous land rights, and even resource extraction (Bashford 2023, 279–281, 293–295; Patankar 2016). These geological scripts are frequently contested, as they are for Selby Hearth, who explains how Tsumeb copper played a different role in Haillom, Ndonga, Herero, and Anglo-German societies during a long period of colonial struggle over land and resources in Southern Africa. Not only does this "mineralogic" perspective show how Tsumeb copper shaped the history of colonialism in South West Africa-then-Namibia, but it also helps draw attention back to the Native people who mined and worked the copper and those who stewarded the land (Hearth 2021). Place-based approaches have also helped expand the cast of actors whose intellectual worlds constitute earth science (chapter ▶ "Indigenous Knowledge and Perspectives"). The "cosmovision" of the Mapuche people of south-central Chile, for example, has recently been linked to the tectonic environment in which they live (Bastías et al. 2021). Just as southern geologists such as du Toit took their cues from the materials and conditions that surrounded them, it is clear that Mapuche geomythology is intimately related to the physical setting of the South American continental margin. Earth science has always dealt with discrete materials like Tsumeb copper and even much larger objects such as geographic or tectonic boundaries, but these things have often been absorbed (or not) within northern institutions or intellectual traditions. A history of the earth sciences from the south might resist these tendencies by developing a place-based approach that recognizes the significance of materials, knowledges, and objects over a range of scales.

Since the emergence of modern geology, those who have sought to think methodically about the planet, and then their historians, have embraced time as the critical axis and application of earth science. James Hutton's 1788 *Theory of the Earth* laid the foundations for this obsession with what the contemporary geologist and writer Marcia Bjornerud calls "timefulness," the twists and turns of which have become the great drama of earth science history (Bjornerud 2018). The temporal utility of geology and geological thinking within the complex of the paleo-sciences and within Western modernity more broadly is clear (Rudwick 2014). The interest in polytemporality remains useful; however, it has led to a neglect of the other axis through which earth scientists have made sense of the world: place. Thinking about the earth and its past provided more than just a timeline within which different disciplines could affix their objects of inquiry. Recent developments in thinking about earth science from the south and from other marginal perspectives suggest that there is an emerging place line through earth science history that scholars might productively follow and extend. These developments are fortifying a place-focused tradition within the subfield that has waxed and waned since the 1970s; however, only recently have historians of science been joined by historians of empire, the environment, geographers, and anthropologists in pursuing these grounded dimensions of earth science history. And when scholars start with the conditions of colonial worlds, the highly (geo)political nature of the earth sciences become inescapable (Bobbette and Donovan 2019). Writing about the colonial color lines of the Anthropocene, the geographer Kathryn Yusoff insists that the narratives about the world that earth science generates have always been political. Not only has the extraction of minerals historically been a "motivation and mode of dispossession" and colonial subjugation, but the innovative language of stratigraphy has often smoothly moved into biopolitical, cultural, and racial realms (Yusoff 2018, 103–108; 79). As a colonial science, geology reified and reinforced whiteness (Anderson 2006), but it did so unevenly across different geographies. Such perspectives indicate that not only do horizons of possible thought within earth science change from time to time, as traditional histories of science would have it (Kitts 1986, 5), but they also differ from place to place, depending on the power and material relations that define them.

If there is a place line that new histories of the earth sciences are following, then it runs directly through the colonial world. Colonists have become adept at making use of what Jeremy Schmidt calls "geologic forms of reasoning – surprise observations, hypothesis generation, and probable explanations" to settle the problems of dispossession and exploitation. Schmidt analyzes how these forms of reasoning operate to make sense of seeping bitumen in contemporary Alberta (Schmidt 2020), but they have a long history and an expansive geography. Historian of science Londa Schiebinger perceives a whole colonial science complex that extended from those "Europeans working in a colonial context" and their intermediaries all the way to distant figures receiving information, resources, samples, or problems (Schiebinger 2005, 52). Examinations of this complex that focus on colonial power relations essentially return scholars to the work of Anis Alam and Deepak Kumar, who suggested that earth science was an imperial relation that inflicted epistemic and material violence over 40 years ago (Alam 1978; Kumar 1980). Connecting the dots, this approach offers, per Gayatri Spivak, "an account of how an explanation or narrative of reality was established as the normative one" (Spivak 1988) and how it was grounded in particular colonial worlds. Here, any work that we might refer to as "postcolonial" refers less to periodization and more to a critical posture within the history of imperial science that might pursue the "decentred, diasporic or 'global' rewriting of earlier, nation-centred imperial grand narratives" that Stuart Hall nominates as a key characteristic of the postcolonial turn (Hall 1996, 247). It is only belatedly, and as a product of the new forms of situated and interdisciplinary inquiry explored above, that the history of the earth sciences has turned to this postcolonial critique.

As these postcolonial perspectives on the history of the earth sciences are developing, they draw productively from other fields where questions of temporality, spatiality, and justice are also highly important. Australian historians such as Ann McGrath and Marry Ann Jebb, for instance, have been at the forefront of the development "deep histories," which adopt expansive timescales to balance different disciplinary claims on the distant past as well as the intimate and ongoing connections to Country of First Nations people in Australia (McGrath and Jebb 2015; Way 2022). Sometimes pursuing a decolonial agenda, historians working in this mode aim to put these different forms of thinking about place and time into conversation, revealing how different "historicities might be understood as coexisting and in relation," rather than as competitors within a single universal narrative (Rademaker and Silverstein 2022). However, the suite of disciplines that make up the earth sciences – geology in particular – are not just competing bodies of knowledge and method that "coexist." Rather, as anthropologists and other social scientists interested in the geological have noted, they are key legitimating forces for colonial, capitalist, and racist operations of power via the "ontological distinction between life and nonlife" (Oguz 2022; Povinelli 2016). The geos is also, in and of itself, a powerful political force that human societies must reckon with as it reshapes the material conditions of life through such seismic or volcanic events (Bobbette and Donovan 2019). It may well be possible to reframe the history of geological universalisms – per the deep historians – by exploring their contexts and coexistences, but it is difficult to square this with the very force of the lithospheric processes that have altered or reshaped worlds throughout earth history. "New Earth Histories" – the program set out by Bashford, Kern, and Bobbette – charts a different, if overlapping path into the history of the earth sciences that is careful to balance the particular claims and histories of north Atlantic geological thinking, the array of coexisting knowledges that swirl around the geos, and the active forces of the planet itself (Bashford et al. 2023b). Notably, these perspectives have important and deep links to the south, where the fallacies of Western geo-universalism have often failed to hold.

4 Conclusion

Over the course of its existence as a subfield within the history of science, the history of the earth sciences has remained heavily preoccupied with the accumulation of knowledge about the earth within a set of north Atlantic institutions and networks. Tracing the emergence of the modern earth sciences via its origins in eighteenth century European naturalism, through a period where comparative global study and fieldwork made geology the "ultimate cosmopolitan science" (Secord 2018), and into an era of sophisticated geophysical modeling, many scholars have embraced a view from the north. In this chapter, I have argued that the most promising future directions for the history of the earth sciences lay in a neglected tradition of scholarship that inverts these dominant models and embraces a southern perspective. When historians have considered the history of the earth sciences from a position that focuses on southern archives, materials, knowledges, and networks, they have found that the emergence of the modern thinking about the planet is about so much more than just the European revelation of deep time. Important itineraries twist and turn between the enchanted geomythologies of Indigenous peoples, through imperial and colonial struggles over land and resources, and into a globalizing world where different investments in the earth and its pasts are held concurrently (Bashford et al. 2023a). Many of these facts have traditionally been marginalized within highly technical histories of the earth sciences that focused primarily on the piecing together of a new, deep temporality. However, if we take the spatial axis of earth science as seriously as we take the temporal one, these multiple relationships and new scales become key issues for the subfield. Historians are well aware that horizons of thinking change from time to time; however, my suggestion in this chapter is that historians of the earth sciences might consider inverting their traditional focus on time and temporality and focus place and spatiality. Embracing the place line that partly defines the history of the earth sciences from the south can help scholars construct a more accurate picture of the development of modern thinking about the planet that better accounts for the material, political, anthropological, and possibly postcolonial, dimensions of earth science.

Acknowledgments Research for this chapter was conducted with the support of the Australian Research Council project "Antipodean Geology: A Modern History of Southern Hemisphere Earth" (DP210101217), led by Alison Bashford and Alessandro Antonello. It draws on work originally funded by the State Library of New South Wales David Scott Mitchell Fellowship and presented at the New Earth Histories conference in 2019. My thanks to Alison Bashford, Lisa Ford, Juliette Gout, Naomi Parkinson, Stephen Pascoe, Emma Thomas, Joel Wing-Lun, Amy Way, and the editors of this volume for their comments on earlier versions of the chapter.

References

Alam, M. Anis. 1978. Science and imperialism: What is science? *Race & Class* 19: 239–251.

Anderson, Warwick. 2006. *The cultivation of whiteness: Science, health, and racial destiny in Australia*. Durham: Duke University Press.

———. 2009. From subjugated knowledge to conjugated subjects: Science and globalisation, or postcolonial studies of science? *Postcolonial Studies* 12 (4): 389–400.

Anthony, Patrick. 2018. Mining as the working world of Alexander von Humboldt's plant geography and vertical cartography. *Isis* 109 (1): 28–55.

d'Avignon, Robyn. 2020. Spirited geobodies: Producing subterranean property in nineteenth-century Bambuk, West Africa. *Technology & Culture (Special Issue: Africanizing Technology)* 61 (2): 20–48.

———. 2022. *A ritual geology: Gold and subterranean knowledge in Savanna West Africa*. Durham: Duke University Press.

Bashford, Alison. 2023. Gondwanaland fictions: Modern histories of an ancient continent. In *New earth histories: Geo-cosmologies and the making of the modern world*, ed. Alison Bashford, Emily M. Kern, and Adam Bobbette. Chicago: University of Chicago Press.

Bashford, Alison, Pratik Chakrabarti, and Jarrod Hore. 2021. Towards a modern history of Gondwanaland. *Journal of the British Academy* 9 (6): 5–26.

Bashford, Alison, Emily M. Kern, and Adam Bobbette, eds. 2023a. *New earth histories: Geo-cosmologies and the making of the modern world*. Chicago: University of Chicago Press.

———. 2023b. Introduction: New earth histories. In *New earth histories: Geo-cosmologies and the making of the modern world*, ed. Bashford, Kern, and Bobbette. Chicago: University of Chicago Press.

Bastías, C.A., R. Charrier, C.V. Millacura, L. Aguirre, F. Hervé, and M.A. Farías. 2021. Influence of geological processes in the cosmovision of the Mapuche native people in south central Chile. *Earth Sciences History* 40 (2): 581–606.

Beinart, William, and Saul Dubow. 2021. *The scientific imagination in South Africa: 1700 to the present*. Cambridge: Cambridge University Press.

Benton, Michael J., and Andrey Sennikov. 2022. The naming of the Permian System. *Journal of the Geological Society* 179 (1):1–33.

Bjornerud, Marcia. 2018. *Timefulness: How thinking like a geologist can help save the world*. Princeton/Oxford: Princeton University Press.

Bobbette, Adam. 2022. Processions: How the spiritual geographies of Central Java shaped modern volcano science. *Indonesia* 113: 49–64.

———. 2023. *The pulse of the earth: Political geology in Java*. Durham/London: Duke University Press.

Bobbette, Adam, and Amy Donovan, eds. 2019. *Political geology: Active stratigraphies and the making of life*. London: Palgrave Macmillan.

Cakrabarti, Pratik. 2019. Gondwana and the politics of deep past. Past and Present 242 (1): 119–153.

Chakrabarti, Pratik. 2020. *Inscriptions of nature: Geology and the naturalization of antiquity*. Baltimore: Johns Hopkins University Press.

Chakrabarti, Pratik, and Michael Worboys. 2020. Science and imperialism since 1870. In *Modern science in national, transnational, and global context*, ed. Slotten, Numbers, and Livingstone, 9–31. Cambridge: Cambridge University Press.

Chakrabarty, Dipesh. 2000. *Provincializing Europe: Postcolonial thought and historical difference*. Princeton: Princeton University Press.

Chen, Zhenghong, Guifang Yang, and Robert Wray. 2014. Shiyan Tao and the history of indigenous meteorology in China. *Earth Sciences History* 33 (2): 346–360.

Chetty, Surykanthie. 2021. *'Africa forms the key': Alex Du Toit and the history of continental drift*. London: Palgrave Macmillan.

Comaroff, Jean, and John L. Comaroff. 2012. *Theory of the south or, how Euro-America is evolving toward Africa*. London: Routledge.

Comyn, Sarah, and Porscha Fermanis, eds. 2021. *Nineteenth century literary culture and southern settler colonies*. Manchester: Manchester University Press.

Connell, Raewyn. 2007. *South theory: Social science and the global dynamics of knowledge*. Cambridge: Polity.

Daggett, Cara New. 2019. *The birth of energy: Fossil fuels, thermodynamics, and the politics of work*. Durham: Duke University Press.

Du Toit, Alexander. 1927. *A geological comparison of South America with South Africa*. Washington, DC: Carnegie Institution.

———. 1937. *Our wandering continents: An hypothesis of continental drifting*. Edinburgh: Oliver and Boyd.

———. 1939. *The geology of South Africa*. Edinburgh: Oliver and Boyd. (Original work published 1926).

Dubow, Saul. 2004. Earth history, natural history, and prehistory at the cape, 1860–1875. *Comparative Studies in Society and History* 46 (1): 107–133.

Figueirôa, Silvia. 2007. Geological surveys in the tropics: The Brazilian experience (1875–1934). *Earth Sciences History* 26 (1): 151–171.

———. 2016. Brazilian geology for Brazilian students: The general geology textbook published by John Casper Branner in 1906. *Earth Sciences History* 35 (2): 375–386.

Finney, Vanessa, Jarrod Hore, and Simon Ville. 2022. Chains of custody, oceans of instability: The precarious logistics of the natural history trade. *Journal of World History* 33 (1): 103–137.

Fleetwood, Lachlan. 2022. *Science on the roof of the world: Empire and the remaking of the Himalaya*. Cambridge: Cambridge University Press.

Frankel, Henry. 1985. The biogeographical aspect of the debate over continental drift. *Earth Sciences History* 4 (2): 160–181.

———. 2012. *The continental drift controversy*. 4 vols. Cambridge: Cambridge University Press.

Friedman, Gerald M. 1990. Editorial: Earth sciences history: A change in style and format. *Earth Sciences History* 9 (2): 93–94.

Gould, Stephen Jay. 1987. *Time's arrow, Time's cycle: Myth and metaphor in the discovery of geological time*. Cambridge, MA: Harvard University Press.

Griffiths, Tom, and Libby Robin, eds. 1997. *Ecology and empire: Environmental history of settler societies*. Edinburgh: Edinburgh University Press.

Hall, Stuart. 1996. 'When was 'the postcolonial'? Thinking at the limit' in Chambers, Iain, and Curti, Lidia. In *The postcolonial question: Common skies, divided horizons*, 242–261. London: Routledge.

Hallam, Anthony. 1973. *A revolution in earth sciences: From continental drift to plate tectonics*. Oxford: Oxford University Press.

———. 1983. *Great geological controversies*. Oxford: Oxford University Press.

Hearth, Selby. 2021. The "world's greatest mineral locality"; Haillom, Ndonga, Herero, and the early colonial histories of Tsumeb, Namibia. *Earth Sciences History* 40 (2): 433–460.

Hore, Jarrod. 2022. Settlers in earthquake country: Apprehending instability in New Zealand and California. *Pacific Historical Review* 91 (1): 1–32.

———. 2023. The voices of an eloquent earth: Tracing the many directions of colonial geo-theology. In *New earth histories: Geo-cosmologies and the making of the modern world*, ed. Bashford, Kern, and Bobbette. Chicago: University of Chicago Press.

Hutton, James. 1788. Theory of the earth; or an investigation of the laws observable in the composition, dissolution, and restoration of land upon the globe. *Transactions of the Royal Society of Edinburgh* 1, Part 2: 209–304.

Kitts, David. 1986. The logic of discovery in geology. *Earth Sciences History* 1: 1–6.

Kuhn, Thomas. 1962. *The structure of scientific revolutions*. Chicago: University of Chicago Press.

Kumar, Deepak. 1980. Patterns of colonial science in India. *Indian Journal of History of Science* 15 (1): 105–113.

Laudan, Rachel. 1978. The recent revolution in geology and Kuhn's theory of scientific change. In *PSA: Proceedings of the Biennial Meeting of the Philosophy of Science Association*, Vol. 2: Symposia and invited papers 1978, 227–239.

———. 1982. Tensions in the concept of geology: Natural history or natural philosophy? *Earth Sciences History* 1: 7–13.

———. 1987. *From mineralogy to geology: The foundations of a science, 1650–1830*. Chicago: University of Chicago Press.

LeCain, Timothy. 2017. *The matter of history: How things create the past*. Cambridge: Cambridge University Press.

LeGrand, Homer. 1988. *Drifting continents and shifting theories: The modern revolution in geology and scientific change*. Cambridge: Cambridge University Press.

Leviton, Ann, and Michele Aldrich. 2004. The impact of travels on scientific knowledge: William Thomas Blanford, Henry Francis Blanford, and the Geological Survey of India, 1851–1889. *Proceedings of the California Academy of Sciences* 55 (9): 117–137.

———. 2012. Contributions of the geological survey of India, 1851–1890, to the concept of Gondwana-Land. *Earth Sciences History* 31: 247–269.

Lopcs, Maria Margarct, and Irina Podgorny. 2001. The shaping of Latin American museums of natural history, 1850–1990. *Osiris* 15 (1): 108–118.

MacGregor, Arthur, ed. 2018. *Naturalists in the field: Recording and preserving the natural world*. Boston: Brill.

MacLeod, Roy. 2000. Of men and mining education: The School of Mines at the University of Sydney. *Earth Sciences History* 19 (2): 192–295.

Marvin, Ursula. 1973. *Continental drift: The evolution of a concept*. Washington, DC: Smithsonian Institution Press.

McGrath, Ann, and Mary Ann Jebb. 2015. *Long history, deep time: Deepening histories of place*. Canberra: ANU Press.

Mount, Dana, and Susie O'Brien. 2019. Postcolonialism and the environment. In *The Oxford handbook of postcolonial studies*, ed. Huggan. Oxford: Oxford University Press.

Oguz, Zeynep. Introduction: Geological anthropology. Theorizing the contemporary. *Fieldsights*, September 2022.

Oldham, Thomas. 1860. On the geological relations and probable geological age of the several systems of rocks in Central India and Bengal. *Memoirs of the Geological Survey of India* 1: 299–311.

Oldroyd, David. 1996. *Thinking about the earth: A history of ideas in geology*. Athlone: The Athlone Press.

Oreskes, Naomi. 1999. *The rejection of continental drift: Theory and method in American Earth Science*. New York/Oxford: Oxford University Press.

Patankar, Mayuri. 2016. 'Gondwana'/'Gondwanaland' as a homeland of the Gonds: Story-telling in the paintings of Gond Pilgrims. *Summerhill* 22 (2): 39–48.
Plotkin, Howard. 2005. Henry A. Ward and the recovery of the Santa Rosa, Colombia, Meteorite. *Earth Sciences History* 24 (1): 63–80.
———. 2014. The Iron Creek Meteorite: The curious history of the Manitou Stone and the claim for its repatriation. *Earth Sciences History* 33 (1): 150–175.
Povinelli, Elizabeth A. 2016. *Geontologies: A requiem to late liberalism*. Durham: Duke University Press.
Rademaker, Laura, and Ben Silverstein. 2022. Deep historicities. *Interventions: International Journal of Postcolonial Studies* 24 (2): 137–160.
Raj, Kapil. 2013. Beyond postcolonialism … and postpositivism: Circulation and the global history of science. *Isis* 104 (2): 337–347.
Ramaswamy, Sumathi. 2004. *The lost land of Lemuria: Fabulous geographies, catastrophic histories*. Berkeley/Los Angeles: University of California Press.
Rieppel, Lukas. 2019. *Assembling the dinosaur: Fossil hunters, tycoons, and the making of a spectacle*. Cambridge, MA: Harvard University Press.
Rudwick, Martin. 1976. The emergence of a visual language for geological science, 1760–1840. *History of Science* 14: 149–195.
———. 1985. *The great Devonian controversy: The shaping of scientific knowledge among gentlemanly specialists*. Chicago: Chicago University Press.
———. 2014. *Bursting the limits of time: The reconstruction of geohistory in the age of revolution*. Chicago: Chicago University Press.
Safier, Neil. 2010. Global knowledge on the move: Itineraries, Amerindian narratives, and deep histories of science. *Isis* 101 (1): 133–145.
Sahni, Birbal. 1936. Wegener's theory of continental drift in the light of palaeobotanical evidence. *Indian Botanical Society Journal* 15: 319–332.
Schiebinger, Londa. 2005. Forum introduction: The European colonial science complex. *Isis* 96: 52–55.
Schmidt, Jeremy. 2020. Settler geology: Earth's deep history and the governance of *In Situ* oil spills in Alberta. *Political Geography* 78: 102132.
Secord, James. 1982. King of Siluria: Roderick Murchison and the imperial theme in nineteenth-century British geology. *Victorian Studies* 25 (4): 413–442.
———. 1986a. *Controversy in Victorian geology: The Cambrian-Silurian dispute*. Princeton: Princeton University Press.
———. 1986b. The geological survey of Great Britain as a research school, 1839–1855. *History of Science* 24 (3): 223–275.
———. 2018. Global geology and the tectonics of empire. In *Worlds of natural history*, ed. H. Curry, N. Jardine, J.A. Secord, and E.C. Spary, 401–417. Cambridge: Cambridge University Press.
Seth, Suman. 2009. Putting knowledge in its place: Science, colonialism, and the postcolonial. *Postcolonial Studies* 12 (4): 373–388.
Seward, Albert. 1929. Botanical records of the rocks: With special reference to the early Glossopteris Flora. *British Association for the Advancement of Science Report* 97: 199–216.
Snow, C.P. 1959. *The two cultures and the scientific revolution*. Cambridge: Cambridge University Press.
Spivak, Gayatri Chakravorty. 1988. Can the subaltern speak? In *Marxism and the interpretation of culture*, ed. C. Nelson and L. Grossberg, 271–313. Basingstoke: Macmillan Education.
Stafford, Robert. 1984. Geological surveys, mineral discoveries, and British expansion, 1835–71. *The Journal of Imperial and Commonwealth History* 12 (3): 5–32.
———. 1989. *Scientist of empire, Sir Roderick Murchison, scientific explorations and Victorian imperialism*. Cambridge: Cambridge University Press.
Suess, Eduard. 1904. *The face of the earth (Das Antlitz der Erde)*. Trans. Hertha B.C. Sollas. 2 vols. Oxford Clarendon Press. (Original work published 1885).

Vetter, Jeremy. 2016. *Field life: Science in the American West during the Railroad*. Pittsburgh: University of Pittsburgh Press.
Visher, Glenn. 1986. A history of geological thought. *Earth Sciences History* 5 (2): 137–143.
Waligora, Mariana. 2020. Researchers following the Glossopteris trail: Social context of the debate surrounding the continental drift theory in Argentina in the early 20th century. *Colligo* 3 (3): 1–15.
Way, Amy. 2022. Displacing history, shifting paradigms: Erasing aboriginal antiquity from Australian anthropology. *History Australia* 19 (4): 710–730.
Wegener, Alfred. 1966. *The origin of continents and oceans*. Trans. J. Biram. New York: Dover Publications. (Original work published in 1915).
Yusoff, Kathryn. 2018. *A billion black Anthropocenes or none*. Minneapolis: University of Minnesota Press.

Land, Mines, and Bones: African History and Deep Knowledges of the Earth

Chris Manias

Contents

Abstract

This chapter considers the connections between histories of the earth sciences (and knowledges of the earth more generally) and African history. It argues there is great potential synergy between them, both fields providing cross-disciplinary and long-durational means of considering the history of knowledge and engagement with the earth, especially through different forms of conceptual, political, and economic power. This is followed across two sections. The first is the history of mining and use of mineral resources in Africa, which has been a dynamic field, aiming to recover the agency and importance of African actors, processes, and institutions, while engaging with the importance of mineral extraction for the formation of colonial systems and maintenance of global inequalities. And secondly, the chapter examines the historiography of the sciences of ancient life as related to Africa, especially palaeontology and palaeoanthropology, and how these more conceptual fields have been entangled with values and engagements with African nature, and citations of the importance of the continent to the whole of humanity.

C. Manias (✉)
King's College London, London, UK
e-mail: chris.manias@kcl.ac.uk

E. Aronova et al. (eds.), *Handbook of the Historiography of the Earth and Environmental Sciences*, Historiographies of Science,
https://doi.org/10.1007/978-3-031-40799-4_28

Keywords

Mining · Minerals · Extraction · Geology · Palaeontology · Palaeoanthropology · Colonialism · Inequality

1 Introduction

When thinking about the historiography of Africa and the earth sciences, we are simultaneously confronted by variety and extent in some areas, but relative under-development in others. The size and diversity of the African continent means that a summary discussion of "Africa and the Earth sciences" runs the risk of problematic generalizations. This chapter therefore thinks specifically about possible dialogue between the history of the earth sciences, and African history as two dynamic, yet currently quite separate, historical fields. How can these different areas of history speak to one another, and what insights does bringing them together potentially offer?

As the other contributions in this volume indicate, the history of the earth sciences is one of the liveliest areas in the history of science, examining changing conceptions of time and the world, economic and political power, and the social and economic structures of knowledge. Histories of the earth sciences have traditionally had a North American and European center of gravity, but the field has also consistently been aligned to international and transnational processes. It is therefore difficult to speak of any current "global turn" in the historiography of the earth sciences, as the field has always been globally orientated. Where matters have shifted over recent years is an increased desire to connect the earth and deep-time sciences with political and economic power relations, as well as thinking about how categories like "geology," "palaeontology," and "palaeoanthropology" are rooted within Western intellectual and institutional contexts (Chakrabarti 2020; Chakrabarti and Bashford 2021, and chapter ▶ "History of the Earth Sciences from the South" by Hore). Studies of these processes however have tended to focus on the Americas, Australia, and South Asia, with relatively little consideration of Africa, apart from some work on South Africa. This is possibly surprising, given the importance of African developments to the history of the earth sciences, and the multiple ways African history engages with conceptual, economic, and political power.

African history is fundamentally based on linking different approaches, disciplines, and sources, with great attention to the relationship between periods and timescales – and the still present legacies of historical processes. Vincent Hiribarren, in a recent review article, discusses how African history "was conceived as a radical project from the onset in the 1950s. Its main ambition was not only to discuss the history of a widely ignored continent but also to use methods which were not deemed to be academic or serious enough to analyse the past of newly independent nations" (Hiribarren 2020: v). As will be discussed in specific examples below, African histories frequently adopt a deep perspective, tracing continuities, changes, and disjunctions from early periods (Africanists increasingly resist the term "precolonial" as problematic and flattening), the colonial period (as perhaps the only common experience across the whole continent), and the postcolonial. Within this, there is often a self-consciously emancipatory project, aiming to demonstrate the importance of African actors, and recovering the agency of people and societies who have persistently been marginalized in colonial and postcolonial eras. This in itself requires distinct methodologies, cutting across different fields.

African history is therefore highly interdisciplinary, bringing perspectives from anthropology, literature, archaeology, ethnography, and climate and environmental studies, even if this interdisciplinarity is not always fully recognized or theorized (Semley et al. 2022) and can even lead to Africanists being dismissed for not being sufficiently "historical" (Brizuela-Garcia 2008). Nevertheless, at a time when many scholars are turning toward interdisciplinarity, African history is a field where crossing disciplines has long been crucial. Oral testimonies and ethnographic approaches have for example been utilized extensively by African historians, given the importance of orality in many African societies, and the varying times at which they have used written documents. Of course, stereotypes of African "orality" or "non-literacy" do need to be questioned: Egypt and Sudan contain some of the oldest written records of humanity; other areas, like Mali and Madagascar, have long literary traditions; while in other African societies, widespread use of written documents was a nineteenth and twentieth-century phenomenon. African history has also fruitfully connected with archaeology, bringing in material culture, histories of everyday life, and broader chronologies – although methodological and institutional barriers between history and archaeology, and the colonial implications of many archaeological projects and concepts, pose problems. The observations of outsiders have also been crucial for African history, with travelers and commentators from the Middle East, Europe, and China all reporting on African societies. As Wole Soyinka (2012:27) has argued Africa was never "discovered" by outsiders, but outsiders have persistently constructed images of the continent. African nature, antiquity, and culture were imagined in numerous ways – pristine wilderness, primordial savagery, deep mysterious antiquity, and utopian wonder have all been important motifs. These imaginings became a form of colonial and epistemological dominance over the continent, and moreover, these stereotypes still condition current views of Africa.

The interdisciplinarity of African history also means that Africanists often use the earth and environmental sciences, especially in the developing literature in African environmental history. Works like McCann's *Green Land, Brown Land, Black Land* (1999) contain a great deal of geological discussion. Yet while the colonial background and heritage of geography and geology are widely acknowledged by historians, this is often effaced in environmental histories, which can use methods and concepts from the modern sciences uncritically (raised in Tilley 2018: 17). This is part of a tension discussed by Hersey and Vetter (2019): while the history of science and environmental history often examine similar issues, they can have very different attitudes to the truth claims and authority of the sciences themselves. Additionally, environmental histories of Africa have not widely engaged with the role of geology, minerals, and other products of the earth. The overwhelming focus so far has been on development, farming, deforestation, engagement with animals, land degradation, and increasingly how climate change and "the Anthropocene" relate to Africa (Austin 2017). These are of course not separate issues from the history of the earth sciences: mining and mineral exploitation have been a key part of human engagement with the earth, and (as will be discussed) a place where the earth sciences are particularly entangled with power, economics, and culture.

The wider position of Africa within the history of science, technology, and medicine has seen some consideration over the past few years. Helen Tilley (2018), in an important historiographic review of the relations between African history and the history of science, argues that African history enables us to consider long continuities over time in knowledge systems (even with the disruptions of the colonial period), and the position of African knowledges within wider global trends (something often ignored in the history of science, where African developments are frequently presented as a "field" for work by outsiders). This agenda, drawing on historiographic trends to see all knowledge as mobile and co-constituted (Secord 2004; Raj 2010), while nevertheless continuing to draw attention to power imbalances developing in the colonial and postcolonial periods, represents an important move in the history of science.

And of course, there are also structural issues in African history. It is frequently cited that many Africanists are not Africans themselves, and works of African history produced by European or North American scholars frequently (and quite rightly) begin by acknowledging their outsider status. This is also the position that this chapter is written from: I am a historian of the earth sciences based in a British university. These structural issues have led to attempts to amplify "history from Africa," as African scholars must overcome significant barriers (Semley et al. 6–7): including lack of funding for historical and archaeological research in Africa; difficulties securing resources for the preservation, maintenance, and cataloging of African archival records (highlighted by the British Library's Endangered Archives Programme[1]); and lack of funds, restrictive visa regimes and often also recognition blocking participation in Northern scholarly networks.

Considering connections between the history of the earth sciences and African history allows us to discuss larger issues. In her crucial book, *A Billion Black Anthropocenes – or None* (2018), Kathryn Yusoff draws attention to the deep connections between definitions of Blackness, the development of extractive mining-based economies, and the formation of subjects like geology and palaeontology, arguing how "the racial categorization of Blackness shares its natality with mining the New World, as does the material impetus for colonialism in the first instance" (2). These connections were nevertheless hidden by "the sleight of hand of the Janus-faced discipline of geology (as extractive economy and deep-time paleontology of life-forms) is to naturalize (and thus neutralize) the theft of extraction through its grammars of extraction" (12). Yusoff raises the importance of linking two often-separated fields – geology as leading to colonial extraction, and constructing

[1] https://eap.bl.uk/

histories of the earth. Existing histories of the deep-time sciences, and knowledges of the earth in Africa, offer ways of thinking beyond this binary. The history of the earth sciences in Africa can be about how we can link different historiographies and methods to understand power and economics in the modern world, and uncover new and obscured perspectives of the earth and its history.

The remainder of the chapter will take Yusoff's contention of the "Janus-faced discipline of geology" into its structure, examining the two largest areas where histories of Africa and the earth sciences have been connected, which together illuminate the potential of linking African history with the history of knowledge of the earth. The first will be one of the largest areas of African historiography – the history of mining and exploitation of mineral resources. This has been a huge field, simultaneously showing the crucial importance of the earth for African history, while enabling us to consider long continuities and disjunctions in human relations with the earth and its products. The second will focus on a more minority pursuit, but which has developed massively in conceptual importance – the history of understandings of past life, both human and non-human. As the sciences of palaeontology and palaeoanthropology consolidated over the nineteenth and twentieth centuries, they marked Africa as a crucial region, while often marginalizing Africans themselves. But views of the history of life also offer scope to take these debates further, to a topic where Africa is now considered central to understanding the whole of humanity.

2 Mining

If we talk about Africa, the earth and historiography, then the history of mining dominates. This is an old literature on an even older phenomenon. African mining traditions go back thousands of years, and studies of African mining link political, economic, cultural, spiritual, and anthropological perspectives. Mining has been central to much of African history and is a clear point of intersection with histories of the earth sciences. The wide use of metals like copper, gold, and iron has been a persistent feature for many African societies – see especially Herbert (2003) for copper, and the works of Chirikure (such as 2015 and 2018) for iron. In some social evolutionary contexts, this use of metals was highlighted to belittle African societies, through describing Bantu peoples as living in "the iron age" in contrast to some imagined later modernity – a clear example of "the denial of coevalness" discussed by Fabian (2014). Nevertheless, the extent, sophistication, aesthetics, and complexity of African metallurgy have been widely valued. Indeed, brass-works from West Africa, most notably the Benin Bronzes, have been especially resonant focusses on demands for the restitution of objects looted during the colonial period (Hicks 2020; Gundu 2020).

One old strand in the literature has been based around searches for "origins" of African mining and metal-smelting, which has moved between ideas of independent innovation and diffusion – and not in a linear manner (Alpern 2005). While nineteenth and twentieth century diffusionists often saw Sub-Saharan Africa as receiving metal-working from the Phoenicians (Mauny 1952), late-nineteenth century German and French figures like Gaston Maspero, Felix von Luschan, and Ludwig Beck all argued for the independent invention of iron-metallurgy in Sub-Saharan Africa, with Maspero even arguing that Egypt itself received iron-working from southern neighbors (Maspero 1891). The association of iron-working with the Bantu-speaking peoples has also drawn on diffusionist currents. Indeed, Herbert (2001) has criticized romanticized "grand theories about Bantu expansion that featured the smith as culture hero: Bantu agriculturalists overspread Africa from the Cameroon Highlands to the Indian Ocean and south to the Cape of Good Hope thanks to their possession of iron technology" (48). These debates are particularly strong because the chronologies of metal-working in Africa disrupt norms derived from European contexts of a linear movement through stone, bronze, and iron ages. Egypt was one of the first metal-smelting and mining societies, especially in gold, silver, and copper, but was a relative latecomer to iron technology. Meanwhile, iron-mining and smelting seems to have been adopted extremely rapidly by many central and southern African societies by at least 1000 BCE – which the Senegalese archaeologist Hamady Bocoum (2004) pushes back even further to the second millennium BCE. Debates on dates and origin are still unresolved, and while the balance of interpretations is currently on independent innovation, the pendulum has been persistently swinging, and is doubtful to rest permanently.

Possibly a more fruitful line of inquiry going beyond the question of where it first started (and the diffusionist baggage associated with this debate) has been on the significance of mining and metallurgy for African societies. Chirikure (2018) has used archaeological, historic, ethnographic, and linguistic sources, and science and technology studies, to examine the wide-ranging importance of mining and metallurgy in numerous African societies, discussing how "the chaîne opératoire of metallurgy and mining was imbued with religious, sociocultural, and symbolic attributes, all of which were inseparable from the associated technology (11)." These histories show a mixture of commonality and diversity, with both alluvial and underground mining practiced across the continent, and mining and metallurgy having persistent ritual importance, often associated with gender and reproduction. Metallurgy and mining are also variably connected with politics and state-formation, sometimes highly controlled in centralized politico-economic systems, but more diffuse in other regions. As with much of African history, the history of metallurgy simultaneously shows dynamism; connections between culture, technology, political, and spiritual forces; and tremendous diversity across time and place.

The study of mining, metals, and minerals has also placed African history at the center of numerous global relationships. The paradox of early African history – that the archaeological record and observations of outsiders indicate extremely powerful, sophisticated, and wealthy societies, for which we often lack written records – is highlighted by François-Xavier Fauvelle, who describes Africa as "the cradle of civilizations so radiant, and yet so obscured in the surviving documentation" (Fauvelle 2018: 4). However, the study of metals and minerals provides an important entry point. Extracted minerals gave Africans a huge amount of importance in global commodity trades. Toby Green (2016, 2020) shows how many currencies, including some based on metals, made West African economies and people crucial players in a developing global economy in the medieval and early modern period, even if the value of their commodities depreciated over the centuries, especially through the distorting effects of the trade in enslaved people. Similar work for East Africa, especially the Swahili Coast (Horton et al. 2021 and Seland 2014), shows the importance of Africans in Indian Ocean trading networks, with iron, gold, and gemstones being crucial resources. While these studies generally do not say much about African knowledge of the earth or mining techniques, they nevertheless indicate the need for investigations of how these minerals were extracted, the knowledge and expertise which developed around them, and their persistence into later centuries.

African knowledges of the earth are of course not just about metals, as a wide literature especially in anthropology and archaeology can tell us. Earth, clay, and geological features have been crucial for many African societies in many different ways – clay for pottery, stone for tools and buildings, megaliths erected to stand in the landscape, rocks and the earth for health and healing, and deep knowledge of soils deployed in agriculture (summarized in Insoll 2015). In some cases, this traditional knowledge has been identified in a modern utilitarian manner, being seen to remedy or avoid environmental degradation (Diallo and Diallo 2016; Ngailo and Nortcliff 2007). But there are also important cultural dimensions. Traditions of humans being born from the earth are found across numerous African societies. In parts of western Africa, "geophagy" has had important healing functions, and earth shrines have a significant ritual and political role (Lentz 2013). Once again, histories of the earth in Africa show how impossible it is to disentangle the economic from the cultural and social, and long durational histories of culture and society.

Mining in the nineteenth and twentieth centuries meanwhile becomes closely tied with imperial and colonial histories. This association dates to the colonial period itself, when the idea that colonialism in southern and central Africa was driven by mineral exploitation was widely discussed, and mining interests were entangled with the formation of colonial states (a critique going back at least to J.A. Hobson 1902). The history of colonial mining clearly shows the dispossession and erasure of African systems of ownership, as colonizers seized mines and deposits which had long been worked by Africans (Fauvelle 2018: 126; Dumett 1998). Mines and minerals also consolidated colonial systems. The search for gold and diamonds was a motor of colonial expansion, and a key part of colonialist imaginaries. Rider Haggard's novel *King Solomon's Mines* (1885) was just one cultural production seeking to define the continent (McClintock 1995; Chrisman 2011). The earth sciences, which were essential for prospecting, imagining territories, and converting of resources into commodities, were part of new systems of dominance. African minerals became central within international rivalries. That the 1899–1902 South African War was essentially a "war for gold" has been a persistent source of debates since the war itself (Smith 2000). The historiography has also shown colonial mining as socially transformative, when historical attention has focused on mine-workers rather than the mine-owners (see for example Johnstone 1976 for a pioneering Marxist-inspired work). Mines required miners after all, and studies of labor have emphasized deep engagement with the earth, but also the modernity of African societies in the nineteenth and twentieth centuries, defined by class-formation, new working patterns, migration, urbanization, and the formation of new communities and identities.

The history of mining has been dominated by two regions, both in southern Africa. South Africa has a distinct mining history, as the largest manifestation of settler colonialism on the continent, and racist state policy across the colonial period and Apartheid regime. Gold-mining on the Witwatersrand, and diamond mining in Kimberley, saw a huge amount of study in the 1980s, often following Marxist perspectives (Worger 1987; Turell 1987; Phimister 1988). These works illustrated the transformative impact of South African gold and diamond mines. Gold rushes sparked mass migration from Africa, Europe, and Asia, and mining organization of structured twentieth-century racial discrimination, class, and identity formation (Moodie and Ndatshe 1994; James 1992). As argued by Worger (1987) and Turell (1987), the monopolistic policies of diamond-mining companies, and attempts to depress worker wages and maintain control of diamond supplies, led to extremely coercive measures over Black workers, including draconian punitive regimes and being confined in compounds – presaging and possibly informing the policies of the Apartheid state. The South African context therefore shows the continued, long-term, and persistent connections between minerals and power, something of which historians of the earth sciences should be mindful.

The second large historiographical focus has been the Copperbelt in the north of modern Zambia. This region saw mining activity in earlier African history (Musambachime 2016; Nikis and Livinstone-Smith 2017), which then expanded dramatically, as the Copperbelt became one of the most densely populated and industrialized regions of the world over the early and mid-twentieth century, with Zambia itself producing 12% of the world's copper in 1964 (Sikomo et al. 2016). This work focuses on mining as an engine of industrial and economic development, moving starkly against stereotypes of African "traditionalism" and subsistence economies to examine urbanization and complex social and labor relations. The Copperbelt was also a region of theoretical innovation. The Rhodes-Livingstone Institute, founded in 1937, pioneered new forms of anthropology focusing on Copperbelt mining communities, which moved the discipline away from a stereotyped focus on "primitive societies," to more complex ones based on class and migration. These forms of anthropology have themselves been shown as dependent on African experts, who conducted much of the fieldwork (Schumaker 2001; Tembo 2014), and simultaneously subverted, but also worked within, colonial structures.

There are enough studies of the Copperbelt for an essay of their own, which have shown the region to be one of the most economically and socially changed parts of Africa, but also locked within particular myths, of the African as a miner, and representing a particular African "urban modernity." Recent studies have attempted to shift frameworks somewhat. Miles Larmer has attempted to decenter the "mythic modernist narrative" around the Copperbelt, bringing out alternative perspectives, drawing from environmental history, bringing women more strongly to the narrative, and seeking cross-border and globalized connections. His use of the "obvious but nonetheless irresistible analogy of geological time for our understanding of historical change" to think about the sedimentation of experience, cultural forms, social groups, and economics (Larmer 2021, 26) again shows how African history and thinking about the earth are linked.

For the postcolonial period, mining looms similarly large. Continuing with the Copperbelt, Musasa (2021), who herself grew up in the region, has written about the social destabilizations caused by the later privatization of the mines, simultaneously continuing this history into the present while highlighting the experiences of Zambian mining communities today. Recent work in South Africa meanwhile has followed shifts in mineral production across the later twentieth century, tied to new questions around political and legal frameworks in the post-Apartheid period. Platinum mining in the north-west of the country has generated an especially large literature and lively debates, as a prism to examine relations between landrights, private companies, chieftainship, and government (Manson and Mbenga 2012; Capps 2016; Phillips 2022). These studies have shown the importance of mining for postcolonial states, the social embeddedness of mining, and the continued dominance of external corporations and interests.

In terms of new directions for the history of African mining, and relations between African history and the earth sciences, two authors stand out. Recently, Robyn d'Avignon (especially 2022, but also 2018 and 2020) has examined links between African gold-mining practices in the Birimian greenstone belt in West Africa, conventionally described as "orpaillage" in colonial-era documents. Self-consciously "part history and part ethnography" (2022, 26) d'Avignon's work is based on fieldwork in Senegal and archival research in Dakar, Conakry and Aix-en-Provence, and traces the long histories of West African mining, especially in Maninka-speaking regions of Mali, Senegal, and Gambia. Orpaillage is examined in terms of understanding of the earth, forms of African expertise, communities around mining and mines, and as a form of "ritual geology" shared across a wide region. Gold mining is tied with understandings of the minerals and the earth belonging to spirits which can in some respects be regarded as "actors" in scientific processes (d'Avignon 2020), and is connected to climatic and environmental change, with mining increasing during drought and famine as a crisis resource. And d'Avignon makes a wider point: we should regard African knowledges of the earth as "geology," and not just "traditions" or "vernacular knowledges."

A different, although complementary, approach has been taken by Gabrielle Hecht, in her work on "nuclearity," thinking about uranium extraction across twentieth-century Africa (2012). Hecht shows how Africans were – in multiple ways – denied the status of "nuclearity," while nevertheless being central for the development of the global atomic age – something often ignored in the voluminous work on the Cold War and nuclear culture. African states like Niger, Madagascar, and Gabon were denied the status of being nuclear powers, despite being major producers of uranium, and African miners were denied recognition that health problems were caused by working with radioactive materials. The history of nuclearity in Africa links the personal and geopolitical. Hecht's methods are also worth noting, pulling material from a huge range of collections (26 archives in national, business and nuclear agency archives in seven countries, as well as oral histories), highlighting silences and gaps as well as formal statements in the documents, and with constant recognition that the story she tells is only partial.

D'Avignon and Hecht's work shows directions in which histories of African mining, and knowledge of the earth more generally, can go, and the potential of linking the history of the earth sciences with African history. Continued interdisciplinarity (in varied constellations), highly creative use of sources, and simultaneously regarding mining in Africa as deeply embedded in local contexts, but also enmeshed in global practices – even if Africans themselves are often marginalized. Western-trained geologists are present in these histories, but are only one set of actors (and usually secondary to the African miners themselves). Historians of the earth sciences ought to be mindful of this. For all the apparent conceptual power of the geologists, and their entanglement with colonial states, they were always only one part of much larger stories, and we should be attentive to other actors and knowledges, and the potential of the diverse sources and methods used in African history.

And of course, histories of mining and extraction are not just histories, and it is no surprise that recent histories of mining by d'Avignon, Hecht, and others continue up until the present. The twenty-first century has indeed been presented as a new struggle for Africa's resources, with companies and economies pouring money, personnel and political capital into operating in Africa. As of 2018, 19 African countries gain more than 50% of their export revenues from mineral resources (Jessell et al. 2018). The deep history of mining shows how economics, geology, geopolitics, and the state continue to intermix. And we should be aware of powerful yet distorting cultural constructions. Hecht's opening example of Western panics over "yellow cake from Niger" being used in the Iraqi nuclear program, or media sensationalism over the trade in "blood diamonds" discussed by Cleveland (2014), certainly draws on a reality of inequality and instability in some parts of Africa. But, they also play into stereotypes of Africa as a lawless and dangerous place, working against alternate African perspectives, and African agency over mines and resources. Highlighting these stereotypes and showing how they can be deconstructed, and amplifying what minerals mean to different Africans, are important current directions for African history, and important links with histories of the earth sciences.

3 The Deep History of the Earth and Life

While palaeontology, and related subjects like palaeobotany and palaeoanthropology, have been minority pursuits in compared to the gigantic resources devoted to economic geology and mining, they have nevertheless had huge significance. The construction of deep-time chronologies for life and the earth has been acknowledged (particularly from a Euro-American perspective) as being one of the largest shifts in human engagement with the natural world (Rudwick 2014). And of course, fossil work is not separate from mining and mineral exploitation. Mines, quarries, and other geological diggings have been some of the most productive sites for fossil exploitation, and mining and fossil work present entangled histories.

It is increasingly common in discussions of the history of palaeontology to highlight Indigenous traditions regarding fossils and landscapes, often through the prism of "geomythology." There is no equivalent work to Adrienne Mayor's *Fossil Legends of the First Americans* (2005) for African societies, although African examples do feature in some of the Greek and Roman cases discussed in *The First Fossil Hunters* (Mayor 2011). Nevertheless, there are certainly important areas of investigation here, and Charles Helm and Julien Benoit have – partly with Mayor's encouragement and guidance – begun to work in this area (Helm et al. 2019; Helm and Benoit 2019). They have highlighted examples primarily from South Africa around the collecting and ritual use of what geologists would regard as fossils by many African peoples (both past and present), as well as the association of San rock art with fossilized dinosaur trackways in Mokhali Cave in Lesotho. These depict the extremely bird-like creature which might have made them, possibly one called //Khwai-hemm and Khoumolumo in oral traditions. These studies indicate the extent of African engagement with fossils and the need for more investigation of African knowledge of the earth.

However, there are of course caveats here. The instrumentalization of African oral traditions and genealogies by colonial-era scholars has made many (although by no means all) Africanists skeptical of the recording of long-durational knowledge of geological history in oral traditions. David Henige (2009) highlights the longstanding methodological debates on using oral history as a record of factual information in African historiography to raise doubts on recent claims that some Aboriginal Australian and Native American oral histories illustrate the retention of information on geological processes for thousands, possibly tens of thousands of years (engaging especially with Mayor 2007; Nunn 2018; McGrath and Jebb 2015). Henige argues against the likelihood of such deep transmission but also raises the epistemological point that "it can operate *only* by supposing an ideal – and idealized set of worlds in which every conceivable constraint on change is presumed to have operated effectively, continuously and universally" (225). More widely, the extent it is possible or even desirable to take broad cosmological traditions, extract where they discuss what Western scientists call fossils, and then define that small subset as "geomythology" is something which needs further consideration – a point raised in Australian (Rademaker 2021) and South Asian contexts (Chakrabarti 2020: 104–9).

We know from the literature on enslavement and the mass displacement of Africans to the Americas that knowledge was also transferred: enslaved people bought rice cultivation techniques, healing practices, and botanical knowledge from Africa to the Americas (Carney 2002; Carney and Rosomoff 2009; Schiebinger 2017). How and whether this also included African knowledge of the earth is a further question. There are hints this was significant, especially in the now frequently repeated anecdote from Mark Catesby's *The Natural History of Carolina, Florida and the Bahama Islands* (1731–43) that enslaved Africans in North America were able to identify mammoth teeth as being those of elephants (Mayor 2000). There is clearly more work to be done here, both in terms of the spread of African knowledge, and how it became incorporated into new scholarly disciplines – with all the issues around dispossession (both physical and epistemological) that this implies.

During the colonial period, Holger Stoecker has argued that the "scramble for Africa" was accompanied by a "scramble for dinosaurs" (Stoecker 2018: 25–6). However, it must be said that any "scramble" for African dinosaurs was far more ad-hoc and sporadic than the mass expropriation and exploitation of the continent by colonial systems. Stoecker himself worked on a project supported by the Museum für Naturkunde in Berlin, on possibly the largest attempt to expropriate African fossil material – the exploitation of Tendaguru in modern Tanzania, which involved the excavation of 225 tons of fossils by huge numbers of African workers, mobilized by the German colonial state in the aftermath of the Maji Maji Rebellion (1905–7). Tendaguru has been researched before: Gerhard Maier's *African Dinosaur's Unearthed* (2003) provided an in-depth narrative account of excavations in both the German and British colonial periods. The *Dinosaurs in Berlin* project, culminating in a lavishly illustrated volume, followed a more thematic and varied approach (Heumann et al. 2018). A particularly relevant study for African history was written by Mareike Vennen (2018), who discussed how photographs from Tendaguru codified visions of African labor, feeding into colonial and racial ideologies while highlighting the crucial dependence on African work and expertise. A further study by Michael Ohl and Holger Stoecker (2018) on the naming of the Tendaguru fossils shows how fossil animals were often given Swahili names by the excavators. However, these names were then effaced in the metropole by German scientists, who gave them Latin and Greek names, or often expressly "Germanized" them, most dramatically shown by an iguanodontian named *Dysalotosaurus lettowvorbecki* after the German colonial general – a direct example of the erasure of African expertise and knowledge, and the nationalization of palaeontology.

While we know a great deal about Tendaguru, the history of palaeontology in other parts of Africa is limited. As with mining and so with fossils, South Africa has seen some work, primarily through the works of Saul Dubow (especially 2004 and 2006), who shows the importance of South African geology and prehistory within the British empire, while also being domestically important for the formation of settler society and the later segregationist state. Other works have followed this lead, such as Chetty (2021) which examines the career of the South African geologist Alex Du Toit as simultaneously being central to the formation of ideas of continental drift, and deeply enmeshed within the political and social debates in twentieth-century South Africa.

There has also been some work on the history of interpretations of fossils from Madagascar – potentially a huge field, given the island's geological diversity, the recent extinction of megafauna like *Aepyornis* and *Megaladapis*, and its definition in Western discourses as a strange place of biogeographic isolation. Thomas Anderson (2018) has written on how natural history, geology, ethnography, and palaeontology in nineteenth-century Madagascar intermixed with colonial discourses, and in a more specific article (Anderson 2013) examined the transfer of the eggs and bones of giant extinct flightless birds in Madagascar and New Zealand to examine global relations in the history of science, following the vogue of questioning centers and peripheries. While the Malagasy people are themselves largely invisible in these studies (a particularly noteworthy omission, given that the Malagasy language has an extensive written tradition), this still illustrates the diversity of currents feeding into the construction of African nature. Scientists have also approached the topic, most recently Alison Richard's *The Sloth Lemur's Song* (2022), examining the deep history of the island and its inhabitants, and using this to think about issues of time, diversity, and scientific work.

Even this limited work on palaeontology in South Africa and Madagascar is exceptional. We still know little about the field's history in other parts of Africa. With the exception of some popularizing works (Nothdurft and Smith 2002), the extent of palaeontological excavation in North Africa, which interacted in complex ways with the much more studied history of Egyptology, has barely been touched. The expansion of palaeontology in the Karoo in South Africa, where fossils of what were problematically termed "mammal-like reptiles" became a major source of interest across the twentieth century, has also been unexamined by historians of science. However, there have been works by palaeontologists, most notably Bruce Rubidge (2013) who noted the scale of these projects, and Elsa Panciroli (2021) has highlighted how histories of fossil mammal research allow us to recover a greater variety of actors within palaeontological work. Meanwhile, the deep connections between histories of oil exploration in West Africa, and invertebrate fossils and micropalaeontology, have barely been acknowledged at all. These are not just gaps in the literature, but investigating these in-depth would reframe the history of knowledge and work on fossils in Africa in major and important ways, and draw further connections between ways of knowing the earth, nature, and humans, and the political, economic, and conceptual structures around them.

The primary exception to the lack of a literature on fossil work in Africa has been the history of human origins research in southern Africa. The history of palaeoanthropology – more than many branches of the history of science – has an old seam focusing on "larger than life" characters like Raymond Dart, Robert Broom, and Louis Leakey, debating "bones of contention" (Lewin 1997). Older accounts of the discipline's history are often litanies of fossils and finders, and this is still continued in some popular works, with one recent example explicitly being titled "*Fossil Men*" (Pattison 2020). However, like most branches of the history of science, there has been a drive to deepen consideration of the history of palaeoanthropology, especially as connected to colonialism and debates over humanity, and the range of people involved in palaeoanthropological research (for a summary, see Goodrum 2009).

The broad trend in the history of palaeoanthropology has been an acknowledgment that, over the twentieth century, a general "Asiacentric" model of human development, seeking human origins in Central Asia in particular, gradually and unevenly shifted to either multiregional or Afrocentric views (Regal 2002; Yen 2014; Schweighöfer 2018). We have a good literature on the role of South African excavations for these processes, with several studies examining two especially prominent "fossil men" – Raymond Dart and Robert Broom (Richmond 2009a, 2009b; Schweighöfer 2018; Madison 2020). These works often examine the contested history of the Taung Skull and other southern African hominin finds in the early twentieth century. Partly this is a story around imperial hierarchies of power, as South African finds were questioned and doubted by scientific establishments in metropolitan Britain – often for not conforming to the features expected of a large-brained human ancestor, as had been epitomized by the (forged) Piltdown skull. However, these specimens were also significant for promoting the idea that South African science had crucial global contributions, through controlling and dominating a primordial antiquity.

On the history of human origins research in Africa, a particular mention must be made of Christa Kuljian's *Darwin's Hunch* (2016), which examines the field's close connection with society and politics from the beginning of the twentieth century to the present. This connects Dart and Broom's work with colonial and racist policies in a considered way, and the stealing of human remains of modern people is a constant and disturbing theme. The example of a woman named /Keri-/Keri, whose remains and body-cast were held in Dart's collections after her death from pneumonia in 1939, is a consistent thread. Likewise, the complexities of human origins research in South Africa in the context of Apartheid and the post-Apartheid period are dealt with in a sensitive and nuanced manner, with the field being shown as simultaneously useful for building an international profile for South African institutions, while jostling uneasily with domestic issues, including struggles over Apartheid and Creationist strands in various South African religious denominations. As well as building a complex and engaging picture of science in Africa, the book also admirably crosses the boundary between academic and popular history of science and should be a model for future work on these topics.

Kuljian's work is also important for leading the history of human origins research from the early twentieth century into more recent decades, when – perhaps surprisingly – the range of history of science studies on palaeoanthropology in Africa either dry up, or move away from Africa itself. This is surprising given the number of high-profile finds from the late-1940s, most notably associated with the Leakeys in Kenya (ranging from "Nutcracker Man" or *Zinjanthropus*, to *Homo habilis*, the "handy man" privileged as the first tool-using human). Research on Australopithecines likewise expanded to fully regard them as the "first" human ancestors, with the iconic find of "Lucy" or "Dinkinesh" in 1974 in Hadar in Ethiopia (the subject of some retrospections, such as Johannson and Wong 2009, but no history of science-based account, as of yet). And more recently, genetic research has fully entrenched a "recent out of Africa" hypothesis, dislodging both out-of-Asia models and multiregionalism, and leading to the promotion of the notion that "we are all African apes" (with the "we" being all humans).

The expanding history of palaeoanthropology has shown it as one of the most prominent and culturally resonant sciences of the twentieth century. However, this literature has overwhelmingly focused on how palaeoanthropology was deployed in the United States and Europe. We know for example from Nadine Weidmann's studies of the relations between Raymond Dart and Robert Ardrey (Weidmann 2011) about the complexity of ideas of the "Killer Ape" and "African Genesis" of humanity, and the reception of these ideas in Cold War culture (not least through the famous monolith sequences in Stanley Kubrick's *2001: A Space Odyssey*). These threads have been taken up even more fully by Erika Milam in *Creatures of Cain* (Milam 2019), examining palaeoanthropology as a "vernacular science" in the USA, and how African finds (and stereotypes of Africa) were integrated into conceptions of humanity, science, nature, and aggression in late-twentieth century North America. Marianne Sommer (2016) has shown how genetics research used African palaeoanthropology to argue for pluralistic visions of humanity. These show the tremendous significance of Africa as a field of scientific work and succeed in what they set out to do: examine the role of African palaeoanthropology in discourses in North America and Europe. They do however indicate a parallel with Hecht's work on "nuclearity": African material is crucial for concepts of human origins in the North, and conceptions connected to the Cold War and Atomic Age. But African actors – with a very few exceptions, largely from South Africa – were generally invisible.

The contribution of Africans is hinted at in some works. Kamoya Kimeu (1940–2022) has been noted for his role in Kenyan human origins research in popular accounts and received obituaries in *Nature* and *The New York Times*. Another example is Saul Sithole (1908–1997), who worked as a preparator at the Transvaal Museum. Sithole initially worked on bird specimens, but also played an important role in fossil work, especially around the Sterkfontein site. As well as encouraging us to pay attention to people in ostensibly subordinate museum roles (also highlighted by Wylie 2021 and this volume), Jacobs discusses how while Sithole was accorded "respectability," structures of contemporary scientific work and prejudice meant he was never accorded the status of being a scientist (Jacobs 2016, 180–210). Kuljian also discusses Sithole and engages with the promotion of palaeoanthropology in post-Apartheid South Africa, especially as providing depth for the project of the "African Renaissance." Likewise, we know how "Lucy" is named "Dinkinesh" in Ethiopia (meaning "You are marvelous" in Amharic) and is a national icon, with the fossils being held at the National Museum in Addis Ababa and with casts displayed in museums around the world (Hochadel 2009). And Emily Kern has discussed the role of African archaeology within debates over human origins in the latter twentieth century (Kern 2018). However, these are still relatively short excurses, and a focused study like Sigrid Schmalzer's *The People's Peking Man* (2008) for any African context would be a key future project to fully take these African developments into account.

African perspectives have been investigated in museum studies – a further academic discipline which could easily be brought into these discussions. Scott's *Envisioning African Origins* (2007) discusses human evolution displays in British and US museums, regarding their importance for fixing particular images of human evolution, social development, and African-ness in the USA and Britain (often re-emphasizing ideas of linear evolution from African "savagery" to Western modernity). However, it also includes two fascinating later chapters, examining the responses of Black visitors to displays of human evolution, and National Museums Kenya in Nairobi. These show how Black audiences express surprise at the lack of engagement with modern Africa (and the "Leakey-centric" perspective in the

Nairobi museum), and how human origins can be a source of pride, with displays reread in "Africa-centric" counter-narratives. Yet there is still much more work to be done here. Scott concludes that "for too long modern-day African people, including those of the African diaspora, have been invisible in 'out-of-Africa' scenarios, invisible as representatives of modern humanity and invisible as members of the museum audience" (150) – something which is still true today, both in the historiography and museum displays, over a decade after the work's publication.

The idea that humanity has African origins has also had a strong impact on African history itself. Paul Zeleza has written how the linked projects of decentering Eurocentric histories and categories "requires not only continued vigilance against Eurocentric conceptions of history and categories of analysis, but also vigorous reconstructions of history that (re)centre African history by deepening and globalizing it in its temporal and spatial scope. After all, Africa is the cradle of humanity, the continent where humans have lived the longest" (2006: 203). That Africa is now regarded as the "birthplace" of humanity has been highlighted as a reason for the importance of African history, showing the continent as a crucial center, rather than somewhere peripheral or cut-off from key human stories. Historians of science will of course recognize that this too is an ideological use of the past, albeit one which connects with the liberatory projects of African history, which makes this an area and set of claims where these two fields would have many possible points for discussion.

A final aspect around African palaeontology is the work by palaeontologists acknowledging how their discipline has been deeply shaped by colonialism and extraction. These include Raja et al. (2022), which has used statistical methods to construct a "parachute index" based on how scientific communities in particular countries have been subjected to various inequalities within palaeontological research. Four of the top ten countries in this index were African. Even more starkly, North et al.'s survey of 3,573 "high-impact geoscience articles" (2020) found that only 3.9% dealt with Africa, and of those, only 30% contained an African author (80% of whom came from either South Africa, Egypt, Morocco, Ethiopia, Algeria, and Cameroon). These legacies of colonialism and exclusion are not surprising to historians of science, who have long engaged with the imperial and colonial connections of these fields. But they have dramatically highlighted these issues to earth scientists and raise the potential for socially engaged cross-disciplinary work between historians of science and working scientists.

The public prominence of palaeontology has also made it an important field in diversifying science and building African scientific institutions. Fossils have become an important focus to move debates on restitution – which have traditionally focused on human remains and human-made objects – to objects of a more "natural" provenance. These have included the Tendaguru fossils still held in Berlin, and other specimens like Kabwe 1 (historically termed "the Broken Hill skull" or "Rhodesia Man") held at the Natural History Museum in London, which has been requested by Zambia since the 1970s (Stewens et al., 2022). That palaeontology also provides an opportunity to inspire African involvement in science have been raised by science communicators. One notable figure is Sibusiso Biyela, who has worked to develop new terms and modes of communicating science in Zulu, particularly through writing reports and coining terms around palaeontology, following the long trend of using dinosaurs as "ambassadors for science" (Biyela 2019). The public resonance of palaeontology has made it a central part of wider debates over the role and conduct of science, and attempts to diversify and expand it to traditionally excluded audiences.

4 Conclusion

To then return to the question posed in the opening of this piece: how can African history and the history of the earth sciences speak to one another, and what insights does bringing them together potentially offer? This chapter can of course only give a partial answer, and one which is strongly informed by my own position as a European primarily trained in the history of science. And given that this piece is part of a potential dialogue, it certainly is not the final word, and responses from Africanist and African positions would be most welcome. Nevertheless, three main points could be suggested. The first is the crucial, yet often hidden and ignored, contribution of African actors and processes to the history of human engagement with the earth, in all its manifestations and entanglements, including the economic, political, conceptual, and ecological. African history can inform the drive in the history of the earth sciences to broaden the range of actors involved in understanding of the earth and regard western geology as just one mode of knowing. The second is a related point that African history in dialogue with the earth sciences can underscore and develop a fundamental contention in both fields – that knowledges of the earth depend on various forms of power. One strand of this story is an expected one – the power exerted through colonial and neocolonial processes, in which the earth sciences and mineral extraction were central to exploitation and domination. But works like those of Hecht, d'Avignon, and the hints around the role of Africans in the history of palaeontology, can also broaden our understanding and recover a greater plurality of voices, actors, and perspectives. And finally, this is a history deeply relevant to the present. Whether discussing legacies and persistence of mineral exploitation, inequalities in research in palaeontology and palaeoanthropology, and the relative marginalization of African scholarship today, the history of the earth sciences in dialogue with African history can both illustrate the roots of these processes and potential means of redressing them in the present and the future.

References

Alpern, Stanley. 2005. Did they or didn't they invent it? Iron in sub-Saharan Africa. *History in Africa* 32: 41–94.

Anderson, Thomas. 2013. Aepyornis as moa: Giant birds and global connections in nineteenth-century science. *British Journal for the History of Science* 46: 675–693.

———. 2018. *Reassembling the strange naturalists, missionaries, and the environment of nineteenth-century Madagascar*. Lanham: Lexington Books.

Austin, Gareth, ed. 2017. *Economic development and environmental history in the Anthropocene: Perspectives on Asia and Africa*. London: Bloomsbury.

Biyela, Sibusiso. 2019. Decolonizing science writing in South Africa. The Open Notebook: https://www.theopennotebook.com/2019/02/12/decolonizing-science-writing-in-south-africa/
Bocoum, Hamady. 2004. *Aux origines de la métallurgie du fer en Afrique: une ancienneté méconnue, Afrique de l'Ouest et Afrique centrale*. Paris: Éditions UNESCO.
Brizuela-Garcia, E. 2008. Towards a critical interdisciplinarity? African history and the reconstruction of universal narratives. *Rethinking History* 12: 299–316.
Capps, Gaving. 2016. Tribal-landed property: The value of the chieftaincy in contemporary Africa. *Journal of Agrarian Change* 16: 452–477.
Carney, Judith. 2002. *Black Rice: The African origins of rice cultivation in the Americas*. Cambridge, MA: Harvard University Press.
Carney, Judith, and Richard Nicholas Rosomoff. 2009. *In the shadow of slavery: Africa's botanical legacy in the Atlantic world*. Berkeley: University of California Press.
Catesby, Mark. 1731–43. *The natural history of Carolina, Florida and the Bahama Islands*. 2 vols. London: Printed at the expence of the Author.
Chakrabarti, Pratik. 2020. *Inscriptions of nature: Geology and the naturalization of antiquity*. Baltimore: Johns Hopkins University Press.
Chakrabarti, Pratik, and Alison Bashford. 2021. Towards a modern history of Gondwanaland. *Journal of the British Academy* 9: 5–26.
Chetty, Suryakanthie. 2021. *'Africa forms the key:' Alex du Toit and the history of continental drift*. Basingstoke: Palgrave Macmillan.
Chirikure, Shadreck. 2015. *Metals in past societies: A global perspective on indigenous African metallurgy*. Heidelberg: Springer.
———. 2018. Precolonial metallurgy and mining across Africa. In *The Oxford research encyclopedia of African history*, ed. Thomas Spear: https://doi.org/10.1093/acrefore/9780190277734.013.148.
Chrisman, Laura. 2011. *Rereading the imperial romance: British imperialism and South African resistance in Haggard, Schreiner, and Plaatje*. Oxford: Oxford University Press.
Cleveland, Todd. 2014. *Stones of contention: A history of Africa's diamonds*. Athens: Ohio University Press.
D'Avignon, Robyn. 2018. Primitive techniques: From 'customary' to 'artisanal' mining in French West Africa. *The Journal of African History* 59: 179–197.
———. 2020. Spirited Geobodies: Producing subterranean property in nineteenth-century Bambuk, West Africa. *Technology & Culture* 61: 20–48.
———. 2022. *A ritual geology: Gold and subterranean knowledge in Savannah West Africa*. Durham: Duke University Press.
Diallo, Drissa, and Bocar Allaye Diallo. 2016. Ethnopedology for solving problems of soil management and sustainable agriculture implementation in West Africa Savannah regions. *African Journal of Agriculture Research* 11: 812–817.
Dubow, Saul. 2004. Earth history, natural history, and prehistory at the cape, 1860–1875. *Comparative Studies in Society and History* 46: 107–133.
———. 2006. *A commonwealth of knowledge: Science, sensibility, and white South Africa*. Oxford: Oxford University Press.
Dumett, Raymond. 1998. *El Dorado in West Africa: The gold mining frontier, African labor, and colonial capitalism*. Athens: Ohio University Press.
Fabian, Johannes. 2014. *Time and the other: How anthropology makes its object*. New York: Columbia University Press.
Fauvelle-Aymar, François-Xavier. 2018. *The Golden rhinoceros: Histories of the African middle ages*. Princeton: Princeton University Press.
Goodrum, Matthew. 2009. The history of human origins research and its place in the history of science: Research problems and historiography. *History of Science* 47: 337–357.
Green, Toby. 2016. Africa and the Price revolution: Currency imports and socioeconomic change in west and west-Central Africa during the 17th century. *Journal of African History* 57: 1–24.

———. 2020. *A fistful of shells: West Africa from the rise of the slave trade to the age of revolution*. London: Penguin.
Gundu, Zacharys. 2020. Looted Nigerian heritage: An interrogatory discourse around repatriation. *Contemporary Journal of African Studies* 7: 47–66.
Hecht, Gabrielle. 2012. *Being nuclear: Africans and the global uranium trade*. Cambridge, MA: MIT Press.
Helm, Charles, and Julien Benoit. 2019. Geomythology in southern Africa. *The Digging Stick* 36: 15–19.
Helm, Charles, Julien Benoit, Adrienne Mayor, Hayley C. Cawthra, Cameron R. Penn-Clarke, and Renee Rust. 2019. Interest in geological and palaeontological curiosities by southern African non-western societies: A review and perspectives for future study. *Proceedings of the Geologists' Association* 130: 541–558.
Henige, David. 2009. Impossible to disprove yet impossible to believe: The unforgiving epistemology of deep-time Oral tradition. *History in Africa* 36: 127–234.
Herbert, Eugenia. 2001. African metallurgy: The historian's dilemma. *Mediterranean Archaeology* 14: 41–48.
Herbert, Eugenia. 2003. *Red gold of Africa: Copper in precolonial history and culture*. Madison: University of Wisconsin Press.
Hersey, Mark, and Jeremy Vetter. 2019. Shared ground: Between environmental history and the history of science. *History of Science* 57: 403–440.
Heumann, Ina, Holger Stoecker, Marco Tamborini, and Mereike Vennen, eds. 2018. *Dinosaurierfragmente: Zur Geschichte der Tendaguru-Expedition und ihrer Objekte, 1906–2018*. Göttingen: Wallstein.
Hicks, Dan. 2020. *The Brutish museums: The Benin bronzes, colonial violence and cultural restitution*. London: Pluto Press.
Hiribarren, Vincent. 2020. African history will make us breathe. *Journal of West African History* 6: v–viii.
Hobson, J.A. 1902. *Imperialism: A study*. London: George Allen and Unwin.
Hochadel, Oliver. 2009. Das Postergirl der Paläoanthropologie. Lucy zwischen Wissenschaft und Öffentlichkeit. In *Mythos-Helden-Symbole. Legitimation, Selbst- und Fremdwahrnehmung in der Geschichte der Naturwissenschaften, der Medizin und der Technik*, ed. Siegfried Bodenmann and Susan Splinter, 216–231. München: Martin Meidenbauer.
Horton, M., N. Boivin, and A. Crowther. 2021. Eastern Africa and the early Indian Ocean: Understanding mobility in a Globalising world. *Journal of Egyptian History* 13: 380–408.
Insoll, Timothy. 2015. *Material explorations in African archaeology*. Oxford: Oxford University Press.
Jacobs, Nancy. 2016. *Birders of Africa: History of a network*. New Haven: Yale University Press.
James, Wilmot Godfrey. 1992. *Our precious metal: African labour in South Africa's gold industry, 1970–1990*. Bloomington: Indiana University Press.
Jessell, Mark, David Baratoux, Luc Siebenaller, Kim Hein, Anthony Maduekwe, François Morou Ouedraogo, Lenka Baratoux, Mababa Diagne, Joe Cucuzza, Adele Seymon, and El Hajdi Sow. 2018. New models for geoscience higher education in West Africa. *Journal of African Earth Sciences* 148: 99–108.
Johannson, Donald, and Kate Wong. 2009. *Lucy's legacy: The quest for human origins*. New York: Harmony Books.
Johnstone, Frederick. 1976. *Class, race and gold: A study of class relations and racial discrimination in South Africa*. Lenham: University Press of America.
Kern, Emily. 2018. *Out of Asia: A global history of the scientific search for the origins of humankind, 1800–1965*. Ph.D. Thesis, Princeton University.
Kuljian, Christa. 2016. *Darwin's Hunch: Science, race and the search for human origins*. Auckland Park: Jacana.
Larmer, Miles. 2021. *Living for the City: Social change and knowledge production in the central African Copperbelt*. Cambridge: Cambridge University Press.

Lentz, Carola. 2013. *Land, mobility, and belonging in West Africa: Natives and strangers*. Bloomington: Indiana University Press.

Lewin, Roger. 1997. *Bones of contention: Controversies in the search for human origins*. Chicago: University of Chicago Press.

Madison, Paige. 2020. *Discovering human origins: Fossils, practices, and controversies*. Ph.D. Thesis, Arizona State University.

Maier, Gerhard. 2003. *African dinosaurs unearthed: The Tendaguru expeditions*. Bloomington: Indiana University Press.

Manson, Andrew, and Bernard Mbenga. 2012. Bophuthatswana and the North-West Province: From Pan-Tswanaism to mineral-based ethnic assertiveness. *South African Historical Journal* 64: 96–116.

Maspero, Gaston. 1891. Les forgerons d'Horus. *L'Anthropologie* 2: 405–407.

Mauny, Raymond. 1952. Essai sur l'histoire des métaux en Afrique occidentale. *Bulletin de l'Institut français d'Afrique noire* 14: 545–595.

Mayor, Adrienne. 2007. *Fossil legends of the first Americans*. Princeton: Princeton University Press.

———. 2011. *The first fossil hunters: Dinosaurs, mammoths, and myth in Greek and Roman times*. Princeton: Princeton University Press.

McCann, James. 1999. *Green land, Brown land, black land: Environmental history of Africa, 1800–1990*. Portsmouth: Heinemann.

McClintock, Anne. 1995. *Imperial leather: Race, gender, and sexuality in the colonial contest*. New York: Routledge.

McGrath, Ann, and Mary Ann Jebb, eds. 2015. *Long history, deep time: Deepening histories of place*. Canberra: ANU Press.

Milam, Erika. 2019. *Creatures of Cain: The hunt for human nature in cold war America*. Princeton: Princeton University Press.

Moodie, T. Dunbar, and Vivienne Ndatshe. 1994. *Going for gold: Men, mines, and migration*. Berkeley: University of California Press.

Musambachime, Mwelwa. 2016. *Wealth from the rocks: Mining and smelting of metals in pre-colonial Zambia*. Lexington: Xlibris.

Mususa, Patience. 2021. *There used to be order: Life on the Copperbelt after the privatisation of the Zambia consolidated copper mines*. Ann Arbor: University of Michigan Press.

Ngailo, Jerry, and Stephen Nortcliff. 2007. Learning from Wasukuma Ethnopedology: An indigenous well-established system for transfer of agro-technology in Tanzania. *Indilinga: African Journal of Indigenous Knowledge Systems* 6: 64–75.

Nikis, Nicholas, and Alexandre Livingstone-Smith. 2017. Copper, trade and polities: Exchange networks in southern Central Africa in the 2nd millennium CE. *Journal of Southern African Studies* 43: 4–5.

North, Michelle, Warwick Hastie, and Lauren Hoyer. 2020. Out of Africa: The underrepresentation of African authors in high-impact geoscience literature. *Earth-Science Reviews* 208: 1–12.

Nothdurft, William, and Josh Smith. 2002. *The lost dinosaurs of Egypt*. New York: Random House.

Nunn, Patrick. 2018. *The edge of memory: Ancient stories, oral tradition and the post-glacial world*. London: Bloomsbury.

Ohl, Michael, and Holger Stoecker. 2018. Taxonomien am Tendaguru. Wie die Berliner Saurier ihre Namen bekamen. In *Dinosaurierfragmente*, ed. Heumann et al., 232–253. Göttingen: Wallstein.

Panciroli, Elsa. 2021. *Beasts before us: The untold story of mammal origins and evolution*. London: Bloomsburg Sigma.

Pattison, Kermit. 2020. *Fossil men: The quest for the oldest skeleton and the origins of humankind*. New York: William Morrow.

Phillips, Lauren. 2022. Below the land deals: The making of mineral property in Ga-Mphahlele, South Africa, 1880–1994. *The Journal of African History* 63: 4–18.

Phimister, I.R. 1988. Historians and the big hole: Kimberley's historiography reviewed. *South African Historical Journal* 20: 105–113.

Rademaker, Laura. 2021. A history of deep time: Indigenous knowledges and deep pasts in settler-colonial presents. *History Australia* 18: 658–675.
Raj, Kapil. 2010. *Relocating modern science: Circulation and the construction of knowledge in South Asia and Europe, 1650–1900*. New York: Palgrave Macmillan.
Raja, Nussaïbah, Emma M. Dunne, Aviwe Matiwane, Tasnuva Ming Khan, Paulina S. Nätscher, Aline M. Ghilardi, and Devapriya Chattopadhyay. 2022. Colonial history and global economics distort our understanding of deep-time biodiversity. *Nature Ecology & Evolution* 6: 145–154.
Regal, Brian. 2002. *Henry Fairfield Osborn: Race and the search for the origins of man*. Aldershot: Ashgate.
Richard, Alison. 2022. *The sloth Lemur's song: Madagascar from the deep past to the uncertain present*. New York: HarperCollins.
Richmond, Jesse. 2009a. Design and dissent: Religion, authority, and the scientific Spirit of Robert Broom. *Isis* 100: 485–504.
———. 2009b. *Experts and australopithecines: Credibility and controversy in the science of human evolution, 1924–1959*. Ph.D. Thesis, University of California San Diego.
Rubidge, Bruce. 2013. The roots of early mammals lie in the Karoo: Robert Broom's foundation and subsequent research progress. *Transactions of the Royal Society of South Africa* 68: 41–52.
Rudwick, Martin. 2014. *Earth's deep history: How it was discovered and why it matters*. Chicago: University of Chicago Press.
Schiebinger, Londa. 2017. *The secret cures of slaves: People, plants, and medicine in the eighteenth century Atlantic world*. Palo Alto: Stanford University Press.
Schmalzer, Sigrid. 2008. *The People's Peking man: Popular science and human identity in twentieth-century China*. Chicago: University of Chicago Press.
Schumaker, Lynn. 2001. *Africanizing anthropology fieldwork, networks, and the making of cultural knowledge in Central Africa*. Durham: Duke University Press.
Schweighöfer, Ellinor. 2018. *Zum Neanderthal nach Afrika: Der Streit um den Ursprung der Menschheit im 19. Und 20. Jahrhundert*. Göttingen: Wallstein.
Scott, Monique. 2007. *Rethinking evolution in the museum: Envisioning African origins*. Abindgon: Routledge.
Secord, James. 2004. Knowledge in transit. *Isis* 95: 654–672.
Seland, E.H. 2014. Archaeology of trade in the Western Indian Ocean, 300 BC–AD 700. *Journal of Archaeological Research* 22: 367–402.
Semley, Lorelle, Teresa Barnes, Bayo Holsey, and Egodi Uchendu. 2022. Editors' introduction: African history's interdisciplinary roots, ruts, and routes. *History in Africa* 49: 1–8.
Sikamo, J., A. Mwanza, and C. Mweemba. 2016. Copper Mining in Zambia – History and future. *Journal of the Southern African Institute of Mining and Metallurgy* 116: 491–496.
Smith, Iain. 2000. A century of controversy over origins. In *The South African war reappraised*, ed. Donal Lowry, 23–49. Manchester: Manchester University Press.
Sommer, Marianne. 2016. *History within: The science, culture, and politics of bones, organisms, and molecules*. Chicago: University of Chicago Press.
Soyinka, Wole. 2012. *Of Africa*. New Haven: Yale University Press.
Stewens, Paul, Nussaibah Raja, and Emma Dunne. 2022. The return of fossils removed under colonial rule. *Santander Art and Culture Law Review* 8: 69–94.
Stoecker, Holger. 2018. Maji-Maji-Krieg und Mineralien. Zur Vorgeschichte der Ausgrabung von Dinosaurier-Fossilien am Tendaguru in Deutsch-Ostafrika. In *Dinosaurierfragmente*, ed. Heumann et al., 24–37. Göttingen: Wallstein.
Tembo, Alfred. 2014. The Rhodes-Livingstone Institute and interdisciplinary research in Northern Rhodesia (Zambia), 1937–1964. *Strategic Review for Southern Africa* 36: 90–99.
———. 2018. The history and historiography of science. In *The Oxford research encyclopedia of African history*, ed. Thomas Spear. https://doi.org/10.1093/acrefore/9780190277734.013.353.
Turrell, Robert Vicat. 1987. *Capital and labour on the Kimberley diamond fields, 1871–1890*. Cambridge: Cambridge University Press.

Vennen, Mareike. 2018. Arbeitsbilder – Bilderarbeit. Die Herstellung und Zirkulation von Fotografien der Tendaguru-Expedition. In *Dinosaurierfragmente*, ed. Heumann et al., 56–75. Göttingen: Wallstein.

Weidman, Nadine. 2011. Popularizing the ancestry of man Robert Ardrey and the killer instinct. *Isis* 102: 269–299.

Worger, William. 1987. *South Africa's City of diamonds: Mine workers and monopoly capitalism in Kimberley, 1867–1895*. New Haven: Yale University Press.

Wylie, Caitlin. 2021. *Preparing dinosaurs: The work behind the scenes*. Cambridge, MA: MIT Press.

Yen, Hsiao-Pei. 2014. Evolutionary Asiacentrism, Peking man, and the origins of sinocentric ethno-nationalism. *Journal of the History of Biology* 47: 585–625.

Yusoff, Katrin. 2018. *A billion black Anthropocenes, or none*. Minneapolis: University of Minnesota Press.

Zeleza, Paul. 2006. The disciplinary, interdisciplinary and global dimensions of African studies. *International Journal of African Renaissance Studies* 1: 195–220.

Historiography of Earth Sciences in Latin America

Silvia Fernanda de Mendonça Figueirôa
and Maria Margaret Lopes

Contents

Abstract

The geoscientific knowledge of the territories of the Americas was at the core of colonizing projects and international disputes over these territories. The historical studies of various aspects of Earth sciences bring countless references. They have been present since the sixteenth and seventeenth centuries in the writings of the Portuguese Jesuits or chroniclers. Latin America continues to participate in the historiographical production of Earth sciences at the international level, in tune with mainstream trends, but with varying intensity depending on each country's specific contexts and intellectual traditions. New perspectives suggest counterpoints to the unidirectional views of the old diffusionist proposals in the History of Sciences. They emphasize that governments' individual, professional, institutional, political, economic, scientific, and theoretical interests are not dissociated and are linked to rocks, fossils, soils, mineral and energy resources, and landscapes. There is much to be explored along with the perspective of circulation of

S. F. d. M. Figueirôa (✉)
Depto. de Ensino e Práticas Culturais/DEPRAC, Faculdade de Educação-FE/UNICAMP, Campinas, SP, Brazil
e-mail: silviamf@unicamp.br

M. M. Lopes
Programa Interunidades em Museologia-PPGmus, Museu de Arqueologia e Etnologia-MAE/USP, São Paulo, Brazil

E. Aronova et al. (eds.), *Handbook of the Historiography of the Earth and Environmental Sciences*, Historiographies of Science,
https://doi.org/10.1007/978-3-031-40799-4_27

ideas, practices, and objects, and scientific cooperation in the continental context, without losing the dimension that the phenomena associated with Earth sciences, due to their spatial and temporal dimensions, are not subordinated to geopolitical frontiers that, in Latin America and various regions of the world, as a result of colonizing processes, wars, and territorial disputes, have changed over time.

Keywords

Latin America · Earth Sciences · Geosciences · History

1 Introduction

The geological and geographic characteristics and specific mineral and energy resources contributed to outlining the profile of the historiographical production of each country in Latin America. Gold, silver, mineral resources in general, and expectations of inexhaustible wealth attracted the interests not only of the Portuguese and Spanish empires since the early years of conquests in the late fifteenth century but also of explorers, naturalists, traders, adventurers, exploration companies, and several European countries. Moreover, the geoscientific knowledge of the territories of the Americas was at the core of colonizing projects and international disputes over these territories. For this reason, the historical studies of various aspects of Earth sciences bring countless references. That is because they have been present since the sixteenth and seventeenth centuries in the writings of the Portuguese Jesuits or of the chroniclers of the Indies – as America was called. It includes information on minerals used by the inhabitants of these regions, occurrences of earthquakes and volcanism, and also information on bones attributed to giants, extending from Argentina and Chile, further south of the continent and to the more northern territories, then Spanish, incorporated by force or by purchasing by the United States.

These are themes that continue to be emphasized in current studies, such as mining, in which Argentina is highlighted in classical studies due to its very name closely related to mineral exploration (Alonso 2010), to dreams of incalculable riches, as also critically affirmed in the works on the Brazilian 'El dorados' and the 'visions of paradise' (Figueirôa 1997).

At the international level, the history of Earth sciences has been consolidated for some decades. The International Geological Congresses (IGCs) increasingly include the theme in specific sessions and symposia, since the meeting in India, in 1964. In 1967, the INHIGEO – International Commission on the History of Geological Sciences[1] – was created, linked to two important international organizations: the IUGS (International Union of Geological Sciences) and IUPHST (International Union of History and Philosophy of Science and Technology). The INHIGEO brings

[1] For more details, access: http://www.inhigeo.com/, https://www.iugs.org/, http://iuhpst.org/

together researchers interested in a wide range of topics in the history of geological sciences, which currently has approximately 300 members from 57 countries.

Since its early years, the INHIGEO has had the participation of Latin American researchers dedicated to the history of geology in their countries, such as Othon Henry Leonardos (1889–1977), professor at Geosciences Institute, Federal University of Rio de Janeiro, since 1968, and Telasco Garcia Castellanos (1912–2008), professor of economic geology at the National University of Cordoba, since 1970, in addition to Horácio Homero Camacho (1922–2015) and Juan Carlos Manuel Turner (1918–1979), professors of paleontology and geology both from the University of Buenos Aires.

However, Latin American participation became more present at INHIGEO from the 1980s onward. A new generation of young researchers started to mobilize and participate, since 1985, with colleagues from Brazil and Venezuela and, in the following years, with colleagues from other Latin American countries (Costa Rica, Colombia, Bolivia, Argentina, Mexico, and Chile). As a result, initiatives around the history of geological sciences in this region gained momentum. An expressive set of international events that took place in Latin America since the end of the 1980s, relying not only on local researchers but also on several European countries, the United States, Canada, Australia, and Japan, constituted landmarks in this process.[2] The interest of these events also resides, in addition to the strengthening of internal exchanges within the Latin American and international region, in the productions that have been published, where one can find a vibrant and varied production, either from the point of view of themes or approaches, revealing the dynamism of the area and the quality of local historiographic production on the Earth sciences.[3]

The institutions and events mentioned marked essential inflection points in research and, consequently, in the historiographical production on Earth sciences at the international level. Latin America continues to follow and participate in these movements, in tune with mainstream trends, but with varying intensity depending on

[2]As an example, we mention the first Brazilian Colloquium on the History and Theory of Geological Knowledge (University of Campinas, São Paulo, Brazil, July 1988); the 18th International INHIGEO Symposium (Campinas, SP and Ouro Preto, MG, Brazil, July 1993) (Lopes and Figueirôa 1990; Figueirôa and Lopes 1994). The Boletín de Historia de las Geociencias de Venezuela since 1984; the organization of the Sociedad Venezolana de Historia de las Geociencias 1988; the first Venezuelan Meeting on the History of geological sciences, 1991; the establishment of the Committee on the History and Philosophy of the Geosciences and Geotechnologies – COHIFIGEO, founded in San Jose, Costa Rica in 1989. Since 2007, Congresses on the History of Argentinean Geology (CAHGEO) have been held (Aceñolaza 2008); see Revista de la Facultad de Ciencias Exactas, Fisicas y Naturales, Universidad Nacional de Córdoba, vol. 7, Supl. 1, 2020 (https://revistas.unc.edu.ar/index.php/FCEFyN/issue/view/2083). The First Chilean Symposia on History of Geology was held in Chile in 2010 and the INHIGEO 43rd Meeting was held in Mexico DF, in 2018. http://www.inhigeo.com/record/51.pdf

[3]Also, within the framework of scientific associations, the History of Earth Sciences Society (HESS) founded in 1982 deserves emphasis. Since then, this society has published the journal *Earth Sci*ences History (ESH) is one of the few, recognized international journals focused on this field of knowledge, which has had the participation of Latin American authors and publishers. For more details, access: http://www.historyearthscience.org/

the specific contexts and intellectual traditions of each country. An indisputable and comprehensive milestone, however, was the creation of the Sociedad Latinoamericana de Historia de las Ciencias y la Tecnología (SLHCT) in 1982. Its strong impact lies less in its foundation itself and in its international design, which brought together researchers from Mexico, Brazil, Colombia, Venezuela, Peru, Ecuador, Cuba, Costa Rica, Argentina, Ecuador, and Spain, but mainly in the new intellectual frameworks launched to think about the sciences in this region, escaping the trap of conceiving it as eternally peripheral and, above all, not allowing to subordinate intellectual and scientific contexts to economic and political ones (Saldaña González 1986). Alongside the historiography on various scientific fields, Earth sciences were present in the SLHCT, especially in its official journal *Quipu*[4] (started in 1984), including various approaches to different themes. It is from then on that the concept of 'empire' becomes part of academic production, either as a reference framework for analysis or in its concreteness, as manifested in the scientific practices themselves and in the lives of those who made them happen. Before that, the expressions "colonial" or "imperial," when present, referred to a period of historical time, without a more strictly political connotation. Furthermore, the concern with constructing a "national" science, in dialogue with but not subordinate to the North Atlantic centers and full of native and/or local knowledge, permeates many of the works.

Consequently, the historiographic production of Earth sciences in Latin America can be roughly aggregated into two large groups. Considering the themes, in general, there is more continuity than rupture. The history of mining, cartography, biographies, educational and research institutions, disciplinary fields, and foreign researchers comprise the main focus of the works produced in one or another set. However, in addition to some innovative themes, as will be mentioned later, it is the new approaches that constitute the most distinctive aspect among these sets of articles, produced either by geologists and/or historians of science and other professionals. We will present below a non-exhaustive synthesis of some characteristics of each of these sets in order to situate the readers, offering them a textual compass to navigate the historiography of Earth sciences in Latin America.

2 Episodic-narrative perspectives

The first set, likely produced before the 1980s, is characterized by the primacy of works with a strong episodic-narrative characteristic, not very analytical and interpretive, and with lists of facts and famous names, almost always male. The biographies, quite frequently, have a flattering and hagiographic tone. The preferred theme brings broad views of specific disciplinary fields in specific countries. In the texts,

[4]The journal owes its name to an ancient mathematical computing system used by the Incas in the Andes. This fact represented a symbol of the guiding ideas that were intended to be presented and implied the recognition of the scientific and technological past of Latin America.

people and institutions are almost always mentioned in chronological sequence but not always interconnected with each other or with contexts (local, national, or international). There is also an intense appreciation of foreigners, to the detriment of the national ones, who are generally presented in secondary positions.

Along these lines, in 1895, geologist Orville A. Derby (1851–1915), whose professional career was entirely built in Brazil, published what appears to be the first article specifically on the history of the Earth sciences in a Latin American country (Derby 1895). Derby stated that the science practiced in Brazil was not characterized by including research and, furthermore, that some reputations had been built without original and worthy work. At the same time, it devalued local production and exalted the contributions of foreigners, like him, a perspective that would be kept in works by the nationals, continued until the 1980s, as previously mentioned. In the year following Derby's article, José G. Aguilera published, in the Boletín del Instituto Geológico de México, a bibliography of the engineer Antonio del Castillo, founder and director of the Instituto Geológico de México, accompanied by a short summary of the Instituto Geológico's initiatives, organized in 1886 to introduce the country's geology and maps (Aguilera 1896).

The exaltation of foreigners as founders is seen in several works, whether in Colombia, with emphasis on the Germans (Ramírez 1955), or in Brazil, addressing the British and German contributions to geology (Leonardos 1970, 1973). In fact, various historical works of the engineer and geologist Othon H. Leonardos (1899–1977), editor of the only Brazilian publication specifically dedicated to "Mining and Metallurgy," which he created in 1936, deserve mention, as well as his co-authorship with Avelino Ignácio de Oliveira (1891–1970), from the book "Geologia do Brasil." This book began with a first chapter on the history of Earth sciences in Brazil since the beginning of colonization, highlighting the production of the first geological map prepared by nationals and the scientific cooperation of geologists from neighboring countries (de Oliveira and Leonardos 1940). On the other hand, the work of Alvarez Conde (1957), despite highlighting foreigners, already incorporated local contributions to emphasize, in broad panoramas, the importance of knowledge on soil, hydrology, cartography, paleontology, mining, and geophysics, including archaeological studies in Cuba.

Among these works – with a strong episodic-narrative nature and full of information – the work of Aguilera (1904) brings the origins of the geology of Mexico, in addition to initial derogatory references to the original peoples, mentions to Spanish historians, to von Humboldt, the Real Seminario de Minería of 1783 and the work of Andrés Manuel del Rio, "Elementos de Orictognosia," published from 1795 to 1805, considered to be one of the first books on mineralogy to appear in New Spain. These mentions are repeated up to the moment of what is considered to be one of the first outlines of Mexican geology history carried out by the men of science themselves, such as Zoltan de Cserna (1968, 1990), and already in a critical way, a work that appreciated their own practices (Azuela-Bernal 2016).

Earth sciences were included in the general works about the History of Sciences in several Latin American countries. The Argentine tradition dates back to the initiatives of Aldo Mieli (1879–1950) – a collaborator of George Sarton

(1884–1956) – and of Argentine historians (Papp et al. 1961) and, more recently, to the works of Garcia Castellanos (Riccardi 2020) and Camacho (2001), among others. In Mexico, Gortari's book (De Gortari 1963) covers the art of metals from the cultures of ancient Mexico, mentions indigenous science, and makes brief references to local geological institutions, societies, and publications, up to the 20th International Geological Congress (IGC) in 1956, a landmark in the geological sciences in Mexico, as was the tenth International Geology Congress, held in Mexico in 1906. Also, within the framework of perspectives considered externalist and Eurocentric, but already evaluated as more innovative and more critical, Elias Trabulse's voluminous work (1985) refers to the period 'of the arrival of European science' in Mexico until the nineteenth century. This work drew attention to the lack of historians with scientific preparation to deal with themes from several scientific areas, which was then the reality. Classical texts by authors of the enlightened period regarding geography, geodesy, cartography, by José Antonio Alzate (1737–1799), or Andrés Manuel del Rio's classification of minerals (1764–1849), and the benefits of exploring silver minerals by Faust de Elhuyar (1755–1833) and others, have been reproduced, as well as by nineteenth-century authors on mineralogy, paleontology, geography, hydrography, tidal studies, or Eduardo Noriega's 'progress in the geography of Mexico.' In Brazil, the extensive work organized by sociologist and educator Fernando de Azevedo (1894–1974) included wide-ranging chapters, prepared by professors of respective disciplines at the universities of São Paulo, on geology and paleontology (Leinz 1955), mineralogy, and petrography (Leonardos 1955), and geography (da Costa Pereira 1955). A review and expansion of this work was undertaken by botanist Mário Guimarães Ferri (1918–1985) and science historian Shozo Motoyama (1940–2021) between 1979 and 1981, addressing mineralogy and petrography (Franco et al. 1981) and paleontological research (Mendes 1981), where this time the field of Earth sciences deserved the title of "Geosciences" and focused on climatology, meteorology, oceanography, and geomorphology (Ab'Saber and Christofoletti 1980).

In Colombia, the "Apuntes para la Historia de las Geologicas Investigaciones en Colombia" (Botero 1978) and "Reseña Histórico de la Geologia en Colombia" (Durán 1973) stand out among the wide-ranging contributions. In Bolivia, the Instituto Panamericano de Geografía e Historia and the Academia Nacional de Ciencias were concerned about redeeming the history of science in that country. In the book "Contributions to the History of Science in Bolivia," Ismael Montes de Oca presents "Historia de la ciencia geologica" and a brief historical review of seismology (Montes de Oca 2001), José Luis Tellería-Geiger addresses geophysics in Bolivia (Tellería-Geiger & Pabón 2001), Waldo Ávila Salinas writes about mineralogical and petrological investigations (Ávila Salinas 2001a, b), and Estela Minaya has a biography on "vidas dedicadas a la ciencia de la Tierra" (Minaya 2001).

The emphasis in studies on the lives of geoscientists deserves to be mentioned, also due to its constant presence in the literature. Both foreigners, such as Charles Darwin (1809–1882) (Sagredo Baeza & Allamand 2011), and native characters were the object of attention. Alexander von Humboldt (1769–1859) is without a doubt a frequent presence in works from Argentina, Colombia, Mexico, or Venezuela

(Aceñolaza 2008; Aguilera 1904; De Terra 1956), as he traveled widely and extensively through the territories now occupied by these countries. Others, such as Orville Derby, mentioned above, also deserved texts and tributes honoring their birth centenaries (Lamego 1951). But local characters, creoles, or figures from imperial metropolises are also on the list of those entitled to have their lives counted since colonial times. The brothers Fausto and José d'Elhuyar, Tulio Ospina, José Maria Cabal, Enrique Hubash, Manuel Ferreira da Câmara, José Bonifácio d'Andrada e Silva, to mention just a few, clearly illustrate what we have just said.

In short, a large part of the production prior to the 1980s (but not exclusively), whether its authors were aware or not, strongly share – if not completely – the positivist historiographic perspectives, that is, the coming of science (in the singular, because it assumes a unity in this endeavor) is continuous and progressive; science is a collection of results and a set of methods; what counts, in fact, are the contents and concepts, and the people who developed them, leaving aside the broader relations in which the sciences are inserted and established with the other spheres of society. In addition, of course, to the strong Eurocentrism that permeates and pervades it.

3 Contemporary perspectives

The production of the second set, frequent from the 1980s onward, is meant to map geoscientific activities in their different times and spaces. Although the perspectives of the first phase can still be found occasionally and are still valuable due to the amount of compiled information, the works of the second set track identify and situate practices, processes, and actors, strongly supported by primary documentation, international bibliography, and contemporary theoretical references that, in many cases, we could also say decolonial. Scientific exchanges, knowledge circulation, collections, and people acquire thematic relevance and new explanatory perspectives, articulating contexts at different scales. Less generalist panoramas and more specific thematic ones, deepening local and regional histories, appear in approaches to mining, paleontology, oil, scientific institutions (mainly scientific associations, museums, and geological/geographical surveys), impactful geological phenomena, such as volcanism, earthquakes, and meteorites, just to mention some themes.

A relevant theme that crosses the entire historiography is that of mining. In this second set of works, the intertwining of economic, political, cultural, and technical-scientific issues is highlighted in the interpretation and acquires explanatory substance (Uribe-Salas et al. 2016). Changes are followed on a smaller scale, even at the level of practices, which occur within the Earth sciences in their articulation with the mining activity and their repercussions in other disciplines. Some texts exemplifying this theme, published in the same book, reflect a new phase in which new perspectives circulate more widely. The work of military engineers in California in the eighteenth century is analyzed related to the dynamics of the survey of natural resources (including minerals) in this territory, with the interests of several States coming together, such as France, England, Russia, and, later, the United States, in addition to Spain (Moncada Maya 1995). The quick scientific exchanges of the last quarter of the eighteenth century, which brought technical-scientific innovations applied to socially relevant sectors such as the mining industry, commerce, and public works, are analyzed simultaneously with the specialization and institutionalization of various areas of knowledge (Molina Martínez 1995; Aceves Pastrana 1995). Another good example is Azuela's (2005) article that connected the introduction of experimental laboratory practices to mining activity in the construction of geology in an educational space in Mexico since the colonial period. Similarly, Figueirôa (1997) examined in detail the investments of the Portuguese metropolis in Earth sciences and Natural History since the Pombaline reforms, whose consequences, especially in mining, infrastructure works, and the training of technical staff, resulted in the institutionalization of geological sciences in Brazil. In Mexico, the institutionalization process of the Earth sciences in the sixteenth century was investigated in depth by Omar Moncada (2003). Márcia Alvim (2010) made a relevant contribution by associating sources traditionally used by history, but not really present in the history of sciences, such as the missionary chronicles and codices from the period of occupation, to rescue Mesoamerican knowledge about the natural world. As for the Mexican nineteenth century, it is important to highlight the works of Luz Azuela (1996), Mendoza Vargas (1999), and Morelos Rodríguez (2012).

At the same time, this production takes the concepts of "science" in each period, which means abandoning the present and ahistorical frameworks, common in previous works that continuously (and artificially) organized information about the Earth as if they had always been part of the same corpus of knowledge. Accordingly, for example, we find texts that connect geography, natural history, and physics to marine expeditions, bringing the richness and dynamics of the imbrications of these fields in their respective historical time (Capel 1995). In addition to valuing native knowledge, its circulation is considered a two-way street, breaking one-way and reductionist perspectives that only considered circulation from Europe to America or from Metropolis to Colonies. It is understood that "foreign" knowledge – especially European – goes through local "acclimatization" processes, almost always "amalgamating" with original knowledge and generating new knowledge that, in due course, returns to Europe to continue being used (Serrano Bravo 1995).

Production is abundant. Citing a few more themes, since the 1980s and 1990s, pre-Hispanic mining in Mexico (Langescheidt 1985; Sánchez-Flores 1988), neogranadine metallurgy (Espinosa Baquero 1987) or Peruvian pre-Hispanic (Grinberg and Palacios 1992), metallic mining in Central America (Soto 2011), or colonial mining in northwestern Argentina (Aceñolaza and Alonso 2018) have been treated in detail in specific localities, regions, or countries, addressing technology transfers, exchanges, and comparisons between America and Europe (Sánchez Gómez and Mira 1993; Contreras and Mira 1993), or mining representations in the scientific and technical press (Azuela and Morelos 2011).

Not intending to rebuild all mining-metallurgical activity in the region and opening new research perspectives for innovative approaches, Serrano Bravo (2004) relates mining in current Bolivia to agriculture, to climate processes and their effects on the environment, and implications regarding the health of populations. The mining of gold, silver, copper, and tin in Bolivia also invited the interest of foreigners, at various times, attracted by the dream of riches. As cited by the author, it also favored the exchange of knowledge between inhabitants of different regions of the territory and colonizers, in contexts where the performance of mineral policies in interculturality cannot be abstracted, in terms of practices in exploitation, and processing, and smelting of minerals. Such publications addressed aspects of the life of miners and entrepreneurs, the magnitude, intensity, and duration of certain companies (Fernández 2005), and opened perspectives for more critical analysis of those who did not find wealth. These aspects were only part of the producing mechanism and went unnoticed. Along these lines, historical reflections strongly resonate today, including issues of the environmental effects of mining or the exploitation of sea mineral resources (Coll-Hurtado et al. 2002; Uribe-Salas 2010; Machado and Figueirôa 2020).

To extract minerals, it is necessary first to carry out exploratory surveys and map the territory. Thus, another significant portion of historiographic production was dedicated to the history of Cartography and its multiple spatial approaches. An excellent example is the work "Historias de la Cartografía de Iberoamérica. Nuevos caminos, viejos problemas," organized by Héctor Mendoza Vargas and Carla Lois (2009). This book discusses cartographic representations and their technical aspects, the relations between the State, national maps, and the professionals involved in their creation, passing through case studies that expand the theme in its various connections. As a rule, the institutions responsible for surveying territories were Geographical and Geological Commissions and Services, mostly inspired by the international model of "geological surveys." Present practically all over the world, they constituted the institutional mark of developing the Earth sciences from the nineteenth century onward and Latin America did not overlook these actions. In Brazil, there were three initiatives in the independent and republican periods, two of them lasting until the present and inspired by the North American model. In each country, these institutions also focused on specific issues, from droughts in the Northeast of Brazil to the search for oil (Figueirôa 2007), also a common attribution to the Colombian Geological Service, the beneficiary of several institutions since the Chorographic Commission of New Granada (Restrepo Forero 1984). Organized in the twentieth century, in addition to prospecting for mineral resources, it also dealt with environmental disasters related to volcanism, earthquakes, landslides, and avalanches (Espinosa Baquero 2016). In the Mexican case, it was not until the last third of the nineteenth century that an official attempt was made to put together a geological survey, which had to overcome a series of local obstacles to become a reality (Azuela 2017).

The theme of Geological Surveys allows us to consider other types of institutions that were also studied, such as Academies, Museums, Research Institutes, Universities, among others, but outside the laudatory and encomiastic perspectives that are typically used during celebrations of festive dates such as anniversaries, silver or gold jubilees, or centenaries. Such studies generally used analytical categories involving aspects like "successes" or "failures," "number of publications," and "scientist's administrative capacity," for example, which concealed more than revealed the scientific activities that, concretely and daily, were carried out in this part of the world (Figueirôa 2019). Furthermore, Hebe Vessuri (1987) pointed out that the history of scientific institutions allows us to enter the strategic *loci* of science construction. Currently, there are very rich studies on scientific societies and museums that reveal their complex relationships as mediating institutions between science and society. From the perspective of the geography of institutional spaces, trying to put science in its place, historiographical studies show that geological, cartographic, geographic, mineralogical, and paleontological practices were enabled in different institutional spaces (Figueirôa 1992).

Among these institutions, we highlight only the museums, which, for example, were some of these seminal *loci*. In the changing perspectives of the 1980s and 1990s, mineral resources, particularly in Colombia (Rodríguez-Prada 2013), and the geographic and geological exploration of the territory in Chile (Sanhueza 2014) were addressed in studies on the museums that were created as legacies of illustrated colonialism in the first decades of the nineteenth century (Lopes and Podgorny 2000; Azuela et al. 2009; Lopes 2020). Such studies have shown a much more complex panorama about the earth sciences, until recently mainly focused on metropolitan collections only. They brought new analytical perspectives in works that concentrated on the difficulties of fieldwork in the nineteenth century, the circulation and commerce of objects and collections, correspondence, and ideas (Farro 2009). These exchanges were not only with researchers from institutions in Paris, Berlin, Padua, and Washington, but within the Latin American context between researchers from museums in Buenos Aires, La Plata, São Paulo, Rio de Janeiro, Montevideo, Valparaiso, and Santiago de Chile, Bogotá, and México City. They showed how the fossils, bones, minerals, and rocks connected people, crossing national, local, disciplinary, or epistemological borders. Many of these analyses highlighted and demystified the recurrent ideas of natural history museums as symbols of invented nationalities and nationalized natures, highlighting the richness of their daily practices that are far removed from national political projects, whether implemented or not.

In the spaces of the countless museums organized since the end of the eighteenth century in Latin America, areas of knowledge such as paleontology have become institutionalized. Latin American historiography has devoted considerable attention to paleontology and stratigraphy, portraying the roused interest since the colonial period regarding the first findings of Latin American megafaunas; studies on fossils, fundamental for establishing geochronology and geological mapping of territories, or for coal surveys in the nineteenth century; or micropaleontological studies fundamental to oil research (Peyerl 2019). Comprehensive panoramas about Mexico or Costa Rica (Gío-Arsgáez et al. 2003; Alvarado Induni 1989) coexist with strategies of following the collections to identify the articulation of international communication networks among specialists (Waligora 2020) or to resume an unconventional biography by Florentino Ameghino (1853–1911), an unavoidable character in Latin American paleontology (Podgorny 2021), to reconstruct fieldwork, disputes over classification priorities, competition for publications, or the trajectories of the fossil trade.

In addition to paleontology, much of this historiography paid attention to and continues to address the consolidation processes of the various disciplinary fields and sub-fields of earth sciences throughout the eighteenth and nineteenth centuries, such as mineralogy and geography (García and Morelos 2011; Azuela 2011), but one of the recent trends is a shift to the more current themes of the twentieth century (La Podgorny 2005). Some examples are the aforementioned environmental approaches to mining, oil investigations, the inseparability of catastrophic geological processes, and their social implications, which have deserved the attention of various science historians and professionals who incorporated them into their specific themes.

Historical studies on oil show that explorations in different Latin American countries have gained prominence since the end of the nineteenth century and especially in the twentieth century, improving expertise in stratigraphy, sedimentology, and geophysical research (Peyerl and Figueirôa 2020). In Venezuela, where the oil hunt started in 1912, location-specific approaches turn their attention to the violent confrontations of oil initiatives with communities that defended their ancestral lands, as well as projects considered significant economic failures (Urbani Patat 2020), and in countries such as Mexico, Argentina, Bolivia, or Brazil (Lopes 2019). In the context of the Argentinean agro-export society, for example, it was the Dirección de Minas, Geología y Hidrología, created in 1904 and heir to previous agencies and currently the Servicio Geológico Minero Argentino, that led oil research in the country since the beginning of the twentieth century (Riccardi 2016), as well as the most recent petrochemical investigations (Matharan and Feld 2016). These studies evidence not only the scientific and technological aspects, the industrial diversification, but also its impacts on political, economic, social, cultural, environmental, and diplomatic issues of history (Figueirôa et al. 2019).

However, geological realities differ from country to country. Unlike countries such as Brazil, Bolivia, Mexico, Colombia, or Chile, where mining and the search for energy sources have been the fundamental promoters of geological knowledge, in Costa Rica, the seismic and volcanic phenomena have been the underlying foundation of geological investigation, and therefore historical. Contemplating relationships such as art and culture since pre-Columbian times related to volcanoes and various topics on the evolution of philosophical and scientific ideas and technology related to volcanology, a recent publication focusses on volcanological research throughout history in Costa Rica, addressing themes on geothermal energy development, volcano observatories, and the historical study of volcanic hazards (Alvarado Induni et al. 1991; Alvarado Induni 2021).

Throughout Latin America, as was done in Europe, previously unknown documents were compiled in the archives, since colonial times, on the occurrence of earthquakes, directly related to volcanism to understand the past. Incorporating anthropological perspectives, the authors of the 'History of Seismological

Assessment in Venezuela' highlight the centrality of documentary research, including the churches, to unravel the characteristics of each earthquake in different times and the understanding of the social context for a broad understanding of earthquakes, as did other authors (Podestá and Olson 1987; Altez 2006; Altez et al. 2004; Poirier 2018; Alonso 2016).

In the perspectives that have understood earthquakes as hybrid entities, in the interface between nature and society, even in Brazil, a country known for the absence of intense earthquakes and active volcanism, Brazilian authors have also called attention to the fact that Brazil has been living under the impact of severe environmental, cultural, and political disasters for centuries. Retrieving centuries-old earthquake records, they highlight the urgency of taking innovative environmental, legal, and political steps to safeguard and preserve archives and material collections. Similarly, in Venezuela, debates about current risks and natural disasters are discussed while incorporating the historical perspective. These studies have shown that the key to understanding and solving today's most pressing environmental challenges lies in the fabric that interconnects the Earth sciences and their historical record (Urbani Patat 2018; Lopes and Figueirôa 2020).

Biographies continue to be present in the historiography of Latin American Earth sciences, but without the hagiographic characteristic that has usually marked them. Instead, they seek to bring aspects of the life trajectories of local builders not prioritized by the previous historiography, such as José Vieira Couto (1752–1827) (da Silva 2002) and Guilherme Schüch de Capanema (1824–1908) (Figueirôa 2005), pivotal to the institutionalization process of Earth sciences in Brazil, as well as Germán Avé-Lallemant (1835–1910), a polymath whose work ranged from mineralogy to education in Argentina (Ferrari 2000). Or propose new ways to understand celebrated characters, such as Spanish-Mexican Andrés Manuel del Río (1764–1849), who was revisited within the context of the circulation of ideas between Saxony and New Spain (Escamilla-González and Morelos-Rodríguez 2020), José Bonifácio d'Andrada e Silva (1763–1838), revised in the inseparable dimension between a man of politics and science (Varela 2006), or Ignacio Domeyko (1802–1889), analyzed within the institutional consolidation of geological sciences in Chile (Hervé and Charrier 2016). These works have demonstrated how dialectically interconnected and interdependent scientists, institutions, and contexts are, also highlighting the role of individuals in history. Without, however, ignoring the fact that they relate to the collective and to their peers, who are part of a community, producing science and technology in the conflict between the individual and collective *praxis*.

Furthermore, particularly in the context of biographies, an emerging trend of work focused on gender relations in science is identified. These approaches, although present in an extensive historiographical production in several Latin American countries (Lopes et al. 2014), only recently have they started to receive greater attention from the historiography of Earth sciences, whether in presentations at symposiums (Morelos Rodríguez 2019), specific publications, or references to women's life stories and work processes, discussing the inclusion and impact measures of women in the mining industry, in the search for more significant gender equity in Chile[5] (Narváez 2018) and Brazil (Carrilho 2021). Another focus has been the career of women in scientific disciplines, both in academia and in their profession, marked by gender prejudices and stereotypes, in Chilean (Mora-Stock et al. 2018), Mexican (Morelos Rodríguez 2019), Costa Rican, Cuban (Figueredo et al. 2017), or Argentinian productions (Herbst and Anzótegui 2016; Guereschi et al. 2021).

Studies on the histories of twentieth-century space and ocean explorations have also begun to gain greater expression in cross-fertilization with other areas of knowledge. Some investigations conceive marine scientific cultures as strategic areas for geological sciences by mining marine mineral resources (Lopes 2021). Others addressed the negotiations on the International Geophysical Year in 1957 as a model in the Cold War years on how to conceive international relations through science, aggregating different branches of geology, and geophysics, including oceanography or Antarctic treaties in the 1960s, that proposed preservation status of this region as an international reserve for science (García 2016).

4 Conclusion

In the historiography of Earth sciences in Latin America, the priority of analysis continues to be given to each country. However, new perspectives suggest counterpoints to the unidirectional standpoints of the old diffusionist proposals in the history of sciences. They emphasize that the individual, professional, institutional, political, economic, scientific, and theoretical interests of governments are not dissociated and are imbricated to rocks, fossils, soils, mineral and energy resources, and landscapes. They point to approaches about the spatiotemporal locations of countries, institutions, individuals, histories, and complex descriptions of specific details of contingent situations, favoring ideas that the production of knowledge in circulation can transgress scales, settings, and territories. The emphasis on places can help to better understand the processes of circulation and transfer of knowledge, technologies, and expertise in the context of international networks, even if they are marked by asymmetric relationships, dependent on economic trades, exchanges of favors, patronage, friendship, and obligation. For a comprehensive and global historiography of the Earth sciences based on such ideas, it is important to know what happened in each country and, especially, what happened in specific institutions or places, and in the activities of particular agents within international contexts, also contingent. Several local and regional distinctive aspects of disciplinary areas, which the present article would not be able to carry out, suggest new perspectives in which other unknown and ignored agents may emerge.

[5]Presentation of the book '100 mujeres y sus historias en la minería chilena' by Claudia Monreal, president of Women In Mining Chile https://www.reporteminero.cl/noticia/mujer-minera/2017/08/video-women-in-mining-chile-presenta-el-libro-que-regala-reporte-minero

A large body of works continues to prioritize, given their historical relevance, the contributions, and collaborations between researchers and European countries and the United States in the development of Earth sciences in Latin America. Some studies open innovative perspectives for historical reflection regarding the exchanges within the continent itself, focusing on the practices of nineteenth-century museums, geological surveys, and oil exploration in the twentieth century or environmental perspectives of the twenty-first century.

There is still much to be explored under the framework of the circulation of ideas, practices, objects, and scientific cooperation in the continental context. One must be aware, however, that the phenomena associated with Earth sciences, due to their spatial and temporal dimensions, are not subordinated to geopolitical frontiers that, in Latin America and various regions of the world, as a result of the colonizing processes, wars, and territorial disputes, have changed over time. At the same time, our Earth is only one, and therefore, a new path also arises: the need to widen the historical dialogue with the other fields of earth sciences, such as the history of soil sciences, meteorology, or oceanography, among others.

References

Ab'Saber, Aziz N., and Antonio Christofoletti. 1979. Geociências. In *História das ciências no Brasil*, ed. Mario. G. Ferri and Shozo Motoyama, vol. 2, 117–238. São Paulo: EPU: Edusp.

Ab'Saber, Aziz N., and Antonio Christofoletti. 1980. Geociências. In *História das ciências no Brasil*, ed. Mario G. Ferri and Shozo Motoyama, vol. 2, 117–238. São Paulo: EPU: Edusp.

Aceñolaza, Florencio G., ed. 2008. *Los Geólogos y la Geología en la Historia Argentina*. San Miguel de Tucumán: Instituto Superior de Correlación Geológica (INSUGEO). Serie Correlación Geológica.

Aceñolaza, Florêncio G., and Ricardo N. Alonso. 2018. *Historia de la Geología del Noroeste Argentino*. Salta: Mundo Gráfico Editorial.

Aceves Pastrana, Patricia. 1995. Intercambios científicos en la Nueva España: una nueva lectura de la Comisión Minera. In *De la Ciencia Ilustrada a la Ciencia Romántica*, ed. Alejandro Díez Torre, Tomáz Mallo, and Daniel Pacheco Fernández, 359–375. Madrid: Doce Calles.

Aguilera, José G. 1896. *Bosquejo geológico de México. Boletín del Instituto Geológico de México*. México: Oficina Tip. de la Secretaría de Fomento. http://xcaret.igeofcu.unam.mx/bogeolpdf/geo1896/geo1896-1.pdf.

———. 1904. Reseña del Desarrollo de la geología en México. *Boletín de la Sociedad Geológica Mexicana* 1: 35–117. https://doi.org/10.18268/BSGM1904v1n1a3.

Alonso, Ricardo. 2010. *Los Antiguos Mineros. Ensayos para uma Historia de la Minería de Hispanoamérica*. Salta: Mundo Editorial.

Alonso, Ricardo N. 2016. Catastrofes. In *Diccionario histórico de las ciencias de la Tierra en la Argentina*, ed. Irina Podgorny et al., 96–102. Rosario: Prohistoria Ediciones; La Plata: Museo de La Plata; Buenos Aires: CONICET.

Alonso, Ricardo, N.G. Solís, E. Silva de Cruz, and N.H. Robles. 2018. *Historia Minera Colonial. Documento Minero Inédito del Noroeste Argentino en el Archivo General de Indias (1760)*. Salta: Mundo Gráfico Editorial.

Altez, Rogelio. 2006. *El desastre de 1812 en Venezuela: sismos, vulnerabilidades y una patria no tan boba*. Caracas: Universidad Católica André Bello, Fundación Empresas Polar.

Altez, Rogelio, Jose Rodriguez, and Franco Urbani. 2004. *Historia del pensamiento sismológico en Venezuela: ... Una mirada inquieta*. Caracas: Ediciones de la Biblioteca, Universidad Central de Venezuela.

Alvarado Induni, Guillermo. 1989. Historia de la paleontología de los vertebrados en Costa Rica. In *Historia de la Ciencia y la Tecnología: El Avance de una Disciplina*, ed. Ángel Ruiz et al., 273–289. Cartago: Editorial de Costa Rica.

Alvarado Induni, Guillermo E. 2021. *Costa Rica y sus volcanes*. San José: Editorial UCR: EUNA: Editorial Tecnológica de Costa Rica.

Alvarado Induni, Guillermo, Luis Diego Morales, and Gerardo José Soto. 1991. Historia de las ciencias geológicas y sismológicas. In *Ciencia y tecnología: estudios del pasado y del futuro*, ed. Z. Ángel Ruiz, 121–142. San José: Guayacán. http://www.centroedumatematica.com/aruiz/libros/CienciayTecnologia/HistoriadelasCienciasGeologicasySismologicas/GuillermoAlvarado.html.

Alvarez Conde, José. 1957. *História de la geologia, mineralogia y paleontologia en Cuba*. La Habana: Publicaciones de la Junta nacional de Arqueologia y Etnografia.

Alvim, Márcia. 2010. *Dos céus e da Terra: Conhecimentos sobre o mundo natural nas histórias missionárias do século XVI novo-hispano*. São Paulo: Annablume.

Ávila Salinas, Waldo. 2001a. Breve reseña de las investigaciones mineralógicas en Bolivia. In *Contribuciones a la Historia de la Ciencia en Bolivia*, ed. Instituto Panamericano de Geografía e Historia y Academia Nacional de Ciencias de Bolivia, 185–192. México: IPGH.

———. 2001b. Sinopsis de los estudios de Petrología ígnea en Bolivia. In *Contribuciones a la Historia de la Ciencia en Bolivia*, ed. Instituto Panamericano de Geografía e Historia y Academia Nacional de Ciencias de Bolivia, 193–198. México: IPGH.

Azuela, Luz Fernanda. 1996. *Tres sociedades científicas en el Porfiriato. Las disciplinas, las instituciones y las relaciones entre la ciencia y el poder.* México: Sociedad Mexicana de Historia de la Ciencia y de la Tecnología: Instituto de Geografía: UNAM.

———. 2005. *De las minas al laboratorio. La demarcación de la geología en la Escuela Nacional de Ingenieros (1795–1895)*. México: Instituto de Geografía: Facultad de Ingeniería: UNAM.

———. 2011. La emergencia de la geología en el horizonte disciplinario del siglo XIX. In *La Saga de la Ciencia Mexicana. Estudios sociales de sus comunidades. Siglos XVIII al XX*, ed. Jorge Bartolucci, 55–77. Instituto de Investigaciones sobre la Universidad y la Educación.

———. 2016. Zoltan de Cserna y la historia de la geología. Una interpretación historiográfica. *Boletín del Instituto de Geología de la UNAM* 120: 71–80.

———. 2017. Towards a nationwide geological survey in nineteenth century Mexico. In *History of geoscience. Celebrating 50 years of INHIGEO*, ed. Wolf Mayer, Rénée Clary, Luz Azuela, Teresa Mota, and Stanislaw Wolkowicz, 289–301. London: Geological Society of London. https://doi.org/10.1144/SP442.11.

Azuela, Luz, and Lucero Morelos. 2011. Las representaciones mineras en la prensa científica y técnica (1860–1904). In *La geografía y las ciencias naturales en el siglo XIX mexicano*, ed. Luz F. Azuela and Rodrigo Vega, 163–177. México: UNAM, Instituto de Geografia.

Azuela, Luz Fernanda, Rodrigo Vega, Ortega Baez, and Raúl C. Nieto García. 2009. Un edificio científico para el Imperio de Maximiliano: El Museo Público de Historia Natural, Arqueología e Historia. In *Geografía e Historia Natural: Hacia una historia comparada. Estudio desde Argentina, México, Costa Rica y Paraguay*, ed. Celina Lértora, vol. 2, 101–124. Buenos Aires: Ediciones FEPAI.

Botero, Gerardo. 1978. Apuntes para la historia de las investigaciones geológicas en Colombia. In *Ciencia y Tecnología en Colombia*, ed. Instituto Colombiano de Cultura, 147–180. Bogotá.

Camacho, Horacio. 2001. Las ciencias geológicas en la Argentina, hasta 1939. *Saber y Tiempo* 12: 177–220.

Capel, Horacio. 1995. Ramas en el árbol de la ciencia: Geografía, Física e Historia Natural en las expediciones náuticas del XVIII. In *De la Ciencia Ilustrada a la Ciencia Romántica*, ed. Alejandro Díez Torre, Tomáz Mallo, and Daniel Pacheco Fernández, 503–535. Madrid: Doce Calles.

Carrilho, Anabelle. 2021. *Invisíveis, mas necessárias: mulheres trabalhadoras da mineração*. Curitiba: Ed. Apris.

Coll-Hurtado, Atlántida, María Teresa Sánchez-Salazar, and Josefina Morales. 2002. *La minería en México: geografía, historia, economía y medio ambiente*. México: UNAM: Instituto de Geografía.

Contreras, Carlos, and Guillermo Mira. 1993. Transferencia de tecnología minera de Europa a los Andes. In *Mundialización de la ciencia y cultura nacional*, ed. Antonio Lafuente, Alberto Elena, and Ma. Luísa Ortega, 235–249. Madrid: Doce Calles.

da Costa Pereira, José Veríssimo. 1955. A Geografia no Brasil. In *As ciências no Brasil*, ed. Fernando de Azevedo, 349–462. São Paulo: Melhoramentos.

da Silva, Clarete P. 2002. *O Desvendar do Grande Livro da Natureza. Um Estudo da Obra do Mineralogista José Vieira Couto, 1798–1805*. São Paulo: Annablume.

De Cserna, Zoltan. 1990. La evolución de la geología en México (1500–1929). *Revista mexicana de ciencias geológicas* 9 (1): 1–20. https://www.geologia.unam.mx:8080/igl/publs/simposio/DeCserna_historia.pdf.

De Gortari, Eli. 1963. *La ciencia en la História de Mexico*. México/Buenos Aires: Fundo de Cultura Económica.

de Oliveira, Avelino I., and Othon Henry Leonardos. 1940. *Geologia do Brasil. Com um mapa geológico do Brasil e parte dos países vizinhos*. Rio de Janeiro: Comissão Brasileira dos Centenário de Portugal.

De Terra, Helmut. 1956. *Humboldt, su vida y su época*. México: Biografías Gandesa.

Derby, Orville A. 1895. As investigações geológicas no Brasil. *Revista Brazileira* II (9): 1–25.

Durán, L.G. 1973. Reseña histórica de la geología en Colombia. In *Historia extensa de Colombia*, ed. Academia Colombiana de Historia, Vol. 20, 289–307. Bogotá: Editorial Lerner.

Escamilla-González, Francisco, and Lucero Morelos-Rodríguez. 2020. Bringing Werner's teachings to the New World: Andrés Manuel Del Río and the chair of mineralogy in the School of Mines of Mexico (1795–1805). *Earth Sciences History* 39 (2): 246–261.

Espinosa Baquero, Armando. 1987. El descubrimiento del platino y su metalurgía en la Nueva Granada en el siglo XVIII: Nuevos documentos de los archivos colombianos. *Quipu* 4 (3): 401–412.

———. 1997. Fuentes y Estudios sobre desastres históricos en Colombia. Retrospectiva y estado actual. In *Historia y Desastres en América Latina*, ed. Virginia Garcia Acosta, vol. II, 289–315. Lima: La RED; CIESAS; ITDG.

———. 2016. *El Servicio Geologico Colombiano. 1916–2016. Cien años de vida institucional del Servicio Geológico Colombiano*. Colección Centenario del Servicio Geológico colombiano. Bogotá: Servicio Geológico Colombiano, Vol. 1.

Farro, Máximo. 2009. *La formación del Museo de La Plata. Coleccionistas, comerciantes, estudiosos y naturalistas viajeros a fines del siglo XIX*. Rosario: Prohistoria Ediciones.

Fernández, José Antonio. 2005. *Mercado, empresarios y trabajo: la siderurgia en el Reino de Guatemala*. San Salvador: Biblioteca de Historia Salvadoreña.

Ferrari, Roberto. 2000. Germán Avé-Lallemant e sua obra científica e Tecnológica na Argentina (c. 1869–1910). In *Um olhar sobre o passado: história das ciências na América Latina*, ed. Silvia Fernanda de M. Figueirôa, 231–252. Campinas: Unicamp.

Figueirôa, Silvia Fernanda de M. 1992. Associativismo científico no Brasil: o Instituto Histórico e Geográfico Brasileiro como espaço institucional para as ciências naturais durante o século XIX. *Interciência* 17(3):141–146.

———. 1997. *A formação das ciências geológicas no Brasil: uma história social e institucional*. São Paulo: HUCITEC.

———. 2005. Ciência e tecnologia no Brasil Imperial: Guilherme Schüch, Barão de Capanema (1824–1908). *Varia História* 21:437–455.

———. 2007. Geological surveys in the tropics: The Brazilian experience. *Earth Sciences History* 26(1):151–171.

———. 2019. Algumas questões metodológicas relativas à história de instituições científicas. *Terra Brasilis (Nova Série)* 12:1–20. https://doi.org/10.4000/terrabrasilis.4164.

Figueirôa, Silvia Fernanda de M. and Maria Margaret Lopes, eds. 1994. *Geological Sciences in Latin America: Scientific relations and exchanges*. Papers presented at the XVIII INHIGEO Congress. Campinas: Instituto de Geociências/UNICAMP.

Figueirôa, Silvia Fernanda de M., Gregory Good, and Drielli Peyerl. 2019. *History, exploration & exploitation of oil and gas*. Cham: Springer.

Figueredo Figueredo, M. Ana Luisa, Aliana Corona Rodríguez, Matos Figueredo, and Maritza de la Caridad. 2017. Presencia de las geólogas en revistas científicas cubanas a partir del estudio de su producción científica. Bibliotecas: Revista de la Escuela de Bibliotecología, Documentación e Información 35, no 3 (noviembre 2017): 1. https://doi.org/10.15359/rb.35-3.4.

Franco, Rui R. A mineralogia e a petrologia no Brasil. História das ciências no Brasil Mario G. Ferri and Shozo Motoyama, 1–42. São Paulo: EPU: Edusp, 3, 1981.

García, Susana V. 2016. El mar en la ciencia argentina: las ciencias marinas a mitad del siglo XX. In *Contra viento y marea. Emergencia y desarrollo de campos científicos en la periferia: Argentina, segunda mitad del siglo XX*, ed. Pablo Kreimer, 61–86. Buenos Aires: CLACSO. http://biblioteca.clacso.edu.ar/clacso/gt/20161215020729/ContraVientoyMarea.pdf.

García, Alejandro, and Lucero Morelos. 2011. La Geografía y la Mineralogía en El Minero Mexicano (1873–1879). In *Geografía e historia natural, hacia una historia comparada: estudio a través de Argentina, México, Costa Rica y Paraguay, Buenos Aires*, ed. Celina Ana Lértora, 225–257. Buenos Aires: Ediciones FEPAI.

Gío Argáez, Raúl, and Hugo Rodríguez Arévalo. 2003. Panorama general de la paleontología mexicana. *Ciencia Ergo Sum* 10 (1): 85–95. https://www.redalyc.org/pdf/104/10410110.pdf.

Grinberg, D.M.K., and T.A. Palacios. 1992. Hornos prehispánicos peruanos empleados en la reducción de minerales de plata. *Quipu* 9 (2): 149–171.

Guereschi, Alina B., D. Martino Roberto, and Victor A. Ramos. 2021. *La Mujer en la Geología*. Buenos Aires: Asociacíon Geológica Argentina. Serie D. Publicación Especial no. 37. https://drive.google.com/file/d/1tIJP2ptzx8JGhwhLpoybtbkm74HLV699/view.

Herbst, Rafael, and Luisa M. Anzótegui. 2016. Las mujeres en la paleontología argentina. *Revista del Museo de La Plata* 1: 130–137. https://doi.org/10.24215/25456377e024.

Hervé, Francisco, and Reynaldo Charrier. 2016. Legado de Ignacio Domeyko (1802–1889) a la geología y a la institucionalidad científica de Chile. *Revista del Museo de La Plata* 1: 138–148. https://doi.org/10.24215/25456377e025.

La Podgorny, I. 2005. Tierra en el laboratorio: las ciencias de la Tierra en el siglo XX. In *Filosofía de las ciencias naturales, sociales y matemáticas*, 129–159. Madrid: Ed. Estany, CSIC. (Enciclopedia Iberoamericanas de Filosofia 28).

Lamego, Alberto. 1951. Derby, o sábio incompreendido. *Boletim Geográfico* 9 (103): 720–722.

Langescheidt, Adolphus. 1985. Bosquejo de la minería prehispánica de México. *Quipu* 2 (1): 37–57.

Leinz, Viktor. 1955. A geologia e a paleontologia no Brasil. In *As ciências no Brasil*, ed. Fernando de Azevedo, 243–263. São Paulo: Melhoramentos.

Leonardos, Othon Henry. 1955. A mineralogia e a petrografia no Brasil. In *As ciências no Brasil*, ed. Fernando de Azevedo, 265–313. São Paulo: Melhoramentos.

———. 1970. *Geociências no Brasil: a contribuição britânica*. Rio de Janeiro: Forum.

———. 1973. *Geociências no Brasil: a contribuição germânica*. Rio de Janeiro: Forum: Porto Alegre: Sulina.

Lopes, Maria Margaret. 2019. Petroleum: New energy perspectives for Brazil in 1922. In *History, exploration & exploitation of oil and gas*, ed. Silvia Fernanda de M. Figueirôa, Gregory A. Good, and Drielli Peyerl, 25–36. Cham: Springer. https://doi.org/10.1007/978-3-030-13880-6_3.

———. 2020. Cataloguing the fauna of deep time: Paleontological collections in Brazil in the beginning of the 20th century. *Colligo. Revue dédiée de naturalia & artificialia* 3. https://revue-colligo.fr/index.php/vol-3-num-3?id=42.

———. 2021. Culturas científicas sobre os oceanos na historiografia das ciências no Brasil. *Varia História* 37 (74).

Lopes, Maria Margaret, and Silvia Fernanda de M. Figueirôa, eds. 1990. *O Conhecimento Geológico na América Latina: Questões de História e Teoria. Trabalhos do 1o Colóquio*

Brasileiro de História e Teoria do Conhecimento Geológico. Campinas: Instituto de Geociências/UNICAMP.

———. 2020. The history of geology meets disasters: A Brazilian perspective. *Isis* 111: 104–111. https://www.journals.uchicago.edu/doi/10.1086/707820.

Lopes, Maria Margaret, and Irina Podgorny. 2000. The shaping of Latin American Museums of natural history, 1850–1990. *Osiris* 15 (1): 108–118. https://doi.org/10.1086/649321.

Lopes, Maria Margaret, Rebeca B. Feltrin, Maria C.F. Alencar, and Bruna Vasconcelos. 2014. Intersecções e interações: Gênero em Ciências e Tecnologias na América Latina. In *Perspectivas latinoamericanas en el estudio social de la ciencia, tecnología y conocimiento*, 233–243. México: Siglo 21.

Machado, Iran F., and Silvia Fernanda de M. Figueirôa. 2020. *História da Mineração Brasileira*. Curitiba: CRV.

Matharan, Gabriel, and Adriana Feld. 2016. La constitución del campo de la investigación petroquímica en la Argentina (1966–1983). In *Contra viento y marea. Emergencia y desarrollo de campos científicos en la periferia: Argentina, segunda mitad del siglo XX*, ed. Pablo Kreimer, 61–86. Buenos Aires: CLACSO. http://biblioteca.clacso.edu.ar/clacso/gt/20161215020729/ContraVientoyMarea.pdf.

Mendes, Josué C. 1981. A pesquisa paleontológica no Brasil. In *História das ciências no Brasil*, ed. Mario G. Ferri and Shozo Motoyama, vol. 3, 43–71. São Paulo: EPU: Edusp.

Mendoza Vargas, Héctor, ed. 1999. *Lecturas geográficas mexicanas, siglo XIX*. México: UNAM.

Mendoza Vargas, Héctor, and Carla Lois, eds. 2009. *Historias de la Cartografía de Iberoamérica: Nuevos caminos, viejos problemas*. México: UNAM.

Minaya, Estela. 2001. Pierre Marie Descotes S.J. y Ramón Cabré S.J.: vidas dedicadas a la ciencia de la Tierra. In *Contribuciones a la Historia de la Ciencia en Bolivia,* ed. Instituto Panamericano de Geografía e Historia y Academia Nacional de Ciencias de Bolivia, 205–207. México: IPGH.

Molina Martínez, Miguel. 1995. Ciencia y minería en las expediciones a América. In *De la Ciencia Ilustrada a la Ciencia Romántica*, ed. Alejandro Díez Torre, Tomáz Mallo, and Daniel Pacheco Fernández, 221–231. Madrid: Doce Calles.

Moncada Maya, Omar. 1995. Los ingenieros militares en California, siglo XVIII. In *De la Ciencia Ilustrada a la Ciencia Romántica*, ed. Alejandro Díez Torre, Tomáz Mallo, and Daniel Pacheco Fernández, 477–500. Madrid: Doce Calles.

———. 2003. *El nacimiento de una disciplina: la Geografía en México (Siglos XVI a XIX)*. México: UNAM: Instituto de Geografía.

Montes de Oca, Ismael. 2001. Historia de la ciencia geológica. In *Contribuciones a la Historia de la Ciencia en Bolivia,* ed. Instituto Panamericano de Geografía e Historia y Academia Nacional de Ciencias de Bolivia, 115–119. México: IPGH.

Mora-Stock, C. et al. 2018. Asociaciones de Mujeres en las Ciencias: desde una conversación de pasillo hacia la equidad de género, p. 183. https://congresogeologicochileno.cl/wp-content/uploads/2018/12/Libro-de-Actas-XVCongresoGeologicoChileno2018-2.pdf

Morelos Rodríguez, Lucero. 2012. *La geología mexicana en el siglo XIX: una revisión histórica de la obra de Antonio del Castillo, Santiago Ramírez y Mariano Bárcena*. Morelia: Secretaría de Cultura del Estado de Michoacán.

———. 2019. *Women in the earth sciences*, 137. Mexico City: National Congress on Geology. http://www.inhigeo.com/record/52.pdf.

Narváez, Patricia. 2018. Breve Historia del desarrollo profesional de Geólogas en Chile, con énfasis en las "pioneras". https://congresogeologicochileno.cl/wp-content/uploads/2018/12/Libro-de-Actas-XVCongresoGeologicoChileno2018-2.pdf

Papp, Desidério, José Babini, and Aldo Mieli. 1961. *Ciencia de la Tierra y Técnica del Siglo XIX*. Buenos Aires: Espasa Calpe.

Peyerl, Drielli. 2019. *The oil of Brazil*. Cham: Springer.

Peyerl, Drielli, and Silvia Fernanda de M. Figueirôa. 2020. Applied geophysics in Brazil and the development of a national oil industry (1930–1960). *History and Technology* 36 (1): 1–22. https://doi.org/10.1080/07341512.2020.1765618.

Podestá, Bruno, and Richard Olson. 1987. Predicción de un sismo, su trama y manejo. *Quipu* 4 (1): 33–52.
Podgorny, Irina. 2021. *Florentino Ameghino y Hermanos. Empresa Argentina de Paleontología Ilimitada*. Buenos Aires: Editorial Edhasa.
Poirier, Jean-Paul. 2018. Saints as protectors against earthquakes in popular culture in Italy and Latin America. *Earth Sciences History* 37 (1): 157–164.
Ramírez, Jesús Emilio. 1955. Los alemanes y las ciencias geológicas y geográficas en Colombia. *Boletín de la Sociedad Geográfica de Colombia* 13 (47–48): 213–218.
Restrepo Forero, Olga. 1984. La Comisión Corográfica: un acercamiento a la Nueva Granada. *Quipu* 1 (3): 349–368.
Riccardi, Alberto C. 2016. Petróleo–Investigaciones. In *Diccionario histórico de las ciencias de la Tierra en la Argentina*, ed. Irina Podgorny et al., 308–322. Rosario: Prohistoria Ediciones; La Plata: Museo de La Plata; Buenos Aires: CONICET.
———. 2020. Telasco García Castellanos: Promotor de la Geología y de las ciencias. *Revista De La Facultad De Ciencias Exactas, Físicas y Naturales* 7 (1): 95–114. https://revistas.unc.edu.ar/index.php/FCEFyN/article/view/28662.
Rodríguez-Prada, María Paola. 2013. *Le Musée National de Colombie, 1823–1830. Histoire d'une création*. Paris: L'Harmattan.
Sagredo Baeza, Rafael, and Francisco Hervé Allamand. 2011. Introducción. Un geólogo en terreno. Darwin en América del Sur. In *Charles Darwin, Observaciones geológicas en América del Sur*, ed. Rafael Sagredo, 13–50. Santiago: Editorial Universitaria.
Saldaña González, Juan José, ed. 1986. *El perfil de la ciencia en América*. México: Sociedad Latinoamericana de Historia de las Ciencias y de la Tecnología.
Sánchez Gómez, Julio, and Guillermo Mira. 1993. Minería Americana y minería europea, 1750–1820: una perspectiva comparada. In *Mundialización de la ciencia y cultura nacional*, ed. Antonio Lafuente, Alberto Elena, and Ma. Luísa Ortega, 105–110. Madrid: Doce Calles.
Sánchez-Flores, Ramón. 1988. Tecnología minera en México. Referencias a Zacatecas. *Quipu* 5 (1): 85–113.
Sanhueza, Carlos. 2014. El Gabinete de Historia Natural de Santiago de Chile (1823–1853). In *Museos al detalle: Colecciones, antigüedades e historia natural, 1790–1870*, ed. Miruna Achim and Irina Podgorny, 201–227. Rosario: Prohistoria Ediciones.
Serrano Bravo, Carlos. 1995. Intercambio tecnológico en la amalgamación entre los centros mineros de América con sus homólogos europeos. *Boletín de la Sociedad Geográfica y de Historia Potosí* 16: 63–82.
———. 2004. *Historia de la Minería Andina Boliviana (Siglo XVI–XX)*. https://www2.congreso.gob.pe/sicr/cendocbib/con4_uibd.nsf/6EF6AA797C1749E905257EFF005C493F/$FILE/Historia_de_Miner%C3%ADa_Andina_Boliviana.pdf
Soto, Gerardo J. 2011. Sobre la minería metálica en América Central entre los años 1500 y 1900. In *Geonaturalia. Geografía e Historia Natural: hacia una historia comparada. Estudio a través de Argentina, México, Costa Rica y Paraguay*, ed. Celina Lértora Mendonza, 43–50. Buenos Aires: Ediciones FEPAI.
Tellería-Geiger, José Luis and Reinaldo Salgueiro Pabón. 2001. Geofísica en Bolivia. In *Contribuciones a la Historia de la Ciencia en Bolivia*, ed. Instituto Panamericano de Geografía e Historia y Academia Nacional de Ciencias de Bolivia, 121–132. México: IPGH.
Trabulse, Elias. 1985. *La historia de la ciencia en México*. México: Fondo de Cultura Económica: Consejo Nacional de Ciencia y Tecnología.
Urbani Patat, Franco. 2018. La cordillera de la costa de Venezuela: evolución de su conocimiento desde el siglo XVI al XXI. https://www.researchgate.net/profile/Franco-Urbani/publication/326834888_EVOLUCION_DEL_CONOCIMIENTO_GEOLOGICO_DE_LA_CORDILLERA_DE_LA_COSTA_VENEZUELA_1/links/5b6784b6299bf1b9303ca1bf/EVOLUCION-DEL-CONOCIMIENTO-GEOLOGICO-DE-LA-CORDILLERA-DE-LA-COSTA-VENEZUELA-1.pdf

———. 2020. The Petroleum of the Lora River Basin: Perijá Range, 1917–1926. *Boletín de la Academia Nacional de la Historia Venezuela,* 524–602. https://www.researchgate.net/publication/347441621_The_Petroleum_of_the_Lora_River_Basin_Perija_Range_1917-1926_%2D%2D_EL_PETROLEO_DE_LA_CUENCA_DEL_RIO_LORA_SIERRA_DE_PERIJA_1917-1926

Uribe-Salas, José Alfredo. 2010. *Historia económica y social de la Compañía y Cooperativa Minera 'Las Dos Estrellas', en El Oro y Tlalpujahua (1898–1959)*. Morelia: Universidad Michoacana de San Nicolás de Hidalgo: CSIC-España.

Uribe-Salas, José Alfredo, I. Herrera-Canales, A. Parra-Campos, F.O. González-Escamilla, and L. Morelos Rodríguez. 2016. *Economía, sociedad y cultura en la historia de la minería latinoamericana*. Morelia: Instituto Nacional de Antropología e Historia, Universidad Michoacana de San Nicolás de Hidalgo.

Varela, Alex Gonçalves. 2006. *"Juro-lhe pela Honra de Bom Vassalo e Bom Português": Análise das Memórias Científicas de José Bonifácio de Andrada e Silva (1780–1819)*. São Paulo: Annablume.

Vessuri, Hebe. 1987. Introducción. El papel de las instituciones científicas en la sociedad. In *Las Instituciones Científicas en la historia de la ciencia en Venezuela*, ed. Hebe Vessuri, 7–11. Caracas: Fondo Editorial Acta Científica Venezolana.

Waligora, Mariana F. 2020. Researchers following the Glossopteris trail: social context of the debate surrounding the continental drift theory in Argentina in the early 20th century. *Colligo. Revue dédiée de naturalia & artificialia* 3 (3) https://revue-colligo.fr/index.php/vol-3-num-3?id=45.

Regional Geopolitics, French Geography, and Climatology in "Dry Spain," Eighteenth to Twenty-First Centuries: Birth of the "Mediterranean Climate"

Alberto Celis and Lino Camprubí

Contents

Abstract

While its limits and definitions are still widely discussed, the Mediterranean region is a firm unit of analysis both in the earth and environmental sciences and in history. The connections between the scientific and the historical approaches to the Mediterranean region are widely recognized in the literature, at least since the works of Fernand Braudel. French and German physical geography first defined a botanic and climatic region which then human geographers and historians interpreted as a space partly determining a shared human past. This chapter contributes to this historiography by highlighting continuities between early modern scientific and political interest in "dry Spain" and later climatological definitions of the "Mediterranean climate" as characterized by summer droughts—a definition which became global through Wladimir Köppen's extension of the Mediterranean climate to places like California, South Africa, and southern Australia. This conceptualization, however, contradicts current approaches to defining the Mediterranean climate by seasonal rains induced by the Mediterranean Sea.

A. Celis (✉) · L. Camprubí
Universidad de Sevilla/DEEPMED, Seville, Spain
e-mail: pozuelo@us.es; lcamprubi@us.es

E. Aronova et al. (eds.), *Handbook of the Historiography of the Earth and Environmental Sciences*, Historiographies of Science,
https://doi.org/10.1007/978-3-031-40799-4_26

Keywords

Mediterranean climate · Mediterranean history · French imperialism · Mediterranean region · Environmental history · French geography · Climate theory

1 Introduction

While the historical literature on the Earth and environmental sciences is rapidly growing, the focus remains on the development of global concepts—say "fossil record," "climate change," and "Anthropocene" (Heymann and Dahan 2019). Nonetheless, a few historians have insisted that local scales are an essential step in building global earth-system concepts (Hulme 2010; Coen 2016; Camprubí and Lehmann 2018). Historians of polar science have been particularly successful in locating the significance of regional histories of geopolitics and geoknowledge (Howkins and Roberts 2023). This chapter enquires about the history of a paradoxical regional climatic concept: the "Mediterranean climate." The paradox we refer to is dual. First, the "Mediterranean climate" refers to regions as far from the Mediterranean Sea as California and Australia. Second, until recent decades, the "Mediterranean climate" did barely consider the role of the "Mediterranean Sea" in shaping atmospheric dynamics. These are two sides of the same coin: the detachment of the "Mediterranean climate" from the Mediterranean Sea. In our day, when climatologists are questioning this detachment, the question of how the "Mediterranean climate" came to be as a scientific concept becomes particularly pressing. We want to contribute to its answer by paying attention to regional geopolitics and the notion of "dry Spain."

The link between geopolitics and geohistory in the Mediterranean is a well-established theme. In 1949, historian Fernand Braudel presented the first version of *The Mediterranean,* which in 1966 grew into the famous three volumes which composed the second edition. The environmental *longue durée* occupied most of the work, and it provided a sort of stable background capable of explaining social, economic, and political history. According to Braudel, writing the history of the Mediterranean region consisted first and foremost in identifying the two *longue-durée* environmental structures determining Mediterranean history: a common "Mediterranean climate" and surface sea currents enabling connectivity between Mediterranean peoples (Braudel 1949). This (contested) connection between the status of the Mediterranean as an object in both history and in the earth and environmental sciences has been widely recognized in the historical literature (Bourguet et al. 1998; Paris 2002; Wick 2016; Cazzato 2017). In the mid-nineteenth century, Élisée Reclus was perhaps among the first in the French tradition in bringing together physical and human geography when describing the Mediterranean region (Ruel 1991). Braudel himself acknowledged his debts to Italian, French, and German, physical geography and climatology (Paris 2002). Later works would complete and problematize this tradition. In his *The Mountains of the Mediterranean World*, environmental historian John McNeil recognized the

Mediterranean as a distinct ecological region, defined by the mountain chains that run parallel to the coast both north and south of the Sea (McNeill 1992: 12). Unlike his predecessors, McNeill's approach drew from plate tectonics to argue for geological unity at the base of the environmental resemblances of Mediterranean mountains. Focusing his attention on five of these mountains, McNeill argued that while their deep history shows human activities did not degrade the Mediterranean mountain environment, industrialization marked an irreversible ecological decline visible through deforestation, soil erosion, and depopulation. Later, the geographer and climatologist Alfred T. Grove paired with the environmental historian Oliver Rackham and objected to McNeill that "degradation" had been a common but false rhetorical trope since early discussions of the Mediterranean region (Grove and Rackham 2003). Diana K. Davis (2018) pointed toward more recent origins of the idea of dryness as degradation in French and German colonialism in North Africa.

In this chapter, we contribute to the historicization of the Mediterranean region as an object of geographic and historiographic analysis. We argue that the notion of a "Mediterranean climate" characterized by dryness partly stems from early modern northern European imperial interests in the region. We do so by looking at Braudel's far precedents in France: the seventeenth- and eighteenth-century advocates of French expansionism toward the south who bound together what they perceived as the Spanish landscape with the idleness of its peoples. From the early nineteenth century onward, French botanists and geographers working within French imperialism in North Africa developed two new and interconnected concepts: the Mediterranean region and the Mediterranean climate. In 1826, Napoleon's captain and botanist Jean Baptiste Bory de Saint-Vincent applied these concepts back to Spain, identifying much of this country as having a Mediterranean climate. Our thesis is that unveiling its early modern precedents sheds new light both on the history and the structure of the concept of the "Mediterranean climate," which is still debated today by climatologists largely unaware of its political origins. We argue that the nineteenth-century identification of the early modern idea of a "dry Spain" with the new concept of "Mediterranean climate" is one of the sources of the sustained (and largely contested) characterization of the Mediterranean climate with aridity.

Other factors have contributed to defining the Mediterranean region—among others, diet, anthropology, and art. However, the emergence of climatology in the early nineteenth century was a key development in the formation of this concept.[1] Within this discipline, the Mediterranean region became identified with a climatic zone characterized by summer droughts (Ruel 1991). There are surely sources other than early climatology for the idea of the Mediterranean climate as an arid. After all, Ancient philosopher Plato already lamented what he perceived as the decline of Greek soil (Grove and Rackham 2003: 288).[2] Likely, further histories of the concept

---

[1] In his isotherm map, Alexander von Humboldt was the first to divide the planet into different climatic regions according to its temperature, improving the classical, traditional division distinguishing between the torrid, temperate, and frigid zones (Klein 2018).

[2] We thank Paul Edwards for bringing this point to our attention.

of the Mediterranean climate should take into account developments in Greece and Italy, as well as the British and German scientific and political relationships with the Ottoman empire, the middle East, and North Africa. We think, however, that focusing on the Mediterranean increasing role in defining "dry Spain" illuminates some unknown precedents of the French geographic tradition, whose import in the historiography of the Mediterranean region is widely recognized.

There are important studies on the origins of the association of the Mediterranean region with a characteristic climate. According to these works, the "scientific invention of the Mediterranean" was the job of early nineteenth-century imperial French human geography and biogeography developed by Élisée Reclus and others (Ruel 1991; Deprest 2002). Paul Vidal de la Blache, widely considered the founder of French regional geography, drew on this tradition to establish the presence of a dry summer season as the defining characteristic of the Mediterranean region in his 1886 paper "Des rapports entre les populations et le climat sur les bords européens de la Méditerranée." Vidal de la Blache emphasized how the Mediterranean climate profoundly influenced the character and way of life of the people inhabiting these areas (Claval 1988). Fernand Braudel emphatically embraced Vidal de la Blache's regional take on the Mediterranean. In his 1949 *The Mediterranean and the Mediterranean World in the Age of Philip II*, Braudel pointed to "Drought: [as] the scourge of the Mediterranean" (Braudel 1975: 238).

Characterizing the Mediterranean climate as dry enabled Braudel to situate the regional environment as an active stage, partly shaping human history. Later environmental approaches to Mediterranean history have had Braudel's notion of a shared climate defining a common region (Greene 2017). And yet, as the later section of this chapter shows, there is nothing self-evident about the idea of the Mediterranean region as a dry one. New definitions of the "Mediterranean climate" point to autumn and winter rainfall as the specific characteristics of the Mediterranean region. The Mediterranean Sea now emerges as a main driver of the Mediterranean climate through cyclone genesis. (Re)situating the Sea at the center of the Mediterranean has opened new questions about the unity of the region, stressing subregional systems of equal atmospheric pressure and dynamics. The question we ask can thus be reframed: Why did definitions of the Mediterranean climate erase the Mediterranean Sea for so long? The notion of the Mediterranean region, we show, has a long political history linked to French and Spanish plans of transformation of the Spanish landscape through engineering works directed to tame a wild nature and culture. The next two sections turn to those plans and their associated notions of aridity.

2 Spanish and French "Climate theory": Between the Decline of the Habsburgs and the Rise of the Bourbons

The personification of a dry Spain has long been a prominent literary motif. Take, for instance, Cervantes' Don Quixote, nicknamed "de La Mancha" after the largest arid plain in the Iberian Peninsula. French travelers who arrived in the Iberian Peninsula in the seventeenth century took up this trope, often comparing the dryness of the southern landscape to the green they were used to at home. In both cases, the insistence on aridity reflected the concern for the decline of the Spanish Empire. This section tracks the origins of this idea through Spanish and French sources, emphasizing its changing social and political interpretations, including "climate theory"—which held that climate determines the character of a population. While often used interchangeably by the actors, in what follows we try to reserve "aridity" for references to bioclimatic conditions and grant "dryness" a broader reach, including the effects of human inaction (lack of public works) and action (deforestation).

In 1599, Jesuit Scholastic and historian Juan de Mariana discussed the problem of drought and how to counter it. He advocated policies of reforestation and irrigation of fields as measures for economic improvement and public health, "so as to reduce the air's dryness." Modifying the climate might be possible with the arrival of more precipitation thanks to the evaporation of water from the canals (Mariana [1599] 1845: 347). Drought and hydraulic works were becoming common themes among Spanish thinkers. In 1627, for instance, economic thinker Caxa de Leruela voiced a complaint by Castilian farmers about the lack of irrigation projects, pointing out that, meanwhile, the Spanish Crown spent large sums of money on the construction of artificial lakes in Flanders (Caxa de Leruela 1627: 18–19).

The debate focused on who should undertake these hydraulic works. In the sixteenth century, Spanish economic thought favored private initiative over state intervention in agriculture (Fernández Álvarez 2014). Along the seventeenth century, however, demand for the king's intervention increased (Fernández Delgado 2003). The call for larger state intervention in the economy characterized the so-called "arbitristas." This group of thinkers plead for reforms that would strengthen the king's power in the economy, particularly by promoting field irrigation and reforestation (Hermann 1989). Only in this way, they argued, could the crown prevent the depopulation of territories, which they associated to decline and lack of revenue. They peppered their writing with words like aridity, sterility, and scarcity (Lorenzo 2014: 83). In a time of imperial decline, arbitristas advanced internal colonization as a response to the metropole's lack of population.

Those political ideas coincided with a growing interest in climate change among early modern European thinkers and governments (Parker 2013). The need to alter climatic conditions to boost agricultural production and improve the quality of life for the population was a defining feature of rulers' actions in the Early Modern age. This approach is what some authors have termed "climate theory": the argument that the climate could determine both the wealth and the character of nations (Morgan and Miglietti 2017). A late exponent was Charles de Montesquieu. In *The Spirit of the Laws* (1748), he explored how climate influenced the character and customs of people in different regions, including Spain. According to this French political philosopher, the intense heat in southern European countries, like Spain, affected the temperament and behavior of its inhabitants. Montesquieu believed that heat made people more passive, less industrious, and more prone to laziness or leisure. He considered that warm climates fostered an inclination toward a contemplative or reflective life, in contrast to the colder climates of northern Europe, which, in his view, encouraged hard work, constant activity, and greater physical strength (Montesquieu 1748: 300).

Earlier authors were often less fatalistic: Climates do determine national characters, but humans can affect climates. While these ideas commonly referred to heat, they often extended to dryness, its human-made causes, and how to counter them. In France, the idea that the government could improve the climate to improve the nation took hold through several important public works. Raphaël Morera referred to the drainage and agricultural transformation of wetlands as "governing with the environment," creating "microclimates" conducive to the growth of the French population. After the ascension to the throne of Henry IV of France in 1589, transforming a "desertic and uninhabited country " became a fundamental objective to the crown (Morera 2017). The house of Bourbon also framed drainage of wetlands as a legitimizing exercise of good governance. Soon, French travelers began to apply these ideas to Spain, often drawing on the abovementioned Spanish sources and often on their own experience and political interests. In 1655, as Antoine Brunel approached the border after a long travel around to Spain, he compared both countries:

> Without exaggeration or hyperbole, I concluded that, in one place, Judgment Day would have little to destroy, while in the other, the flames that will mark the end of the world would be the last to extinguish. It seemed as though fire had already fallen from the sky in the former, leaving the land half-baked, while in the latter, only a pleasant and life-giving warmth was received. (Brunel 1666: 320–321)

A few days earlier in Pamplona, located in northern Spain near the French border, he had already distinguished the Navarrese from other Spaniards based on geographical characteristics:

> We observed that, as we gradually moved away from the dryness of Castile and the austerity of Aragon, we arrived in a more sociable country, with greater communication among its inhabitants, as well as with those who were merely passing through. (313)

He and other French travelers compared the successes achieved in France under the Bourbon dynasty in producing Morera's "microclimates" to what they perceived as Spanish inaction and even active destruction through deforestation. The political motivation behind some of these negative views was the war between the kings of France and Spain. Brunel's more positive views on Navarre, for instance, were perhaps linked to the fact that Henry IV, the first Bourbon to reign in France (1589–1610), had also been king of Navarre.

The discourse of climate theory and aridity could nonetheless be adapted to new political situations. In 1669, some decades after his compatriot Brunel, François Bertaut also described a depopulated and barren country: Spaniards were to blame for the depopulated dryness of the Castilian land, because with their laziness and pride they refused to cultivate it. "The disordered behaviour of its men and women starts from childhood," Bertaut argued in the tone of climate theorists, "resulting in less kids being born and those born not being as vigorous as in colder and more moderate countries" (Bertaut 1669: 274). He left, however, ample room for hope regarding Spain's economic prospects. The two countries had now signed the Peace of the Pyrenees (1659), and Bertaut traveled to Spain together with King Louis XIV's ambassador to reinforce that alliance. He thus had grounds to be more optimistic than Brunel had been. But canalization and reforestation could recover an allegedly lost wealth. Discussing the example of the Islamic hydraulic works in Granada, "where the water flows through small canals like those that irrigate the lands in France over the meadows," he proposed "that if the Spaniards were willing to do the same everywhere, this would be the most fertile country in Europe" (276). Heavy taxes imposed by a corrupt dynasty, Bertaut concluded making his political case, were the only obstacle to these reforming prospects.

The idea of a dry Spain was thus not only linked to the aridity of its climate but also to the perceived laziness of its inhabitants (Schaub 2003: 211). Some authors argued that the heat determined this behavior, as supported by "climate theory," but some others emphasized that indolence and bad government hindered the reforms that were necessary to address the dryness caused by the climate and the resulting depopulation. In both cases, aridity became the hallmark of the Spanish and a powerful image to explain the loss of imperial hegemony.

These ideas played a pivotal role in the broader diplomatic strategy of the French Bourbons in Spain. As Charles II of Spain had no heir, the French Bourbons devised plans to replace the Habsburg dynasty as rulers of Spain (Schaub 2003: 337). Louis XIV's diplomats, travelers, and ambassadors promoted a propagandistic view of the overall poor state of the Spanish monarchy compared to a France prospering thanks to political and technical reforms. The connection between decline and dryness came to the front of the political discourse. Cardinal Pierre de Bonzi, Louis XIV's ambassador to Spain, wrote to his king in 1679 that the Spanish would rather drown themselves than build canals or divert water (Alvarez López 2008: 273). This, of course, contrasted with important works by Spanish engineers and technicians both in Spain and in the Americas (Lucena Giraldo and Fernández-Armesto 2022). But the narrative of idleness and the need for transformative projects became the usual legitimization for the French king's political ambitions. A group of courtiers consolidated in Madrid who associated good governance with the reforms implemented in France (Alvarez López 2008: 114).

When the Bourbon dynasty finally seized power in 1700 with Philipp V, they moved to implement some of the promised reforms. Philip V found in his grandfather Louis XIV of France not only a model of good governance but also a roadmap for executing his policies. The king Louis XIV sent Jean Orry to Spain to assist his grandson in building an absolutist crown there (Dubet 2008). Increasing tax revenue was a first step. Orry noted that depopulation was an essential obstacle for this. The king received absolutely nothing from "waters and forests," he complained, while the extension of "uninhabitable Spain" kept increasing.[3] Orry's model soon became the 1669 *Ordonnance sur les Eaux et Forêts*, promoted by Louis XIV's minister

[3] Archives diplomatiques du ministère des affaires étrangères, *Moyens de remédier à l'état présent des affaires du Roy d'Espagne*... CPE 119, p. 132–213

Colbert. Mimicking its emphasis on state control over natural resources, Orry advocated for the role of the intendant as a territorial representative of the king to implement reforms.[4] Finally, in 1718, an ordinance was approved for the creation of intendants, whose responsibilities included river canalization, the use of underground waters, and reforestation.[5] However, the Bourbons war of succession against the Habsburgs left little room for these hydraulic works to materialize.

Philipp's son, Ferdinand VI, became king in 1746 and, short after, initiated an interventionist policy aimed to end aridity and depopulation. His most powerful minister led the reform: The Marquis of Ensenada. Ensenada had managed to gain the king's trust through a series of letters known as "representations" (Delgado Barrado 2002). The most famous is the one written in 1751, where he associated "bad government" with dry Spain and "good government" with what was accomplished by Louis XIV:

> There is no land in Europe drier than Spain (...) its rivers have not been made navigable to the extent possible, nor are there canals for irrigation and transportation (...) if in his territory the great Louis XIV ordered rules and regulations that were followed and continued, why can't they adopt and practice the same in Spain while you are the king? (Rodríguez Villa 1878: 47–48)

Ensenada's government was the first to design an ambitious hydraulic plan aimed at transforming the territory, anticipating the hydraulic policy followed during the second half of the eighteenth century (Gil Olcina 1992: 174): The Castile Canal or the Imperial Canal of Aragon, as well as the drainage of lagoons and meadows to create new arable land, received a definitive boost at that time with the desire to turn dry Spain into the stronghold of the Spanish empire.

3 Equating "dry Spain" to the "Mediterranean region" from French Imperialism to Spanish Regenerationism

Napoleon's botanists and geographers introduced the scientific concept of a Mediterranean region as a natural unit of analysis. They did so in the context of imperial travels to Egypt and attempts at settling colonies in North Africa. While this story is somewhat known, the role of these same geographers in Spain has been largely overlooked. In this section, we argue that, in the nineteenth century, French geographers reinterpreted the old concept of "dry Spain" through the newly coined "Mediterranean climate." They did so, moreover, in a context of political and physical transformation of the Spanish landscape which would then continue in the Spanish tradition of engineering, regenerationism, and climate studies.

After the defeat of Trafalgar in 1805, Napoleonic France focused on continental domination. One of Napoleon's dragons, Captain of the fifth Cavalry Regiment Jean

[4]Ibidem, p. 213

[5]Biblioteca Digital Hispánica. *Ordenanza de 4 de Julio de 1718 para el establecimiento é instruccion de Intendentes, y para Tesorero General, Pagadores y Contadores de los Ejercitos y Provincias,* p. 47–48. http://bdh-rd.bne.es/viewer.vm?id=0000219669&page=1

Baptiste Bory de Saint-Vincent, played a prominent role in the identification of dry Spain with the Mediterranean region. Bory had received training as a naturalist and geographer in his youth from the disciple of Count Buffon, Bernard Germain Lacépède, and went on to become a prominent botanist and to participate in some of the French scientific expeditions in overseas provinces (Álvarez and Quirós Linares 2004). There, he was among the proponents of the new concept of a Mediterranean region unified by unique vegetation characteristics (Davis 2018). Moreover, he identified this Mediterranean region with the "Mediterranean climate," characterized by heat (Bory de Saint-Vincent 1826: 46). Bory, however, did not only blame the natural climate for an arid soil and vegetation. He and others adopted a discourse of decay caused by human activity, which became a legitimating common place following the French conquest of Algeria in 1830 (Davis 2018: 131).

We want to argue that Bory de Saint-Vincent's Mediterranean Spain served as a significant precursor to a narrative which he helped create as a member of French scientific expeditions in Algeria. In 1808, Bory entered Spain with the Grande Armée, where he carried out various tasks of geographical reconnaissance, accumulating important information. This work formed the basis of his later works, including "Résumé géographique de la Péninsule Ibérique" (1826) where he first divided the Iberian Peninsula, including a cartographic representation (577), into two large regions differentiated by their "natural climate": the European region, also known as the Oceanic or temperate region, and the Mediterranean region, referred to as the African or hot region (74–76). The idea of a "dry Spain" reemerged in his writings as a powerful rhetorical device to support this classification. This idea retained all the previous political connotations, including the idealization of a remote fertile past which would have been degraded by idle rulers and people:

> Thus, this once fertile land, darkened by the lack of trees and devoured by the intensity of a sun whose heat no foliage tempers, has gradually seen its springs dry up, its lagoons disappear, and its streams run dry. At every step, one encounters dry basins and valleys that must have once been irrigated but where, with every element of freshness now gone, the soil can produce only a sombre dust that the wind scatters in suffocating whirlwinds through a scorching atmosphere. (70)

Echoing commonplaces of seventeenth-century French travelers, Bory linked the aridity of Spain to backwardness, desolation, and the ignorance of taciturn and proud Spaniards. Only the wettest Spain, full of joy and intelligence, escaped this curse (67–70). Bory de Saint-Vincent would return, as in the seventeenth century, to the ruined Arab hydraulic past to point out the lack of hydraulic infrastructure. A good example was the Guadalquivir River, navigable up to Cordoba during Muslim rule and cut from it from the reign of Emperor Charles V onward. According to Bory, the French marshal Soult achieved to reopen this connection "in less than four months" during the Napoleonic Wars (90).

Nevertheless, there were significant differences between seventeenth-century French travelers and Bory de Saint-Vincent. Among the most relevant was new meteorological and biogeographical knowledge. As late as the mid-eighteenth century, Spanish scientists had developed more meteorological knowledge for certain parts of the American empire than for the metropole. Coming back from the French Spanish geodesic expedition to the Equator, naval officer and scientists Jorge Juan convinced the government of the need to modify the Armada through instruments and observations. At least from 1786, the Spanish government began taking more systematic measurements of winds and rain in the peninsula (Safier 2012). Around the same period, a new decree asked city majors to issue meteorological data to the government on a biweekly basis, but it is not clear whether they complied nor with which instrumentation did they count to do so. In the decade of the 1850s, Spanish physician and physicist Manuel Rico Sinobas published the most complete review of Spanish meteorological data to this date, assembling almost everything available from the late eighteenth century onward (Anduaga Egaña 2012: 38). He was particularly worried about the extreme droughts often affecting the Spanish Southeast, for which he suggested that a better understanding of the influences of the Mediterranean Sea could lead to better forecasts and irrigation schemes (Rico Sinobas 1851).

Bory de Saint-Vincent was less interested in meteorological data and on the marine influences upon it. Perhaps because of lack of rainfall and other relevant data in Spain and elsewhere, he relied on biogeographical features as the primary indicator for defining the Mediterranean region, especially agricultural production, vegetation, and animal species. Carlos Cañete (2012: 6) argues that Bory's notion of the Mediterranean region reflected imperial ambitions rather than a proper scientific discourse arising from expeditions. While we would nuance this interpretation to combine it with the scientific interest of French botanical explorations, we would also add that Bory's prior approach to the Iberian Peninsula fitted the colonial interventionist approach. Regarding dry Spain as part of a larger Mediterranean region enabled the French to use a similar legitimizing discourse in the military campaigns of Spain and Egypt: restoring civilization. The Mediterranean region stemmed from a negative perspective that viewed Iberian Peninsula and North African communities as marginal: "the general physiognomy of the Iberian Peninsula hardly allows it to be considered part of Europe" (Bory de Saint-Vincent 1826: 116).

Later in the century, the connection between aridity and colonial ambitions in the Mediterranean would be reinstated by proponents of scientifically refined versions of "climate theory." Within the new discipline of human geography, French and German geographers adapted old notions of the climate's influence upon human behavior. Friedrich Ratzel, the German fore bringer of anthropogeography, had been an ardent proponent of climatic determinism in anthropology and geopolitics. The French geographer Paul Vidal de la Blache reflected Ratzel's ideas and defended that "the exciting influence of the climate is reflected in the body's general appearance" (Vidal de La Blache 1886: 409). As before, this meant that transforming the climate could lead to bettering the people. Projects of climate engineering became bolder, as historian Philipp Lehmann (2022) has shown, including reforesting the Sahara Desert and desiccating the Mediterranean Sea.

Paul Vidal de la Blache is usually credited with founding French human geography. Drawing from the botanic region defined in the first half of the nineteenth century, Vidal de la Blache was also one of the main theorizers of the Mediterranean climatic and human region. One of his earliest works, "Des rapports entre les populations et le climat sur les bords européens de la Méditerranée" (1886), redefined the Mediterranean region based on climatic characteristics and new meteorological data. During the nineteenth century, climatology became an empirical practice grounded in the collection and correlation of meteorological data, forming the foundation upon which climate regionalization was built (Heymann and Achermann 2018: 606). Its development relied heavily on the expansion of meteorological observations, which enabled the collection of data on precipitation, temperature, and atmospheric pressure. This information enabled new global climate classifications and maps. Alexander von Humboldt was among the firsts to combine these new data with mathematics as the basis of his work, following the physicist Pierre-Simon de Laplace (Olcina Cantos 2020: 4). Humboldt aimed to enhance traditional thematic classifications by applying isolines to climate studies in his isotherm maps, which connected points of equal temperature to identify distinct climatic regions. His map became the first of a series of climatic classifications, from Julius von Hann to Wladimir Köppen (Heymann and Achermann 2018: 606).

Köppen admitted that he took inspiration for his classification from a map created by Alfred Grisebach (1866) depicting the world's vegetation zones, which included a Mediterranean flora zone (46). Observing the symmetries among different regions of the planet—such as the Mediterranean, California, Chile, Australia, and South Africa—Köppen sought to unify these areas using a climatic formula. He identified temperate climate zones with the letter C, distinguishing those with dry summers as Cs (from the German *sommer*) (Köppen 1936). Although Köppen never used the term "Mediterranean climate," the fact that he based his work on maps like Grisebach's, linking summer dry seasons with Mediterranean vegetation, led to Cs climates to be identified as "Mediterranean" in various publications, including regions far from the Mediterranean Sea, such as Chile or California (Joffre and Rambal 2001). The global "Mediterranean climate" was thus born.

Vidal de la Blache explicitly drew from these pioneering climatological innovations to define the Mediterranean climate as characterized by a dry summer season and greater abruptness in contrast to the mildness of the oceanic climate. His geography relied heavily on meteorological observations. But it also followed paths previously explored by "climate theory." "This dry climate," he wrote, "tightens the skin tissues, hardens the flesh, and accelerates blood circulation. The blood, poorer in water, acts as an energetic stimulant on the nervous system, exciting its functions. The vivacity of the gaze, the promptness of gestures, the agility of movements, and the sonority of articulated sounds are effects that derive directly from the climate. These traits are part of the common description of Mediterranean coastal populations" (Vidal de La Blache 1886: 408). Vidal de la Blache even revisited the idea of indolence brought about by the climate in Mediterranean populations, which, according to him, had led to a state of decay in certain areas where the land was no longer cultivated (419).

This early work would be continued by other main proponents of French human geography. An important landmark is the work of Vidal's disciple Jean Brunhes. He dedicated his 1902 doctoral thesis to aridity in Iberia and North Africa and how to counter it: "L'irrigation, ses conditions géographiques, ses modes et son organisation dans la Péninsule Ibérique et dans l'Afrique du Nord." There, he distinguished human geography from physical geography in that, while the latter studied how natural phenomena affected populations, human geography focused on the changes brought about by human activity to overcome these constraints (Brunhes 1904: 1–2). His own work on irrigation in arid regions worked as an example. Brunhes chose to focus on the Mediterranean region, which he considered a geographic unit of analysis defined by the Mediterranean climate. This climate, "opposed to that of Western and Central Europe, is mainly characterized by a certain dryness and great serenity. These defining characteristics of the Mediterranean climate are reflected in shrub vegetation forms found around the Mediterranean Sea, giving this region a true climatic identity" (10). As a new insight, Brunhes associated the sea's influence with a moderating effect on temperatures and higher humidity compared to the desert areas further south.

In his thesis, Brunhes dedicated considerable attention to the physical geography of the Iberian Peninsula and then explored the arid and irrigated areas of Spain. He distinguished a dry, Mediterranean Iberia from a humid, Atlantic Iberia (24–33). Although this division had already been suggested by Bory de Saint Vincent, Brunhes now had access to far more meteorological observations, mainly thanks to the data collected by Spain's meteorological station network established in 1850 (Anduaga Egaña 2012: 69). From this data, Brunhes was able to produce an isohyet map that established a more scientific division between dry and humid Iberia. The approximate dividing line was the 600 mm annual isohyet, which extended Spain's dry zone to most of the country (29–33) (Fig. 1a).

This climatic division was further elaborated by other members of the French geography school. Vidal de la Blache's son-in-law, Emmanuel de Martonne, refined the definition of the Mediterranean climate in his renowned *Traité de Géographie Physique*, emphasizing its characteristic dry summer ([1909] 1913: 215). Later, he introduced his "aridity index," based on a detailed analysis of temperature and precipitation data (Martonne 1926) (Fig. 1b). In 1934, Max Sorre (Martonne 1942) utilized this index to map the Mediterranean region in Volume VII of Vidal de la Blache and Lucien Gallois' *Geographie Universelle*. These advancements in climatology and the adoption of mathematical methods solidified the concept of a Mediterranean region defined by summer drought and aridity.

At the same time, the relationship between geography and history gained increasing attention within the *Annales* school. Lucien Febvre's *La Terre et l'évolution humaine* (1922) played a crucial role in shaping this perspective by emphasizing the interconnections between physical environments and human societies. This book significantly influenced Fernand Braudel and later historians by highlighting the long-term relationship between geography and history, particularly in the Mediterranean region. Braudel's seminal work on the Mediterranean would later expand on this idea, demonstrating how geographical constraints—including the characteristic summer drought of the Mediterranean climate—shaped human activities, economies, and political structures over centuries.

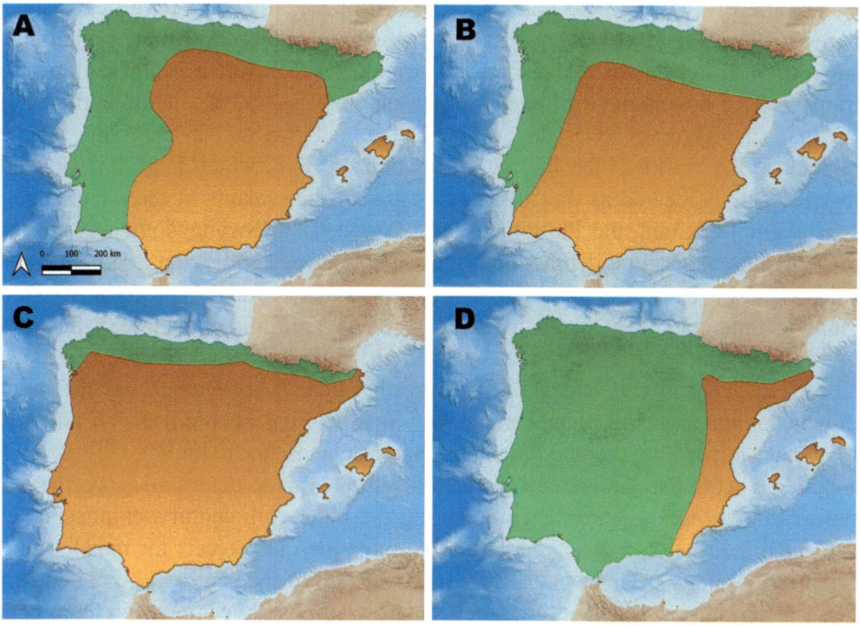

Fig. 1 The extent of the influence of the Mediterranean climate on the Iberian Peninsula (in orange), according to several geographers. Based on our review of the literature spanning from 1902 to 2021, we identify four maps. We utilize meteorological data from the period 2000–2021: (**a**) Brunhes (1902); (**b**) De Martonne (1926); (**c**) Rivas Martínez (1983); and (**d**) Martín-Vide and López Bustins (2006). (From: Celis and Villarín 2024)

Although Brunhes noted in his thesis on several occasions that the binary division between a dry and a humid Iberia might be overly simplistic (Brunhes 1904: 34), he used it to propose ways to overcome the challenges posed by aridity, especially in areas where steppe landscapes dominated. Irrigation could create true oases, which Brunhes called "centres of culture," the true object of study in his human geography (55). From the study of irrigation in the Spanish steppes, he extended his analysis to North Africa, specifically Algeria, Tunisia, and Egypt. As such, "dry Spain" paved the way for overcoming the aridity imposed by the Mediterranean climate on North Africa.

The Brunhes' thesis was both an academic work in human geography and a contribution to the endeavors of French colonialism. Brunhes extrapolated freely from the Spanish example to French Algeria (Deprest 2017). In particular, his definition of *steppes* as a specific biome of the Mediterranean region served to intertwine irrigation with good governance. Steppes traditionally referred to northern European latitudes. In German studies of Russian steppes between Europe and Asia, their association to sterile lands suggested the need for improvement (Coen 2011). In 1852, German botanist Heinrich Moritz Willkomm published a book introducing that biome to characterize the interior plains of Spain (Willkomm 1852). Brunhes then exported it further to French colonial territories. Steppes thus came to demarcate Europe from Africa through the Mediterranean region and to justify the colonial enterprise.

The life of this imported notion in Spanish political life is telling of the links between climate science and political discourse and action. Following Willkomm's suggestion, a debate on the nature and origin of the Spanish steppes followed, triggering the beginnings of Spanish phytosociology as well as the modernization of Spanish geography (Casado de Otaola 2010: 117–133). The notion became popular in the works of the literary Generation of 1898, which largely responded to what they perceived as the "disastrous" loss of the remains of the Atlantic and Pacific empire: Cuba, the Philippines, and Puerto Rico (Swyngedouw 2007). For these writers, the steppe defined both the landscapes of Castile and its inhabitants. José Martínez Ruiz *Azorín,* for instance, in his book titled "The route of Don Quixote" attempted to explain the behavior of the *errant knight* through his geographical setting (Azorín 1905).

Linked to, but partly independent from, this literary movement was the more directly political and technical thought of Spanish "regenerationism," a political movement advocating the retooling of the Spanish landscape to improve the nation's social and economic situation. The rhetoric of geographic determinism permeated the writings of the regeneracionists. During the nineteenth century, engineers had cemented their new political power by reformulating old projects for transforming Spain's hydrology (Martykánová 2023). By the end of the century, the notion had permeated reformist scientists and intellectuals. Geologist Lucas Mallada represents the new link between science and politics well. He blamed the dry climate for the barrenness of large areas of the territory. Moreover, he argued that the arid landscape resulted from centuries of uninformed human action (Mallada 1890: 22–24). The excesses of the Spanish empire became a usual suspect, for instance, through the (exaggerated) blaming of Philipp II's *Armada* for the decimation of Castilian forests (Wing 2012).

The regenerationists promised to end aridity as a catalyst to restore the nation's lost prestige. Brunhes' thesis contributed to creating a framework for Spanish government policies. Spanish geographers would develop and apply Brunhes' ideas. Juan Dantín Cereceda, usually considered to be the modernizer of Spanish geography, firmly adhered to the Vidal de La Blache school (Ortega Cantero 1997). Dantín confirmed in 1922 Brunhes' division between wet Spain and dry Spain, the latter under the influence of the Mediterranean climate (Dantín Cereceda 1922: 17). The regionalization proposed by Brunhes and appropriated by Spanish geographers provided a scientific alibi for Spanish reformist plans. The first major National Hydraulic Works Plan, projected during the government of the Second Republic in 1933 and which would influence hydraulic works ever since, relied on Brunhes' studies (Lorenzo Pardo [1933] 1993: 173). This plan included transferring water from the Tagus River to the Spanish Levant, Spain's most significant water transfer project to date. This transfer, conceived by the regenerationists and projected by the Second Republic (1931–1936), was partially executed in the later part of the Francoist regime (1939–1975). It was presented as key to the growth of what has been pompously called "the orchard of Europe," hundreds of thousands of hectares of irrigation in Alicante, Murcia, and Almería, the driest areas of the Spanish Mediterranean (Bartolomé Rodríguez et al. 2024). The old dream of internal colonization through water works regained momentum in the era of big dams. Exporting water from humid Spain to dry Spain seemed the obvious way forward.

"Spain tormented us for its dryness," General Francisco Franco proclaimed in 1952, "with its misery, with the needs of our villages and towns. We have now redeemed the pain with these great national hydraulic works (...) beautifying this once desolate landscape while creating a liquid gold that is the foundation of our independence." This discourse by dictator Franco inaugurating a dam on the Ebro, the most abundant river in the Iberian Peninsula, shows what it meant for different Spanish governments to erase the aridity of Spain from the map: independence (Camprubí 2014: 1). As we have shown, this nationalist narrative of redemption drew from a scientific and political tradition linking the old trope of dry Spain to newer notions of a degrading Mediterranean climate.[6] This identification, in turn, led geographers to associate dryness to the Mediterranean climate and region, an assumption that would be contested in the second half of the twentieth century.

4 The "Mediterranean climate" Has Changed: Putting the Sea Back in the Mediterranean

We have shown that Max Sorre, in 1934, gave mathematical form to the old idea of a dry Mediterranean climate. Nevertheless, definitions of the Mediterranean climate underwent important modifications shortly after Sorre's publication. Improvements in modern meteorological forecasts and the development of aviation facilitated important changes in atmospheric sciences and climatology (Fleming 2016). The United States of America played a significant role after World War II. Funding for the US Army's weather service contributed to developing models of atmospheric general circulation (Byers 1958). Advances in meteorology, but also in oceanography, have prompted a reexamination of the traditional boundaries of the Mediterranean region. The introduction of maritime masses as a fundamental factor in atmospheric configuration led to revising prior understandings of the Mediterranean Sea. The reference point to define the Mediterranean climate would no longer be the lack of summer rains caused by atmospheric stability—as in climate theory and French and Spanish geography—but rather the role of the sea in intense autumn and winter precipitations.

In the second half of the twentieth century, the Mediterranean Sea emerged as a key factor in defining the Mediterranean climate, particularly through the atmospheric disturbances caused by the evaporation of water masses: the cyclones. In the 1950s, Sverre Pettersson, known for his accurate meteorological predictions that facilitated the Allied landing in Normandy, depicted the Mediterranean Sea as a significant global cyclone genesis area (Lionello et al. 2006: 328). The key was to distinguish cyclones formed in the sea from others with a different origin. A new

[6] Interestingly, the idea of "Spain, more African than Africa" also played a role in international projects of nature conservation during Francoism (Camprubí 2020)

breakthrough came from Yugoslavia, the other side of the Iron Curtain. In the 1960s, Djuro Radinović (1965) identified subregional areas where cyclone genesis occurred, such as the Gulf of Genoa and the Aegean Sea. These cyclones resulted from differences in sea and atmospheric temperatures: The Mediterranean Sea acted as a source of latent heat that favored cyclone formation.

The existence of an inland sea that favored the formation of local cyclones gradually came to define the "Mediterranean climate" in different disciplines. French geographers, for instance, began to use atmospheric pressure and weather maps as reference points in the 1950s (Billaut et al. 1956). This, in turn, led to a greater emphasis on regional differences within the Mediterranean region. The uniform character provided by the existence of a dry season in summer gave way to high complexity of the different atmospheric patterns converging in the Mediterranean basin (Lionello et al. 2006).

To be sure, older approaches to the Mediterranean region coexisted with the new climatic one. French botanist and phytogeographer Louis Emberger produced in 1962 the first Bioclimatic Map of the Mediterranean Zone as part of UNESCO's Arid Zone Program (Davis 2018: 139). Inspired by Emberger's work, in 1983 the Spanish geobotanist Salvador Rivas Martínez defined "Mediterranean Spain" through a climatic index based on the presence or absence of summer drought (Rivas-Martínez 1983). He concluded that practically the entire territory pertained to the Mediterranean region, including inland areas and coastal areas bathed by the Atlantic Ocean (Fig. 1c). Today, Rivas Martínez's index remains the main reference in Spanish geography for delineating the country's two main natural regions: the Mediterranean and the Eurosiberian.[7]

In the late twentieth century, however, different groups of meteorologists focused their efforts on linking rainfall on the Mediterranean coast with these cyclones. Subregional differences gained increasing relevance as the dynamics of the Mediterranean Oscillation (MO) became the focus of study (Martin-Vide and Lopez-Bustins 2006: 1456). The MO refers to differences in surface atmospheric pressure configurations over long periods: An anticyclone in the western Mediterranean and Iberia opposes the lower pressures of the eastern Mediterranean. Scientists further emphasized subregional differences by studying more local pressure gradients. Understanding these regional and subregional dynamics provided a better understanding of rainfall patterns in the Mediterranean basin. For instance, the climatology group at the University of Barcelona created the Western Mediterranean Oscillation index (WEMOi) (Martin-Vide and Lopez-Bustins 2006). The index showed that the complex orography surrounding the Mediterranean isolates it from northern hemisphere atmospheric dynamics, giving it unique climatic characteristics due to local factors. These unique characteristics would explain rainfall and drought patterns. For Spain, between 1910 and 2000, winter precipitation on the eastern slope of the Iberian Peninsula, especially in the Gulf of Valencia, significantly

[7] "Atlas Nacional de España, Instituto Geográfico Nacional," accessed March 1, 2024, https://www.ign.es/web/ane-area-ane

increased with negative WeMOi values. These correlations led to a new regionalization of the Iberian territory: Only the eastern coast received most of its precipitation from the Mediterranean Sea, much of it in the form of torrential rains, while the rest depended on the Atlantic Ocean (Fig. 1d).

In the lapse of a century, attention shifted from the Mediterranean's dry season to its rainy season. The Spanish geographer Antonio Gil Olcina (2007) argued that a dry season in summer is typical of all temperate dry-season climates. What is specific to the Mediterranean climate is the presence of an inland sea that determines precipitation in coastal areas. With insight, it seems that, paradoxically, older definitions of the Mediterranean climate through drought were overlooking the geographical feature that gives the region its name: the Mediterranean Sea! Introducing sea-atmospheric interactions in a dynamic system led to a redefinition of the Mediterranean climate and of the areas under its climatic influence. For instance, "dry Spain" and the Spanish steppes no longer fall into the "Mediterranean" area of influence in current indexes and maps. Figure 1 captures the historicity of the reach of the Mediterranean climate in the Iberian Peninsula by applying current rainfall data to the divergent indexes defined by conflicting definitions of the Mediterranean climate.

5 Conclusion

As defined by the reach of the Mediterranean climate, the Mediterranean region is historical in at least two senses. First, the indexes used to define it have changed through time. For us, this historicity opens the door for cultural and political explanations of the changing definitions. In particular, we have argued that a long-standing concern about drought in Spain was a key source for early modern ideas of climatic influence on human behavior and for modern associations of the Mediterranean region to a dry climate. We have shown how, through time, these ideas were politically situated in imperial disputes and reform plans.

According to historian Diana K. Davis, some of the main assumptions of Emberger Mediterranean map for UNESCO were deeply rooted in French colonial perceptions of a degraded North African landscape. We hope to have shown that Rivas' Mediterranean index was no less rooted than Emberger's in imperial struggles and political projects of landscape transformation. The old characterization of Spain as a dry and backward country has been one of the sources of definitions of the Mediterranean region as characterized by a dry climate. A more recent example is the Intergovernmental Panel on Climate Change (IPCC), which references Köppen's climatic classification to define the Mediterranean region by the presence of a dry season in summer.[8] This characterization has even been extended to define the climate of areas far from the Mediterranean, such as California, Chile, South Africa, and Australia. This detachment of the notion of the Mediterranean climate from the specificities of the

[8] "Mediterranean Region, Intergovernmental Panel on Climate Change (IPCC)," accessed April 14, 2024, https://www.ipcc.ch/report/ar6/wg2/downloads/report/IPCC_AR6_WGII_CCP4.pdf

Mediterranean Sea is increasingly contested. As the devastating floods in Valencia in October 2024 show, Mediterranean regions in Eastern Spain suffer intense rainfall produced by the intersection of isolated masses of cold air left behind by the polar jet stream and warm and humid masses of air coming from the Mediterranean sea—what Spanish scientists called DANA and German ones Kaltlufttropfen, cold drop. As the Mediterranean climate changes, recognizing the changes to the very notion of "the Mediterranean climate" becomes more urgent. As Fig. 1 illustrates, regional mappings of the Mediterranean influence in the Spanish climate are historical and contested. This demonstrates that scientific definitions of the Mediterranean region had also extra-scientific sources. They were part of larger political and geopolitical conflicts, often with enduring and invisible legacies.

Competing Interest Declaration The author(s) has no competing interests to declare that are relevant to the content of this manuscript.

References

Alvarez López, A. 2008. *La fabricación de un imaginario: los embajadores de Luis XIV y España*. Madrid: Ediciones Cátedra.

Álvarez, J. C., and F. Quirós Linares. 2004. La contribución de Bory de Saint-Vincent (1778–1846) al conocimiento geográfico de la Península Ibérica. Redescubrimiento de una obra cartográfica y orográfica olvidada. *Ería* 64–65:177–205. https://doi.org/10.17811/er.0.2004.177-205.

Anduaga Egaña, Aitor. 2012. *Meteorología, ideología y sociedad en la España contemporánea*. Madrid: Consejo Superior de Investigaciones Científicas, CSIC.

Azorín. 1905. *La ruta de Don Quijote / Azorín*. Madrid: Leonardo Williams.

Bartolomé Rodríguez, M. I., M. Rubio Varas, and D. Sesma Martín. 2024. Water for whom?: Unravelling the allocation of water storage capacity between irrigation and electricity uses in Spain during the 20th century. *Historia Agraria: Revista de Agricultura e Historia Rural* 94:165–201.

Bertaut, F. 1669. *Journal du voyage d'Espagne; contenant une description fort exacte, de ses royaumes, & de ses principales villes; avec l'estat du gouvernement, & plusieurs traittés curieux, touchant les regences, les assemblées des Estats...* Paris: Claude Barbin. https://gallica.bnf.fr/ark:/12148/bpt6k318124j.

Billaut, M., P. Birot, D. Cavalier, and P. Pédelaborde. 1956. Problèmes climatiques sur la bordure nord du monde méditerranéen. *Annales de géographie* 65 (347): 15–39. https://doi.org/10.3406/geo.1956.14340.

Bory de Saint-Vincent, J.B. 1826. *Résumé géographique de la péninsule ibérique, contenant les royaumes de Portugal et d'Espagne, par M. le colonel Bory de Saint-Vincent.* Paris: Ambroise Dupont et Roret, Éditeurs. https://gallica.bnf.fr/ark:/12148/bpt6k62258595

Bourguet, Marie-Noëlle, Bernard Lepetit, Daniel Nordman, and Maroula Sinarellis. 1998. *L' invention scientifique de la Méditerranée. Égypte, Morée, Algérie*. Paris: EHESS.

Braudel, F. 1949. *La Mediterranee et le Monde mediterraneen a l'epoque de Philippe II*. Paris: Librairie Armand Colin.

Braudel, F. 1975. *The Mediterranean and the mediterranean world in the age of Philip II*. New York: Harper and Row Publishers.

Brunel, A. 1666. *Voyage d'Espagne, curieux, historique, et politique: fait en l'année 1655 dedié a son altesse royale mademoiselle*. Paris: Robert de Ninville. https://gallica.bnf.fr/ark:/12148/bpt6k8727043s?rk=21459;2.

Brunhes, Jean. 1904. *L'irrigation dans la Péninsule Ibérique et dans l'Afrique du Nord: ses conditions géographiques, ses modes et son organisation*. Paris: C. Naud. https://gallica.bnf.fr/ark:/12148/bpt6k5496542c.

Byers, H. R. 1958. Carl-Gustaf Arvid Rossby (1898–1957). *Deep Sea Research (1953)* 5 (1): 83–84. https://doi.org/10.1016/S0146-6291(58)80015-6.

Camprubí, Lino. 2014. *Engineers and the making of Francoist Regime*. Cambridge MA: MIT Press.

Camprubí, Lino. 2020. Birds without borders: Ecological diplomacy and the WWF in Franco's Spain. *Historical Studies in the Natural Sciences* 50 (4): 433–455.

Camprubí, L., and P. Lehmann. 2018. The scales of experience: introduction to the special issue experiencing the global environment. *Studies in History and Philosophy of Science Part A* 70:1–5. https://doi.org/10.1016/j.shpsa.2018.05.003.

Cañete, C. 2012. The difficult unity of the Mediterranean in the works and expeditions of Bory de Saint-Vincent (1778–1846). *Culture & History Digital Journal* 1 (2): m105. https://doi.org/10.3989/chdj.2012.m105.

Casado de Otaola, S. 2010. *Naturaleza patria. Ciencia y sentimiento de la naturaleza en la España del regeneracionismo*. Madrid: Marcial Pons.

Caxa de Leruela, M. 1627. Discurso sobre la principal causa y reparo de la necessidad comun, carestia general y despoblacion destos Reynos. https://digibuo.uniovi.es/dspace/handle/10651/27747

Cazzato, L. 2017. *Sguardo inglese e Mediterraneo italiano. Alle radici del meridionismo*. Milano: Mimesis.

Celis Pozuelo, A., and M.C. Villarín Clavería. 2024. *Dataset for defining Mediterranean Climate zones in the Iberian Peninsula using various climate indices*. idUS (Depósito de Investigación de la Universidad de Sevilla). https://hdl.handle.net/11441/164275.

Claval, P. 1988. Les géographes français et le monde méditerranéen. *Annales de géographie* 542: 385–403.

Coen, D. R. 2011. Imperial Climatographies from Tyrol to Turkestan. *Osiris* 26 (1): 45–65. https://doi.org/10.1086/661264.

Coen, D. R. 2016. Big is a thing of the past: Climate change and methodology in the history of ideas. *Journal of the History of Ideas* 77 (2): 305–321. https://doi.org/10.1353/jhi.2016.0019.

Dantín Cereceda, J. 1922. *Ensayo acerca de las regiones naturales de España: tomo I*. Madrid: J. Cosano.

Davis, D. K. 2018. Between sand and sea: Constructing Mediterranean plant ecology. In *The Palgrave handbook of critical physical geography*, ed. R. Lave, C. Biermann, and S. Lane, 129–151. Cham: Palgrave Macmillan. https://doi.org/10.1007/978-3-319-71461-5_7.

de Mariana, Juan. (1599) 1845. *Del rey y de la institución de la dignidad real*. Madrid: Imprenta de la Sociedad Literaria y Topográfica.

de Martonne, Emmanuel. [1909] 1913. *Traité de Géographie physique*. Paris: Libraire Armand Colin.

de Martonne, Emmanuel. 1926. L'indice d'aridité. *Bulletin de l'Association de Géographes Français* 3 (9): 3–5. https://doi.org/10.3406/bagf.1926.6321.

de Martonne, Emmanuel. 1942. Nouvelle carte mondial de l'indice d'aridité. *Annales de géographie* 288:241–250.

Delgado Barrado, J.M. 2002. Ensenada versus Carvajal: un tópico a debate. In *El Catastro de Ensenada. Magna averiguación fiscal para alivio de los vasallos y mejor conocimiento de los reinos: 1749–1756,* coordinated by I. Durán and C. Camarero, 101–112. Madrid: Ministerio de Hacienda. Centro de Publicaciones y Documentación.

Deprest, F. 2002. L'invention géographique de la Méditerranée: éléments de réflexion. *L'Espace géographique* 31 (1): 73–92.

Deprest, F. 2017. What is a colonial geographer? Jean Brunhes, irrigation and human geography (1894–1911). *L'Espace geographique* 46 (3): 264–284.

Dubet, A. 2008. *Jean Orry et la réforme du gouvernement de l'Espagne (1701–1706)*. Clermont-Ferrand: Presses Universitaires Blaise Pascal.

Febvre, L. 1922. *La Terre et l'évolution humaine: introduction géographique à l'histoire*. Paris: Albin Michel.

Fernández Álvarez, Á. M. 2014. La evolución institucional de la propiedad y de los derechos subjetivos en los siglos XVI y XVII en España. *Esic market* 148:327–388.

Fernández Delgado, R. 2003. La ruptura del pensamiento económico castellano en el siglo XVII: Juan de Mariana y Sancho de Moncada. Doctoral dissertation. Universidad Complutense de Madrid.

Fleming, J. R. 2016. *Inventing atmospheric science: Bjerknes, Rossby, Wexler, and the foundations of modern meteorology*. Cambridge, MA: MIT Press.

Gil Olcina, A. 1992. Las políticas hidraúlicas del reformismo ilustrado. In *Hitos históricos de los regadíos españoles*, coordinated by A. Gil Olcina and A. Morales Gil, 143–82. Madrid: Ministerio de Agricultura, Pesca y Alimentación.

Gil Olcina, A. 2007. Mediterraneidad y subtropicalidad climáticas. *Boletín de la Asociación de Geógrafos Españoles* 44:53–68.

Greene, M. 2017. The Mediterranean Sea. In *Oceanic histories*, ed. A. Bashford, D. Armitage, and S. Sivasundaram, 134–155. Cambridge: Cambridge University Press.

Grisebach, Alfred. 1866. Die Vegetations-Gebiete der Erde, übersichtlich zusammengestellt, *Mittheilungen aus Justus Perthes' Geographischer Anstalt über wichtige neue Erforschungen auf dem Gesammtgebiete der Geographie* 14: 45–53.

Grove, A. T., and O. Rackham. 2003. *The nature of Mediterranean Europe: An ecological history*. New Haven; London: Yale University Press.

Hermann, C. 1989. L'arbitrisme: un autre état pour une autre Espagne. In *Le premier age de l'état en Espagne (1450–1700)*, coordinated by C. Hermann, 239–256. Centre National de la Recherche Scientifique.

Heymann, M., and D. Achermann. 2018. From climatology to climate science in the twentieth century. In *The Palgrave handbook of climate history*, ed. S. White, C. Pfister, and F. Mauelshagen, 605–632. London: Palgrave Macmillan UK. https://doi.org/10.1057/978-1-137-43020-5_38.

Heymann, M., and D. D. Dahan. 2019. Epistemology and politics in Earth system modeling: Historical perspectives. *Journal of advances in Modelling Earth Systems* 11 (5): 1139–1152. https://doi.org/10.1029/2018MS001526.

Howkins, A., and P. Roberts. 2023. *The Cambridge history of the polar regions*. Cambridge University Press. https://doi.org/10.1017/9781108555654.

Hulme, M. 2010. Cosmopolitan climates. *Theory, Culture & Society* 27 (2–3): 267–276. https://doi.org/10.1177/0263276409358730.

Joffre, Richard, and Serge Rambal. 2001. *Mediterranean ecosystems*. eLS. https://doi.org/10.1038/npg.els.0003196.

Klein, Mike. 2018. The first isothermic World Maps | Worlds revealed. The Library of Congress. 25 April 2018. https://blogs.loc.gov/maps/2018/04/the-first-isothermic-world-maps.

Köppen, Wladimir Peter. 1936. Das geographische System der Klimate. In *Handbuch der Klimatologie*, ed. W. Köppen and R. Geiger, vol. 1, Part C. Berlin: Verlag von Gebrüder Borntraeger.

Lehmann, Philipp. 2022. *Desert Edens: Colonial climate engineering in the age of anxiety*. Princeton University Press. https://doi.org/10.2307/j.ctv2j6xds8.

Lionello, P., P. Malanotte-Rizzoli, R. Boscolo, P. Alpert, V. Artale, L. Li, J. Luterbacher, et al. 2006. The Mediterranean climate: An overview of the main characteristics and issues. In *Developments in earth and environmental sciences*, ed. P. Lionello, P. Malanotte-Rizzoli, and R. Boscolo, vol. 4, 1–26. Elsevier. https://doi.org/10.1016/S1571-9197(06)80003-0.

Lorenzo, Darío R. 2014. *Los Arbitristas y el arte de gobernar: Discursos y prácticas políticas en un contexto de crisis*. Doctoral dissertation, Universidad Nacional del Mar del Plata, Facultad de Humanidades.

Lorenzo Pardo, M. (1933) 1993. *Plan Nacional de Obras Hidráulicas. Tomo II–III.* Ministerio de Obras Públicas, Transportes y Medio Ambiente.

Lucena Giraldo, Manuel, and Felipe Fernández-Armesto. 2022. *Un imperio de ingenieros*. Madrid: Taurus.

Mallada Pueyo, Lucas. 1890. *Los males de la patria y la futura revolución española*. Madrid: Tipografía de Manuel Ginés Hernández.

Martin-Vide, Javier, and Joan-Albert Lopez-Bustins. 2006. The Western Mediterranean oscillation and rainfall in the Iberian Peninsula. *International Journal of Climatology* 26 (11): 1455–1475. https://doi.org/10.1002/joc.1388.

Martykánová, Darina. 2023. *Los ingenieros en España: el nacimiento de una élite*. Leioa: Universidad del Pais Vasco, Servicio Editorial.

McNeill, J. R. 1992. *The mountains of the Mediterranean World*. New York: Cambridge University Press.

Montesquieu, Charles. 1748. *Complete works, vol. 1 The spirit of laws*. London: T. Evans.

Morera, Raphaël. 2017. Marshes as microclimates: Governing with the environment in early modern France. In *Governing the environment in the early modern world*, ed. Sara Miglietti and John Morgan, 56–75. New York: Routledge.

Morgan, John Emrys, and Sara Miglietti. 2017. Introduction: Ruling "climates" in the early modern world. In *Governing the environment in the early modern world*, ed. John Morgan and Sara Miglietti, 1–21. London/New York: Routledge.

Olcina Cantos, J. 2020. La aportación a la ciencia climática de A. de Humboldt en el Cosmos. *Scripta Nova. Revista Electrónica de Geografía y Ciencias Sociales* 24 (648): 1–44. https://doi.org/10.1344/sn2020.24.30071.

Ortega Cantero, Nicolás. 1997. Juan Dantín Cereceda y la Geografía española. Ería: Revista cuatrimestral de geografía, n.º 42, 7–34.

Paris, E. 2002. *La genèse intellectuelle de l'œuvre de Fernand Braudel: 'La Méditerranée et le monde méditerranéen à l'époque de Philippe II' (1923–1947)*. Athens: FNRS.

Parker, Geoffrey. 2013. *Global crisis: War, climate change and catastrophe in the seventeenth century*. New Haven; London: Yale University Press.

Radinović, Djuro. 1965. Cyclonic activity in Yugoslavia and surrounding areas. *Archiv Für Meteorologie, Geophysik Und Bioklimatologie, Serie A* 14 (4): 391–408. https://doi.org/10.1007/BF02253485.

Rico Sinobas, Manuel. 1851. *Memoria sobre las causas meteorológico-físicas que producen las constantes sequias de Murcia y Almería, señalando los medios de atenuar sus efectos*. Madrid: Imprenta de D.S. Compagni.

Rivas-Martínez, S. 1983. Pisos bioclimáticos dc España. *Lazaroa* 5:33–43.

Rodríguez Villa, Antonio. 1878. *Don Cenón de Somodevilla, Marqués de la Ensenada ensayo biográfico formado con documentos en su mayor parte originales, inéditos y desconocidos*. Madrid: Librería de M. Murillo.

Ruel, Anne. 1991. L'invention de la Méditerranée. *Vingtième Siècle. Revue d'histoire* 32:7–14. https://doi.org/10.2307/3769993.

Safier, Neil. 2012. *Measuring the new world: Enlightenment science and South America*. Chicago: University of Chicago Press.

Schaub, Jean-Frédéric. 2003. *La France espagnole: les racines hispaniques de l'absolutisme français*. Paris: Éditions du Seuil.

Swyngedouw, Erik. 2007. Modernité et hybridité Nature, "regeneracionismo" et la production du paysage aquatique espagnol. 1890–1930. *Géographie, économie, société* 9 (1): 39–68. https://doi.org/10.3166/ges.9.39-68.

Vidal de La Blache, Paul. 1886. Des rapports entre les populations européennes et le climat sur les bords européens de la Méditerranée. *Revue de Géographie* 10:401–419.

Wick, A. 2016. *The Red Sea: In search the lost space*. Oakland: University of California Press.

Willkomm, Moritz. 1852. *Die Strand und Steppengebiete der Iberischen Habbinsel, und deren Vegatation*. Leipzig: Fleischer.

Wing, John T. 2012. Keeping Spain Afloat: State forestry and imperial defense in the sixteenth century. *Environmental History* 17 (1): 116–145. https://doi.org/10.1093/envhis/emr123.

Mining and Metallurgy in Early Imperial Russia

Anna Graber

Contents

Abstract

The global turn in the history of science enables us to account for the non-Western and transcultural origins of modern science. While the knowledge traditions of some non-Western empires have begun receiving substantial historical treatment, the Russian Empire has been little considered by historians of science and technology. This gap in the literature is particularly surprising in the case of the earth and mining sciences given Russia's dominance in early modern Eurasian metallurgy. The aim of this chapter is to bring together two literatures—the history of Russian imperial mining and the global history of earth science and mining technology—that are not often in conversation with one another. As historians of an empire that was independent yet deeply engaged with the earth and mining sciences of a variety of Wests, Russianists offer novel approaches to the some of the shared issues facing scholars of global science history.

A. Graber (✉)
University of Minnesota, Minneapolis, MN, USA
e-mail: agraber@umn.edu

E. Aronova et al. (eds.), *Handbook of the Historiography of the Earth and Environmental Sciences*, Historiographies of Science,
https://doi.org/10.1007/978-3-031-40799-4_29

Keywords

Mining · Metallurgy · Russia · Soviet Union · Russian Empire · Technology · Colonialism · Technological diffusion · Artisanal knowledge · Local knowledge

1 Introduction

The global turn in the history of science is a welcome development, enabling us to account for the non-Western and transcultural origins of modern science, and to consider seriously nonscientific or Indigenous knowledge systems. But too often, global science is bounded by the footprint of European empire and thus becomes a means of doing British or French history by another name. Historians of global science such as Kapil Raj (2007) and Sujit Sivasundaram (2010) have called for the field to move away from the previous eurocentrism of science and empire studies and for studies of natural knowledge in non-European empires, noting in particular the understudied entanglement of science and imperial rule in the Persian, Ottoman, and Chinese empires. One should add to this list the Russian Empire.

A major world and imperial power since the mid-sixteenth century, the Russian Empire nevertheless has been the subject of few histories of science and technology. Recent work in the history of medicine and botany (Griffin 2022; Romaniello 2016a, b; Koroloff 2018), as well as studies of intellectual life at the early modern court (Collis 2012; Chrissidis 2016) and of natural historical and imperial intelligence (Jones 2014), have made crucial advances in filling this lacuna, but knowledge of the earth, the subject of this volume, in imperial Russia is *terra incognita* in the English-language literature of global science. That this gap exists is particularly surprising given Russia's dominance in early modern Eurasian metallurgy: For most of the eighteenth century, the Ural Mountains were Eurasia's most productive iron and copper mining and metallurgical zones, and the Altai Mountains in southern Siberia yielded more silver than any other region of Eurasia at a time when South American silver production was depressed. These mountain ranges were not only industrial zones: They were theaters in which mining officials and plant owners leveraged multiple mining traditions, governing practices, and knowledges of the earth to subjugate resource-rich, multiethnic borderlands, in the process forging the modern Russian empire.

While the English-language literature of Russian earth knowledges is limited, mining science and technology has played an important role in Russian-language narratives of the empire's transition to modernity. This is especially true for the period under investigation in this chapter, the early seventeenth to early nineteenth centuries, in which most historians place the rise, boom, and relative decline of the mining industry. The question of the role of artisanal versus state-backed institutional expertise in the development of this industry, which was linked to the larger issue of the existence or absence of a native metallurgical and scientific tradition (as opposed to the view of the empire as the recipient of science and technologies diffused from the West), has held important implications for Russia and the Soviet

Union's fitness as a modern state, ability to compete with capitalist nations, and ultimately the legitimacy (in Marxist terms) of the Bolshevik Revolution.

After the revolution, the Soviet historiography of science and technology developed apart from the English-language historiography with its own commitments, methodologies, and preferences in genre. The principal concern of the field has been the positivist (and often propagandistic) identification of past Russian scientists, advances in scientific theory, and technological innovations, developed largely through biographical studies of individual scientists. With the fall of the Soviet Union, Russian scholars of mining and earth science have forged deeper connections with historians based in Scandinavia, Germany, the United Kingdom, and the United States, resulting in scholarship that has begun the project of placing Russia's mining and earth knowledges in European, if not global, history. The future of these connections, however, is unassured due to Russia's isolation from European and North American academia following the February 2022 invasion of Ukraine.

It is the goal of this chapter to continue this project of placing Russian histories of earth science and technology within global context, and to bring together two literatures—the history of Russian mining and the global history of earth science and mining technology—that are not often in conversation with one another. This requires some careful reading of the Russian-language scholarship on mining history for, while earth knowledge and metallurgical expertise has long been an important topic within this literature, the most important works in the field have been written by economic and social historians rather than historians of science and technology. One of the aims of this chapter is to translate this largely Soviet historiography for contemporary historians of the earth sciences, as these studies of Russian mining history have much of interest to offer our field. As Jarrod Hore reminds us elsewhere in this volume, the earth sciences are field sciences, which necessitates close attention to the disparate geographies of earth knowledges (including the still little-studied northern half of Eurasia) in history. Soviet historians of technology have argued since the 1930s that the chemistry of iron and especially of copper ores requires adaptation to local conditions even when using established, foreign technologies (Baklanov 1935). Recent work investigating the environmental history of mining likewise uncovers the ways in which geographical specificities of Russia's mining regions encouraged the development of globally unique mining technologies and organization (Evtuhof 2023).

The Russian industry will be of interest to historians studying the interaction of the earth sciences and empire for the particular dynamics of colonial science at mines and metallurgical plants. Never colonized by a foreign power, Russia and it earth knowledges do not fit into the schemas of colonial and postcolonial science developed for societies of the Global South. But neither were Russian scientists and technical experts free of a (real or perceived) intellectual dependence on Europe, which Russia was both culturally and geopolitically a part of although situated at the margins of scientific society. A complex dynamic between European and Russian knowledge traditions—if these traditions can ever be disentangled enough to be considered separate—was further complicated in Russia's Asian colonies, where imperial agents encountered and mobilized Indigenous knowledge systems as they established indirect and then direct imperial rule. This is particularly true of Russia's most important metallurgical zone, the Urals, which boasted an ancient, non-Slavic mining tradition and a diverse body of prospectors. The Urals and to a lesser extent the silver-bearing Altai Mountains and the Nerchinsk region in southeastern Siberia on the Qing border were geocontact zones in which the science and technology of multiple European cultures interacted with Russian artisanal tradition and the knowledges of Siberian and Uralian peoples. Recent work—including my own—on Russian mining administrators as imperial agents is beginning to consider the dynamics of knowledge regimes and hierarchies at Russian mines and to elucidate what should be an important case study in the global history of science.

2 Artisanal or Institutional? Knowledge and Expertise in Imperial Mining

"Russia," declared the historian and liberal politician Pavel Miliukov in 1896, was a country "without capital, without workers, without entrepreneurs, and without a market," when Tsar (from 1721 Emperor) Peter I (r. 1682–1725) established a host of industries, including flourishing iron and copper industries, in the opening decades of the eighteenth century (1:80; unless stated otherwise, all translations are my own). With these words, Miliukov opened a century-long debate on the role of the state in Russian industry, and historians in both the Soviet Union and abroad have since examined this question with particular focus on the mining and metallurgical industries. While economic and social issues were central preoccupations for this literature (i.e., studies of the origins of industrial capital, the use of compulsory labor in the industry, the formation of a proletariat, the estate and class identity of mine owners, and above all many attempts at identifying Russia's transition from a feudal to a capitalist economy and society), knowledges of the earth and especially of how to work its products were also key subjects for historians invested in this debate. Did the industry have its intellectual origins in the country's blacksmithing tradition, or did it owe its success in the eighteenth century to a network of state mining institutions—most importantly the College of Mines (*Berg-kollegiia*), in operation 1719–1736, 1742–1783, and 1796–1807—modeled on German and Swedish institutions, and deploying central and northern European management practices, metallurgical technologies, and academic mineral science? In asking these questions, both Soviet historians and those working in Miliukov's classically liberal tradition debated whether Russian industry—and hence, for Marxist historians, civilization—organically developed from folk sources or was a foreign institution imposed by the state. This led into even larger questions of whether modern Russian history conformed to or deviated from either liberal or Marxist models built on studies of Western and Central European history. Though the fall of the Soviet Union has taken the urgency out of this debate, recent additions to this literature offer historians of global earth knowledges approaches to moving between and beyond dichotomous narratives of scientific nationalism in non-Western countries versus models of scientific and technological diffusion from and dependence on the West.

Scholars seeking to understand the traditions that the metallurgical boom of the eighteenth century emerged from have looked to the small but dynamic seventeenth-century iron industry for antecedents of Russia's industrialization. There were three centers of early modern artisanal iron production: the oldest near Tula to the south of Moscow; another centered in Olonets in Karelia far to the north, where Old Believers, members of a persecuted, exiled schismatic Russian Orthodox sect, fled in the mid-seventeenth century to poor lands ill-suited to agriculture and so took up bog iron working; and third in the Ural Mountains, where a regional trade developed in iron produced in smithies and at monasteries. At these sites, smiths worked near-surface or bog ores mined in the vicinity and purified them in bloomery furnaces.

These early iron mining and working centers, in particular Tula and Olonets, figure prominently in the work of Soviet scholars committed to a native, Russian metallurgical industry with medieval roots. Early Soviet scholarship of mining and other heavy industry arose from turn-of-the-century debates among leftist groups (agrarian socialists, Mensheviks, and Bolsheviks) as to the maturity of Russian industry and capitalism and the empire's readiness for socialist revolution. For historians after 1917, to emphasize state involvement in industrialization or foreign sources of metallurgical technology was to open oneself to the charge of doubting the organic development of Russian capitalism and ultimately of suggesting that Russian history did not align with Marxist doctrine (Griffiths 1986, 2). By the 1930s, a group of historians led by Stanislav Strumilin argued against what they termed the "Milukov thesis" that industry had been imposed by Peter I's state in the early eighteenth century and instead sought the origin of Russian industry in artisanal activity in the seventeenth century (Strumilin 1932, 10; Kafengauz 1949; Zaozerskaia 1970). This group produced regional studies of industry, like A. P. Glagoleva's *The Olonets Factories in the First Quarter of the Eighteenth Century* (1957). In this work, Glagoleva examines the extent of the northern bog iron trade in the seventeenth century, which she characterizes as a peasant-led, flourishing regional industry that showed early features of capitalism. In her view, the Olonets region was a cradle of national mining and metallurgical expertise, as bog miners and smiths contributed their local knowledge of their ores (which were of low quality and difficult to smelt), charcoal preparation, and dam construction to the first large, foreign-built water-powered iron factories of northwestern Russia, factories which in turn served as a training ground for the first generation of Russian-speaking skilled labor in the later developing but more important Urals mining region.

Strumilin himself, in his *History of Iron Metallurgy in the USSR* (1954), argues that the mining boom of the eighteenth century grew "from a base" of thousands of bloomeries which served local needs (all subsequent citations to the 1967 edition, 153). Elena Zaozerskaia likewise emphasizes the blacksmithing and artisanal origins of the first generation of water-powered factory owners in the first quarter of the eighteenth century in her *Manufacturing in the Reign of Peter I* (1947). The key region for both these historians was Tula, which by the end of the seventeenth century was home to a small but bustling iron mining, smithing, and weapons-making industry using limonite mined by Cossacks at a riverside mine 30 km southeast of the city. Their artisanal manufacturing, though modest in output, featured well-constructed bloomery furnaces that operated efficiently (Zavyalov 2018). By running small workshops, entrepreneurial smiths became versed in every step of the small arms manufacturing process, a knowledge which members of seven smithing families were able to translate into much larger, water-powered enterprises both in central Russia and, from the eighteenth century, the iron-rich Ural Mountains (Pavlenko 1962, 71–162). This literature describes an entangled cluster of knowledges—artisanal, metallurgical, and embodied, but also entrepreneurial, managerial, and even mineralogical—that makes it impossible to separate the history of mining and metallurgy from the history of earth knowledges in Russia.

Whereas Soviet scholars labored to demonstrate the profundity of native Russian metallurgical expertise at the dawn of the eighteenth century, foreign historians examining the Tula and Olonets industries dated the origins of Russian heavy industry to Europeans' entry into the industry, with their construction of the first water-powered plants in the country. In the St. Petersburg-born West German historian Eric Amburger's *The Marselis Family: Studies in Russian Economic History* (1957), the subjects are the small Dutch-Danish-German family groups that received patents from Tsar Michael (r. 1613–1645) to build Russia's first four lasting water-powered iron plants near Tula in 1629–1632. Though Amburger acknowledges the blacksmithing and mining tradition of the region, his work emphasizes the foreign, external character of these water-powered plants: The Marselis, Witsen, and Akkema families he studies had been recruited from abroad by Tsar Michael's court to produce wrought iron for court consumption and to cast iron cannon and small arms for the army using German methods and technologies. These arms, he says, were superior to the simple firearms crafted by Tula smiths in what he terms "primitive" Russian smelting ovens (Amburger 1957, 94). In his study of the seventeenth-century iron industry as a capitalist enterprise, American historian Joseph Furhmann (1972) likewise considers the water-powered iron plants to be a foreign institution, the result of the native economy not being able to compete technologically or militarily with its western neighbors. In these studies as well as the later work of American historian Hugh Hudson, Jr. on the Tula smithing and entrepreneurial family the Demidovs, these European-built factories were important as sites of technology transfer and of the "breaking" of technological barriers: Russian miners and smiths were already skilled enough to be able to follow plans for the construction of water-powered factories and so learned of German innovations like the blast furnace and water-turned hammers through their participation in factory construction (Amburger 1957, 99; Hudson 1986, 34). It is only under a 60-year period of tutelage, in this model of technology diffusion, that Russians learned how to build the water-powered factories needed to arm the navy and the army and become a European great power. As the European metallurgical enterprises deploying water-based technologies were sponsored by and produced arms specifically for the imperial court, historians emphasizing the role of the state in the advancement of Russian knowledge likewise have taken the construction of the Marselis iron plants or court-sponsored prospecting expeditions undertaken in the fifteenth century as the beginning of Russian industry (Pavlenko 1962, 9; Portal 1950, 25).

3 Artisanal and Institutional Knowledges in the Urals

Studies of the seventeenth century set the dominant narratives of state-sponsored technological diffusion versus artisanal and entrepreneurial innovation that continue in work on the eighteenth-century golden age of the Russian iron, copper, and silver industries. Historians writing on this period have also found success, however, in escaping these dichotomous narratives. In addition, their work has examined the links between mining and the production of scientific knowledge of the earth, with a number of studies published after the fall of the Soviet Union on the scientific culture of the upper stratum of mine owners. This endeavor is linked in turn to an at first nationalistic (during the Soviet period), and then regionalist (since 1991), positivist project celebrating the intellectual world of Russia's industrial zones through the identification of scientists, engineers, inventors, writers, and scholars connected to the mining industry.

Iron production expanded rapidly at Tula and Olonets from the end of the seventeenth century as Tsar Peter I granted concessions to foreign and Russian entrepreneurs to build water-powered factories and produce small firearms, cannon, and anchors for his reformed army and newly founded navy (for statistics showing a tripling in iron works construction and a fivefold increase in cast iron production across Peter's reign, see Zaozerskaia 1947, 154–164 and Strumilin 1967, 153). The production of war materiel became essential as Peter launched the Great Northern War (1700–1721) against Sweden, the erstwhile supplier of the majority of Russia's iron and arms. Cut off from European arms markets and with most of his artillery captured in an early loss to Sweden, Peter was obliged to develop his own defense industries, initially in the preexisting centers of Olonets and Tula which were convenient to both the fronts and the capitals (Moscow and the new capital of St. Petersburg, founded 1703).

The war won, in the 1720s metallurgical activity shifted eastward to the Urals. Iron and copper ores there were of far higher quality than those of central and northwestern Russia, with some experimental smelts yielding 25% pure iron. This iron was also much easier to work with than the brittle bog iron of the Olonets region, and ores lay near the surface in massive nests accessible by picks and crowbars (de-Gennin 1937, 70, available in English translation: de Hennin 1992). The region was also abundant in forests, huge reserves of which were necessary for charcoal production, and featured swift, powerful, predictable mountain rivers to power bellows and hammers. The principal rivers also flowed westward into the Volga-Kama system, and hence to Moscow and via canals to St. Petersburg and the Baltic Sea, offering cost-effective transportation to domestic and international markets. From the 1720s, cheap Russian iron flooded European markets, and by 1740 the Urals were the most important metallurgical zone in Eurasia as Russia surpassed English and Swedish iron production. From 1750, over half of all Russian iron was sent to the export market, with England and Scotland, followed by Sweden, as the principal destinations—Urals iron was a key material in England's Industrial Revolution (Strumilin 1967, 173, 177–202; Evans et al. 2002; Evans and Rydén 2007).

Adherents of the antistatist, nativist school of mining history stress the role of Russian artisanal expertise in the success of the Urals industry. Strumilin and Zaozerskaia's colleague B. B. Kafengauz takes the entrepreneurial activity of the Demidov family, the most successful of the eighteenth-century private mine owners, as a case study in the development of Russian capitalism (one of the central objects of study of economic historians in the Soviet Union) in his monumental *History of the Demidovs' Enterprise* (1949). Like his colleagues, he argues against the Miliukov thesis that mining and metallurgy were a state-directed transplantation of European skill, technology, and capital to Russia. Instead, he argues that the Urals industry had its origins in the seventeenth-century expertise and capital of Tula. The Demidov family patriarch, Nikita Demidov, was an arms maker who built one of the first Russian-owned water-powered factories in Tula. He was no mere blacksmith, Kafengauz argues, but among the wealthiest stratum of an already diversified group of Tula arms makers, metallurgists, and smiths. When Demidov was granted by Peter I the patent to a state-owned iron factory in the Urals in 1702, he brought with him his Tula masters, metallurgists, and other skilled labor. This is the kind of knowledge transfer that Soviet economic historians who saw water-powered heavy industry as an organic development of Russian history described: a transfer of skill from one area of Russia to another, rather than of European knowledge to Russia.

But the shift of mining and metallurgical activity from European Russia to the Urals was also the result of the new institutionalization of mining knowledge. The first persisting state institution regulating mining and metallurgy, the College of Mines, was founded in St. Petersburg in 1719 as part of Peter I's reform of central state administration into a Swedish-system collegiate system. The stated aim of this college was to order the mining industry and bring it into a "diligent arrangement" according to "mining art" (*gornoe isskustvo*) (*Polnoe sobranie zakonov Rossiiskoi imperii* 1830, 5: December 10, 1719, no. 3464). This term encompassed the administration of mine and factory systems according to the practices developed in early modern Germany, especially Saxony, undergirded by a mathematized and mechanized understanding of nature. This approach to mining administration was brought to the Urals by a pair of early College of Mines officials, Vasilii Tatishchev and the Nassau-born metallurgist Georg Wilhelm de Hennin, as they established a Siberian Head Mining Office (located on the grounds of a large iron and copper industrial complex they founded in 1723–1724 in the new city of Ekaterinburg), initiated prospecting surveys, developed cadres of skilled labor, reoriented iron production from domestic weaponry to the export market, and oversaw the construction of 21 iron and copper factories in the Urals between 1720 and 1739 (Portal 1950, 67).

From the 1940s, some historians criticized the antistatist position of Kafengauz and Strumilin and attempted to bring state institutions back into the historiography of Russian industry. The most influential figure in this group was the economic and social historian Nikolai Pavlenko, who contends in a series of books and articles on the origins of Russian capitalism that early industry was the creation of the state (Pavlenko 1953; Pavlenko et al. 1969). In a monograph on the formation of an industrial bourgeoisie in the Urals, Pavlenko emphasizes the role that state policy had in shaping the growth and eventual retardation of this industry (1962). He details how the mining administration resolved disputes among owners, permitted or blocked mine and factory construction, managed forest reserves, and provided technical expertise to private factory owners. Additionally, Pavlenko finds the mining administration a far more competent actor in the industry than the high aristocrats who received privatized state factories in the middle decades of the eighteenth century. This work set the stage for a new generation of Soviet historians active in the 1960s who wished to reconsider the role of the state in the empire's mining boom.

In the wake of Pavlenko's rehabilitation of the state as an actor in the mining industry, historians began writing biographies of individual mining administrators as progressive (that is, supporting science, education, and state building) figures in Russian history. The figure who attracted the most detailed biographical study was the aristocrat, historian, and first head of the Siberian Mining Office, Vasilii Tatishchev (Iukht 1985; Kuz'min 1981; Shakinko 1986; Daniels 1973 is an English-language biography). Conrad Grau, a graduate of East Germany's Institute for History of the Peoples of the USSR, examines Tatishchev's role as a "developer" of Russia in his *The Economic Promoter, Statesman, and Scholar Vasilii N. Tatishchev* (1963). He argues that Tatishchev developed the country economically through his work as an early leader of the College of Mines to shift the locus of mining activity from European Russia to the Urals, cofounding the city of Ekaterinburg in the process; and that Tatishchev contributed to the ideological development of Russia as an elaborator of enlightened absolutism. In Grau's account, Tatishchev is a progressive figure struggling against feudal elements (i.e., the widespread use of serf labor, and factory owners who vociferously, even violently, objected to any government interference in the management of their plants) to rationalize private production according to universalized mining sciences and to oversee correct, sustainable use of the Urals' natural resources (28–41, 66–82). Underlying the narratives of Grau, Pavlenko, and other postwar historians who cast state actors as industrial modernizers is a belief that the knowledge tradition these figures mobilized in their construction and management of the mining industry, the "mining sciences" (*gornye nauki, Bergwissenschaften*; understood by members of the College of Mines as those aspects of mineralogy, chemistry, physics, mechanics, surveying, architecture, and hydraulics that could be applied to mining and metallurgy), were universalizable sciences that, though originating abroad, were effective and unproblematic guides to economic, social, and even civilizational development. This narrative of the progressive, even liberating power of science was shared by postcolonial science promoters globally in this period.

4 Industry Leaders as Scientific Figures

With the increasing importance of science and science education to Soviet ideology in the Cold War, historians cast state actors not only as developers of Russia, but also as scientists, inventors, and educators, in the process exploring the connections between daily work in the mines and factories and the generation of scientific knowledge (Kozlov 1981). Tatishchev was here again an important figure, as historians detailed his establishment of a system of schools of reading and arithmetic at state factories, including a higher school in Ekaterinburg in which students were taught physics, mechanics, German, and Latin (Nechaev 1944). Tatishchev also pursued scholarly interests in history, science promotion, and in the fossil curiosities he encountered and collected during his service in the Urals, even publishing some of Russia's first scholarship on mammoths (1725, 1979). He assembled one of the largest libraries in provincial Russia, which he left at the Head Siberian Mining Office when he was recalled from the Urals in 1737 (Safranova et al. 2005). He was far from the only mining official to be studied as a scientist: As Steven Usitalo has argued, scientific biography was the preferred genre of Soviet historians of science, as a country promoting science education needed past role models for students to look up to and to demonstrate the centuries-old roots of Russian science (2013, 14). Naum Raskin was particularly prolific in this genre, writing biographies of mining officer Aleksandr Karamyshev, a Linnaeus student who wrote on the geology of eastern Siberia and on the state of the Nerchinsk lead and silver mines (Raskin and Shafranovskii 1975), and of a Swedish naturalist and mining expert in Russian service, Erik Laxmann (Raskin and Shafranovskii 1978). He also wrote on A. A. Musin-Pushkin, a Vice President of the College of Mines and metallurgical chemist (Raskin 1981). Musin-Pushkin contributed to scholarly chemical journals and witnessed early replications of Lavoisier's experiments on oxygen, and his work on platinum's salts and experimental alloys aided in the chemical identification of this metal as a new element. These biographies limn a network in place by the 1770s of officers of the College of Mines who exchanged mineralogical observations, mineral and other natural curiosities, and regional natural historical and prospecting descriptions with each other across mining centers in Siberia and with college leaders in Moscow and St. Petersburg (Graber 2016, 47–56, 97–137).

With the fall of the Soviet Union, historians embraced new approaches and topics in regional history, business history, and cultural history, often building on formations laid by late Soviet scholars. In a new free market economy, the question of the origin of Russian capitalism morphed into the origin of Russian entrepreneurialism. Now that it was politically permissible to study the culture of elites, a particularly productive area was the study of the rise, management style, and scientific contributions of the great mining families of the eighteenth and nineteenth centuries: the Stroganovs, the Turchaninovs, and especially the Demidovs (Nekliudov 2004). The substantial 1994 edited volume *The Demidov Chronicle: A Historical Almanac* is emblematic of this trend (Cherkasova 1994a). In a sprawling collection devoted to the Demidov family, editors and contributors seek to move beyond Soviet-era economic histories of metallurgical output (such as Kafengauz's work) and instead prosopographically place the members of the family and their circle in focus (Cherkasova 1994a, 7). Akinfii Demidov, of the family's second generation, in particular emerges in this volume as the model mining entrepreneur, who had the technical and mineralogical knowledge to effectively oversee every aspect of the industry from ore prosecting to smelting—an exemplar of the Urals' past to be emulated in post-Soviet economic conditions.

For post-Soviet historians of the Urals, the knowledge and skill of the Demidov family extended beyond their own business interests to increase the cultural and intellectual sophistication of the region and even the empire. Historian of science I. N. Iurkin, for example, has devoted much of his career to the study of the intellectual world of the Demidov family, who he approaches as "scholars, engineers, and organizers of science and industry" (2001). In a country with few universities or research institutes to support science, Iurkin argues, it was mine owners who advanced science in Russia in the eighteenth century (2001, 35). He describes in detail the Demidov family's support of and engagement in natural historical scholarship—Grigorii Akinf'evich Demidov, a member of the third generation of Demidov magnates, planted a botanical garden at his estate in the western Urals that served as an essential repository of local flora and also of exotic East Asian plants collected by academic naturalist members of the Second Kamchatka Expedition to Alaska, who stopped at his home multiple times as they crossed the empire (Iurkin 2001, 114–120). Grigorii Demidov exchanged seeds with Carl Linnaeus and sent his sons to Uppsala to study with him (Pobedimova 2006), while his brother Prokofii created the largest botanical garden in Moscow (Koroloff 2014; Elina 2018). Pavel Demidov, Grigorii's son, amassed one of the largest mineralogical and natural collections in the empire. It became the basis of the Moscow University natural collection when he donated it to that institution in 1806, and it included 2000 Siberian mineral specimens and a library of 3000 titles (Iurkin 2001, 176–182; Fischer 1806, 2). Other chroniclers of the cultural interests of magnate families similarly document how, from the 1760s, the great mine owners assembled mineralogical collections from their mines, invited traveling academicians to tour their plants and mines, and shared specimens with scholarly correspondents (Pirogova et al. 2008; Hudson 1986). Taken together, the literature of the two decades after the fall of the Soviet Union treats mine owners as skilled entrepreneurs and major cultural and even scientific figures that must be reinserted into Russian history and history of science.

5 Mining and Metallurgical Technologies: Beyond Diffusion

The development of Russian industrial technology, especially as compared to Saxon, Swedish, and British mining and metallurgical technologies, has been a particular concern of both Soviet and more recent historians of the mining industry. Historians of the industry have had to explain why Russia, despite having the most productive mining industry of eighteenth-century Eurasia, did not experience the rapid industrialization of late eighteenth-century England's Industrial Revolution until the 1880s. Still, the place of Russia in the literature on the "great divergence" between European and Chinese economies in the eighteenth and nineteenth century is ambiguous and underexamined—the central authors of the literature, such as Kenneth Pomeranz, have little to say on Russia. Pomeranz at times includes Russia as part of Europe, at other points it is a periphery that supplies useful, but not by any means essential, timber, grain, and iron to industrializing Britain

(Pomeranz 2000; Turchin 2013; Rönnback 2010). Others consider Russia "poor and underdeveloped" on the eve of World War I in comparison to the major Western economies (Clark and Feenstra 2003, 294).

Though infrequently discussed by global economic historians, Russia specialists have proposed many reasons for the mining industry's relative decline—reliance on enserfed labor, strangling state control over industry, a late and inefficient transition from charcoal to coal as a fuel source, and the lack of a bourgeoisie or a middle class—but the one most relevant to historians of earth and mining science is that Russia failed to compete technologically with the major Western economies. The consensus among both American and Soviet historians throughout the twentieth century was that private mine owners, especially those of aristocratic origin, clung doggedly and conservatively in the nineteenth century to charcoal as an energy source and resisted new advances in steel making (Blackwell 1968, 56–62, 1970; Pipes 1974, 191–220; Strumilin, 1967; for more on this historiographical position see Blanchard 2005). To counter this argument, more recent historians of Russian mining and metallurgical technology have searched for innovations within the industry. In doing so, some scholars have uncovered ways of escaping the old narratives described above of technological diffusion from the West or of purely native innovation, through work that examines the unique environmental factors and transnational collaborations that shaped Russian technological development.

The foundational work in the history of Russian mining technology is N. B. Baklanov's *Technology of Metallurgical Production in the Urals in the Eighteenth Century* (1935). A sweeping study of every aspect of the mining industry from ore extraction to shipment for export, Baklanov finds innovation arising from how mining officers, assayers, and skilled metallurgists dealt with the chemical and physical makeup of the Urals and its ores. Copper metallurgy, he notes, is highly dependent on the chemistry of local ores, meaning that recipes for copper alloys could not simply be transferred from Sweden or Germany to the Urals, but were developed on site, especially in the chief state plant in Ekaterinburg, over years of experimental smelts. In her work on Akinfii Demidov's experimental iron smelts and tests of what Demidov called "German boasts," Aleksandra Cherkasova continues this argument, writing that, due to variation in local ores and fuel sources, all metallurgy is experimental and thus metallurgical knowledge always local (Cherkasova 1994b, 20).

Water technologies were another area where local climate and ecological conditions resulted in unique techniques and machinery being developed: Baklanov cites in particular the development of a highly efficient system of water wheels used to pump several bellows simultaneously (1935, 178–179). As described in Catherine Evtuhof's recent essay on the environmental history of the Urals industry, the water resources of the Urals differed greatly from those of Saxony and therefore required innovative means to manage them. The canal system used to store water needed to turn hammers and pump bellows in Saxony would freeze through for half the year in the Urals. To maintain the necessary year-round water supply to operate factories, factory managers situated plants immediately next to large reservoirs created by damming rivers. These large "factory ponds" were unique features of the Urals industry, and Russian engineers constructed strings of reservoirs behind particularly high dams which could power multiple factories year-round along a single river (Evtuhof 2023, 52–56). With such reliable water power, Malcom Hill argues in an article-length history of Urals iron, Russia faced far less need for coal or steam technologies, two key markers of the Industrial Revolution, than the United Kingdom did (2006). The Urals' abundant timber resources, plus innovations in charcoal making, and the Demidov-led adaptation of the puddling process to charcoal-fueled iron production, Ian Blanchard argues, allowed iron production to rise in productivity and efficiency across the nineteenth century and meet domestic demand without switching to coal fuel (2005). Similarly, compulsory labor, especially for support tasks like timber felling, charcoal production, carting, road building, and shipping, lessoned the need for labor-saving technologies as found in British industry (a discussion of the historiography of the mining industry's workforce, which included such compulsory laborers as serfs attached to factories, state peasants working part of the year in lieu of cash payment of the poll tax, and convicts and prisoners of war, is beyond the scope of this chapter. For more on bonded labor in the mining industry and its historiography, see Griffiths 1986 and Hudson, DeHart, and Griffiths 1995). Instead of being deficient when compared to British technology, Russian technology was simply different.

Consideration of the environment was one way for historians like Baklanov, Evtuhof, and Hill to evaluate Urals technologies without falling into narratives of either diffusion or of development straight from artisanal, early modern Tula, or Olonets techniques. The work of Andrei Keller studying German and Russian collaboration in metallurgy and arms making offers another way of escaping the diffusion or native innovation dichotomy. Keller compares small-scale weapons production in both western Germany and the Urals to understand the impact of foreign experts in Russian industrial history. He takes as his main case study the early nineteenth-century enterprise of the German metallurgical and weapons-making entrepreneur Andrei Knauf, who recruited two dozen German, Danish, and Swedish metallurgists to the iron plant he acquired at Zlatoust in the southern Urals, as well as over 100 German and Alsatian weapons-making masters for an arms workshop. Russian masters also worked for Knauf, and Russian and foreign masters exchanged ideas, experimented together, and developed both artisanally and industrially scaled technologies of weapons production that complimented each other and satisfied market demand—even through World War I—until exigent state demands for mass-produced weapons in the 1920s–1930s collectivized artisans and ended the craft weapons-making tradition (Keller 2015, 2000, 2022).

These narratives of Russian innovation born of engagement with local ores, the specificities of regional climate and geography, and cross-cultural collaboration do not ignore the fact that Russian metallurgical production did fall behind western European production in the nineteenth century and place some blame for this fact on Urals technologies. While Baklanov stresses areas of innovation, he also finds stagnation in technological development: He argues that mining administrator Georg Wilhelm de Hennin established such an advanced, efficient, and profitable state iron and copper factory in Ekaterinburg in 1723 that for the rest of the century all new factories were but copies of the original, which itself was outmoded in comparison to European plants by the end of the century (Baklanov 1935, 162, 178). Themes found in Baklanov—that there were a number of innovative technologies, especially earlier in the eighteenth century, and that toward the end of the century there was a lag in comparison with the West—have been further developed by subsequent authors. In his study of Russia's silver industry, Ian Blanchard notes technological successes, like the introduction of the Saiger process in the Altai Mountains, in which the local argentiferous copper ores were mixed with the argentiferous lead of the Nerchinsk region in special hearths, that made the Siberian silver industry highly productive from the mid-1740s to mid-1770s. But he also frequently underlines the technological limitations of the Nerchinsk silver industry in southeastern Siberia, where it was economical only to refine extremely pure ores of 12.5 ounces of silver per ton, which caused the Nerchinsk industry to sputter in and out of production (Blanchard 1989, 61–86). V. L. Stepanov (2008–2009), writing on the Ministry of Finance's mining policies, finds inadequate application of Western technologies (the puddling process, cupola shaft furnaces, and rolling equipment) in the opening decades of the nineteenth century, but to him this was a symptom rather than a cause of the industry's decline. He continues by detailing how the industry was revitalized by a new generation of highly educated mining engineers active in the second quarter of the nineteenth century. These men modernized factories, experimented with metallurgical processes, sponsored research into foreign technologies, and produced scholarship in mining science (Stepanov 2008–2009, 28–31). All this activity however, could not compensate for the fundamental brake on Russia's industrialization, the reliance on serf labor (Stepanov 2008–2009, 31). The concern at the root of the literature on eighteenth- and early nineteenth-century technology is that Russia's delayed heavy industrialization was the result of technological backwardness. These authors, however, describe an intellectually vibrant industry whose leaders monitored and adapted foreign technologies, and that also, because of regional ecological and social conditions, evolved in its own, innovative way.

6 The Science of Empire

With an increasingly aggressive and centralizing state under President Vladimir Putin, the state has reentered the historiography in the past 15 years as the central actor in the imperial mining industry. Sidestepping the old debates over whether the state helped or hindered industrial development, current historians examine the mining industry as a site of experimentation in governing styles, bureaucratic sciences, and imperial rule in Russia's transformative eighteenth century. Instead of studying the impact of state policy on industrial development, these historians, drawing at times on the scholarship of German, Swedish, and Habsburg mining administration discussed in Sebastian Felten's contribution to this volume, consider how the rapidly expanding mining industry shaped the development of the imperial state.

In a biography of the Russian-born Dutchman Andrei Vinius, Kees Boterbloem (2013) studies this governor of Siberia active early in Peter I's reign as both a conduit of foreign technology (as the son of one of the first foreigners to found a water-powered ironworks in Russia) and governing practices and as the chief bureaucrat overseeing the transition of Siberia's economy from one founded on fur harvesting to mining. In a series of articles, Mikhail Kiselev charts experimentation, evolution, and ultimately instances of effective state building in the mining administration of the Urals in the first half of the eighteenth century (2015, 2019a, b). Nikolai Petrukhintsev (2014) details the interrelation of politics at the court of Empress Anna (r. 1730–1740) and questions of privatization and reform in the Urals iron industry in his magisterial study of Anna's reign. Chechesh Kudachinova (2019a, b) explores gold, silver, and Indigenous metal objects and tales of precious metals in the imperial imagination of the Altai. E. G. Nekliudov (2018) examines mine administration and reform in the nineteenth century, in particular the privatization of mines as part of Emperor Alexander II's (r. 1855–1881) Great Reforms, in an effort to understand how the major reforms of the 1860s and 1870s were implemented on the ground in the provinces. This new wave of historians, in which I include myself, argues that the mining industry and the Russian Empire coforged each other (Graber 2016).

This imperial turn in the historiography of mining has become especially salient since the 2014 seizure of the Donets Coal Basin (*Donbass*) by Russia from Ukraine. With ample forest cover in the main eighteenth-century metallurgical zones (the Urals and the Altai), Russians faced nowhere near the pressure of the English to find alternatives to charcoal. Still, from at least the 1760s scholars and mining administrators studied coal, pondered its chemistry and origins, and promoted its use, initially more as an alternative to wood in domestic settings than as a new fuel in industry (Shlatter 1760; Lomonosov 1763; Güldenstädt 1776, 52; L'vov 1799). To speed the exploitation of anthracite discovered in geological expeditions to the Donets Basin in 1838, in 1840 the Ministry of Finance seized Cossack-owned lands containing fuel reserves. Over the course of the following decade, coal production quadrupled, to 3,540,000 poods (57,988 metric tons) a year in 1850 (Stepanov 2008–2009, 22). Further development of the region, especially with the 1869 founding of Welshman John Hughes' massive coal and iron complex, the largest in the empire, drew in laborers and settlers from across the empire (Wynn 1992; Kuromiya 1998; Hill 2019; McQuillen 2023). While the history of the Donbass has been studied in detail by Soviet scholars, the contribution of mining to the incorporation of a region that had been until 1764 the Ottoman and Cossack frontier into the industrial heartland of the empire and Soviet Union can be further explored (Novik et al.1960; Shcherban 1969; Bakulev 1955; Nestrenko 1954).

Another aspect of science, empire, and mining in Russia very much worth further development is the study of the multiethnic character of mining and its knowledges in the Urals and Siberia. The favored methodology in the historiography of imperial Russian mining and metallurgy has been comparative, and it has been consistently motivated by a desire to measure Russian industrial performance against that of the West. Even studies of cross-cultural knowledge creation, like Kees' work on Vinius and Keller's on Knauf, are concerned only with the interaction of Russian and European knowledges. Missing from this discussion are the earth knowledges and mining practices of the Urals' and Siberia's Indigenous populations. Bashkirs, a Turkic, Muslim, seminomadic people, served as prospectors in the eighteenth century in the southern Urals and the neighboring copper-rich Orenburg Steppe, and the Demidovs relied on Voguls (now known as Mansi, a seminomadic people speaking an Ugric language) to aid in prospecting work in the western Urals. The richest mines in the Urals and the Altai—Gumashevskii copper mine and a mountain of magnetic iron (*magnitaia gora,* developed by the Soviets into the steel complex of Magnitogorsk) in the Urals, and Zmeiagorsk/Schlangenberg silver mine in the Altai—all originated from veins discovered for the Russians by Indigenous prospectors or from abandoned mines predating Russian settlement. Nikolai Petrukhintsev, Catherine Evtuhof, and I separately have begun studying the interactions of Indigenous (in particular Bashkir) and official knowledges of minerals and geography using mostly Russian- and European-language sources, and these interactions will be an important part of my book manuscript in progress (Graber 2024). But the question of Indigenous contributions to the mining industry's eighteenth-century success is unresolved, and little historical work has been done on their earth knowledges, cosmologies, and mining practices. The field needs speakers of Turkic, Ugric, or Altaic languages to contribute to this project using sources in these languages, and to do the kind of philological recovery of Indigenous earth and metal knowledges that Alison Bigelow (2020) has done for Spanish America.

7 Conclusion

The history of earth sciences and of mining, as subfields in the history of science and technology, have been written overwhelmingly in the traditions of western and central European national and imperial histories. By bringing the historiography of imperial Russian mining into conversation with earth and mining history, I aim to join historians of the Global South and of non-European empire in uncovering the global history of modern science and technology. As historians of an empire that was independent yet deeply engaged with the earth and mining sciences of a variety of Wests, Russianists offer novel approaches to the some of the shared issues facing scholars of postcolonial science. This chapter suggests two topics in the literature that will be of particular interest to historians of the global earth sciences and technologies. First, there is the Soviet-era work of historians engaging with opposing narratives of Russia as either the recipient and beneficiary of western science and technology or as the home of an independent, long-standing artisanal knowledge tradition that developed into Eurasia's first modern metallurgical power. Russianists focusing on the environmental, climate, and chemical specificities of the Urals and its ores offer one way to move beyond ideas of native versus foreign science: The physical conditions of each region demanded that metallurgists and technicians experiment and adapt inherited knowledge traditions to fit local circumstances.

Second, Russianists' (admittedly still unfinished) study of the dynamics of knowledge creation in such a uniquely ethnically, religiously, nationally, and intellectually diverse mining regions as the Urals and the Altai complicates narratives of colonial science written for European empire. As a land empire with a long history of engagement, cooperation, and conflict with the Indigenous peoples of the steppes, the Urals, and Siberia, the case of Russian Empire and its cooption of Indigenous miners and mineral knowledges differs fundamentally from those of the better studied European maritime empires. As mine administrators and owners were embedded in trans-Eurasian epistolary networks, the results of cross-cultural mineral knowledge exchange did not remain locked away in the Urals and Siberia, and work on the circulation of imperial Russian earth knowledges to Europe and beyond will aid historians of earth science in arriving at a more accurate, complex history of global knowledge of the earth and its products.

Acknowledgments I owe thanks to Elena Aronova, Patrick McCray, and the members of the University of California, Santa Barbara History of Science and Technology Workshop; Gregory Afinogenov, Michael David-Fox, and the members of the Washington, DC Russian History Seminar; and Valerie Kivelson and the members of the University of Michigan Russian History Workshop for their comments on earlier drafts of this chapter. Thank you as well to Anna Amramina, who provided research support.

References

Amburger, Erik. 1957. *Die Familie Marselis: Studien zur russischen Wirtschaftsgeshichte*. Gießen: Wilhelm Schmitz.

Baklanov, N.B. 1935. *Tekhnika metallurgicheskogo proizvodstva XVIII veka na Urale*. Leningrad: Sotsekgiz.

Bakulev, G.D. 1955. *Razvitie ugol'noi promyshlennosti Donetskogo baseina*. Moscow: Gos. Izd-vo politicheskoi literatury.

Bigelow, Alison Margaret. 2020. *Mining language: Racial thinking, indigenous knowledge, and colonial metallurgy in the early modern Iberian world*. Chapel Hill: University of North Carolina Press.

Blackwell, William L. 1968. *The beginnings of Russian industrialization: 1800–1860*. Princeton: Princeton University Press.

———. 1970. *The industrialization of Russia: An historical perspective*. New York: Thomas Y. Crowell Company.

Blanchard, Ian. 1989. *Russia's 'age of silver': Precious-metal production and economic growth in the eighteenth century*. London: Routledge.

———. 2005. Nineteenth-century Russian and 'Western' ferrous metallurgy: Complementary or competitive technologies? In *The industrial revolution in iron: The impact of British coal technology in nineteenth-century Europe*, ed. Chris Evans and Göran Rydén, 129–150. Aldershot: Ashgate.

Boterbloem, Kees. 2013. *Modernizer of Russia: Andrei Vinius, 1641–1716*. New York: Palgrave Macmillan.

Cherkasova, A.S., ed. 1994a. *Demidovskii vremennik: Istoricheskii al'manakh. Kniga 1*. Ekaterinburg: Demidovskii institut.

———., ed. 1994b. Chtob zhelezo delat' samym dobrym masterom. In *Demidovskii vremennik: Istoricheskii al'manakh. Kniga 1*, ed. A. S. Cherkasova, 10–29. Ekaterinburg: Demidovskii institut.

Chrissidis, Nikolaos. 2016. *An academy at the Court of the Tsars: Greek scholars and Jesuit education in early modern Russia*. Dekalb: Northern Illinois University Press.

Clark, Gregory, and Robert C. Feenstra. 2003. Technology in the great divergence. In *Globalization in historical perspective*, ed. Michael D. Bordo, Alan M. Taylor, and Jeffrey G. Williamson, 277–321. Chicago: University of Chicago Press.

Collis, Robert. 2012. *The Petrine Instauration: Religion, esotericism and science at the court of Peter the great, 1689–1725*. Leiden: Brill.

Daniels, Rudolph L. 1973. *V. N. Tatishchev: Guardian of the Petrine revolution*. Philadelphia: Franklin Publishing.

de Hennin, G.W. 1992. *Description of Ural and Siberian factories in 1735*. Trans. A.S. Despande, and W.V. Nabar. Washington, DC: Smithsonian Institution Libraries and the National Science Foundation.

De-Gennin, Vil'gem. 1937. *Opisanie ural'skikh i sibirskikh zavodov, 1735*. Moscow: Gosudarstvennoe izd-vo "Istoriia zavodov".

Elina, Olga. 2018. A passion for plants: Collections and power games in botany in the Russian Empire from the eighteenth to the early nineteenth century. *Centaurus* 60: 257–275.

Evans, Chris, and Göran Rydén. 2007. *Baltic iron in the Atlantic world in the eighteenth century*. Leiden: Brill.

Evans, Chris, Owen Jackson, and Göran Rydén. 2002. Baltic iron and the British iron industry in the eighteenth century. *The Economic History Review* 55 (4): 642–665.

Evtuhof, Catherine. 2023. Early Russian industrialization: An environmental perspective. In *Thinking Russia's history environmentally*, ed. Catherine Evtuhof, Julia Lajus, and David Moon, 48–73. New York: Berghan Books.

Fischer, Gotthelf. 1806. *Museum-Demidoff*. Vol. 2, *Contenant la description des Minéraux et des Pétrifications*. Moscow: Imprimerie de l'Université Impériale.

Furhmann, Joseph. 1972. *The origins of capitalism in Russia: Industry and progress in the sixteenth and seventeenth centuries*. Chicago: Quadrangle Books.

Glagoleva, A.P. 1957. *Olonetskie zavody v pervoi chetverti XVIII veka*. Moscow: Izd-vo Akademii nauk SSSR.

Graber, Anna C. 2016. *Tsardom of rock: State, society, and mineral science in Enlightenment Russia*. PhD diss. Yale University.

———. 2024. Depicting expertise and managing diversity in the Urals mining industry (1773–1819). In *Picturing Russian Empire*, ed. Valerie Kivelson, Sergei Kozlov, and Joan Neuberger, 156–163. Oxford, UK: Oxford University Press.

Grau, Conrad. 1963. *Der Wirtschaftsorganisator, Staatsmann und Wissenschaftler Vasilij N. Tatiscev (1686–1750)*. Berlin: Akademie-Verlag.

Griffin, Clare. 2022. *Mixing medicines: The global drug trade and early modern Russia*. Montreal/ Kingston: McGill-Queen's University Press.

Griffiths, David M. 1986. Introduction: The Russian manufactury in soviet historiography. In *The rise of the Demidov family and the Russian iron industry in the eighteenth century*, ed. Hugh D. Hudson Jr., 1–26. Newtonville: Oriental Research Papers.

Güldenstädt, A.J. 1776. *Discours academique sur les produits de Russie*. St. Petersburg: Imprimerie de l'Académie Impériale des Sciences de St. Petersbourg.

Hill, Malcolm R. 2006. Russian iron production in the eighteenth century. *Icon* 12: 118–167.

———. 2019. Russian steel production from the repeal of serfdom to the First World War. *Historical Metallurgy* 51, part 2 for 2017: 99–110.

Hudson, Hugh D., Jr. 1986. *The rise of the Demidov family and the Russian iron industry in the eighteenth century*. Newtonville: Oriental Research Papers.

Hudson, Hugh D., Jr., Bruce J. DeHart, and David M. Griffiths. 1995. Proletarians by fiat: The compulsory Ural metallurgical work force, 1630–1861. *International Labor and Working-Class History* 48: 94–111.

Iukht, A.I. 1985. *Gosudarstvennaia deiatel'nost' V. N. Tatishcheva v 20-kh-nachale 30-kh godakh XVIII v.* Moscow: Izd-vo Nauka.

Iurkin, I.N. 2001. *Demidovy. Uchenye, inzhenery, organizatory nauki i proizvodstva: Opyt naukovedcheskoi prosopografii*. Moscow: Izd-vo Nauka.

Jones, Ryan Tucker. 2014. *Empire of extinction: Russians and the North Pacific's strange beasts of the sea, 1741–1867*. Oxford: Oxford University Press.

Kafengauz, B.B. 1949. *Istoriia khoziaistva Demidovykh v XVIII-XIV vv*. Moscow: Izd-vo Akademii nauk SSSR.

Keller, Andrei. 2000. Zavodchik Andrei Knauf i paradigmy razvitiia ural'skoi promyshlennosti XIX v. In *Ural industrial'ny. Bakuninskie chteniia: materialy XIV Vserossisskoi nauchnoi konferentsii*, 490–504. Ekaterinburg: Izd-vo UMTs UPI.

———. 2015. "Raison d'etat" i "chastnyi interes" v Rossii k. XVIII v. – nach. XIX v.: na primere A. Knaufa v gornozavodskoi promyshlennosti Urala, 1797–1833 gg. *Bylye Gody* 37 (3): 508–518.

———. 2022. Oruzhie, inostrantsy i stal': Gosudarstvennye i partikuliarnye praktiki v metalloobrabatyvaiushchei i oruzhneinoi promyshlennosti v Rossii. *Rossiia XXI* 6: 22–47.

Kiselev, Mikhail. 2015. State metallurgy factories and direct taxes in the Urals, 1700–50: Paths to state building in early modern Russia. Trans. A. Graber. *Kritika* 16 (1): 7–36.

———. 2019a. Perepiska V. N. Tatishcheva s Ia. V. Briusom i Berg-Manufaktur-Kollegiei v 1720–1721 gg. *Istoriia: Fakty i simvoly* 2019 (2): 16–26.

———. 2019b. Vyshnee gornoe nachal'stvo v Solikamske (1722–1724 gg.). In *Rus', Rossiia: Srednevekoe i novoe vremia vyp*, vol. 6, 424–428. Moscow: Izd-vo Moskovskogo universiteta.

Koroloff, Rachel. 2014. *Seeds of exchange: Collecting for Russia's apothecary and botanical gardens in the seventeenth and eighteenth centuries*. PhD diss. University of Illinois, Urbana-Champaign.

———. 2018. Travniki, Travniki and Travniki: Herbals, herbalists and herbaria in seventeenth-century and eighteenth-century Russia. *ВИВЛІОѲИКА: E-Journal of Eighteenth-Century Russian Studies* 6: 58–76.

Kozlov, A.G. 1981. *Tvortsy nauki i tekhniki na Urale XVII-nachalo XX veka. Biograficheskii spravochnik*. Sverdlovsk: Sredne-Ural'skoe knizhnoe izd-vo.

Kudachinova, Chechesh. 2019a. Mapping the Altai in the Russian geographical imagination, 1650s–1900s. PhD diss. Humboldt-Universität zu Berlin.

———. 2019b. The muscovite silver crusade: Power, space, and imagination in early modern Eurasia. *Ab Imperio* 4: 49–72.

Kuromiya, Hiroaki. 1998. *Freedom and terror in the Donbas: A Ukrainian-Russian borderland, 1870s–1990s*. Cambridge, UK: Cambridge University Press.

Kuz'min, A.G. 1981. *Tatishchev*. Moscow: Molodaia gvardiia.

L'vov, Nikolai. 1799. *O pol'ze i upotreblenii ruskago zemlianago uglia*. St. Petersburg: Tip. I. K. Shnora.

Lomonosov, Mikhail. 1763. O sloiakh zemnykh. In *Pervoe osnovaniia metallurgii ili rudnykh del*. St. Petersburg: Imperatorskaia Akademiia nauk.

McQuillen, Colleen. 2023. The Imperial underground: Coal mining narratives of Donbas at the Fin de Siècle. *Slavic and East European Journal* 67 (2): 163–182.

Miliukov, Pavel Nikolaevich. 1896. *Ocherki po istorii russkoi kul'tury*. 2 vols. St. Petersburg: Tip. I. N. Skorokhodova.

Nechaev, N.V. 1944. *Shkoly pri gornykh zavodov Urala v pervoi polovine 18-go stoletiia*. Moscow: Moskovskii Gos. Pedagogicheskii Institut im. V. I. Lenina.

Nekliudov, E.G. 2004. *Ural'skie zavodchiki v pervoi polovine XIX veka: Vladel'tsy i vladeniia*. Nizhny Tagil: Izd-vo NTSPA.

———. 2018. *Gornaia reforma v Rossii vtoroi poloviny XIX—nachala XX veka: Ot zamysla k realizatsii*. St. Petersburg: Nestor-Istoriia.

Nestrenko, A.A. 1954. *Ocherki istorii promyshlennosti i polozheniia proletariata Ukrainy*. Moscow: Politizdat.

Novik, E.O., V.V. Permiakov, and E.E. Kovalenko. 1960. *Istoriia geologicheskikh issledovanii donetskogo kamennougol'nogo basseina (1700–1917)*. Kiev: Izd-vo Akademii nauk Ukrainskoi SSR.

Pavlenko, N.I. 1953. *Razvitie metallurgicheskoi promyshlennosti Rossii v pervoi polovine XVIII veka*. Moscow: Izd-vo Akademii nauk SSSR.

———. 1962. *Istoriia metallurgii v Rossii XVIII veka: Zavody i zavodovladel'tsy*. Moscow: Izd-vo Akademii nauk SSSR.

Pavlenko, N.I., et al. 1969. Perekhod ot feodalizma k kapitalizmu v Rossii. In *Perekhod ot feodalizma k kapitalizmu v Rossii*, ed. V.I. Shunkov et al. Moscow: Nauka.

Petrukhintsev, N.N. 2014. *Vnutrenniaia politika Anny Ioannovny (1730–1740)*. Moscow: ROSSPEN.

Pipes, Richard. 1974. *Russia under the old regime*. New York: Charles Scribners' Sons.

Pirogova, E.P., et al. 2008. *Rod Turchaninovykh*. Ekaterinburg: Sokrat.

Pobedimova, G.A., ed. 2006. *Puteshestvie brat'ev Demidovykh po Evrope*. Moscow: Indrik.

Polnoe sobranie zakonov Rossiiskoi imperii. 1830. St. Petersburg.

Pomeranz, Kenneth. 2000. *The great divergence: Europe, China, and the making of the modern world economy*. Princeton: Princeton University Press.

Portal, Roger. 1950. *L'Oural au XVIIIe siècle. Etude d'histoire économique et sociale*. Paris: Institut d'études slaves.

Raj, Kapil. 2007. *Relocating modern science: Circulation and the construction of knowledge in South Asia and Europe, 1650–1900*. New York: Palgrave Macmillan.

Raskin, N.M. 1981. *Apollos Apollosovich Musin-Pushkin*. Leningrad: Izd-vo Nauka.

Raskin, N.M., and I.I. Shafranovskii. 1975. *Aleksandr Matveeevich Karamyshev: 1744–1791*. Leningrad: Izd-vo Nauka.

———. 1978. *Erik Gustavovich Laxmann, noted traveler and naturalist of the 18th century*. Springfield: Amerind Publishing Company.

Romaniello, Matthew P. 2016a. Humoral bodies in cold climates. In *Russian history through the senses: From 1700 to the present*, ed. Matthew P. Romaniello and Tricia Starks, 23–43. London: Bloomsbury Academic Press.

———. 2016b. True rhubarb? Trading Eurasian botanical and medical knowledge in the eighteenth century. *Journal of Global History* 11 (1): 2–23.

Rönnbäck, Klas. 2010. New and old peripheries: Britain, the Baltic, and the Americas in the great divergence. *Journal of Global History* 5 (3): 373–394.

Safranova, A.M., et al. 2005. *Katalog knig V. N. Tatishcheva i pervoi biblioteki Ekaterinburga v fondakh Sverdlovskogo Oblastnogo Kraevedcheskogo Muzeia*. Ekaterinburg: Izd-vo Ural'skogo Universiteta.

Shakinko, I.M. 1986. *Vasilii Tatishchev*. Sverdlovsk: Sredno-Ural'skoe knizhnoe izd-vo.

Shcherban, A.N., ed. 1969. *Istoriia tekhnicheskogo razvitiia ugol'noi promyshlennosti Donbassa*. Kiev: Naukova dumka.

Shlatter, Ivan. 1760. *Obstoiatel'noe nastavlenie rudnomu delu*. St. Petersburg: Imperatorskaia Akademiia nauk.

Sivasundaram, Sujit. 2010. Sciences and the global: On methods, questions, and theory. *Isis* 101: 146–158.

Stepanov, V.L. 2008–2009. E. F. Kankrin and the development of mining and metallurgy in Russia. *Russian Studies in History* 47 (3): 7–37.

Strumilin, S.G. 1932. Tsarskaia manufaktura XVII veka. In *Krepostnaia manufaktura v Rossii. Chast' III. Dvortsovaia polotnianaia manufaktura XVIII veka*. Leningrad: Izd-vo Akademii nauk SSSR.

———. 1954. *Istoriia chernoi metallurgii v SSSR*. Moscow: Izd-vo Akademii nauk SSSR.

———. 1967. *Istoriia chernoi metallurgii v SSSR*. Moscow: Izd-vo Nauka.

Tatishchev, Vasilii. 1725. Generosiss. Dr Basilii Tatischow Epistola ad d. Ericum Benzelium de Mamontowa Kost, id est, de ossibus bestia Russis *Mamont* dicta. *Acta literaria Sveciae* 2: 36–43.

———. 1979. Skazanie o zvere mamonte. In *V. N. Tatishchev: Izbrannye proizvedeniia*, ed. S.N. Valk, 36–50. Leningrad: Izd-vo Nauka.

Turchin, Peter. 2013. The west and the rest: The science of the great divergence. *Cliodynamics: The Journal of Quantitative History and Cultural Evolution* 4 (1): 67–85.
Usitalo, Steven A. 2013. *The invention of Mikhail Lomonosov: A Russian National Myth*. Boston: Academic Studies Press.
Wynn, Charles. 1992. *Workers, strikes, pogroms: The Donbass-Dnepr bend in late Imperial Russia, 1870–1905*. Princeton: Princeton University Press.
Zaozerskaia, E.I. 1947. *Manufaktura pri Petre I*. Moscow: Izd-vo Akademii nauk SSSR.
———. 1970. *U istokov krupnogo proizvodstva v Russkoi promyshlennosti xvi-xvii vekov*. Moscow: Nauka.
Zavyalov, V. 2018. Modelling of bloomery processes in a medieval Russian furnace. *Archeologické Rozhledy* 70 (3): 450–456.

Index

E. Aronova et al. (eds.), *Handbook of the Historiography of the Earth and Environmental Sciences*, Historiographies of Science,
https://doi.org/10.1007/978-3-031-40799-4

GPSR Compliance

The European Union's (EU) General Product Safety Regulation (GPSR) is a set of rules that requires consumer products to be safe and our obligations to ensure this.

If you have any concerns about our products, you can contact us on ProductSafety@springernature.com

In case Publisher is established outside the EU, the EU authorized representative is:

Springer Nature Customer Service Center GmbH
Europaplatz 3
69115 Heidelberg, Germany

Batch number: 10755214

Printed by Printforce, the Netherlands